SIMULATION MODELING WITH SIMIO: A WORKBOOK V4

Jeffrey Allen Joines

Stephen Dean Roberts

North Carolina State University

Raleigh, North Carolina

November 2015

4th Edition

Published by:
SIMIO LLC
504 Beaver St, Sewickley, PA 15143, USA
http://www.simio.com

Table of Contents

About the Authors

JEFFREY A. JOINES is an Alumni Distinguished Undergradate Professor, an Associate Professor and the Associate Department Head of Undergraduate Programs in the Department of Textile Engineering, Chemistry, and Science at NC State University. He received a B.S. in Electrical Engineering and B.S. in Industrial Engineering, a M.S in Industrial Engineering, and Ph.D. in Industrial Engineering, all from NC State University. He received the 1997 Pritsker Doctoral Dissertation Award from the Institute of Industrial Engineers. He is a member of IEEE, IIE, ASEE, Tau Beta Pi, Etta Kappa Nu, Alpha Pi Mu and Phi Kappa Phi. His research interests include supply chain optimization utilizing computer simulation and computational optimization methods. Dr. Joines teaches graduate and undergraduate classes in computer information systems, computer based modeling in Excel and VBA, Lean Six Sigma, and computer simulation modeling. Dr. Joines has also teaches industry programs in the areas of Design for Six Sigma, Simulation and Six Sigma, and Data Management to Assist in Six Sigma through the Textile Extension programs Six Sigma Black Belt and Master Black Belt program. Dr. Joines served as the Program Chair for the 2005 Winter Simulation Conference (WSC) and the Proceedings Editor for the 2000 WSC as well as developed and maintained the WSC paper management system from 2000-2009. He currently serves on the WSC Board of Trustees representing the IEEE Systems, Man, and Cybernetics and was the 2015 Board Chair. He has also been an author and session chair for several Winter Simulation Conferences. He received the 2014 Distinguished Service Award from INFORMS College on Simulation.

Dr. Joines is involved in utilizing technology in the classroom and how it impacts problem solving. He was awarded the 2012 Alumni Distinguished Undegraduate Professor, the 2006 NC State University Outstanding Teaching Award and College of Textiles Outstanding Teacher which allowed him to become a member of the Academy of Outstanding Teachers. In 2009, Dr. Joines (along with Professor Roberts) was awarded the Gertrude Cox Award for Innovative Excellence in Teaching and Learning with Technology for Transformative Large Scale Projects.

STEPHEN D. ROBERTS is the A. Doug Allison Distinguished Professor in the Edward P. Fitts Department of Industrial and Systems Engineering at NC State University. Professor Roberts received his: Ph.D., M.S.I.E., and B.S.I.E. (with Distinction) from the School of Industrial Engineering at Purdue University. His primary teaching and research interests are in simulation modeling and health systems engineering. He has been a faculty member at NC State University since 1990, serving nine years as Department Head of the Department of Industrial Engineering and three years as Interim Director of the Integrated Manufacturing Systems Engineering Institute. Prior to serving at NC State, he was a faculty member in the Department of Internal Medicine at the Indiana University School of Medicine and the School of Industrial Engineering at Purdue University as well as the Director of the Health Systems Research Group at Regenstrief Institute for Health Care. Previously, he was a faculty member in the Department of Industrial and Systems Engineering at the University of Florida and Director of the Health Systems Research Division of the J. Hillis Miller Health Center, University of Florida. He has had sabbaticals at Wolverine Software and the University of Central Florida/Institute for Simulation and Training.

Professor Roberts is a member of Alpha Pi Mu, Tau Beta Pi, Sigma Xi, Sigma Tau, and Omega Rho and a Fellow of the Institute of Industrial Engineers. He has held Kaiser Aluminum Fellowship and a NDEA Title IV Fellowship. He received the AIIE 1967 Graduate Research Award, Outstanding Paper Award at the 12th Annual Simulation Symposium, the Pritsker and Associates Outstanding Undergraduate Teaching Award in the Purdue School of Industrial Engineering, the CA Anderson Outstanding Teacher in the NCSU Department of Industrial and Systems Engineering, the Outstanding Teacher from the NCSU College of Engineering, membership of the NCSU Academy of Outstanding Teachers, the Gertrude M. Cox Award for Transformative projects from NC State University (with Professor Joines), the Distinguished Service Award from INFORMS College on Simulation, and has served as member, Vice- Chair, and Chair of the Winter Simulation Conference (WSC) Board of Directors representing TIMS (now INFORMS) College on Simulation, and Secretary, Vice-President/Treasurer, President, and Past-President of the WSC Foundation Board. He was the Proceeding Editor for the 1983 WSC, the Associate Program Chair in 1985, and Program Chair in 1986. He has been a WSC presenter, session chair, track coordinator, and keynote speaker.

Preface to the Fourth Edition

This edition of this workbook maintains the successful "participatory" style introduced in the first edition. You don't sit and read the book without a computer loaded with SIMIO (the book was created using version 7.124 of SIMIO™). We expect your active participation in using SIMIO as you turn the pages. We try to carry on a conversation with you. Our belief is that simulation is not a spectator sport. You have to practice to gain skill with it and you develop that skill through modeling practice. This book encourages you to practice and use your skill, and feedback from earlier editions appear to validate the approach. This book retains its focus on simulation modeling with SIMIO and most of the simulation statistical analysis and analytical issues are more thoroughly covered in other books. We strongly suggest that if you are teaching/learning simulation that you also have one of these non-language books available.[1]

We have deliberately tried to keep the price of the book low (i.e., the E-book or the paper copy). A relatively new simulation language like SIMIO is constantly. In fact, the SIMIO developers have a history of new releases (called "sprints") about every three months. Any book that describes SIMIO will go out of date quickly, so we have tried to track new features and update this book fairly often. If the book price is low, maybe you will want to re-buy this book from time to time, so you have the latest information. Also, we feel when learning and teaching, the paper copy allows the learner to write directly in the book.

Part 1: Organization of this Edition

This edition of the workbook has an evolved structure based on use and experience. More emphasis is placed on "why" modeling choices are made, to supplement the "how" in using SIMIO in simulation. In Chapter 1, we present fundamental simulation concepts, independent of SIMIO which can be skipped for those who already understand these fundamentals. In Chapters 2 through 6, concentrates of the use of the Standard Library Objects in SIMIO. You can do a lot of simulation modeling without resorting to more complex concepts. A key part of those chapters is learning to identify/separate the data in a model from the model structure. Chapter 7 introduces the fundamental topic of "processes," which we frequently employ in the following chapters. Chapters 8 and 9 concentrate on the important topics of flow and capacity. Chapter 10 introduces optimization in the context of supply chain modeling. Chapter 11 presents the influence of bias and variability on terminating and steady-state simulation. Chapter 12 introduces SIMIO materials handling features. Chapter 13 extends the use of resources while Chapters 14 and 15 describes the use of workers including the detailed services provided by task sequences and their animation. Chapter 16 details the simulation of call centers with reneging, balking, and cost optimization. Chapters 17 through 20 presents object-oriented simulation capabilities in SIMIO. Chapter 17 builds a model out of an existing model (we call it sub-modeling). Chapter 18 describes the anatomy of an existing SIMIO and in Chapter 19 we build a new object by "sub-classing" an existing object. In Chapter 20 a new object is designed and built from a base SIMIO object and its creation is contrasted with standard SIMIO object. Chapter 21 presents some of the continuous modeling features in SIMIO. Chapters 22 and 23 demonstrates the power of object-oriented simulation in the modeling supply chains and process planning respectively. We include an appendix on input modeling, although SIMIO does not provide software.

The book is designed to be read from chapter to chapter, although it is possible to pick out certain concepts and topics. Some redundancy is helpful in learning. By the time you have finished this book you should be well-prepared to build models in SIMIO and to understand the virtues of different modeling approaches.

---

[1] For example, *Discrete-Event System Simulation* (5th Edition), Jerry Banks, John Carson, Barry Nelson, David Nicol, Prentice-Hall, 2010 (622 pages)

Like SIMIO itself, this workbook has been designed for a variety of student, teacher, and practitioner audiences. For example, if you are interested in manufacturing, you will want to be sure to study data-based modeling in Chapter 5, assemply and packaging in Chapter 6, the workstation in Chapter 9, and material handling in Chapter 12. If you are interested in logistics, don't miss modeling of distances in Chapter 3, flow and capacity in Chapter 8, inventories and supply chains in Chapter 10, and free space travel in Chapter12. If you are interested in healthcare, be sure to review scheduled arrivals in Chapter 8, resource decision making in Chapter 13, mobile workers in Chapter 14, and animated people and task sequences in Chapter 15. If object-oreinted simulation is your interest, make sure to study Chapters 17 through 20, which describes how SIMIO provides composition and inheritance to create objects. Manufacting examples and examples from the service sector are used throughout. Also we pay some attention to input modeling (including input sensitivity) and output analysis (including confidence intervals and optimization). This workbook provides comprehensive and in-depth discussion of simulation modeling with SIMIO.

At the end of most chapters, we offer commentary on topics presented. We will emphasize the strengths and weaknesses of the modeling approach and the language (we have no financial stake in SIMIO). To help insure that everyone participates in this active learning process, we sprinkled questions throughout the chapters. They have short answers and require the student pay some attention to what is going on. You can use these in class. Accordingly, even though you don't officially take attendance, you can give credit to students who turn in their in-class assignment each day. These practices help develop a reputation as a class you need to attend. We can provide you with answers to the questions, lecture notes, homework, and tests through a shared Dropbox™ if you contact us.

Part 2: Specific Changes in the Fourth Edition

- In any type of software book, it is not until the book is actually used do mistakes crop up. This edition corrects several mistakes and typos that occurred in the third edition.
- Since the third edition, SIMIO has gone through many changes and upgrades. The second edition and to some extent the third edition just added the new features to existing chapters or added another chapter later with the same case to highlight those new features (e.g., the original Chapter 7 which introduced related tables based on the Chapter 4 case). In the new book we have combined those two chapters which is now Chapter 5. We feel related data tables are an extremely important modeling concept and needs to be introduced earlier in the development of the simulation model.
- Also, we moved all the material handling to a standalone chapter (i.e., Chapter 12). For those who would like to have material handling sooner, Chapter 12 can be introduced any time after the Chapter 7 which introduces processes.
- As mentioned earlier, a new Chapter 1 which introduces simulation fundamentals which caused the original Chapters 1 - 3 to shift to Chapters 2 – 4. Chapter 3 introduces the concept of Input Sensitivity which is further explored in the new Chapter 11 devoted to Simulation Output Analysis. We do a more thorough job of using confidence intervals to examine output.
- Chapter 6 is now the original Chapter 5 which introduced combiners for the memory chip problem. However, the chapter is now followed directly by the processes chapter (i.e. Chapter 7) based on the same memory chip problem rather than waiting several chapters as was done before.
- The new concept of Tokenized processes which allows one to create a generic process that can be passed input parameters to specialize the process are an important concept. No longer do you have to have almost identical processes. Tokenized processes are utilized in Chapters 7, 9, and 15.
- In the first three editions, the original Chapter 9 and 10 utilized a simple clinic to introduce concepts of flow and capacity where chapter 9 was a little obsolete by the third edition. These two chapters were combined into the new Chapter 8 improving the flow along with a different case (i.e., the DMV).
- Some chapters were just refreshed and updated to the new versions but have moved within the book. For example, Chapter 9 and 10 are just refreshed versions of Chapter 8 and 11 from the third edition. However, Chapter 9 does utilize Tokenized processes when producing the materials in the JIT section.
- Chapters 13 and 14 were combined into one chapter (Chapter 13) improving the flow. Another example was added to the earlier editions when new features were added.

- Chapter 15 was updated to use the new Reserve option for acquiring resources. Also, SIMIO has introduced more process capability beyond simple processing times by using the task sequences. Chapter 15 explores the various features related to tasks utilizing the bank problem.
- Chapter 20 is now the new object chapter and was updated to include states and different animation pictures along with capacity.
- Consistent with the third edition, the continuous chapter, Chapter 21 is located after the discussion of objects. However, a new section was added that utilizes the new container entity, filler, and emptier objects.
- The more advanced Chapters 22 (Shirt Folding Line), 25 (Lean Systems), and 26 (Multiple product) were not included in this version for several reasons. Many of the new features of SIMIO have made some of the tricks used since the first edition obsolete. Also we needed to limit the length of the book. If anyone would like access to these chapters, we can provide you with the pdf and models of these chapters not included.

Part 3: Styles Used in this Book

Certain styles have been used in this book to illustrate objects, names and parameters and to make it easier to distinguish these types of parameters. Standard SIMIO objects will be set in small caps using a Courier New font (e.g., `SERVER`) while objects that are created by the modeler will be also bolded (e.g., **`DELAYOBJECT`**). Properties associated with these objects will be italicized (e.g., *Processing Time*). Process names will be italicized and placed in quotes, as "*OnEnteredProcessing*", while Add-on process triggers (e.g., *Exited*) will be will be only italicized since they are properties. Process steps like *`Assign`* will be set in italicized in Courier New font. SIMIO uses lots of expressions. These are set in Courier New font (e.g., `SrvOffice.Contents >0`). Names of all objects specified by the modeler will be bolded (e.g., Insert a `SERVER` named **SrvOffice)**. Values associated with properties will be set in a fashion similar to expressions or in quotes for strings (e.g., "True").

Part 4: Acknowledgements

We wish to thank our students, who have added much to our understanding and who let us often display our ignorance. Also we thank the many colleagues and friends (new and old) who have read, commented as well as used the book in their classes – hopefully they know who they are. In our writing this book, we appreciated the response of SIMIO developers and SIMIO Support to our endless stream of questions and doubts, especially Dave Sturrock and Dennis Pegden. Finally we thank our families for understanding and patience as we often spend more time talking with each other than with them.

Please let us know how we can improve this workbook and how it can better meet your needs.

Jeff Joines (JeffJoines@ncsu.edu)
Steve Roberts (roberts@ncsu.edu)
North Carolina State University

Raleigh, North Carolina

Chapter 1
Introduction to Simulation: The Ice Cream Store

This chapter will give the novice in simulation an introduction to the terminology and mechanics of simulation, as well as statistics used in performing a simulation model. If you are already familiar with these basic simulation concepts, feel free to start with Chapter 2 which starts the simulation modeling in SIMIO.

Part 1.1: What is Simulation?

The word "simulation" has a variety of meanings and uses. Probably its most popular usage is in games. A video game "simulates" a particular environment, maybe a formula one race, a battle, or a space encounter. The game allows the user to experience something similar to what it is like to drive in a race, maybe with other people participating. In the military, commanders create a battlefield simulation where soldiers act according to their training perhaps defending a base or assaulting an enemy position. In aeronautical engineering, an engineer may take a model airplane to a wind tunnel and test its aerodynamics. Most simulations have elements of reality with the intention that the participant will learn something about the environment through the simulation or perhaps only learning to play the game better.

We employ simulation to study and improve "systems" of people, equipment, materials, and procedures. We use simulation to mimic or imitate the behavior of systems like factories, warehouses, hospitals, banks, supermarkets, theme parks – just about anywhere a service is provided or an item is being produced. Our simulations are different from gaming and training simulations in that we want the simulation to model the real system, so we can investigate various changes before making recommendations. In that sense, simulation is a *performance improvement tool*. The simulation acts as an experimental laboratory, except that our laboratory is not physical but instead a computer model. We can then perform experiments on our computer model.

Modeling

Many performance improvement tools (i.e., Lean, Six Sigma, etc.) rely on models. For example value stream maps, spaghetti diagrams, process flow charts, waste walks, etc. are useful conceptual/descriptive models for performance improvement based on direct observations of the system. These models provide a wide-range of vehicles for describing and analyzing various systems. More formal models employ mathematical and statistical methods. For example, linear regression is a popular modeling technique in statistics. Queuing models offer a means for describing an important group of stable stochastic processes. Linear programming is a formal optimization method of finding the values of variables that minimize or maximize a linear objective function subject to linear constraints. Nevertheless, all these methods require a variety of assumptions about the system being modeled. For example, the variables are constants (i.e., real or integer valued) and often related linearly. If the variables have statistical variation, they are assumed to be normally or exponentially (i.e., Markovian) distributed.

Simulation is a *model-based* improvement tool. However, few assumptions need to be made to build the model. The model can be non-linear, described by arbitrary random variables, have a complex relationship, and change with time (i.e., dynamic). In fact, the simulation model is limited only by your imagination and the nature of the system being considered. You determine the nature of the model, based on what you think is important about the system being studied.

Computer Simulation Modeling

Instead of a formal model, our simulations are *computational*. The model is essentially a logical description of how the components of a system interact. This description is translated into a computational structure within the computer using a simulation language. The computer simulation of the system is executed over and over to generate statistical information about the system behavior. We use the statistics to describe the system performance measures. Based on what we learn about the system, we modify the computer simulation models to study alternative systems (i.e., experiment). By comparing these alternative systems statistically, we are able to offer performance improvement recommendations.

Of course, we could just experiment directly with the system. If we thought an additional person on the assembly line would improve its production, we could try that. If we thought a new configuration of the hospital emergency room would provide more efficient care, we could create that configuration. If we thought a new inventory policy would reduce inventory, we could implement it. But now the value of a computer simulation model becomes apparent. A change in a computer simulation model is clearly cheaper and less risky as compared to changing the real system. It is also faster to use a computer simulation to determine if the changes are beneficial. Also, it might be safer to try a change in a simulation model than to try it in real life. In general, it is much easier to try changes in a computer model than in a real operating system, especially since changes disrupt people and facilities.

A computer simulation model allows us to develop confidence in making performance improvement recommendations. We can try out a wide range of alternatives before disrupting an existing system. You can now see why many companies and many managers require that a simulation study be done before making substantial changes to any working system, especially when there are potential negative consequences of performance changes as well as costs that don't provide improvement.

Verification and Validation

Since simulation models often have serious consequence and are developed with a minimum set of assumptions, the validity of a simulation model needs to be carefully considered before we believe in the recommended benefits. In simulation, we often use the words "*verification*" and "*validation*" and we have specific definitions for each as they relate to simulations.

The word "verification" refers to the model and its behavior. We most often develop our simulation models with a simulation language. This language translates our modeling "intent" into a computation structure that produces output statistics. Most simulation languages can be used to describe a complex operating system and the language provides a framework of components for viewing that system. As our models become more complex, we employ more complex simulation language constructs to model the behavior. This relationship is critically dependent on our understanding of the simulation language intricacies. We may or may not fully understand computational code and thus we confront the key question in verification: *Does the simulation model behave the way we expect?* Is it possible that we may have made a mistake in employing the simulation language or perhaps the simulation language creates an unexpected behavior in the computational code? Therefore, even after we have created our simulation model, it needs to be tested to see if it behaves as expected. If we make the processing time longer, does it result in longer times in the system? If we reduce the arrival rate to a queue, does the waiting time decrease? The answers to any of these questions are specific to the system and your model of it. But, above all else, we need to be sure that our model is behaving the way we expect without error.

The word "validation" refers to the relationship between the model and the real system. *Does the simulation produce performance measures consistent with the real system*). Now, we are assuming our model has been verified but can we validly infer to the real system? This question strikes at the heart of our modeling effort because without a valid model we cannot legitimately say much about performance improvement. Many people new to simulation may want the model to be a substitute for the "real system" and that is a limitless task. After all, the only "model" that is perfectly representative of the real system is the system itself!

We must always remember in simulation that our model is only an *approximation* of the real system. So the most relevant way to validate our model is to concern ourselves with its approximation. How do we decide what to approximate? It depends on our performance measures. If our performance measure is time in the system, then we concern ourselves with those factors that impact time in system. If our performance measure is production, then we focus on those factors that impact production. Usually we are interested in several performance measures, but those measures will be the focus of our concern and will limit our modeling activity. Otherwise without a clear set of performance objectives, we are left with a search for reality in our model, which is a never-ending task.

Generally, people who have only a general familiarity with simulation, think simulation models can mimic anything and often drive the development of a model that is needlessly complex. One of the difficult responsibilities of anyone engaged in simulation modeling is the education of the stakeholders on the benefits and limitations of a simulation model as well as the simulation modeling activity.

Computer Simulation Languages

The process of creating a computer simulation model varies from programming your own model in a programming language such as C++, C#, or Java to using a spreadsheet like Excel. There are a variety of simulation languages including Simula™, GPSS™, SIMSCRIPT™, Arena/SIMAN™, SLX™, ProModel™, Flexsim™, AutoMod™, ExtendSim™, Witness™, AnyLogic, ™ among many others. Each of these simulation languages differ in the way they require users to construct a simulation model. An important distinction is the degree to which computer programming is required. Also many of the simulation languages have evolved from a particular industry or set of applications and are especially useful in that context.

We have chosen to use the simulation language SIMIO which is a relatively new language, having been developed over the past several years. The developers of the language had previously developed the Arena simulation system. SIMIO benefits from more recent developments in object-oriented design and agent modeling. SIMO is a "multi-modeling" language having agents as well as discrete event and continuous language components. SIMIO was developed to provide visual appeal through its 3-D animation and graphical representations. SIMIO provides a wide range of extensions, from direct modification of executing processes to user developed objects. SIMIO also provides interoperation with various spreadsheets and databases. Finally, SIMIO has become widely adopted in industry as well as academic institutions. Learning SIMIO provides you with one broadly-based simulation modeling tool which can help you learn to use others if the need arises.

Part 1.2: Simulation Fundamentals: The Ice Cream Store

A fundamental understanding of simulation will be beneficial throughout your study of simulation. It's easy to get caught up in the creation of a computer model using a simulation language and miss important basic principles of simulation modeling. Too many people associate simulation with a simulation language and for them, simulation is simply learning a simulation language. Learning a simulation language is necessary to using simulation, but it does not substitute for understanding at a fundamental level. If you understand the fundamentals of simulation, then you establish a basis for understanding any simulation language and any simulation model. In fact, this understanding will be a key to learning almost everything else in this text.

The Ice Cream Store

It is helpful if the discussion of simulation is done in the context of a problem – albeit a simple problem. Using this simple problem, we will be able to describe simulation elements of modeling, execution, and analysis. Our problem is the common ice cream store since everyone loves ice cream and probably has visited an ice cream store. As seen in Figure 1.1, customers arrive to the ice cream store where they obtain an ice cream cone. It's a simple store where there is only one attendant and people will wait in a single line to order and receive their ice cream cone. The attendant waits on each customer, one at a time in the order they arrive.

Question 1: If you owned or managed the ice cream store, what might be your operational concerns?

____

Likely one of your most prominent concerns would be this store's operation and how you can improve it. For example, could you buy a new cone making machine to make cones faster or should you hire someone to help service customers? Should you resize the waiting space? These are performance improvement concerns.

Figure 1.1: The Ice Cream Store

Question 2: If you made one of these changes to the ice cream store, how would you decide if it improved the store's operation?

____

In other words, what are the performance measures? Here are some possible measures: number of customers served per day, time customers spent in the store, the waiting time of customers, the number of customer waiting, and the utilization of servers. We will discover these are common performance measures and will often be your first performance measurement choices.

Question 3: How can you expect to improve performance if you don't know what is going on inside the ice cream store?

____

You need to employ all the descriptive tools you know to understand what is going on inside the system. At this stage, one might create a value stream map, a flow chart, a relationship chart, a spaghetti diagram, etc. These techniques will greatly improve your understanding of what happens to customers and the attendant during the sale of an ice cream cone. Perhaps you develop the conceptual model (i.e., flow chart) presented in Figure 1.2.

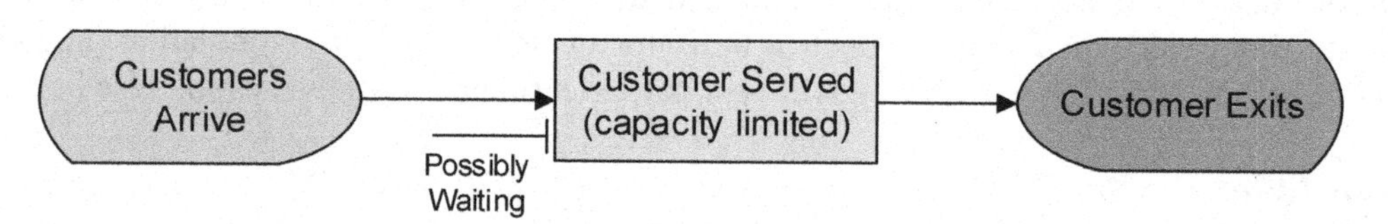

Figure 1.2: Flow Chart of the Ice Cream Store

The conceptual model shows how the customers are served. Notice, we have added the possibility of waiting due to the fact that service is limited by the availability of the single attendant.

Gathering Data about the Ice Cream Store

Descriptive information helps us understand the service process in the ice cream store; however it doesn't give us the performance measures. For that we need to document what is happening (i.e., we need to do a "*present systems analysis*"). Suppose we decide to do a "time study" of the store operations which is shown in Table 1.1.

Table 1.1: Direct Observation Time Study (Event View) of Ice Cream Store

Time	Customer	Process
0		Store Opens – Server Idle
0	1	Arrives –
		Start service on Customer 1 – Server Busy
8.36	1	Departs service – Server Idle
9.01	2	Arrives –
		Start service on Customer 2 – Server Busy
9.98	3	Arrives – Customer 3 waits
13.5	4	Arrives – Customer 4 waits
19.36	2	Departs Service – Customer 2 leaves:
	3	Start service on Customer 3 – Server Busy
23.07	5	Arrives – Customer 5 waits
27.22	3	Departs service – Customer 3 leaves:
	4	Start service on Customer 4 – Server Busy
33.82	4	Departs service – Customer 4 leaves:
	5	Start service on Customer 5 – Server Busy
38.18	6	Arrives – Customer 6 waits
40		End observations

In our time study, we simply record all the "events" and the event time that occurs in the ice cream store. An event occurs when something operationally happens in the store, like an arrival of a customer or the departure of a customer from service. Also we recorded when the attendant (i.e., server) becomes busy or idle. Here we only observed the first 40 minutes of the store's operation. We recognize this isn't really enough time to gain a full understanding, but we are not intending to solve a problem at this time – only to demonstrate a method.

The time study is sufficient for us to compute some performance measures, but first it will be helpful to re-organize our time study data. We are not going to add (or subtract) any information. We are simply going to re-organize our data from an *event view* to an *entity* (i.e., customer) *view*. The event view recorded events but the entity view allows us to follow our customers. The re-organized data is shown in Table 1.2 which shows when the customer arrives to the store, enters service, and leaves the store. Notice, customers five and six service and exiting store happen after the 40 minutes.

Table 1.2: Re-organized Event Data in Customer View From

Customer	Arrives to Store	Enters Service	Leaves Store
1	0.00	0.00	8.36
2	9.01	9.01	19.36
3	9.98	19.36	27.22
4	13.50	27.22	33.82
5	23.07	33.82	??
6	38.18	??	??

Performance Measure Calculations

From the entity view of the time study data, we can easily compute a number of performance measures.

- **Production**: Number of people served
  The number of people served is easily calculated by looking at the number of customers that have exited the system in the simulation time. In our example, four people were served in 40 minutes or the system had a production rate of six per hour (i.e., 4 customers/40 minutes * 60minutes/hour).
- **Flowtime, Cycle Time**: Time in System

The time in system is calculated by averaging the difference when the customer entered the system from when they exited the system. Again, only four customers exited the system and contributed to the average time in system in our simple ice cream system as seen in Table 1.3.

Table 1.3: Calculating Average Time in System for the Ice Cream Store

Customer 1	Customer 2	Customer 3	Customer 4	Average Time in System
8.36-0.0	19.36-9.01	27.22-9.98	33.82-13.5	56.27/4 = 14.07 minutes

- **Non-Value Added Time**: Waiting Time in Queue
  The time in queue (i.e., the waiting or holding time) represents the time customers waited in line before being seen by the ice cream attendant. It is calculated by averaging the difference when the customer entered the system from when they entered service. During our observation window, five customers entered service as seen in Table 1.4.

Table 1.4: Calculating Average Waiting Time in the Queue

Customer 1	Customer 2	Customer 3	Customer 4	Customer 5	Average Waiting Time
0.0-0.0	9.01-9.01	19.36-9.98	27.22-13.50	33.82-23.07	33.85/5 = 6.77 minutes

- **Number Waiting on In Queue:**
  This measure determines the average number of customers one would expect to see waiting in line to receive service. Unlike the previous metrics, number waiting in queue or number in the system measures are *time-persistent statistics* or *time-weighted statistics.* You may have never computed a time-persistent statistic which is a statistic for which we need to know the amount of time a value was observed. We weight the value of the observation by the amount of time that value persists. A graphical depiction of the number in the queue at the ice cream store is in Figure 1.3.

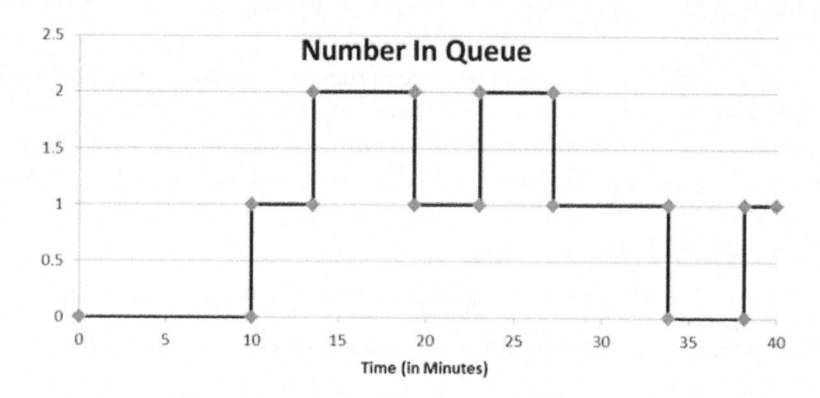

Figure 1.3: Graph of the Number in Queue in the Ice Cream Store

Consider, computing the average number in queue. Suppose we observed the number in queue to be two one time and ten another time. Would you say that on average number we would expect to see six in line? Of course "not"! You need to know how long the value of two was observed and how long the value of ten was observed. Suppose we observe two in queue for ten minutes of the time and ten in the queue for only one minute. So the queue was observed a total of 11 minutes. As a result, the value of two was observed 10/11ths of the total time while ten was observed 1/11th of the total time. Our average then is 2*(10/11) + 10 *(1/11) or = 2.73 customers. Or another way of computing is to realize that the total waiting time[2] observed was 20 + 10 over a total of 11 minutes which also yields 2.73 people. Now, looking at the data from the ice cream store, we need the percentage of time (40 minutes) that there was zero waiting, one waiting , two waiting , three and so forth[3]. In our case, a maximum of two was observed and the time waiting would be computed as seen in Table 1.5.

---

[2] Total waiting time is also the "area" under the curve in Figure 1.3.

[3] We don't include 0 since it makes no contribution to the total waiting time.

Table 1.5: Calculating Average Number in Queue

Number In Queue	Time Period	Time Spent (min)
0	0.00 – 9.98	0*9.98 = 0.000
1	9.98 – 13.5	1*3.52 = 3.520
2	13.50 – 19.36	2*5.86 = 11.72
1	19.36 – 23.07	1*3.71 = 3.710
2	23.07 – 27.22	2*4.15 = 8.300
1	27.22 – 33.82	1*6.60 = 6.600
0	33.82 – 38.18	0*4.36 = 0.000
1	38.18 – 40.00	1*1.82 = 1.820
	Total Waiting Time	35.67 minutes
	Average Number in Queue	35.67/40 = 0.89 Customers

- **Utilization:**
  This performance is the percentage time the server is busy servicing customers which is calculated by dividing the time spent servicing customers divided by the time available. For our example, the attendant is only idle during the time period between finish servicing customer one and the arrival of customer two (i.e., 9.01 – 8.36 = .65 minutes). Therefore, the attendant's utilization will be 39.35/40 or 98.4%:
- Other possible performance measures are maximum values, standard deviations, time between departures, etc.

Table 1.6: Types of Performance Simulation Measures/Metrics

Type	Description/Examples
Counts	The number of parts that exited, entered, etc. (e.g., production).
Observation-based Statistics	These measures typically deal with time like waiting time, time in system etc. SIMIO calls observation-based statistics "Tally" statistics.
Time-Persistence/Time-Average Statistics	These measures deal with numbers in system, queue, etc. when the values can be classified into different states. SIMIO calls these "State" statistics.

Question 4: What is another example of an observations-based performance measure?

____

Question 5: What is another example of a time-persistence performance measure?

____

Question 6: Is "inventory" a time-persistent or observation-based statistics?

____

Question 7: Is the amount of time that a job is late a time-persistent or observation-based statistic?

____

Terminology of a Queuing System

Figure 1.4 shows some elements of a "common" queueing system with the normal terminology. In general we will refer to the arriving objects as "model entities" and the counter/attendant as a "server". You can see the members of the queue and the customer in service. The input processes for this system are the arrival process and service processes. We will provide details on these inputs later.

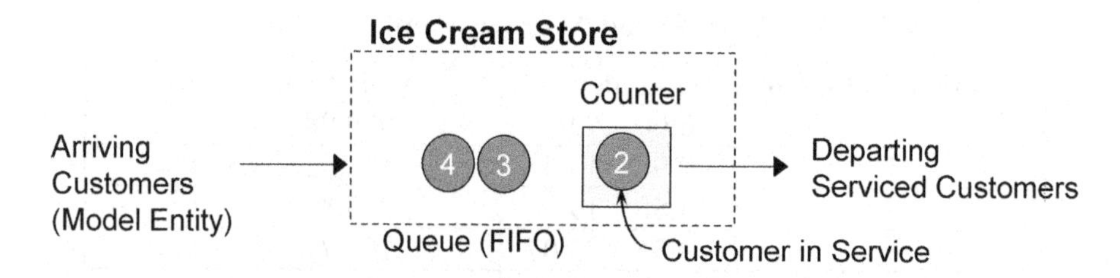

Figure 1.4: A Breakdown of Terminology

Fundamental input processes: arrival times and service times

Can we simulate it?

We want to be able to reproduce the time-study data collection exercise that we used for the ice cream store. But we want to synthesize it numerically as opposed to observing it – in essence simulate it! If we can re-create the time study data, we can compute the performance measures. Notice that the time study records are centered on "**events**". Recall these events are points in time when the system changes it state (i.e., status). A quick review of that data reveals the following three events occur.

Table 1.7: Events of the Ice Cream Store

Number	Type	Event Description
1	Arrival	Indicates an entity will arrive at this time.
2	End of Service	An entity will be departing from service (i.e., service has finished)
3	End	The simulation will terminate and end of all observations.

To facilitate our simulation we need a method to keep track of our events. Although there are other ways to keep track of events, it is convenient to keep them in an "**event calendar**". An event calendar contains the records of future events (i.e., things we think are going to happen in the future), ordered by time (with the earliest event first). Simulation can be executed by removing and inserting events into the event calendar.

Let's add some specifics to our simulation problem (i.e., the ice cream store). We will use "minutes" as our base measure of time. The input data that we currently have is shown in Table 1.8.

Table 1.8: Input Data

Customer	Arrival Time	Interarrival Time	Processing Time
1	0.00	0.00	8.36
2	9.01	9.01	10.35
3	9.98	0.97	7.86
4	13.50	3.52	6.60
5	23.07	9.57	8.63
6	38.18	15.11	10.33
7	42.08	3.90	10.46
8	48.80	6.72	7.96

In the table, the arrival time has been re-stated as an "interarrival time" – namely time between arrivals. Doing this re-statement does not change the arrival times but simply changes how they are presented. Such a representation also requires the time of the first arrival from which the interarrival times are sequentially computed. Table 1.8 provides data on only the first customers. We may need more data, but this is a later discussion.

The Simulation Algorithm

To synthesize the system as presented in the time study, we need a systematic means of moving through time by removing and inserting events on the event calendar. Consider the simulation algorithm shown in Figure 1.5.

Algorithm Step 1: Setup the simulation (i.e., initialize the system)
Algorithm Step 2: Remove the next event from the event calendar
Algorithm Step 3: Update simulation time (i.e., *TimeNow*) to the time of that event
Algorithm Step 4: Execute the processes associated with the event (i.e., adding additional events as needed and collecting statistics)

Arrival Event Process (see Figure 1.6)
Departure Event Process (see Figure 1.7)
End Event

Algorithm Step 5: Repeat *Algorithm Steps 2-4* until complete

Figure 1.5: Simple Simulation Algorithm

Remember that the event calendar is ordered according to the next most recent event. So we can move through time by removing the "next" event from the event calendar, updating time to the time of that event, and executing whatever processes are associated with that event. This simple method is then repeated until we reach some terminating time or condition.[4] In our case, the "events" are the arrival of an entity, the service (departure) of the entity, and the end of the simulation period. The arrival and departure of the entities are the most important. Consider now how the arrival (see Figure 1.6) and the departure (see Figure 1.7) are processed within the context of our simple single queue, single server system. The word "schedule" means to insert this event into the event calendar.

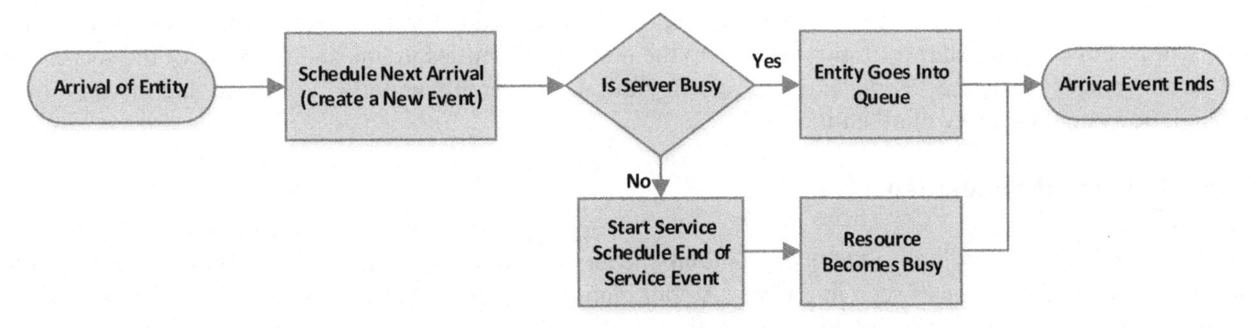

Figure 1.6: Event Process Associated with Arrival of Entity

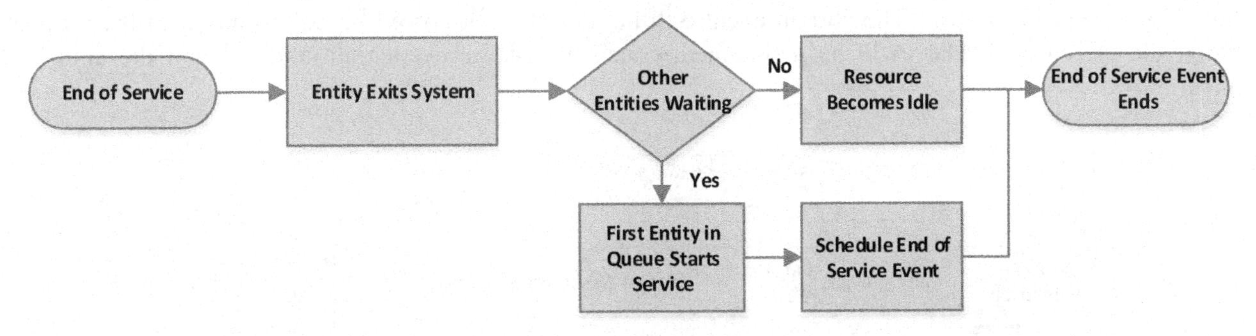

Figure 1.7: Event Process Associated with End of Service

The arrival of an entity creates a new event because we know the next entity's arrival time since we have the interarrival times of entities. The newly arriving entity either must wait because the server (i.e., resource) is busy or that entity can engage the server and start service. If the entity can start service, we now know a new future event, namely the service departure because we know the processing time. So if the removed event is an arrival, we insert a next arrival into the event calendar and may insert the service departure event in the event calendar provided the entity can start service immediately. If the event removed from the event calendar is a service departure, then the entity that has finished service will exit the system. If the waiting queue is empty, then the resource (server) becomes idle. On the other hand, if there is at least one customer in queue,

[4] The practice of removing the "next" event has caused some to refer to our simulation as a "next or discrete event simulation".

then the first customer is brought into service and the resource remains busy and a new future event, namely the service departure is inserted into the event calendar because we have the processing time. Note that a service departure causes the entity to depart the system regardless of what happens to the server.

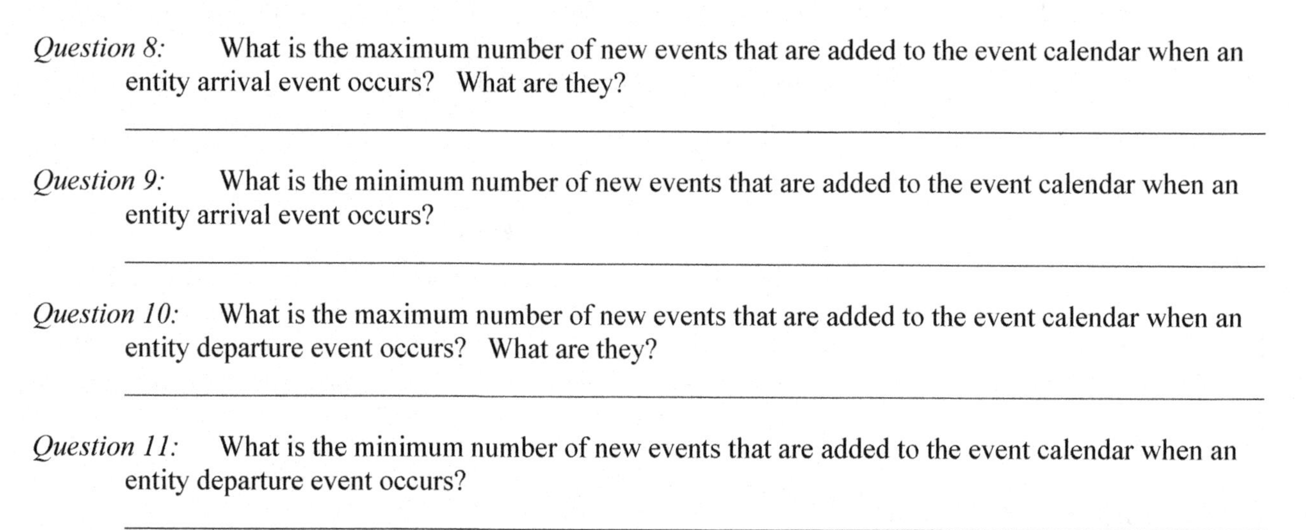

Question 8: What is the maximum number of new events that are added to the event calendar when an entity arrival event occurs? What are they?

____

Question 9: What is the minimum number of new events that are added to the event calendar when an entity arrival event occurs?

____

Question 10: What is the maximum number of new events that are added to the event calendar when an entity departure event occurs? What are they?

____

Question 11: What is the minimum number of new events that are added to the event calendar when an entity departure event occurs?

____

Finally, we note that the very first step of the simulation algorithm calls for the system to be "initialized." In other words, how will we start operation relative to the number of entities in line and the state of the server? It will be convenient to use the "empty and idle" configuration. By "empty and idle" we mean that the server is idle and the system is empty of all entities.

Part 1.3: Manual Simulation

To further understand how a discrete event simulation operates, a simple data structure will be employed to execute a manual simulation as seen in Figure 1.8. Our manual simulation will consist of a "system animation" graphical representation of what is going on in the system with the current customer being serviced shown inside the square and other customers waiting outside the square. The current simulation time (i.e., called "Time Now") will be shown. The current event will be identified, followed by a description of the process. Finally, the "Event Calendar" will be maintained, consisting of the event time and typ and the entity ID number.

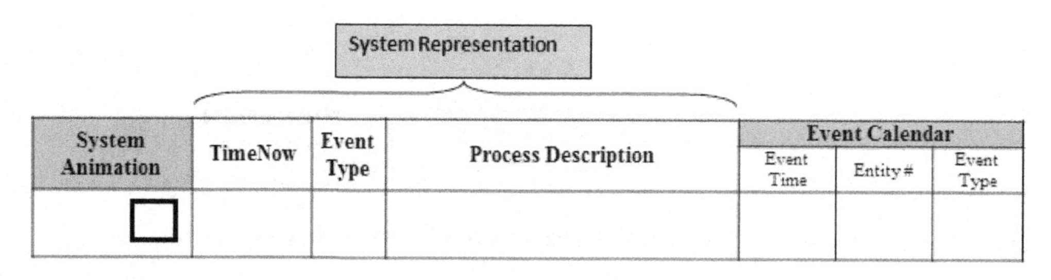

Figure 1.8: Manual Simulation Data Structure

Step 1: Algorithm Step 1 of the simulation as defined in Figure 1.5 is used to "initialize" the system as seen in executing structure of Figure 1.9. Note, that we have inserted into the event calendar the time of the first entity arrival and we have inserted the end event or the time to stop making observations.

System Animation	Time Now	Event Type	Process Description	Event Calendar		
				Event Time	Entity #	Event Type
□	0.0		Initialize simulation	0.0 40.0	1	Arr End

Figure 1.9: System Initialization

Step 2: Executing the simulation algorithm, we remove the next event (Algorithm Step 2) from the event calendar, update simulation time to the time of that event (Algorithm Step 3), and execute the appropriate processes (Algorithm Step 4). The next event is the "Arrival of Entity #1" at time 0.0. The arrival of this entity allows us to schedule into the *Event Calendar* the "Arrival of Entity #2" to occur since the interarrival time between Entity #1 and Entity #2 is 9.01 minutes and thus the *Arrival* event for Entity #2 is at time 9.01 (i.e., 0.0 + 9.01). Because Entity #1 arrives when the server is idle, that entity enters service and we can now schedule its *Departure* event because we know the processing time for Entity #1 as 8.36 minutes thus the event time is 8.36 (i.e., 0.0 + 8.36). The result is that our data structure now appears as Figure 1.10.

System Animation	Time Now	Event Type	Process Description	Event Calendar		
				Event Time	Entity #	Event Type
[1]	0.0	Arr	Store opens and Entity 1 arrives and goes into service	8.36 9.01 40.0	1 2	Dep Arr End

Figure 1.10: Processing Time 0.0 Event

Question 12: We inserted the arrival of Entity #2 event before the departure of the Entity #1 event, so why is the departure event ahead of the arrival event in the event calendar?

____

Question 13: Why do we compute the new event times as 0.0 +? What does the 0.0 mean?

____

Step 3: We are done with the processes associated with time 0.0. Now, the next event in the event calendar is the "Departure of Entity #1" from the system at time 8.36. Since there are no entities in the queue, the server is allowed to become idle and no new events are added to the event calendar as seen in Figure 1.11.

System Animation	Time Now	Event Type	Process Description	Event Calendar		
				Event Time	Entity #	Event Type
□	8.36	Dep	Entity 1 departs service and server become idle	9.01 40.0	2	Arr End

Figure 1.11: At Time 8.36

Step 4: The next event is the arrival of Entity #2 at time 9.01. So we remove it from the event calendar and update the simulation time to 9.01. We can schedule the arrival of Entity #3 at the current time (9.01) plus the interarrival time from Entity #2 to Entity #3 (0.97) which is event time 9.98. Next, Entity #2 can start service immediately since the server is idle, so its departure event can be scheduled as the current time (9.01) plus the processing time for Entity #2 of (10.35), which yields an event time of 19.36. The new status of the simulation is shown in Figure 1.12. We also now have added animation of the status of the entities and the server.

System Animation	Time Now	Event Type	Process Description	Event Calendar		
				Event Time	Entity #	Event Type
2	9.01	Arr	Entity 2 arrives and enters service	9.98	3	Arr
				19.36	2	Dep
				40.0		End

Figure 1.12: At Time 9.01

Step 5: The next event occurs at time 9.98 which is the arrival of Entity #3. Its arrival allows us to schedule the next arrival (i.e., Entity #4) at time 13.50. But now the arriving entity must wait which is indicated in the "System Animation" section of the data structure in Figure 1.13.

System Animation	Time Now	Event Type	Process Description	Event Calendar		
				Event Time	Entity #	Event Type
3 2	9.98	Arr	Entity 3 arrives and must wait	13.5	4	Arr
				19.36	2	Dep
				40.0		End

Figure 1.13: At Time 9.98

Question 14: How did we get 13.50 for the arrival of Entity #4?

____

Step 6: The next event is the arrival of Entity #4, which will schedule the arrival of Entity #5, but will have no other actions since the server remains busy. The new simulation time is 13.50 and the updated status is shown in Figure 1.14

System Animation	Time Now	Event Type	Process Description	Event Calendar		
				Event Time	Entity #	Event Type
4 3 2	13.50	Arr	Entity 4 arrives and must wait	19.36	2	Dep
				23.07	5	Arr
				40.0		End

Figure 1.14: At time 13.50

Step 7: Finally, we see that Entity #2 finishes service at 19.36 and departs. Entity #3 can go into service and we can schedule the service departure of Entity #3 at 27.22 as shown in Figure 1.15

System Animation	Time Now	Event Type	Process Description	Event Calendar		
				Event Time	Entity #	Event Type
4 3	19.36	Dep	Entity 2 departs and Entity 3 goes into service	23.07	5	Arr
				27.22	3	Dep
				40.0		End

Figure 1.15: At Time 19.36

Question 15: How did we compute the service departure for Entity #3 to be 27.22?

____

Question 16: What is the next event?

____

Question 17: What new events are added as a result of this event? (Give the event time, the entity, and the event type?

___

Step 8: From Figure 1.15, we see the next event is the arrival of Enity #5 at time 23.07 which triggers the addition of the next arrival (Entity #6).

System Animation	Time Now	Event Type	Process Description	Event Calendar		
				Event Time	Entity #	Event Type
5 4 [3]	23.07	Arr	Entity 5 arrives and must wait	27.22 38.18 40.0	3 6 	Dep Arr End

Figure 1.16: At Time 23.07

Step 9: Entity #3 completes service at 27.22 which allows Entity #4 to enter serviced which schedules the end of service event for Entity #4 at 33.82 as shown in Figure 1.17.

System Animation	Time Now	Event Type	Process Description	Event Calendar		
				Event Time	Entity #	Event Type
5 [4]	27.22	Dep	Entity 3 departs and Entity 4 goes into service	33.82 38.18 40.0	4 6 	Dep Arr End

Figure 1.17: At Time 27.22

Question 18: What is the next event?

___

Question 19: What new events are added as a result of this event? (Give the event time, the entity, and the event type)?

___

Step 10: The result of this next event is shown in Figure 1.18.

System Animation	Time Now	Event Type	Process Description	Event Calendar		
				Event Time	Entity #	Event Type
[5]	33.82	Dep	Entity 4 departs and Entity 5 goes into service	38.18 40.0 42.46	6 5	Arr End Dep

Figure 1.18: At Time 33.82

Question 20: What is the next event?

___

Question 21: What new events are added as a result of this event? (Give the event time, the entity, and the event type?

___

Step 11: We are getting closer to the time to quit observing the system; however we have one more event and one more time update. The result of this next event is shown in Figure 1.18.

System Animation	Time Now	Event Type	Process Description	Event Calendar		
				Event Time	Entity #	Event Type
(6) [(5)]	38.18	Arr	Entity 6 arrives and must wait	40.0 42.08 42.46	 7 5	End Arr Dep

Figure 1.19: At Time 38.18

Step 12: Finally, the next event calls for the "End" of the simulation at time 40.0. Only time is updated as the event calendar is unchanged. The final state is given in Figure 1.20.

System Animation	Time Now	Event Type	Process Description	Event Calendar		
				Event Time	Entity #	Event Type
(6) [(5)]	40.00	End	Simulation Ends	40.0 42.08 42.46	 7 5	End Arr Dep

Figure 1.20: At Time 40.0

Our simulation approach is referred to as a "*Discrete-Event Simulation*" because the system only changes states at defined event times and the state changes are discrete in that the number in queue changes discretely as well as the number in system. If our model included variables that changed continuously with time, say like the water in a tank, then we wouldn't have a discrete-change system and we would have to consider all points in time, not just those when the system changes state. Simulations that contain continuous variables are called "Continuous Simulations." And we can have combinations of the two types of simulations. In fact, SIMIO can model both kinds of systems together – a multi-method simulation language.

Part 1.4: Input modeling and Simulation Output Analysis

Since we now have simulated the ice cream store time study, we can re-organize this information from an entity viewpoint, and compute the *same performance statistics* that we computed previously!

Question 22: Do we have enough information to draw conclusions about the present system?

____

Question 23: What can we do to extend the information available?

____

If we had a total of 45 interarrival times and 43 processing times, we could simulate a total of 480 minutes of time. For example, we may obtain the information given in Table 1.9. We have included the minimum, maximum, and average values over the 480 minutes. Currently we simulated "only" one day of 480 minutes.

Question 24: Now do you have enough information for a "present systems analysis?

____

Question 25: Is a one-day of simulation long enough?

____

Table 1.9: Final Performance Measures for 480 Minutes

Performance Measure	Value
Total Production	43
Average waiting time in queue	9.59
Maximum waiting time in queue	35.65
Average total time in system	17.87
Maximum total time in system	42.65
Minimum total time in system	7.46
Time-average number of parts in queue	0.88
Maximum number of parts in queue	4
Ice Cream Attendant utilization	78%

One day doesn't give us much information about the day so more "days" of information is needed. But that means more interarrival times and more processing times. If we did a simulation of ten days, we would need approximately 450 interarrival times and 430 processing times. If that information came from an electronic record, then getting more data may not be a problem.[5] But if our only alternative is time studies, then we will need to observe 10 days and that may be a greater intrusion into the actual operation than we can expect.

An alternative to more data collection is for us to find a "model" of this input data (i.e., interarrival and processing times). Perhaps we can use a statistical representation since processing times and interarrival times are most certainly random variables. We can often match the processing times and interarrival times to standard statistical distributions. First we make a histogram of the data we observed. Second, we try to pick out a statistical distribution, like a Gamma, Lognormal, Weibull, or Pert, to match the data. In fact, there is a wide range of software that can help with this undertaking. An example of this activity is shown in Figure 1.21.

Statistic	Value
Sample Size	44
Range	4.31
Mean	8.6277
Variance	1.3242
Std. Deviation	1.1507
Coef. of Variation	0.13337
Std. Error	0.17348
Skewness	0.37355
Excess Kurtosis	-0.85959

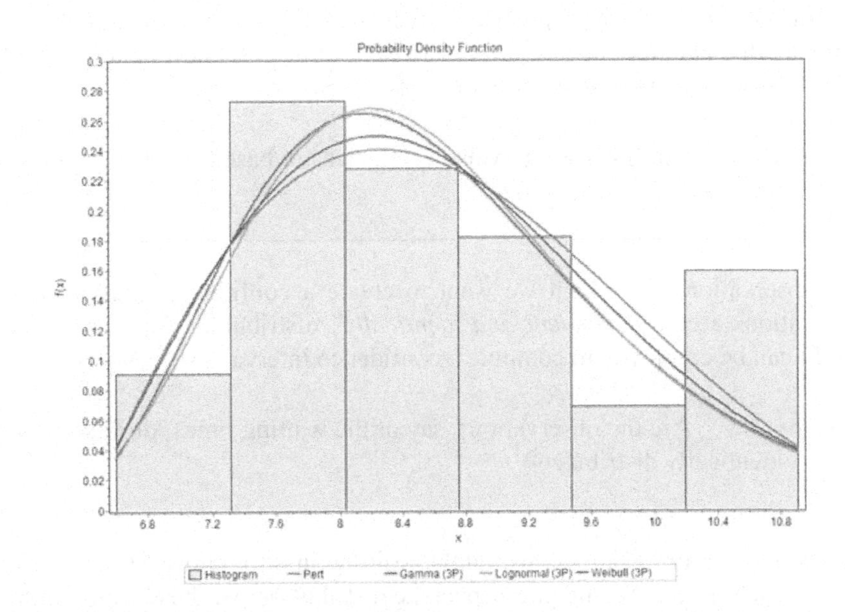

Figure 1.21: Matching Observed Processing Time to a Standard Statistical Distribution

As a result of this input modeling, we can characterize a random processing time or a random interarrival time by a statistical model. For example, we might use a Pert distribution which has three parameters as: Pert (6, 8, 12) for a processing time or maybe a Gamma distribution which has two parameters as: 6.3 + Gamma(3.1, 0.7) where the 6.3 is the "location" of the distribution's origin.

[5] An implication of this need for more data is that performance improvement should be one of the bases on which an information system is designed. The information should not be limited to accounting and reporting, but also performance improvement.

Once we characterize an input with a statistical model, there is simulation technology[6] that allows us to "sample" from that statistical model. Repeated sampling will statistically reproduce the input model, but more importantly give us an "infinite" supply of input data. With that extended data, we can simulate many days of operation. Having many days of simulation provides us with the opportunity to gain an estimate of the precision of the average performance measures we compute. For example suppose we simulate our system for five days. Using simulation terminology, we performed five "replications" or "runs" of our simulation model. Each replication obtains new "samples" for its interarrival times and processing times. To compute the average and the standard deviation, we use the averages from each day (one day yields one average). The results are shown in Table 1.10.

Table 1.10: Results from Five Replication of 480 Minutes Each

Replication	1	2	3	4	5	Avg.	Std.Dev.
Avg. time in queue	9.59	15.26	12.98	8.08	21.42	13.47	5.26
Avg. no. in queue	0.88	1.62	1.17	0.76	2.31	1.35	0.63
Machine Utilization	0.78	0.86	0.75	0.78	0.81	0.80	0.04

Question 26: How many observations are used to compute the average and standard deviation in Table 1.10?

____

Question 27: Why only the averages instead of all the queue waiting times within a given replication?

____

We know that these statistics, like the average and the standard deviation, are themselves random variables. If we looked at another five days (either simulated or real) we wouldn't get the exact same results because of the underlying variability in the model. We need to have some idea about how precise these summary statistics are. In order to judge how precise a given statistic is, we often use a *confidence interval.* For example, we computed the average waiting time as 13.47, but does this estimate have a lot of variability associated with it or are we pretty confident about that value?

Question 28: Confidence intervals in statistics are based on what famous distribution as well as famous theorem?

____

Any observations for which we want to create a confidence interval must satisfy the assumptions that the observations are "*independent and identically*" distributed. As a consequence the **Central Limit Theorem (CLT)** can be employed to compute a confidence interval.

Question 29: Are the observations, say of the waiting times, during a simulated day independent and identically distributed?

____

Question 30: For example, would the waiting time for entity #1, entity #2, and entity #3 be independent of each other? Would you expect the distributions of these waiting times to be identical?

____

Question 31: Would the number waiting in the queue at 9am be independent of the number waiting at 8:45am? Would the distribution of the number waiting be identical?

____

[6] Random number and random variate generation methods.

Almost no statistic (performance measure)[7] computed during a single simulation replication will be independent and identically distributed (e.g., the waiting time of entity #4 maybe dependent on entity #3 waiting and processing times). So instead, we use, for example, the average values computed over the day as an observation (i.e., we will determine the average of the averages).

Question 32: Are daily averages independent and identically distributed? Why?

Since these daily averages (or maximums or others) are independent and identically distributed, we can compute confidence intervals for our output statistics (based on the Central Limit Theorem).

Confidence Intervals on Expectations

A confidence interval for an expectation is computed using the following formula:

$$\bar{X} \pm t_{n-1,1-\frac{\alpha}{2}} \frac{\hat{s}}{\sqrt{n}},$$

where $\bar{X}$ is the sample mean, $\hat{s}$ is the sample standard deviation of the data, $t_{n-1,1-\frac{\alpha}{2}}$ is the upper $1-\frac{\alpha}{2}$ critical point from the Student's t distribution with $n-1$ degrees of freedom, and n is the number of observations (i.e., replications). Note, $\frac{\hat{s}}{\sqrt{n}}$ is often referred as the standard error or the standard deviation of the mean. As the number of replications is increased, the standard error decreases. So a 95% confidence interval of the expected time in the queue using the data from Table 1.10 would be computed as the following calculation.

$$13.47 \pm 2.776 \frac{5.26}{\sqrt{5}}, \text{or } 13.47 \pm 5.53 \text{ minutes or the interval of } [6.93, 20.00] \text{ minutes}$$

Question 33: Are you confident in the 13.47 average?

Question 34: Would you bet your job that the "true" average waiting time is 13.47 minutes?

So another way to express the confidence interval is given by the *Mean* ± the *Half-width,* as in 13.47 ± 5.53 which SIMIO will use. The Central Limit Theorem allows the calculation of the confidence interval by asserting that the standard error (or the Standard Deviation of the Mean) can be computed by dividing the standard deviation of observations by the square root of the number of observations. If the observations are not independent and identically distributed then that relationship doesn't hold. As n increases sufficiently large the CLT and the inferential statistics on the mean of the population become valid.

Question 35: Does the confidence interval we computed earlier (i.e., 13.47 ± 5.53) mean that 95% of the waiting times fall in this interval?[8]

---

[7] Of course there are exceptional cases.

[8] Prediction intervals are used for observations.

Question 36: Does it mean that if we simulated 100 days that 95% average daily waiting times would fall in this interval?

___

Question 37: Or looked at another way. Can we say that there is a 95% chance that the "true, unknown" overall mean daily waiting time would fall in this interval?

___

So the confidence interval measures our "confidence" about the computed performance measures. A confidence interval is a statement about the mean (i.e., the mean waiting time), not about observations (i.e., individual waiting times). Confidence intervals can be "wider" than one would like. However, in simulation, we control how many days (i.e., replications) we perform in our analysis thus affecting the confidence on the performance measures.

Question 38: Using simulation, how can we improve the precision of our estimates (reduce the confidence interval width?

___

So we can run more replications in our simulation if we want "tighter" confidence intervals.[9]

Comparing Alternative Scenarios

In simulation, we usually refer to a single model as a simulation "scenario" or a simulation "experiment". Our model of the present ice cream store is a single scenario. However using simulation as a performance improvement tool, we are interested in simulating alternative models of, for example, our ice cream store. We would refer to each model as a scenario. So, for example, if we added a new ice cream making machine, this change would constitute a different simulation scenario. If we started an ad campaign and expected an increase in business in the ice cream store, we would have yet another scenario. In general, we would typically explore a whole bunch of scenarios, expecting to find improvements in the operations of the ice cream store. The simulation is our experimental lab.

So let's reconsider a different scenario for our ice cream store. What would happen if the arrival rate was increased by 10% (i.e., more customers arrive per hour owing to an ad campaign)? We could reduce the interarrival time by 10% and make five additional replications of the 480 hour day. The results of the original and this alternative scenario are shown in Table 1.11 along with the confidence intervals.

Table 1.11: Results of Original and Added Time Scenarios

Replication	**1**	**2**	**3**	**4**	**5**	**Avg.**	**Std.Dev.**	**LCL**	**UCL**
Avg. time in queue	9.59	15.26	12.98	8.08	21.42	13.47	5.26	6.93	20.00
Avg. no. in queue	0.88	1.62	1.17	0.76	2.31	1.35	0.63	0.56	2.13
Machine Utilization	0.78	0.86	0.75	0.78	0.81	0.80	0.04	0.74	0.85

Increased Customer Arrival

Replication	**1**	**2**	**3**	**4**	**5**	**Avg.**	**Std.Dev.**	**LCL**	**UCL**
Avg. time in queue	14.77	24.42	18.39	12.10	31.18	20.17	7.69	10.62	29.72
Avg. no. in queue	1.50	2.77	1.80	1.18	3.74	2.20	1.05	0.90	3.50
Machine Utilization	0.85	0.92	0.82	0.82	0.85	0.85	0.04	0.80	0.90

Variability in the outcome creates problems for us. We recognize that we simply cannot compare scenarios of only one replication, but even with five, it's hard to know if there are any real differences (although it appears so). Once again we need to rely on our statistical analysis to be sure we are drawing appropriate conclusions.

[9] Note that we can also create a smaller confidence level by using is a larger significance level or α value)

When comparing different sets of statistics like this, we would resort to the Student's t-test. A way to conduct the t-test is to compare confidence intervals for each of the two scenarios. If the confidence intervals overlap then we will fail to reject the null hypothesis that the mean time in system for the original system equals the new system.

The 95% confidence interval for the original five days was [6.94, 20.00] while for the increased arrival rate, the confidence interval is [12.48, 29.76]. Now comparing the confidence intervals, we see that they "overlap", meaning *we cannot say there is a statistical difference in the average waiting time.* Without a statistical difference, any statement about the practical difference[10] is without statistical foundation.

Question 39: What can we do to increase our chance of obtain a statistical difference?

____

Question 40: It is necessary to have a statistical foundation for our recommendations?

____

Although it is probably unnecessary to have a statistical foundation for every recommendation, we strive to do so to avoid the embarrassing situation of making a claim that is later shown to be erroneous – especially since our job may be at risk. By striving to have a statistical foundation for our recommendations, we take advantage of our entire toolbox in decision-making and promote our professionalism.

Part 1.5: Elements of the Simulation Study

The entire simulation study is composed of a number of elements, which we present here. Although these are presented in a sequential manner, rarely is a simulation study done without stopping to return to an earlier issue whose understanding has been enhanced. In many instances we work on several of the elements at the same time. But regardless of the order, we usually try to complete all the elements.

- *Understand the system*: Getting to know and understand the system is perhaps the most intense step and one that you will return to often as you develop a better understanding of what needs to be done.
- *Be clear about the goals*: Try to avoid "feature or scope creep". There is a tendency to continue to expand the goals well beyond what is reasonable. Without clear goals, the simulation effort wanders. During this phase identify the performance measures of interest that will be used to evaluate the simulation.
- *Formulate the model representation*: Here you are clearly formulating the structure and input for your model. Don't spend a lot of time doing data collection at this point because, as you develop the model, its data needs will become clearer. Also be sure to involve the stakeholders in your formulation, to avoid missing important concerns.
- *Translate your conceptual representation into modeling software*, which in our case is SIMIO. A lot of time is spent learning SIMIO so you have a wide range of simulation modeling tools with which to build this and other models.
- *Determine the necessary input modeling*: At some point data will need to be collected on the inputs identified during the formulation and translation of the system into a computer model. The initial simulation model can be used to determine which inputs are the most sensitive with regard to the output which need to be collected. Fitting distributions is generally better but expert opinion can be used to get the model up and running.
- *Verify the simulation*: Be sure the simulation is working as it is expected without errors. Do some "stress tests" to see if it behaves properly when resources are removed or when demand is increased. Explain any

[10] A practical difference means that the difference is important within the context of the problem. When we are concerned with a practical difference of unimportant or cheap items, then a statistical difference is not necessary.

"zeros" that appear in the output. Don't assume you are getting counter-intuitive results when they may just be wrong.

- *Validate the model*: How does the model fit the real world? Is the simulation giving sufficient behavior that you have confidence in its output. Can you validly use the model for performance improvement?
- *Design scenarios*: Determine which alternatives you think will improve the performance of the present system and create the alternative simulation models and associate the models with scenarios.
- *Make runs*: do the simulation experiments for the scenarios. Be sure your simulation output generates the appropriate performance measures. Make multiple runs for each scenario.
- *Analyze results and get insight*: Examine carefully the output from the scenarios and begin to develop judgments about how to increase the performance of the system. Be sure the statistical analysis supports your conclusions.
- *Make Recommendation and Document*: Be sure to discuss the results with all the stakeholders and decision-makers. Make sure you have addressed the important problems and developed feasible recommendations. Document your work so you or someone else can return one year later and understand what you have done.

Part 1.6: Commentary

If you have worked carefully through this chapter, you will have a fundamental understanding of simulation that is completely independent of the simulation software or any particular application. Some of the key points have been.

- A simulation model consists of a system structure and input.
- The insertion and removal of events drive a simulation.
- Random variables are used to represent input.
- Simulation statistics include observations and time-persistent values.
- A simulation may consist of many performance measures.
- Verification and validation are important concerns in any simulation.
- By using a confidence interval, we have some measure of variability as well as central tendency of the simulation output.
- Computerization of simulation greatly facilitates its value in the present and in the future.

Chapter 2
Introduction to SIMIO: The Ice Cream Store

Simulation is a very useful tool. Just about everyone who learns about simulation gets excited about its widespread applicability. Practically any manufacturing, production, or service system can benefit from a simulation-based investigation. SIMIO can make the application of simulation easier to do and at the same time provide a powerful approach for addressing complex problems for designing and improving these types of systems.

Also, simulation is fun when you build models that are visually appealing and SIMIO kicks the fun up a notch by offering 3D (i.e., three dimensional) visualization. Not only does the animation allow you can see how the model is behaving, but people you work with can as well which will greatly increase your credibility with them.

Part 2.1: Getting Started

We are assuming you have installed SIMIO on your computer and are now ready to get going.

Step 1: Invoke SIMIO by either opening the Start menu in Windows or by clicking on a SIMIO icon that perhaps you have located on the Windows taskbar or on the Desktop. The opening SIMIO window as seen in Figure 2.1 has the standard Microsoft Office "look and feel".

Step 2: When you open SIMIO for the first time, you will be placed into a modeling environment. The "*Run Tab*" of the "SIMIO ribbon" has been selected and contains the *Run*, *Run Setup*, *Animation Speed*, and *Display* sections.

Step 3: Below the SIMIO ribbon are the SIMIO window tabs. The "*Facility*" window has been selected and is the canvas that will be used to build your simulation models. The SIMIO window consists of the display of "*Libraries*" area on the left, a "*Browse*" section on the right, and the middle section is the modeling canvas.

Step 4: The *Libraries* section displays the [*Standard Library*] which contains the modeling object definition icons. Those definitions are used to create objects by clicking on the particular definition icon, dragging the icon to the modeling canvas to position it, and then clicking to complete the object definition. Most of the icons have names that connote their meaning. Also under the *Libraries* section is the [*Flow Library*] which will show the object definition icons for modeling various "flow" characteristics. Finally, the last portion of the Libraries section is the "*Project Library*" which will contain the objects that become a part of your SIMIO simulation project including the MODELENTITY object.

Step 5: In the *Browse* section, the top section is the "Navigation" panel, which identifies the components of your simulation model. By default, a new project will be named "MySimioProject" and will consist of two defined objects, a "MODELENTITY" and a "MODEL". The lower section of the navigation panel displays the "*Property Inspector*". This panel will display the properties of the object selected in the navigation panel. A specification in the Property Inspector is to "Show Commonly Used Properties Only" which will filter out certain properties. Since we want to learn all of SIMIO, we will not use this screening and should uncheck the option.

Step 6: Finally, notice the "*Help*" options in the upper right-hand corner which can also be invoked through the ***F1*** key. There is lots of information to be found in the help documents, especially after you have some experience with SIMIO concepts and features.

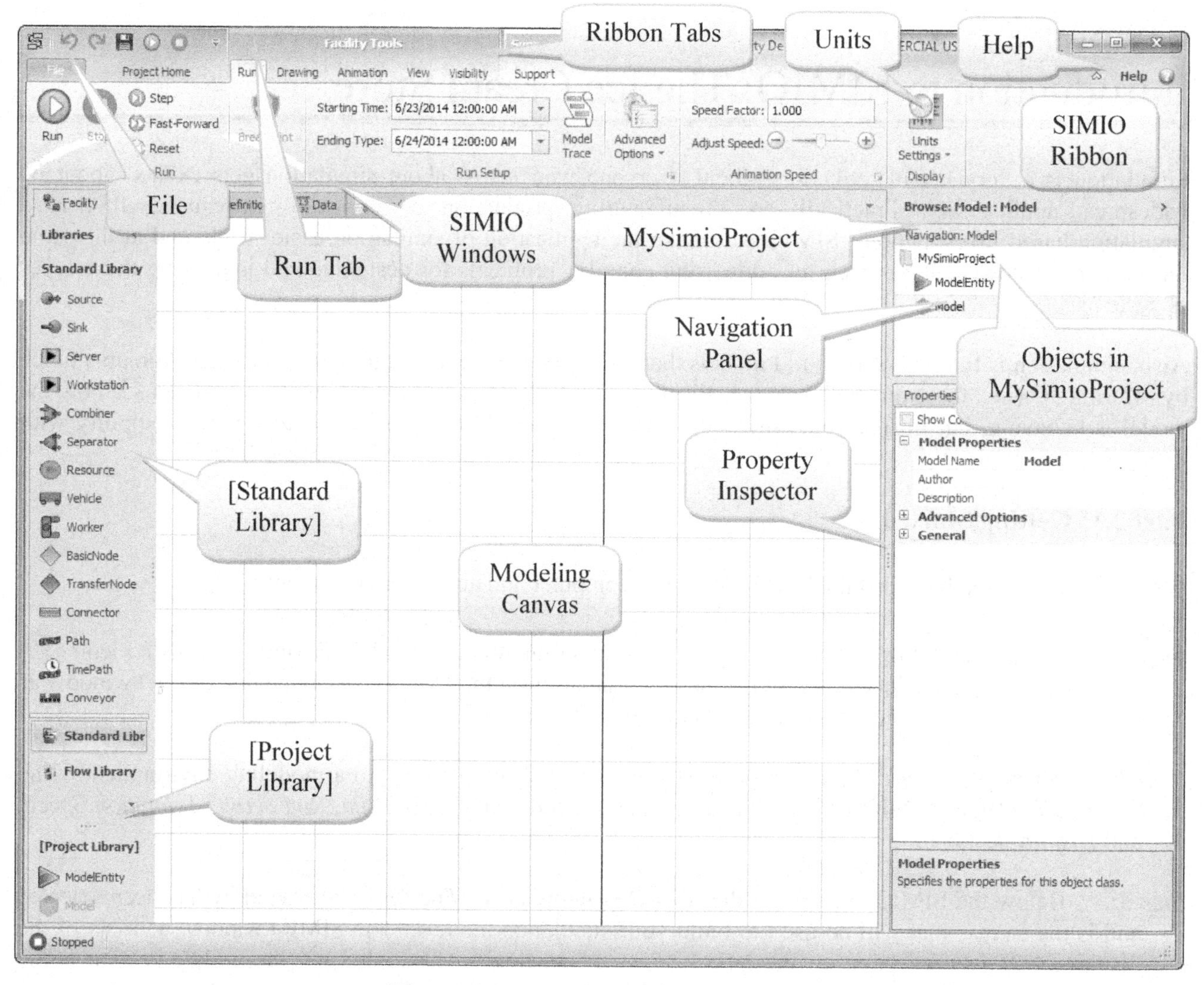

Figure 2.1: Opening SIMIO Window

Other SIMIO resources available to you are found under the SIMO ribbon tab "*Support*". In the "*Learning Simo*" section, the "*SimBits*" will refer you to a library of elemental models (each illustrating a particular modeling structure inside of SIMIO) and the "*Examples*" section will refer you to a library of complete models that provide interesting SIMO models. And there are a variety of Books, Videos, Training, and guides. Also note the SIMIO "version"[11] in the "My Software" section. You will want to keep your version up-to-date.

Step 7: When a new model is created (see Figure 2.1), two objects are automatically defined. The MODELENTITY will create entities that move through our model while the MODEL will contain positioned objects and the flow paths of entities. In a sense, the entities will roam around the positioned or fixed objects.

Step 8: Right-click the mouse on the MODELENTITY and then on the MODEL objects in the [*Navigation*] Panel and select the "Properties".

Question 1: What is the "default" Model Name of the MODELENTITY?

____

[11] SIMIO "Sprint" releases come out approximately monthly, so the software will be changing rather quickly. Be sure you obtain updates with new features and possibly bug fixes.

Question 2: What is the Object Type of the MODEL (look under the "*Advanced Options*")?

____

Step 9: Note the *Undo/Redo* buttons. These can be used to correct mistakes and recover previous modeling components. When an action cannot be undone, SIMIO will issue a warning.

Step 10: The "Units" are a convenient way to set the default specification units for time, length, rate, area, volume and mass.[12]

Part 2.2: The Ice Cream Store

A small ice cream store sells ice cream cones. Customers arrive and wait in line to be served by one of two attendants. These attendants take the ice cream order and give the cone(s) back to the customer who then moves to the separate cashier to pay. After paying, the customers will leave the store. In building a simulation model, it is often important to flowchart the processes. Figure 2.2 shows the four processes of the ice cream store.

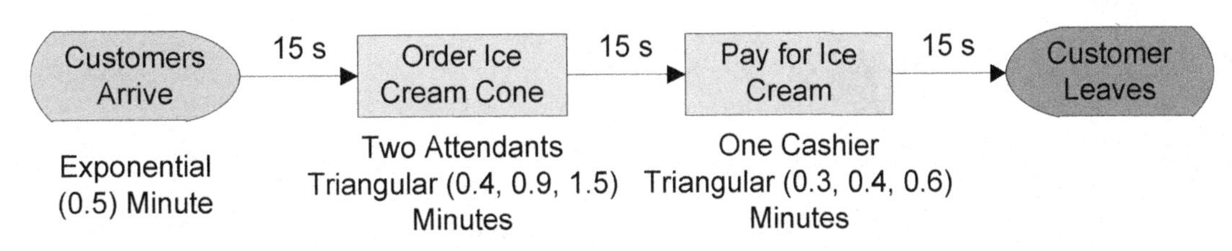

Figure 2.2: Flowchart of Ice Cream Store

In this problem we will assume we know that:

- Customers arrive Exponentially with a mean interarrival time of 0.5 minutes,
- The time an attendant takes to interact with the customer and give them their ice cream cone(s) is modeled with a Triangular distribution with a minimum of 0.4 minutes, a most likely time of 0.9 minutes, and a maximum of 1.5 minutes,
- The time the cashier takes to accept payment for the ice cream is also Triangular with a minimum of 0.3 minutes, a most likely time of 0.4 minutes, and a maximum of 0.6 minutes, and
- The travel time between each process is 15 seconds.

Customers will wait in a single waiting line on a first-come, first-served basis if both attendants are busy. Likewise there is a single waiting line for the cashier. We will also assume there is no limit to the length of the waiting lines.

Step 1: It should be fairly clear that the customers should be modeled by a MODELENTITY while the MODEL will consist of the flowchart of the Ice Cream Store as seen in Figure 2.2. Entities are the objects that will traverse through the network of fixed objects. Add an entity to your model by clicking on the MODELENTITY object in the [*Project Library*] panel and dragging it onto the canvas.

Step 2: Add objects to your model by clicking on the object type in the [*Standard Library*] panel. Drag the object around the modeling canvas and click to drop it into position (you can left-click and drag the objects around to relocate or delete them from the model). When you click on an object, its properties appear in the property inspector (typically on the bottom right side panel.

- Add a SOURCE object, two SERVER objects, and one SINK object – see Figure 2.3.

[12] SIMIO internally keeps all times in hours, distances in meters, rates in per hours, weights in kilograms, areas in square meters and volumes in cubic meters.

- Connect the objects with TIMEPATH objects by clicking on the "output" node (blue diamond) of an object and connecting it to the "input" node (grey diamond) of an object. You can click between the nodes to produce a multi-segment path which allows the path to be more flexible.

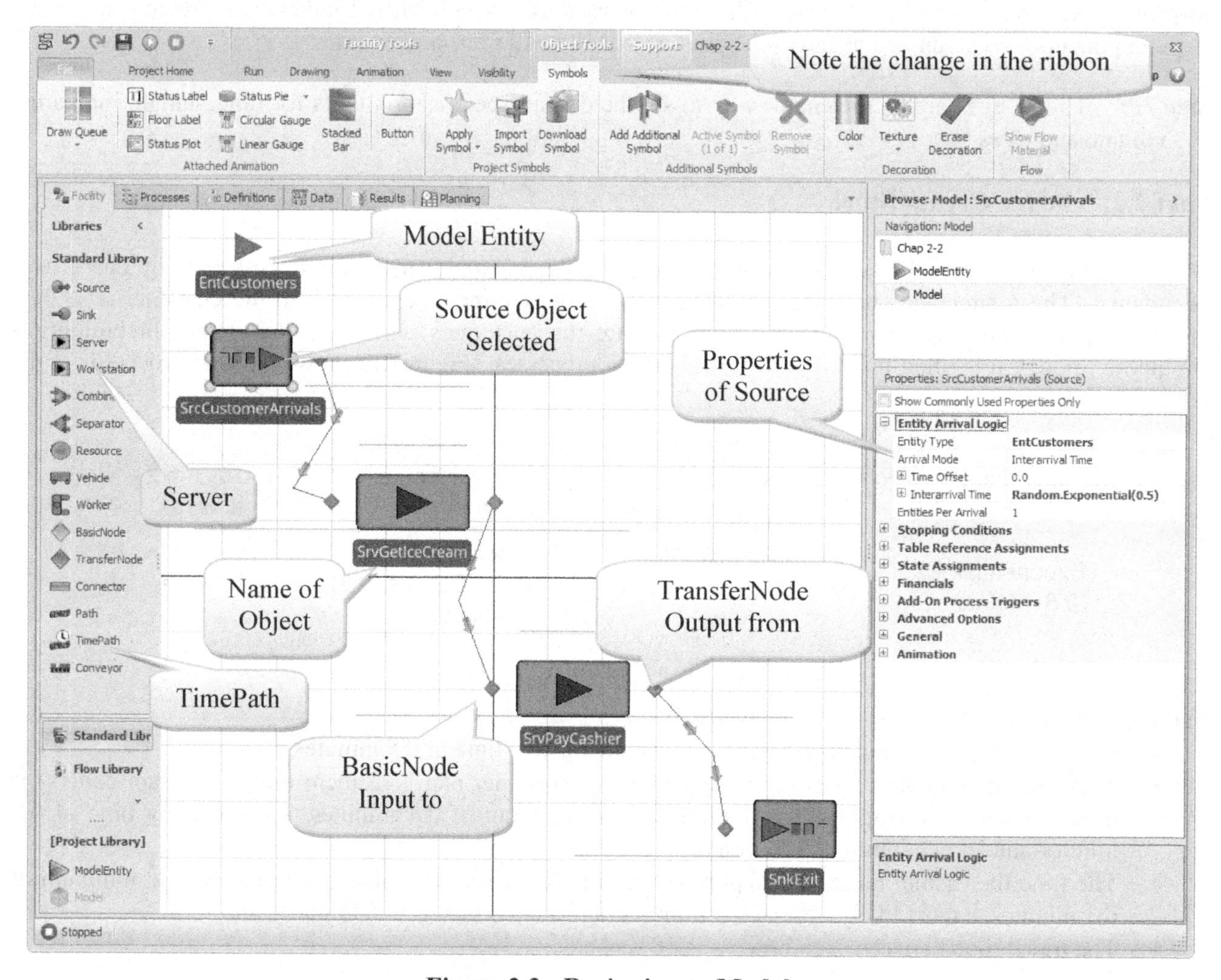

Figure 2.3: Beginning to Model

Step 3: Change the defalut "names" for each object. You can change the name on the object directly[13] or in the "*General*" property section of for each object. Use the names as specified in Figure 2.3.

Question 3: A SERVER object has three "lines" that surround it. What are they called? (note that a "." adds further specification)

____

Question 4: A SERVER has an "input" and an "output" node. These nodes are also [*Standard Library*] objects. What is the standard library name of the input node?

____

Question 5: What is the standard library name for the output node?

____

Step 4: Next, click on each object and fill out its properties according to our assumptions. Figure 2.4 shows the properties associated with the SOURCE object.[14]

[13] Double click or right-click the object to change the name the directly. Do not forget to click enter to save the name change.

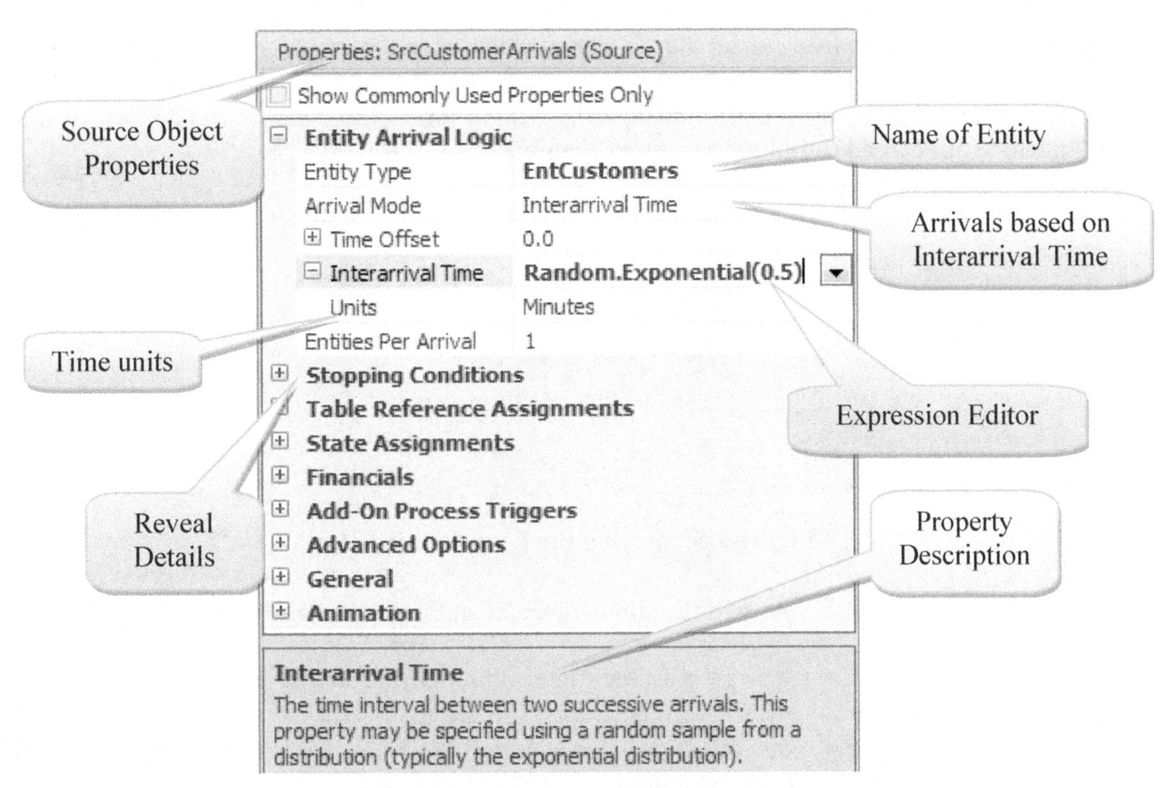

Figure 2.4: Properties of Source

It is important to be sure the time units are correct for expressions involving time and distance. You may need to reveal the details in a dialog by clicking the ⊞ icon to show the additional information. The *Expression Editor* allows you to write expressions which have the following form.

Step 5: `Object.SubObject.SubObject(parameters)`

To begin the expression editor start by specifying a character. Select the object, its subobject(s), and its properties. The expression editor has "tab-completion" which means the tab key will complete the name from its first few characters. It also indicates if subobjects are available for this object.

Question 6: What units of time are available for the interarrival time?

____

Step 6: Complete your modeling by adding the information for the two SERVERS and TIMEPATHs:

- For the **SrvGetIceCream** server object:
  - *Initial Capacity*: 2
  - *Processing time*: `Random.Triangular(0.4, 0.9, 1.5)`
  - *Units*: `Minutes`
- For the **SrvPayCashier** server object:
  - *Processing time*: `Random.Triangular(0.3, 0.4, 0.6)`
  - *Units*: `Minutes`
- For the TIMEPATHS objects:
  - *TravelTime*: `15`
  - *Units*: `Seconds`

[14] Note the default properties that are changed will be bolded to indicate a changed has occurred.

Note that the object names are listed under the "*General*" category in the property inspector.

Step 7: Before we run the model, let's change the ending time in the *Run Setup* section of the SIMIO "*Run*" tab to "Unspecified (Infinite)" as seen in Figure 2.5.

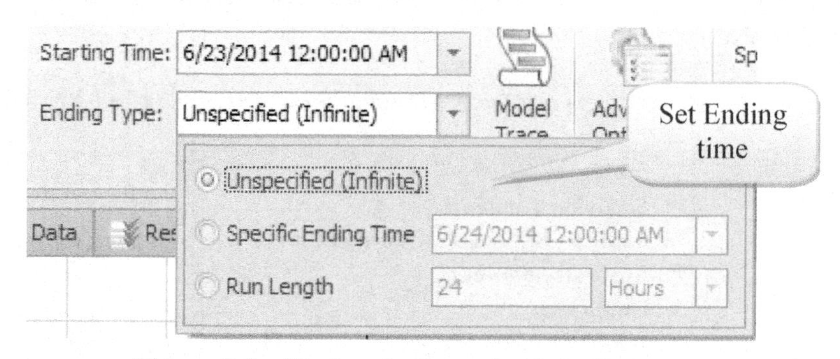

Figure 2.5: Setting the Length of the Run

Step 8: Click the run button to start the simulation and let the simulation "run" for a while.

Step 9: Notice the ModelEntities (▶) as they move through the model across the TimePaths. The entities queue in the `InputBuffer.Contents` and they are in service in the `Processing.Contents`, both shown as "green lines" that surround the Server objects. You can relocate the objects while the model is running.

Step 10: Click the "*View*" tab and select "3-D".

Question 7: Hold down the left mouse button key and move the mouse left and right and up and down. What happens?

____

Question 8: Hold down the right mouse button key and move the mouse left and right and up and down. What happens?

____

Step 11: Switch between 2-D and 3-D using the "2" and "3" as "hotkeys"

Question 9: What happens in 2-D when you hold down the left mouse key and move the mouse left and right and up and down?

____

Question 10: What happens in 2-D when you hold down the right mouse button key and move the mouse left and right and up and down?

____

Step 12: Experiment in 2-D and 3-D by moving stations and changing the "layout" of the model. Change the travel time to the **SrvGetIceCream** station to 50 seconds to see another kind of change.

Part 2.3: Enhancing the Animation

Animation is often what gets people interested in a model. A SIMIO 3-D model, although it might take some work, can bring a lot of attention. In this section, we will give you some instructions on how to create an animation that appears as in Figure 2.6.

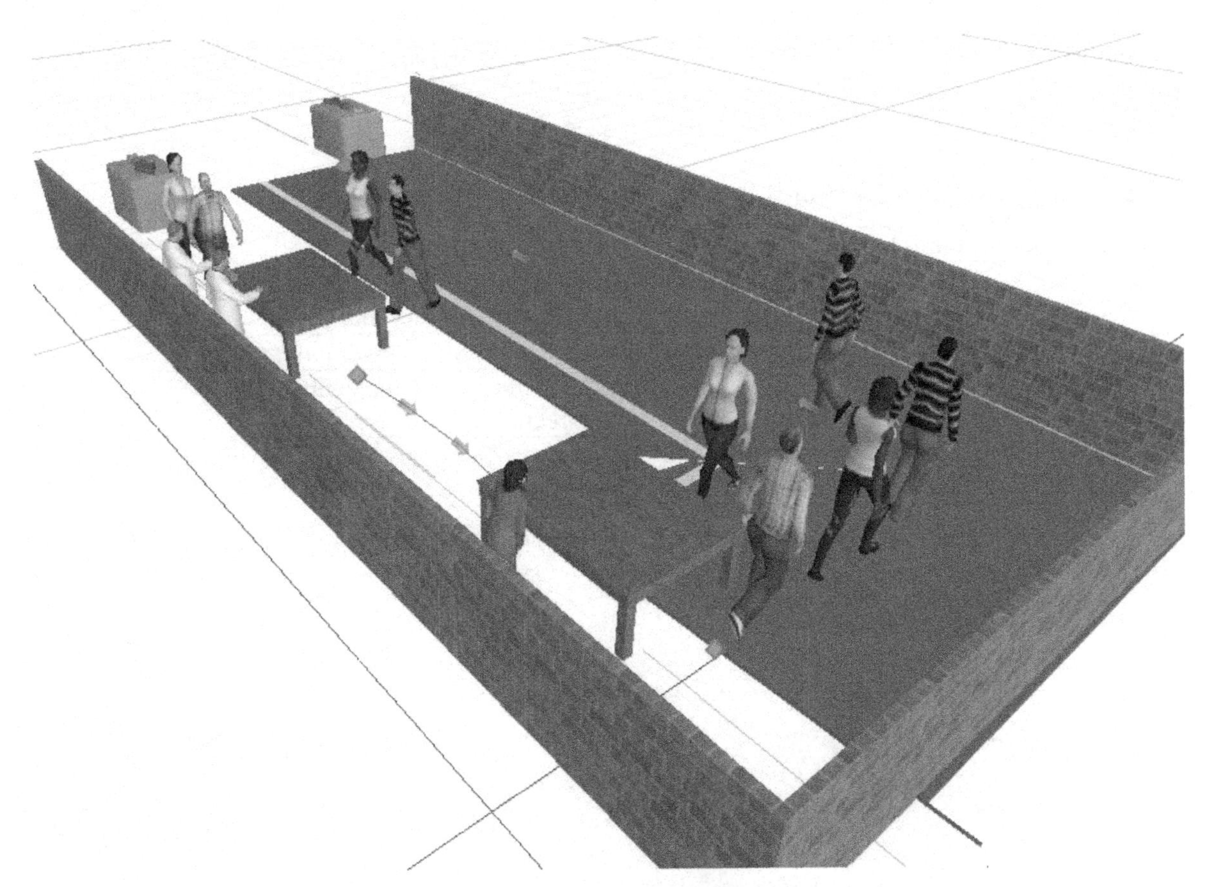

Figure 2.6: The Animation

Step 1: Make the ice cream store customers look like people, rather than triangles. Select the **EntCustomer** entity (i.e., the green triangle) and under the *Symbol→Project Symbols* section, look at all the groupings and pick out a person from the "Library\People\Animated", and click on it to substitute it for the triangle. Look at the picture in 3-D and enlarge it by pulling out on one of the bounding ends. Run the simulation. You may want to change the 3-D perspective, as described previously.

Question 11: What happens to the animated people in the queues when you run the simulation?

____

Step 2: Change what happens to the animated people when they are waiting by clicking on the **EntCustomer** and expand the "*Animation*" details in its ***Properties Inspector***. Change the "Default Animation Action" to "*Moving*" from "*MovingAndIdle*" to eliminate the shaking of the animated people when they are idle.

Step 3: Select the **SrvGetIceCream** SERVER object and substitute a "table" for the SERVER picture by choosing the table under the *Symbol→Project Symbols* section. Repeat the procedure for the **SrvPayCashier** SERVER object. Adjust the size of the SERVER objects to correspond to the persons' size.

Step 4: Next, let's add some stationary people behind the tables to represent the attendants. To do that, switch to 3-D and select the "*Drawing*" tab. Click on the "*Place Symbol*" button and select a person from the "Library\People" to represent the attendant. These non-animated library of people do not move their body parts if used as a symbol for an entity and move by "skating" from point to point. When the person is placed, you can click on one of the corners and hold down the *Ctrl* key to rotate the picture. Place the person behind the table. Duplicate the attendant by *Ctrl-C* and *Ctrl-V*. Now do the same for the **SrvPayCashier** at the cashier station as seen in Figure 2.6.

Step 5: The animated queues (i.e., the three green lines around the server objects) by default have the entities oriented in the same direction as the entities are traveling. Modify the orientation within the queue as specified in Table 2.1 by selecting the `Processing.Contents` queue and click the `Point` or `Oriented Point` button in the *Appearances→Alignment* section as seen in Figure 2.7. You can use `Inline` for the `InputBuffer.Contents` queues.

Table 2.1: Entity Alignment Options for Animated Queues

Queue Alignment Option	Description
`None`	Entities will point from the left to right no matter the queue orientation within the object frame.
`Inline`	Entities will be oriented to point forward along the line
`Point`	Entities only reside on vertices of the queue but will point in the same direction for all vertices
`Oriented Point`	Entities only reside on vertices but can be aligned in different directions for each vertex.[15]

Step 6: Repeat the process for the animated queue for the processing contents for the **SrvPayCashier**.

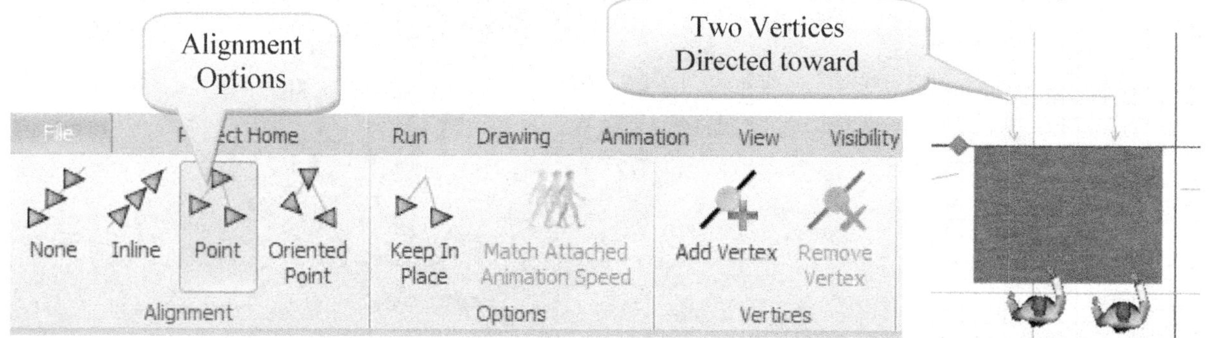

Figure 2.7: Changing the Orientation of Entities in the Animated Queues

Step 7: Move the symbols around on the modeling canvas to produce a better representation of the ice cream store. You can add walls to your store by using the "*Polyline*" tool from the *Drawing* tab. Be sure to set the height of the object in the "*Object*" section of the *Drawing* tab. You should note that the measurements of length are in meters unless you specify the units to be different. You may need to modify the heights of objects to make them consistent.

Step 8: The path from the **SrvPayCashier** object to the **SnkLeave** object can have a "path decorator" added. Path decorators may be added by clicking on the path and selecting a "*Decorator.*" To see people on this path, change the travel time to 300 seconds.

Step 9: Now run the simulation and look at the animation. When you run the simulation, you may need to adjust the way it looks both in 2-D and in 3-D. The various queue symbols (lines) may need to be extended. For instance the line associated with the number of people waiting to get ice cream may be too short, so you might want to extend that line. Same is true for the `Processing.Contents` line. Remember that the length of the animated line has no impact on the actual number in the line (i.e., there may be more entities waiting than can be displayed but just hidden).

Question 12: Show off your animation to your friends. What is their reaction?

____

Step 10: In the prior animation, the default entity triangle was changed to represent a person but the exact same person arrived each time. To allow for different people to arrive, select the model entity and click the *Add Additional Symbol* button under the *Symbols→Additional Symbols* section to create as many types of

[15] Note the default queue alignment is `None`. Also, the `Keep in Place` option will force the entities to remain at the position they enter in the model. This characteristic is seen in queues that represent systems like waiting rooms (i.e., people do not change seats).

people you would like to see arrive. Under the *Active Symbol* dropdown, select each symbol which will be currently identical to the first and change the symbol following the directions in *Step 1* as seen in Figure 2.8 where four total symbols have been added.[16] You will need to adjust the sizes of each of the symbols.

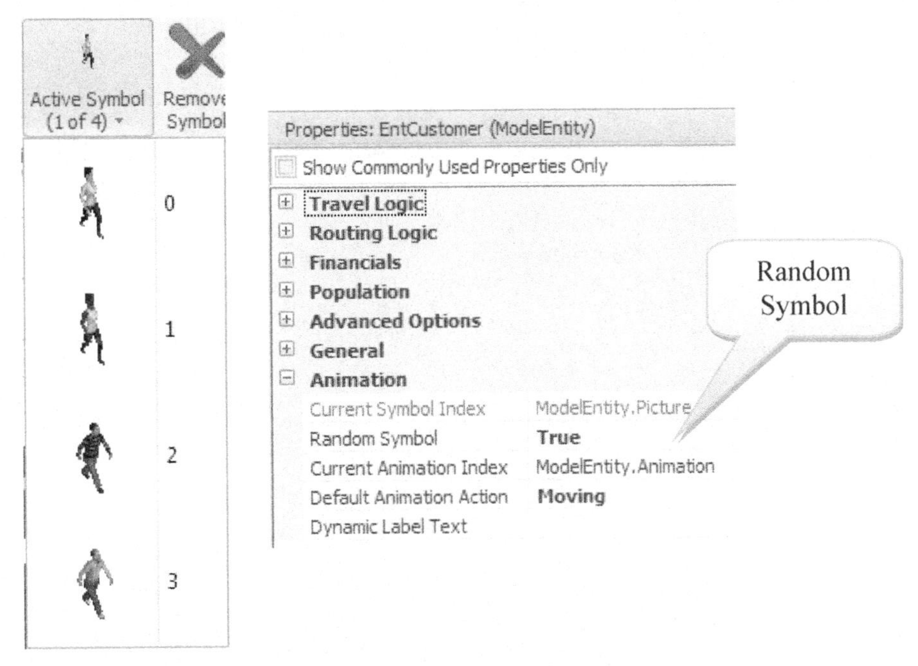

Figure 2.8: Adding Additional Types of People

Step 11: To use the new symbols, change the *Random Symbol* property to **True** under the *Animation* section of the model entity. Now, each time a person is created by the SOURCE, a random symbol will be selected as its picture. Run the new animation and observe what happens as seen in Figure 2.6.

Step 12: Notice that the new symbols have been added to your model in the [*Navigation*] panel.

Part 2.4: Looking at the Results

While the animation brings attention to your model, you will be building the simulation model in order to understand the numerical characteristics of the system being modeled. Typically, you will want to make all the changes to the model until the model accurately represents the system and/or the potential changes have been identified before spending a lot of time on the animation.

Step 1: First, let's look at the results from the basic model as illustrated in Figure 2.9. Perhaps we'll call this the "present system" model. Under the "*Run*" tab, change the "Run Length" to 8 hours and run the simulation. The results of this simulation are found under the "*Results*" tab. Using the "*Fast-Forward*" choice will not display the animation during the simulation. Whenever the simulation is run using the "*Run*" tab, it is being run in "interactive" mode allowing one to pause, step, etc. during the simulation, even extending the run length.

Step 2: Once you run the model, you can select the *Results* tab to access the statistics. Notice the "*Unit Settings*". By default time is in hours, but it is easily changed using the *Unit Settings* button under the *Display* section. The time-based statistics will display the units in parenthesis in the results.

[16] The color of the clothing can be changed by selecting a color from the *Decoration* section and clicking on the part of the symbol you would like colored.

Step 3: The results of a simulation are shown in the form a "pivot table." A pivot table arranges the output data according to attributes that head each column as seen in Figure 2.9.

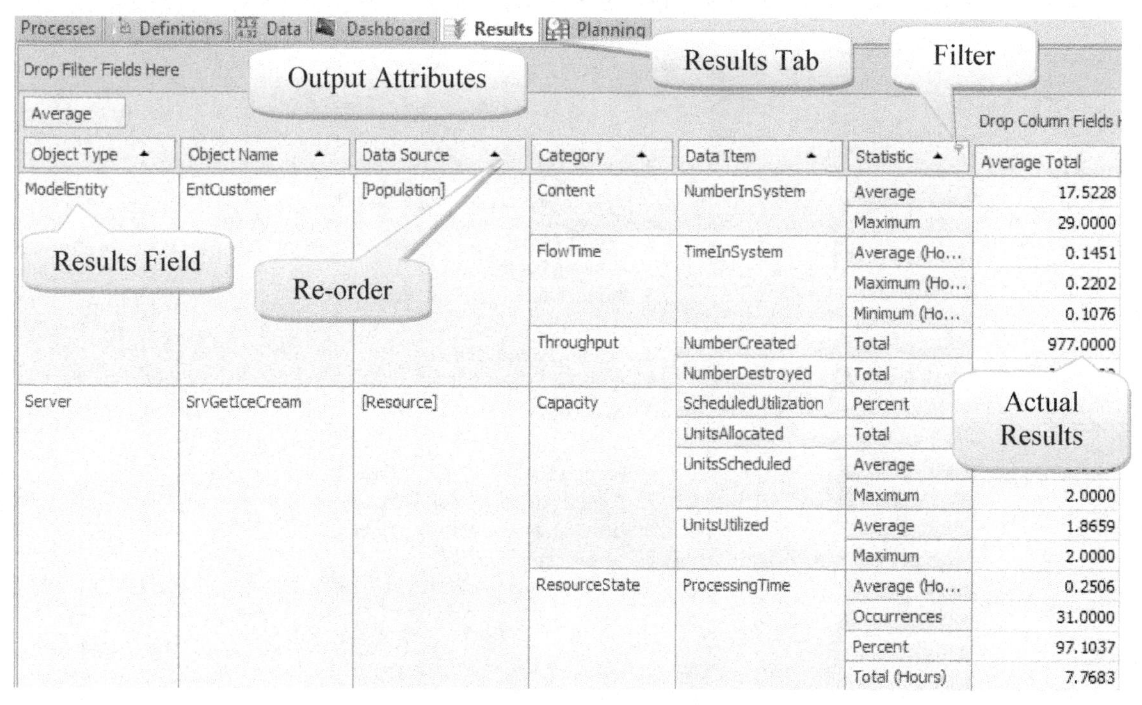

Figure 2.9: Looking at Results

Step 4: Each Output attribute has two symbols – one to re-order the column and the other to "filter" or select the items displayed in the column.

Question 13: Filter the "Statistics" to show only the average values for the MODELENTITY. What is the average *Number in System* and *Time in System*?

Step 5: Attributes can be moved left and right in the display.

Step 6: Right clicking on the "*Results Field*" button (see Figure 2.9), you can display "show field list." You can drag other fields to display by dropping them in columns in the actual results.

Step 7: The "Categories" for statistics are: Content, Throughput, Capacity, FlowTime, ResourceState and HoldingTime. These are applied to each data source as they are relevant.

Step 8: Resource utilization includes "ScheduledUtilization" as well as "UnitsAllocated," "UnitsScheduled" and "UnitsUtilized". These items refer to the actual use of the capacity of the object and the average scheduled capacity. ScheduledUtilization is computed as a ratio of the actual time the resource is utilized divided by the total time its capacity is available. "UnitsUtilized" is the average number of units of the resource that are utilized up to the time this report is generated.

Question 14: What is the ScheduledUtilization of the capacity of the **SrvGetIceCream** object?

Step 9: Notice the waiting at the server object is displayed under the "HoldingTime" of the "InputBuffer" while the "Content" of the **InputBuffer** displays the number in the queue.[17]

Question 15: What is the average number (content) in the **InputBuffer** of the **SrvGetIceCream** server object?

____

Question 16: What is the average waiting time (holding time) in the **InputBuffer** of the **SrvGetIceCream** server object?

____

Question 17: Why does the `Processing Content` average number for the **SrvGetIceCream** server object equal the average `UnitsUtilized` previously?

____

Question 18: What is the utilization of the **SrvPayCashier** server object?

____

Step 10: To switch back to the model, you need to select the "*Facility*" window tab, which is at the same level as the "*Results*" tab.

Part 2.5: Commentary

If you have been paying close attention to the model construction, you may have noticed that:

- The SIMIO undo/redo is very convenient. This feature eases the task of trying different modeling features and then "undoing" and "redoing" various changes. Other features of the model, such as path type can also be easily changed and then changed back by right clicking on the link.

- You can suppress randomness in your simulation by selecting the "*Advanced Options*" within the *Run Setup* section of the "*Run*" tab and "Disable Randomness". Doing this will allow you to follow the behavior without randomness to assist in debugging/verifying your model.

If you are familiar with Object Oriented Design (OOD), the SIMIO [*Standard Library*] could be appropriately called a "class" library. The simulation objects are created from these class definitions. The act of selecting and dropping an object onto the modeling canvas is what "instantiates" the object. Properties, as displayed in the property inspector, define the characteristics of the objects. Later, you will see how to add your own characteristics. Properties are initialized, but cannot be changed. Another type of characteristic, which is called "state variable" in SIMIO, may be added if the characteristic needs to be changed during the simulation.

Generally, running the simulation with only one run (replication) does not generate reliable statistics for results. The next chapter will demonstrate how to obtain multiple replications for a given simulation scenario.

[17] The SIMIO nomenclature of "HoldingTime" and "Content" may seem non-standard. Generally, queuing theory uses the terms "waiting time" and "number in queue". Also "waiting line length" or "number waiting" is used sometimes instead of "number in queue."

Chapter 3
Modeling Distance and Examining Inputs/Outputs

For this chapter, our goal is to model people arriving to an airport and going through the check in process. In the first phase we are only concerned with the amount time it takes the passengers to reach the security checking station so we can determine the needed number of workers at a check-in station. Passengers arrive at the terminal and proceed to the check-in process to get their tickets. After check-in the passengers proceed to the security check point.

Passengers arrive according to an exponential distribution with an average of one arrival per minute. Passengers walk to the check-in station at a rate that is uniformly distributed between two and four miles per hour. Passengers travel 50 yards from the terminal entrance to the check in station. Following the check in process they then must walk 65 yards to the security check point as seen in Figure 3.1. The check-in station currently has four people assigned to process customers, who wait in a single line. The check-in process takes between two and five minutes, uniformly distributed, to completion. The simulation model needs to be run for 24 hours.

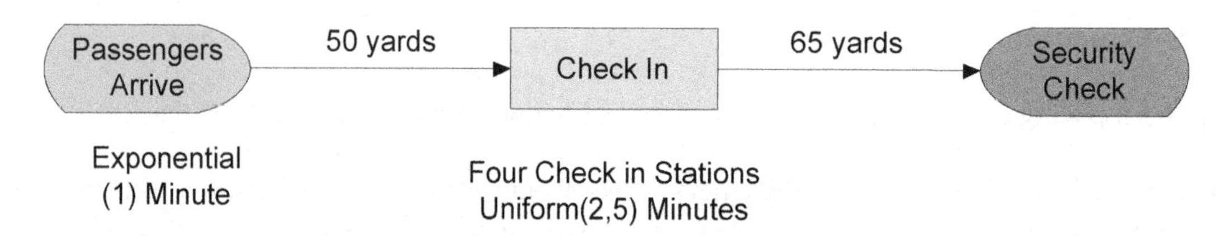

Figure 3.1: Airport Check-in Process

Part 3.1: Building the Model

Step 1: Create a new model, as shown in Figure 3.2. Right-click on the objects in the [*Navigation*] panel and view their properties – you may need to right-click on the [*Navigation*] panel names and select its properties. Call the PROJECT "**AirportProblem.**"

Step 2: Insert a SOURCE named **SrcPassengersArrive**, a SERVER named **SrvCheckIn** and a SINK named **SnkSecurity** from the standard library to the Facility window and connect them using a path linkage (all of this can be done using the "*Add-In*" section on the "*Project Home*" tab).[18]

Step 3: Click and drag an entity into the *Facility* window from the MODELENTITY in the [*Project Library*] panel. Also change the name from **DefaultEntity** to **EntPassenger**. Change the passenger symbol to a person from the people library (either animated or static – we will use the static people pictures ("Woman1") this time so you can see how they behave during the simulation).[19]

[18] To connect objects via links without selecting them from the [*Standard Library*], hold the *Ctrl Shift* keys down while selecting the TRANSFERNODE to start the drawing of the link. After completing the link, choose the correct link.

[19] You can also add more than one symbol for the passenger as was done in Chapter 2.

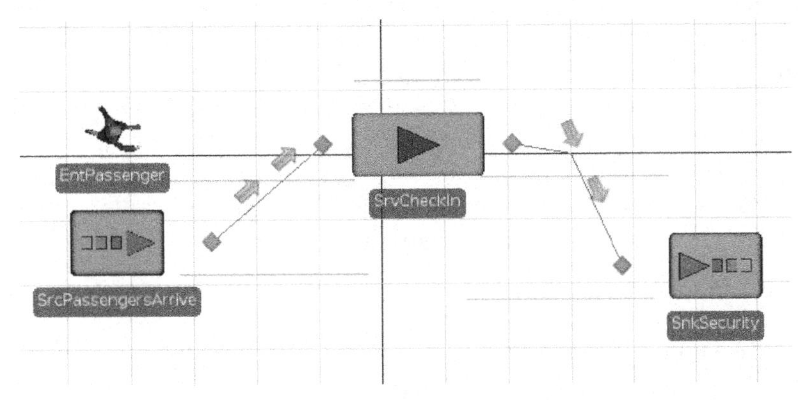

Figure 3.2: The Basic Model

Step 4: Set the speed at which a passenger can walk, namely, `Random.Uniform(2,4)` miles per hour, in the **EntPassenger** object. Set the *Interarrival Time* property, `Random.Exponential(1)` minutes, in the **SrcPassengersArrive** SOURCE object. Set the *Processing Time* in the **SrvCheckIn** object to be `Random.Uniform(2,5)` minutes and make sure to set its initial capacity to four.

Step 5: For the two paths, change the *Drawn to Scale* property of both paths to "FALSE" and set the correct distances for each. Recall that the path from arrive to check-in is 50 yards while the path from check-in to security is 65 yards.

Step 6: Run the simulation and observe the behavior of the "passenger" and the changing color of the check-in.

Question 1: What does the passenger look like as they "move"?

____

Question 2: What color is the check-in when it is idle and when it is busy?

____

The symbol colors for the check-in can be seen by clicking on the *Active Symbol (1 of 5)* down-arrow in the *Additional Symbols* section. It shows the five "default" states of the server.

Question 3: What is the state name and state number when the server is busy?

____

Part 3.2: Using the 3D Warehouse

Although we have a working model, let's change the server symbol. You can either do so by clicking the entity or objects and changing to some of the preloaded symbols in SIMIO, or you can download new symbols from the ***Trimble 3D Warehouse™*** or if you have symbols[20] already on your computer, you can import them into this model.

Step 1: To employ new symbols, select the **SrvCheckIn** object and click one of the choices among Project Symbols. The ***Trimble 3D Warehouse™*** should pop up when you click the "*Download Symbol*" button which will allow you to search for any symbol you want (e.g., try searching for "Airport Baggage"). Once you find a suitable model, simply click "*Download*" button on the top right and SIMIO will import it for you into a window where you can change some of its properties, such as size and orientation before inserting the symbol into your model.

[20] New symbols can be created by packages like Trimble's SketchUp Pro™ as well as other 3-D graphics software.

Step 2: When you download from the 3D warehouse an "Import" window appears. First note you can view the downloaded symbol in 2D and 3D. Also it has length, width, and height that can be specified (in meters). A size "hint" is shown relative to a person, so you can modify size based on that. Finally you can rotate the symbol so it has the proper "orientation"[21].

Step 3: Position the baggage **SrvCheckIn** object so that the **Input** BASICNODE, the **InputBuffer** queue, and the **Processing** queue symbols gives the appearance of a check-in at an airport as shown in Figure 3.3.

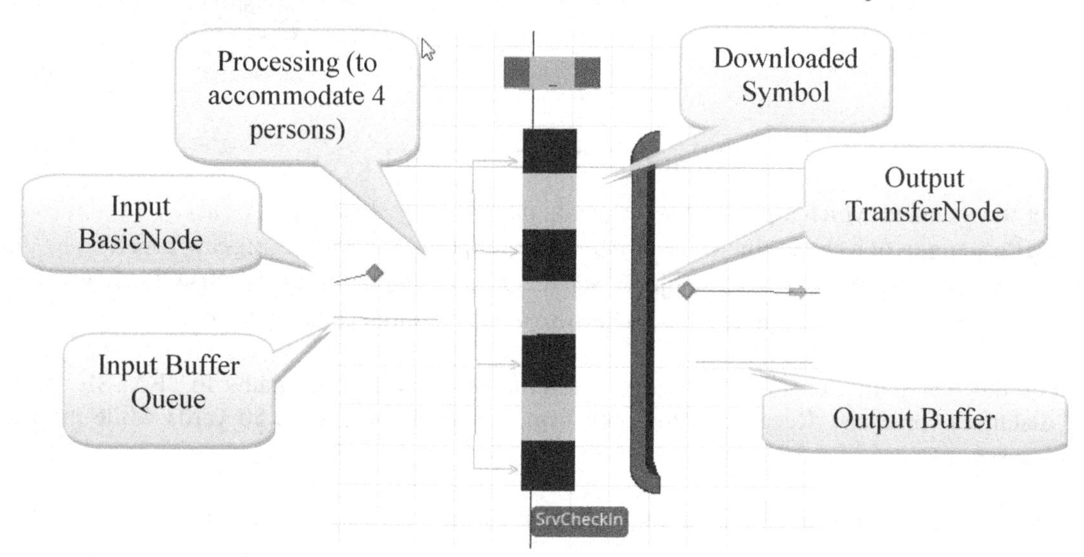

Figure 3.3: Baggage Check-In

Step 4: Run and adjust the animation (in 2D and 3D). Lengthen the **Processing** queue of the **SrvCheckIn** server and modify the symbol using "point alignment" to show that up to four people can be in-process at the check-in station at once by adding two additional vertices.

Part 3.3: Examining Model Input Parameters

Before looking more closely at the output, let's re-examine our "input". We can use SIMIO to understand our assumptions about the interarrival time and processing time using input parameters. Using input parameters instead of directly specifying the expressions allows one to use the same random expression across multiple objects (i.e., five machines all have the same processing time) but more importantly will allow response sensitivity and sample size error analysis to be performed.

Step 1: Click on the "*Data*" tab and select the "*Input Parameters*" icon in the View panel on the left. Three types of input parameters can be defined: *Distribution*, *Table Value*, and *Expression*.[22]

Step 2: Click on "*Distribution*", meaning we are interested in a statistical distribution. By default we are shown a histogram sample of 10,000 observations from a "Normal distribution" whose mean is 1 and whose standard deviation is 1.

Step 3: Let's try some other distributions. Look at a "Triangular" with a minimum of 0.4, a mode of 0.9 and a maximum of 1.5 (that was the "get ice cream" processing time from Chapter 1).

---

[21] Use the hint screen by toggling it by hitting the "*h*" key to get help on rotating, sizing and moving objects in SIMIO.

[22] *Distribution* allows one to specify a single distribution which can be used in both sensitivity and sample size error analysis by specifying the number of samples used to build the distribution.

Question 4: Using the SIMIO "help", what is the mean of a Triangular distribution whose minimum is 1, whose mode is 3, and whose maximum is 8?

____

Step 4: Now define the interarrival time distribution of an Exponential with mean 1 minute (*Unit Type* is "Minutes") having with the name **InpDisInterarrivalTime**. It's a good practice to give each distribution its own "Random Stream", so we will use stream two.[23]

Step 5: Add a second distribution for the Check-In processing time. It is Uniform with a minimum of two minutes and a maximum of five minutes. Name it **InpDisCheckInTime** and give it stream 3.

Step 6: Now go back to your model and replace: (1) the Interarrival Time in the **SrcPassengersArrive** to **InpDisInterarrivalTime**, and (2) the Processing Time in the **SrvCheckIn** to **InpDisCheckInTime**. The easy ways to add these input parameters is to right-click the down-arrow in the specification and select the "Set Input Parameter". From the options, select the name distributions.

Step 7: Run the simulation for a few minutes.

Question 5: Do you notice any change in the behavior of the simulation?

____

Step 8: Of course you could continue to use the direct specification like `Random.Exponential(1)` or you can use the "*Expression*" choice in the "Input Parameters" to give `Random.Exponential(1)` a name to reference.

Step 9: Also you should note that a given input parameter can be used in multiple places within an object, so changing all cases is reduced to changing the input parameter specification.

Part 3.4: Examining Output

The specific output from a simulation is a random variable produced by the entities flowing through the model and receiving services. Their arrivals are random as well as their services. Hence one run/replication of the model produces one experimental value. This value is not precise and to gain precision, we need to make multiple runs/replications.

Step 1: Change the run length to 24 hours and run (i.e., use "Fast-Forward") the simulation. After the simulation is complete, click "Stop", and then click the "*Results*" window tab to access the results window. In the "*Display*" section of the SIMIO ribbon, change the units of time to minutes.

Question 6: What is the average number of people waiting at check in?

____

Question 7: What is the average time in minutes they were waiting?

____

Question 8: Do you think these are "good" results? Why or why not?

____

[23] A random number stream for a distribution can be viewed as a specific series of random observations from that distribution. As such, it is reproduced by choosing that same stream again. This insures that the randomness associated with that stream is reproduced in other scenarios, so its randomness doesn't add additional variance to the output performance measures.

Step 2: To gain confidence in the precision of our results, the model needs to be run for several independent replications. From the "*Project Home*" tab, let's add a new experiment to be able to run the model multiple times by clicking on the "New Experiment" icon.

Step 3: In the "*Design*" window tab, change the number of replications to ten[24] and "Run" the experiment using the icons in the "*Experiment*" section of the SIMIO ribbon. It should run quickly and the progress of each replication is shown in the comment window at the bottom. If you are running with a multi-core computer the runs will not be in order since each replication runs on a different core.

Step 4: You can look at the "*Pivot Grid*" or the "*Reports*" tab to get the results (remember that you can set the "units" in the ribbon for the output). From the *Pivot Grid*, notice that the "Results Fields" now include "Minimum", "Maximum", and "Half-width". The half-width is one half of a confidence interval with the "Confidence Level" as specified in the Experiment Properties (it is 95% by default). Recall that the confidence interval is a statement about the precision of the summary values

Question 9: What did you get for the average, minimum, maximum, and half-width for the number of passengers in the system?

Question 10: What is the average, minimum, maximum, and half-width for time in system (FlowTime) in minutes?

Step 5: With $(1 - \alpha)\%$ as the "confidence level," you may recall that the corresponding statistical confidence interval is given by the following formula.

$$\bar{X} \pm t_{n-1,1-\alpha/2} \frac{s}{\sqrt{n}}$$

$n =$ number of replications

$\bar{X} =$ sample mean

$s =$ sample standard deviation

$t_{n-1,1-\alpha/2} =$ critical value from t statistic

Step 6: Note that:

- The statistical summary values, such as an average, are computed average values from each replication (one observation per replication). The confidence interval will increase in size as the confidence level goes from 90% to 95% to 99% while the confidence interval gets smaller as the number of replications increase.
- The "half-width" is simply one-half the width of the confidence interval
- A simulation run for n_0 replications having a half-width of h_0 will require (approximately) n total replications to obtain a target half-width h according to the following equation.

$$n \cong n_0 \frac{h_0^2}{h^2}$$

$h_0 =$ half width from observation of n_0 replications

Question 11: Suppose you want the average number in the system for Passengers to be within ±5% of its mean? How many replications would be needed given the previous half width calculation?

---

[24] It should have defaulted to ten.

Question 12: Suppose you want the maximum number in the system for Passengers to be within ±2 passengers? How many replications would be needed?

____

Step 7: You can add the standard deviation (actually the standard deviation of the mean or standard error – not the standard deviation of observations within a replication) by right-clicking on of the *Filter Fields* and selecting "Show Field List". Next double-click on the *Std. Dev* field.

Question 13: What did you get for the "Std. Dev." for the number of passengers in the system?

____

Question 14: What part of the confidence interval formulation has the standard error – it's what SIMIO calls the standard deviation here?

____

Step 8: To remove a field, simply move it back to the *PivotGrid Field* List.

Part 3.5: Using Experiments

Suppose now we want to determine the effect a change in capacity of the check-in station will have? We could simply change the capacity and re-run the simulation looking at the results. A better way is to set up a SIMIO "experiment" which contains alternative "scenarios" that can be compared together directly.

Step 1: In the *Facility* view, click on the **SrvCheckIn** object. Right-click on the *Initial Capacity* property and select "*Set Referenced Property*". Use "*Create New Property*" and give it the name **CheckInCapacity**. Doing this creates a new model "property" which is listed in the "*Definitions*" tab of the model under the "Properties" listing.

Step 2: From the "*Project Home*" tab, create a "New Experiment" and view the "*Design*" Tab. Note that this is a second designated "experiment" in the [*Navigation*] panel and our referenced property is now listed in the columns.

Step 3: Add Scenario rows, changing the **CheckInCapacity** to two, three, and four to create an experiment grid as seen in Figure 3.4. You can add rows by clicking on the "*" symbol in the first column. Essentially you are defining three scenarios each having a different check-in capacity.

Design | Response Results | Pivot Grid | Reports | Input Analy

Scenario		Replications		Controls
Name	Status	Required	Completed	CheckInCapacity
ScenarioCap2	Idle	10	0 of 10	2
ScenarioCap3	Idle	10	0 of 10	3
ScenarioCap4	Idle	10	0 of 10	4
*				

Reference Property

Figure 3.4: Experiment Scenarios

Step 4: Now run the simulation performing ten replications of each scenario which is completed very quickly.[25] SIMIO will parallel process the experiments on each core of a multi-core computer. Replications that finish turn green in parallel where the yellow scenarios are currently being run.

---

[25] Run the experiments under the "*Design*" tab in the SIMIO ribbon to employ the scenario controls. If you go back to the Facility window to run the simulation, then the controls employ their original specifications.

Step 5: Notice that the *Pivot Grid* report now includes columns for each scenario.

Question 15: What is the minimum average time in the system for "ScenarioCap4"?

____

Step 6: The "*Reports*" tab is a good way to compare scenarios. In the reports tab, the scenarios are organized under each data item.

Question 16: What are the average and half-width for average time in system for each scenario?

____

Step 7: It is helpful to add "responses" to the experiment to aid in comparing scenarios. Simply click the "*Add Response*" button in the "*Design*" tab. Insert the following two responses with expressions which will look at the average time in system by the passengers and the total number processed specifying units of minutes for the time in system as seen in Figure 3.5.

- TimeInSystem, in Minutes: `EntPassenger.Population.TimeInSystem.Average`
- TotalNumberProcessed: `EntPassenger.Population.NumberDestroyed`

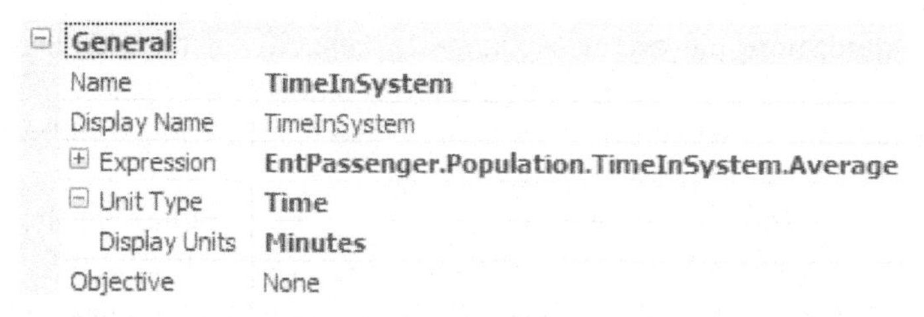

Figure 3.5: Specifying the Time in System Response

Step 8: Reset and run the experiment again looking at the "*Response Result*" charts along with the scenario grid (see Figure 3.6). To arrange the "tabs", click the tab and drag it to the position you desire on the display – note the positioning points. You can also right-click the tab to configure the display of results.

Step 9: The *Response Results* and the SMORE[26] plots contain graphical displays of various summary results from each scenario. Generally, the numerical values are hard to determine from just the graph which is discussed later, but the graph is useful in showing trends and variation across and within scenarios. The confidence interval on the mean, which is most often used, is displayed in the rust color. We know statistically that if confidence intervals between scenarios don't overlap, the scenario results are statistically different.

---

[26] SMORE stands for SIMIO Measure of Risk & Error.

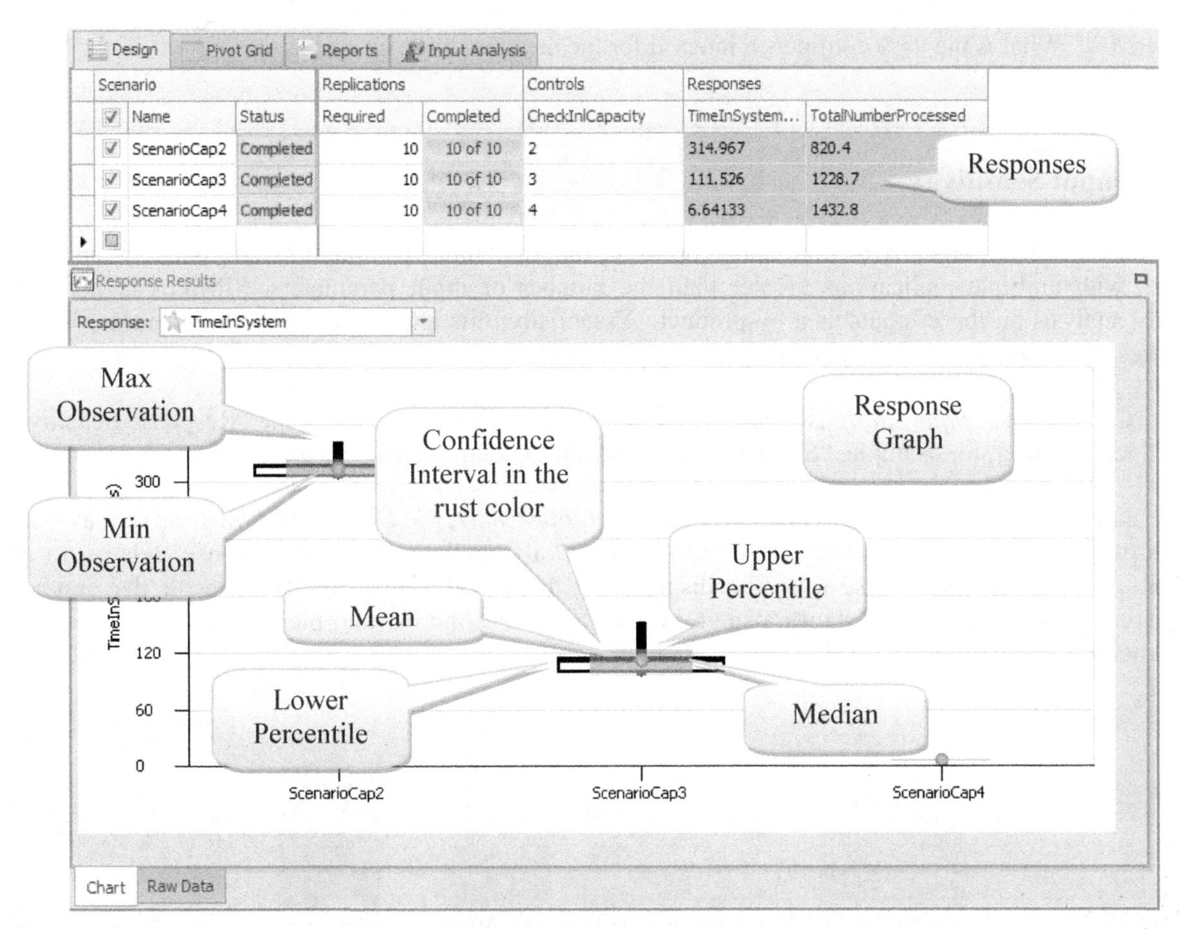

Figure 3.6: Response Results

Step 10: Clearly, as the capacity of the check-in increases, the time in system decreases and the number processed increases. You can modify the graph using the "*View*" section of the "*Response Results*" options in the ribbon. The response graph displays response values (see Figure 3.6) of observations within the simulation. The shading provides their confidence intervals which is specified in the properties of the *Experiment*. Investigate clicking the "*Histogram*" and the "*Means Line*" buttons or try rotating the plot to get a different perspective.

Step 11: If you select the *"Raw Data* tab found at the bottom of the *"Response Results"* tab, you will get the table shown in Figure 3.7, which includes numerical details on the responses, the confidence interval (95%) on the response, and the percentiles (which is 25% and 75% by default).[27]

Name		Values				Mean Confidence			Lower Percentile			Upper Percentile		
Scenario N...	Response ...	Mean	Median	Minimum	Maximum	Half Width	Mean CI Be...	Mean CI End	Lower Value	Lower CI St...	Lower CI End	Upper Value	Upper CI St...	Upper CI End
ScenarioCap2	TotalNumbe...	820.4	822	809	833	5.5222457...	814.87775...	825.92224...	816	809	822	824	NaN	NaN
ScenarioCap2	TimeInSystem	5.2494553...	5.2640565...	5.0566988...	5.7081364...	0.1345578...	5.1148974...	5.3840132...	5.0969630...	5.0566988...	5.2640565...	5.2839316...	NaN	NaN
ScenarioCap3	TotalNumbe...	1228.7	1229	1216	1242	7.0896050...	1221.6103...	1235.7896...	1221	1216	1229	1238	NaN	NaN
ScenarioCap3	TimeInSystem	1.8587716...	1.8337780...	1.5881942...	2.5268071...	0.1914501...	1.6673215...	2.0502218...	1.6516296...	1.5881942...	1.8337780...	1.8985345...	NaN	NaN
ScenarioCap4	TotalNumbe...	1432.8	1425	1391	1517	25.906895...	1406.8931...	1458.7068...	1414	1391	1425	1449	NaN	NaN
ScenarioCap4	TimeInSystem	0.1106887...	0.1055480...	0.0945288...	0.1465531...	0.0111892...	0.0994995...	0.1218779...	0.1031184...	0.0945288...	0.1055480...	0.1101748...	NaN	NaN

Figure 3.7: Raw Data on Graph

Question 17: What is the mean average time in the system for "ScenarioCap3"?

____

[27] You can change the confidence level and percentiles in the EXPERIMENT properties.

Question 18: What is the 95% confidence interval for the mean average time in system for "ScenarioCap3"?

____

Part 3.6: Input Sensitivity

Since we specified the interarrival time and processing time as "input parameters" and our experiment has responses with multiple replications greater than the number of input parameters, SIMIO can perform a sensitivity analysis on these inputs as a by-product. Essentially, this analysis attempts to determine how the inputs affect the responses (i.e., outputs) through simple linear regression.

Step 1: Click on the "*Input Analysis*" window tab for the experiment and select the "Response Sensitivity" panel icon. Select, for example, "ScenarioCap2" and the "TimeInSystem" response.

Step 2: Four lower tabbed displays (i.e., *Tornado Chart*, *Bar Chart*, *Pie Chart*, and *Raw Data*) are available. Each of these charts measure how the particular "input" affects the particular "response" relative to each other. By passing the mouse over the display, the numerical values associated with the graph are displayed. The Tornado chart shows a single response whereas the bar and pie charts show all responses together as seen in Figure 3.8.

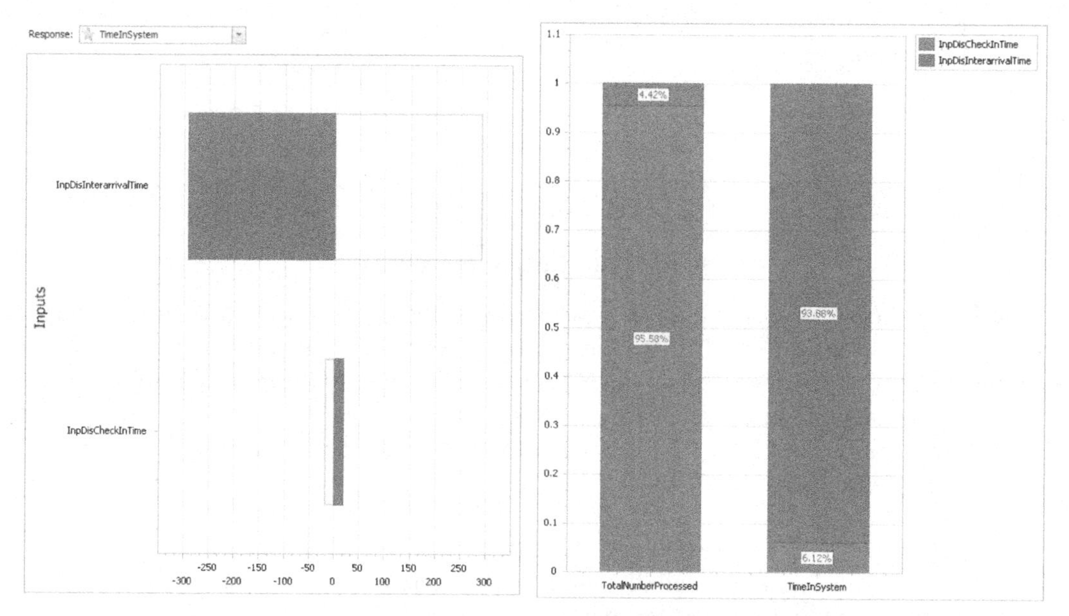

Figure 3.8: Tornado Chart for Sensitivity of Input to Time in System

Step 3: The tornado chart in Figure 3.8 presents the results of the regression of the two inputs, *InpDisInterarrivalTime*(x_1) and *InpDisCheckInTime*(x_2), on the output *TimeInSystem*(y) which can be represented more formally as $y = \beta_1 x_1 + \beta_2 x_2$. By passing your cursor over the bars in the tornado chart you can see that β_1 = -291.261 and β_2 = 18.993. These values have practical interpretation. *TimeInSystem*(y) is negatively related to *InpDisInterarrivalTime*(x_1) and positively related to *InpDisCheckInTime*(x_2). In other words, a unit increase in *InpDisInterarrivalTime*(x_1) causes the *TimeInSystem*(y) to decrease by 291.261 minutes while a unit increase in *InpDisCheckInTime*(x_2) causes the *TimeInSystem*(y) to increase by 18.993 minutes.

Step 4: In the tornado chart the solid line shows the relation between the input parameter and the response while the outlined bar shows the negative. In this case, the lower the interarrival time (i.e., faster arrivals) the higher the time in system. While the higher the processing time the higher the time in system. Clearly, from the figure, the interarrival time has a much greater influence on time in system than the processing time. However, from the bar chart the reverse is true for the total number processed. This information can help modelers determine if more data needs to be collected to verify the input distribution assumptions.

Question 19: Passing your cursor over the bars, what are the sensitivity coefficients for the interarrival time and the processing time for ScenarioCap3?

____

Question 20: When you examine the Bar Chart and Pie Chart for the time in system response, what is the percentage contribution of each input for ScenarioCap3?

____

Step 5: Although the impact of adding capacity to the check-in is clear looking at Figure 3.6, the capacity can be considered an input parameter and be added to the sensitivity displays.

Part 3.7: Commentary

- The standard deviation produced in the output (see the "Results Field" option) is really the standard error (standard deviation of the mean). No standard deviation is computed within one run of the simulation even though it can be computed and can provide some useful modeling options (for instance in establishing inventory policies). The danger is that the standard deviation computed within a single replication cannot be used to compute a legitimate confidence interval, since the observations may not be Normally distributed, nor independent and identically distributed (IID).
- The sensitivity of input parameters on individual responses opens up a useful method to explore the behavior of the simulation model as seen in later chapters.
- Later chapters will describe in detail the SMORE plots, subset selection, ranking/selection, and optimization.
- While other simulation languages may manage the output in a database file, SIMIO manages the output internally but it can be exported externally.

Chapter 4
More Detailed Modeling: Airport Revisited

There is normally more than one type of check-in in an airport, as well as more than one type of passenger (i.e. a curbside check-in or a no checked-bags check-in traveler). Our goal is to embellish the airport model from the last chapter, to handle routing to different check-in stations, varying arrival rates of passengers, and varying types of passengers with rate tables and data tables. We will also introduce a means of giving entities characteristics, called state variables, which can be used to provide various distinctions.

Part 4.1: Choice of Paths

As in the previous chapter, our primary concern is with the amount time it takes the passengers to reach the security checking station and in determining the number of needed check-in stations.

Step 1: Delete the EXPERIMENT object in the prior chapter model by right clicking on the Experiment in the [*Navigation*] Panel.[28] We will embellish the previous airport model to allow a choice of three different routes for passengers arriving at the airport. Passengers who need to check bags and get tickets can either check in at the curbside station (10%) or check in at the main station inside the airport (70%), while some passengers can proceed directly to the security check point (20%). Refer to Figure 4.1 which depicts the modified airline check-in process to be modeled.

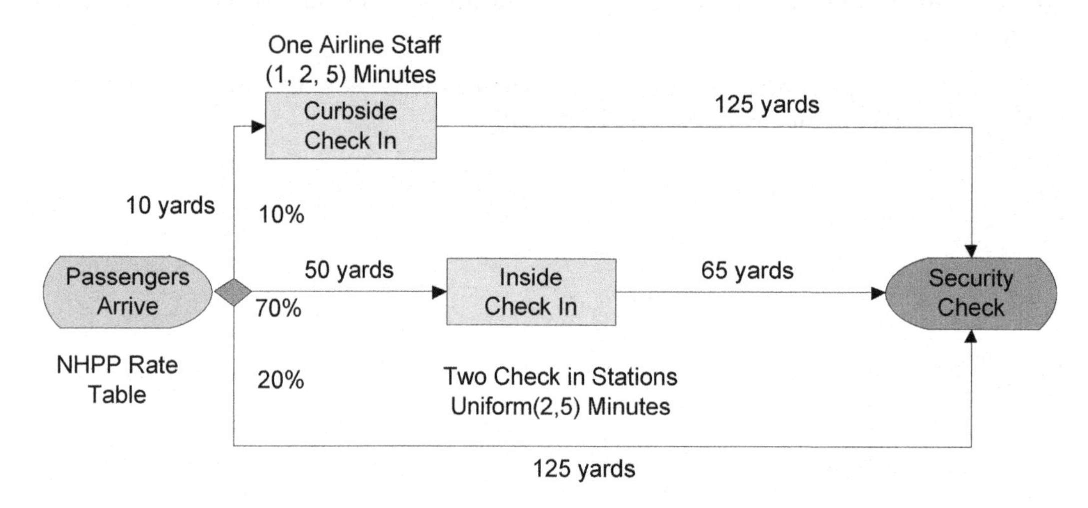

Figure 4.1: Modified Airline Check-in Process

Route1: Passengers who use the curbside check-in need to walk 10 yards to reach the station where a single airline staff member takes between one and five minutes, with a most likely time of two minutes, to check-in a passenger. Once they have checked in, they walk 125 yards to the security check line.

Route2: Passengers who have checked in online and do not need to check any luggage, only need to walk the 125 yards to the security check line once they arrive to the airport.

Route3: Passengers who plan to use the inside check-in first need to walk 50 yards to reach the station where there are currently two airline personnel (instead of four previously) checking passengers into

[28] The [*Navigation Panel*] in the upper right corner contains all of the models, objects, and experiments in the project. You can access the properties, rename the item, or delete the item from the project.

the airport. It takes between three and ten minutes uniformly to process each passenger. Once they have checked in, they then walk 65 yards to reach the security check line.

Step 2: Modify the model from last chapter to include the new curbside check-in (SERVER) as well as the direct path (i.e., Route 2) from the arrivals to the security gate as seen in Figure 4.2. Specify the properties according to description in Figure 4.1.

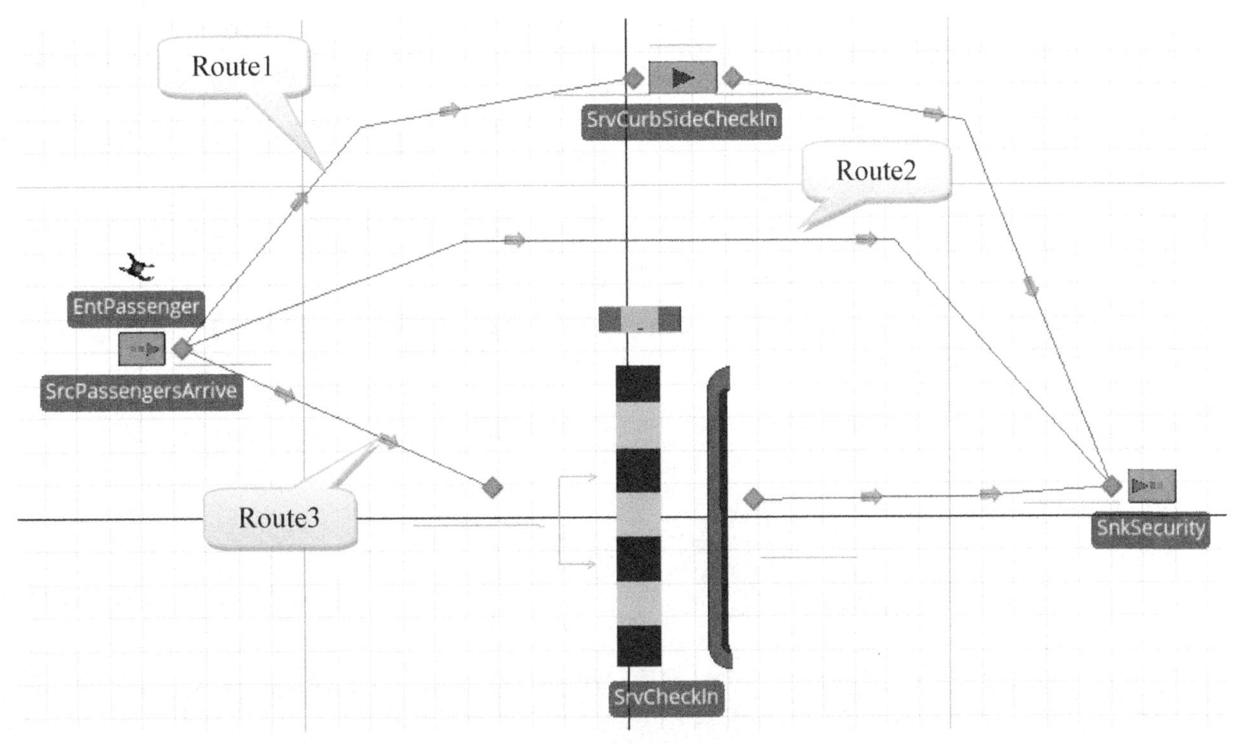

Figure 4.2: Creating Different Routing Paths

Step 3: It has been observed that passengers follow the probabilities in Table 4.1 in determining which route they will follow.

Table 4.1: Passenger Type Percentages

Check-in Type	Percentage
Curbside Check-in	10%
Inside Check-in	70%
No Check-in (Carry on)	20%

Step 4: To model the decision of a passenger to choose their check-in route the Selection Weight property of Path objects will be utilized. The probability of choosing a particular path is that path's individual selection weight divided by the sum of all path selection weights from that node object. Place the proper weights to each different check-in route (see Figure 4.3).

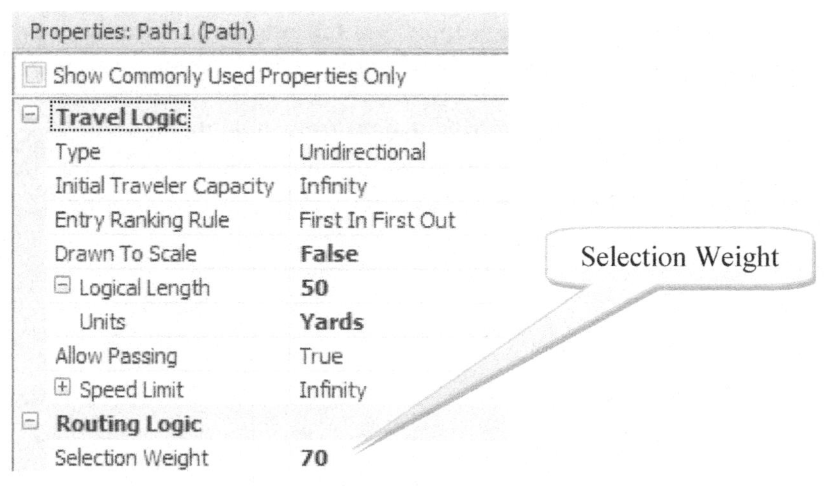

Figure 4.3: Weighting Paths

Step 5: Remember to select the Outbound Link Rule to "By Link Weight" on the TRANSFERNODE **Output@SrcPassengersArrive** (see Figure 4.4).[29]

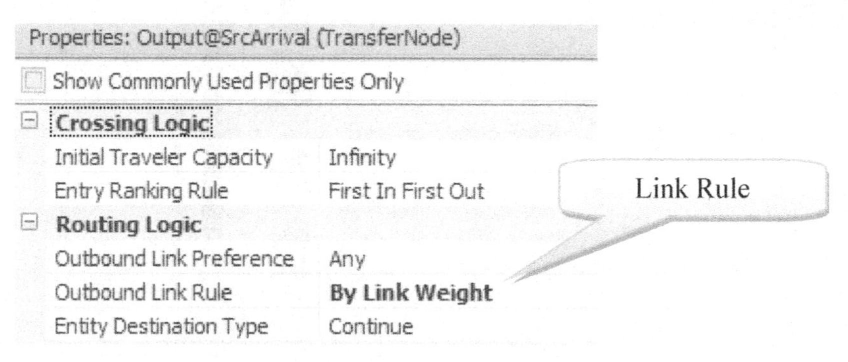

Figure 4.4: Outbound Link Rule

Step 6: Run the model for one, twenty-four hour replication (no "Experiment" needed). You might want to improve on our animation.

Question 1: What is the average number of passengers waiting at Curbside and the Inside Check-in?

____

Question 2: What is the average wait time at each Check-in?

____

Question 3: What is the average time it takes for a passenger to check-in (from entry to security)?

____

Part 4.2: Changing Arrival Rate

Passengers do not arrive at a constant rate (i.e., homogeneous process) throughout the entire day in a real airport (even though they may arrive randomly), so we should expand our model to handle such phenomena. Our approach will be to allow the *arrival rates* to change throughout the day. To model this changing rate, we will use a data object in SIMIO called a *Rate Table* and reference the *Rate Table* we create in the arrival logic of our Source object. It is important to understand that the interarrival times for this rate will be assumed to be Exponentially distributed, but the mean of this distribution changes with time. If an interarrival process has an

[29] Although "Shortest Path" is the default *Outbound Link Rule*, "**By Link Weight**" rule will still be used unless the entity destination has been assigned to a particular node. Therefore, we could have used the default rule.

Exponential distribution, then it is known statistically that the number of arrivals per unit time is Poisson distributed. Since the parameter (its mean) changes with time, it is formally referred to as a "non-homogeneous Poisson" arrival process, also called a NHPP (Non-Homogeneous Poisson Process). Therefore when you use a SIMIO Rate Table you are assuming the arrival process is a NHPP. Table 4.2 shows the hourly rate over the 24 hour day where more passengers on average arrive during the morning and dinner time hours. The zero rates imply no arrivals during that time period.

Table 4.2: Hourly Arrival Rate During Each Hour

Begin Time	End Time	Hourly Rate
Midnight	1am	0
1am	2am	0
2am	3am	0
3am	4am	0
4am	5am	0
5am	6am	30
6am	7am	90
7am	8am	100
8am	9am	75
9am	10am	60
10am	11am	60
11am	Noon	30
Noon	1pm	30
1pm	2pm	30
2pm	3pm	60
3pm	4pm	60
4pm	5pm	75
5pm	6pm	100
6pm	7pm	90
7pm	8pm	30
8pm	9pm	0
9pm	10pm	0
10pm	11pm	0
11pm	Midnight	0

Step 1: Arrivals are only present from 5am to 8pm for a total of 15 hours each day. A simulation run length of 24 hours would cause the time-based statistics like number in station, number in queue, and scheduled utilization to be computed when there were no arrivals, causing lower than expected values for these statistics. Therefore we will set the *Starting Time* within the "*Run Setup*" tab to 5:00 am and the *Ending Type* to *Specific Ending Time* at 8:00 pm as shown in Figure 4.5

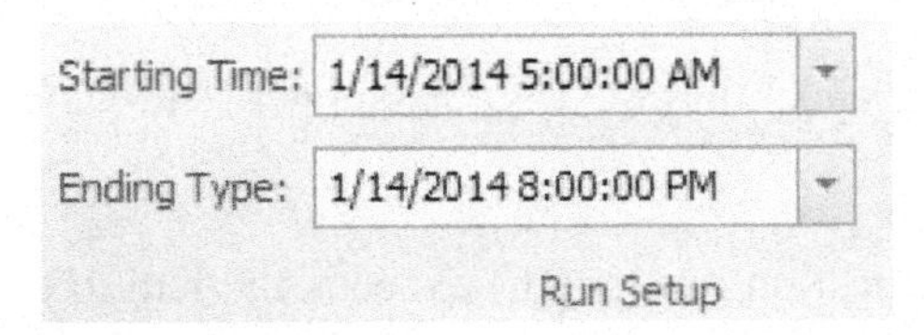

Figure 4.5: Replication Start and Stop Time

Step 2: To create a RATE TABLE, select the "*Data*" tab and then click *Create →Rate Table.* Name the table **PassengerArrivalRate**. Set the *Interval Size* to one hour and the *Number of Intervals* to 15, corresponding to the run length as seen in Figure 4.6. While you can modify the size (e.g., half hour increments) and number of time intervals during which the arrival rate is constant, you cannot modify the rate, which is fixed at *arrivals per hour*.

Step 3: Enter values in the rate table as shown in Figure 4.6.

Starting Offset	Ending Offset	Rate (events per hour)
Day 1, 00:00:00	Day 1, 01:00:00	30
Day 1, 01:00:00	Day 1, 02:00:00	90
Day 1, 02:00:00	Day 1, 03:00:00	100
Day 1, 03:00:00	Day 1, 04:00:00	75
Day 1, 04:00:00	Day 1, 05:00:00	60
Day 1, 05:00:00	Day 1, 06:00:00	60
Day 1, 06:00:00	Day 1, 07:00:00	30
Day 1, 07:00:00	Day 1, 08:00:00	30
Day 1, 08:00:00	Day 1, 09:00:00	30
Day 1, 09:00:00	Day 1, 10:00:00	60
Day 1, 10:00:00	Day 1, 11:00:00	60
Day 1, 11:00:00	Day 1, 12:00:00	75
Day 1, 12:00:00	Day 1, 13:00:00	100
Day 1, 13:00:00	Day 1, 14:00:00	90
Day 1, 14:00:00	Day 1, 15:00:00	30

Properties: PassengerArrivalRate (Rate Table)

Show Commonly Used Properties Only

Basic Logic	
Interval Size	**1**
Units	Hours
Number of Intervals	**15**
General	
Name	**PassengerArrivalRate**
Description	

Figure 4.6: The Rate Table (15 intervals)

Step 4: Modify the **SrcPassengersArrive** source accordingly (see Figure 4.7 for details). *SIMIO refers to the NHPP as a "Time Varying Arrival Rate"* ARRIVAL MODE.

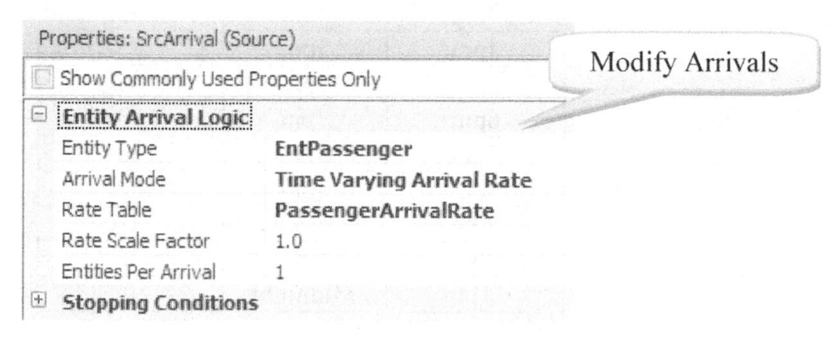

Figure 4.7: Specifying a Time Varying Arrival Rate

Step 5: Run the model again and observe the results.

Question 4: What are the average Wait Times at each check-in location (Curbside and Inside)?

____

Question 5: What is the average check-in time for a passenger (Curbside and Inside)?

____

If the length of the simulation (replication/run length) exceeds the length of the rate table, then the rate table simply repeats until the simulation replication is terminated. So if the run length should be longer and there should be no arrivals during that time, then the rate table should contain zeroes accordingly.

Part 4.3: State Variables, Properties, and Data Tables

Properties and state variables can be characteristics of the "Model" or of the "ModelEntities". A "property" of an object is a part of the object's definition and is assigned at the time an object is created, but it can't be changed during the simulation. Properties can be expressions, so their evaluation and the time of their creation can yield different values. In contrast, a "state variable" is a characteristic of an object that can be changed during the simulation. Generally, when properties and state variables are associated with the model, they are global in the sense that their values are visible throughout the model. When properties and state variable are associated with entities, then you need to have access to that model entity in order to access the state variable value.

In an actual airport more than one type of passenger arrives. Some are continental travelers while others are inter-continental travelers. Still there are some passengers which have handicaps and require special attention or are with families. The different types of passengers result in different processing times. To model these three cases we will use a combination of state variables and properties to store the information and to reference the information when needed.

There is a 33% chance for each type of passenger to arrive to our airport. The processing times at the curbside check-in and the inside check-in varies with the type of passenger. Specifically, see Figure 4.8 for the processing times.

Passenger Type	Curbside Check-In Time (minutes)	Inside Check-In Time (minutes)
1	Triangular(1, 2, 5)	Uniform(2, 5)
2	Triangular(2, 3, 6)	Uniform(3, 5)
3	Triangular(3, 4, 7)	Uniform(4, 6)

Figure 4.8: Check-In Times for Different Passenger Types

The MODELENTITY will need to reference these data as specifications at the check-in counters

Step 1: Since the passenger type is not a general characteristic of SIMIO model entities, we need either to find a SIMIO characteristic we can use to represent the type or define such as characteristic ourselves. User-defined characteristics whose values can be re-assigned are the "STATE VARIABLES." A state variable can be a characteristic of the "Model" (namely global) or a characteristic of the "ModelEntity". Here we are interested in a characteristic of a MODELENTITY.

Step 2: To create a MODELENTITY State Variable, first select the "**ModelEntity**" object in the *Navigation* panel. Click *"States"* panel icon in the *"Definitions"* tab. It is convenient to number our passenger type, so select a DISCRETE "Integer" state from the ribbon. Let's name it **EStaPassengerType**[30]. Next we need to assign each MODELENTITY a value for passenger type.

Question 6: When you define your new state variable, what other state variables are already defined for MODELENTITIES?

____

Step 3: We need to assign a value to **EStaPassengerType** for each entity. To obtain a random assignment of passenger type to entities (i.e., 33% chance for each type), we must use the `Discrete` Distribution (entered in cumulative distribution form).

```
Random.Discrete(1, 0.33, 2, 0.66, 3, 1.00)
```

Step 4: One convenient place is to make this assignment at the **SrcPassengersArrive** source object in the "State Assignments" property. Click on the plus box to reveal "*Before Exiting*". Then click the ellipses box at the end of that specification to bring up the "REPEATING PROPERTY EDITOR"[31], which in this case allows state value assignments.

---

[30] We will use "E" to when denoting "Entity" and "Sta" to denote "State variable".

[31] It's called a Repeating Property Editor, even though, in this case, it is assigning state variables

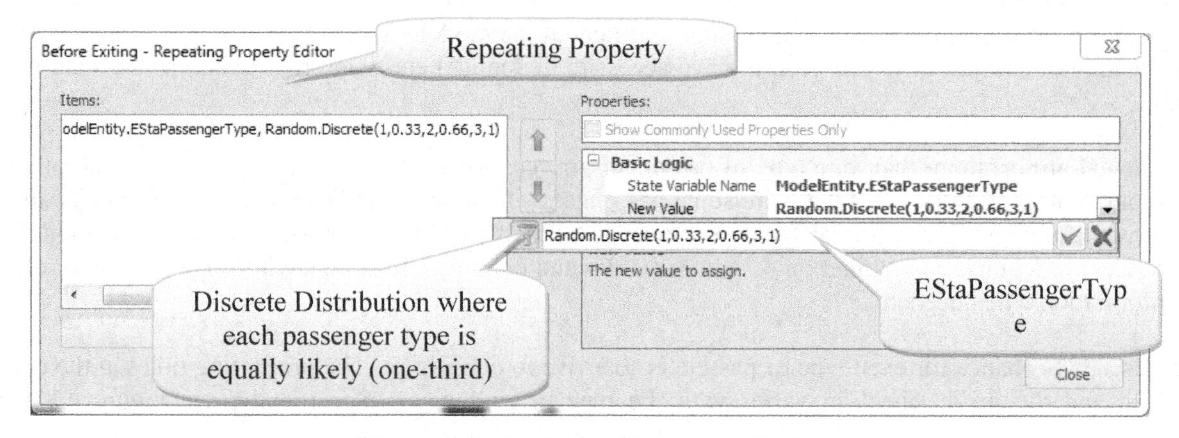

Figure 4.9: Assigning Passenger Type

Now each passenger has a (randomly assigned) state variable that describes its passenger type.

Step 5: Next we need to store the information for the check-in times. The expression of these times (e.g. `Random.Unform2,5))` do not change throughout the simulation. And these check-in times need to be accessed for all entities. The check-in times are not specific to each entity, but are characteristics of the model. So, "properties" are an appropriate way to store these data for the model.

Step 6: A way to organize model properties is through the use of a DATA TABLE. The data table is functionally identical to the table we created for Figure 4.8. To create a DATA TABLE click *"Tables"* in the *"Data"* tab, then *"Add Data Table."* Call it **TablePassenger**. Characteristics, called "properties", can be added to the data tables – these properties are the columns in the table. In our DATA TABLE, the processing times for each of the three types of passengers for each check-in location will be specified. Select the properties from the *"Standard Property"* option in the *Properties* section. **PassengerPriority** will be an *Integer* property whereas the check-in times will be *Expression* properties named **CurbsideCheckInTime** and **InsideCheckInTime.**[32] Priorities 1, 2, and 3 represent Continental, Inter-Continental, and special needs passengers respectively as in Figure 4.10.[33]

Table Passen...

Properties

	Passenger Priority	CurbsideCheckInTime (Minutes)	InsideCheckInTime (Minutes)
▸ 1	1	Random.Triangular(1,2,5)	Random.Uniform(2,5)
2	2	Random.Triangular(2,3,6)	Random.Uniform(3,5)
3	3	Random.Triangular(3,4,7)	Random.Uniform(4,6)

Figure 4.10: The Data Table

Step 7: Since the check-in times are in minutes, we want to specify the "Unit Type" for these properties as seen in Figure 4.11[34].

---

[32] Note that if you first specify the *"General"* name, it will become the default for the *"DisplayName"* property. Also, by using proper case, the table labels as seen in the Figure 4.10 will have spaces between the words.

[33] Even though the check-in times are expression properties, the familiar drop down expression editor is not available in DATA TABLES. Therefore, the easiest way is to put in all three priorities and type in the first value of the first row. Then copy it to the other rows and just modifying the values of the parameters.

[34] It may seem obvious the Unit Type for the time properties should be specified. However, one can leave the units as unspecified then the numbers will take on the units of the property when they are used (i.e., default value of hours).

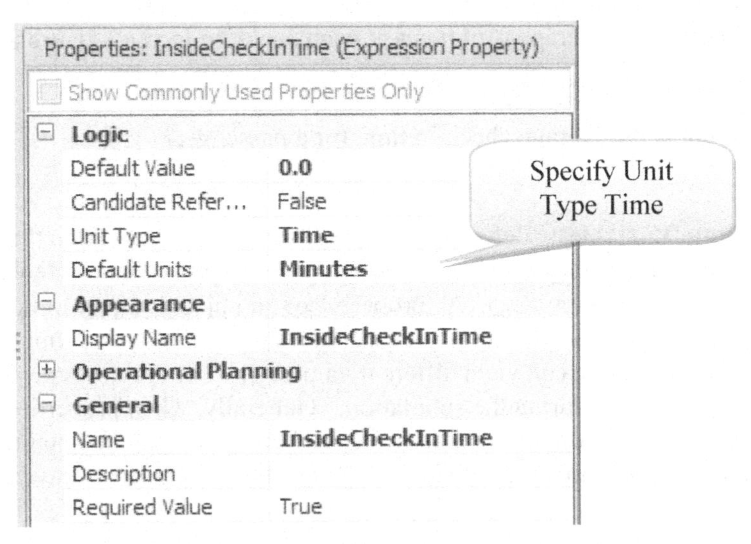

Figure 4.11: Inside Check-In Properties

Step 8: Now we must specify the processing times at the check-in stations to depend on the particular entity's passenger type, shown in Figure 4.12. The passenger type is used as the "row reference" in the table. Thus, the specifications needed for the new processing times use the **EStaPassengerType** state variable as:

- *At* **SrvCurbSideCheckIn**:[35] `TablePassenger[ModelEntity.EStaPassengerType].CurbSideCheckInTime`
  At **SrvCheckIn**:
  `TablePassenger[ModelEntity.EStaPassengerType].InsideCheckInTime`

Notice that the content of the [][36] are used to designate the specific row in the table, similar to arrays in programming languages. Unfortunately, the expression editor does not know about the DATA TABLE properties since they are user-defined, so you must type in the property into the expression. Also be sure that the processing time units in the check-in stations are specified in minutes.

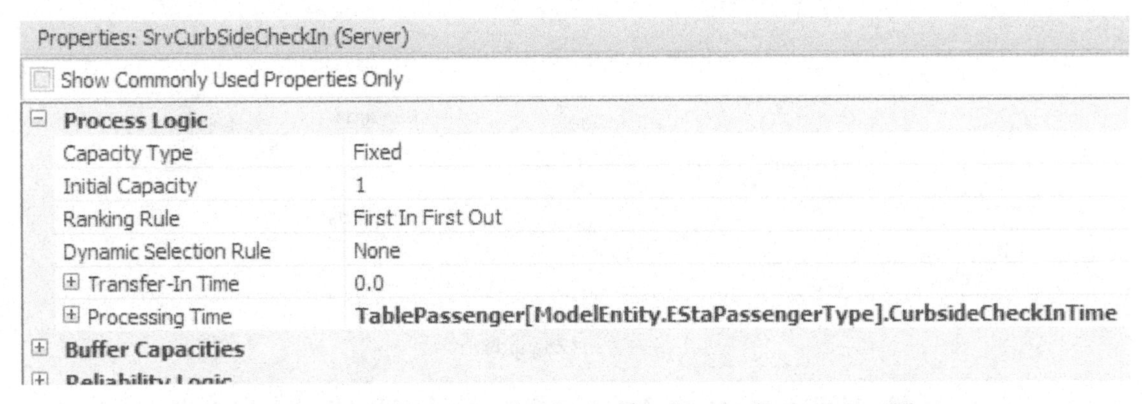

Figure 4.12: Utilizing the Table to Specify the Processing Times

Step 9: Run the model. Remember no arrival occurs before 5am and the simulation is started at 5am so no animation appears before then.

---

[35] To enter this specification so as to employ the expression editor during entry is to first put in the table name and property as `TablePassenger.CurbSideCheckInTime` and then insert `[ModelEntity.EstaPassengerType]`.

[36] Note, table rows start at one (i.e., the first check in time is `TablePassenger[1].InsideCheckInTime`)

Question 7: What is the average wait times at each check-in location (Curbside and Inside)?

____

Question 8: What is the average check-in time for a passenger?

____

Properties and State Variables: Before leaving this section, let's reflect on the characteristics "properties" and "state variables" because they will be in frequent use. Properties and state variables can be characteristics of the MODEL or of the MODELENTITIES. A "property" of an object is a part of the object's definition and is assigned at the time an object is created, but it can't be changed during the simulation. Properties can be expressions, so their evaluation can yield different values. In contrast a "state variable" is a characteristic of an object that can be changed during the simulation. Generally, when properties and state variables are associated with the model, they are global in the sense their values can be obtained almost anytime. When properties and state variables are associated with entities, then you need to have access to that model entity in order to access the value.

SIMIO makes extensive use of properties and state variables, which you can see by looking at their "definitions" and expanding the (Inherited) Properties/States. Some properties and states are related. For example, the "*InitialPriority*" property of a MODELENTITY becomes the initial value of the "*Priority*" state variable. Other states are given their initial values by properties.

Part 4.4: More on Branching

At times, the waiting lines at the inside Check-in station seem overcrowded. We decided to add a second Inside Check-in station that passengers may use during rush hours. Thus, passengers will be routed to whichever Check-in station has the smallest number in queue. The distance from this second Check-in station to the Security Check is 75 yards as seen in Figure 4.13.

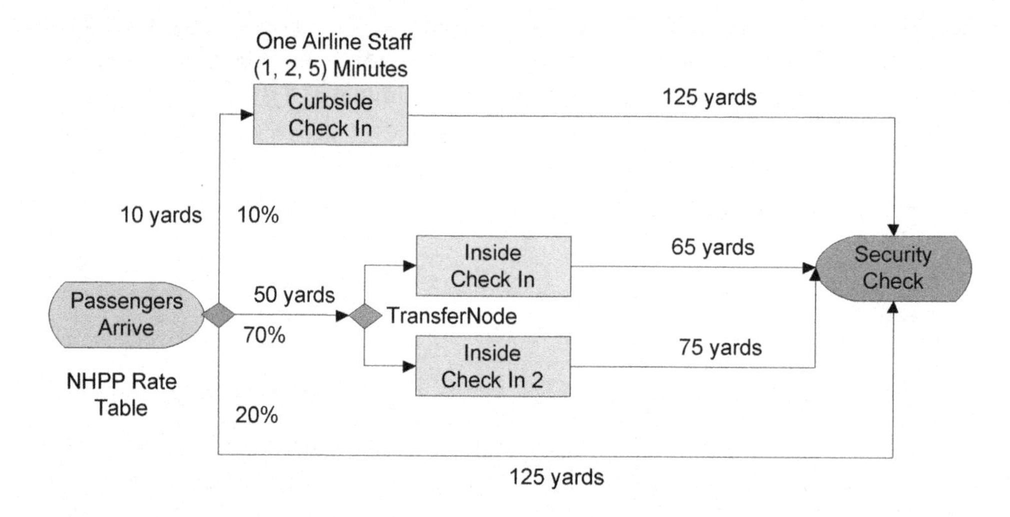

Figure 4.13: Airline Check-in Process with Two Inside Check In Stations

Step 1: Add the second check-in station. Call it **SrvCheckIn2**. Assume it has a capacity of one and it's processing times are the same as those for the regular check-in station. Add the path of 75 yards from this second station to **SnkSecurity**.

Step 2: We need to model the choice made by the "inside" check-in passengers to go to the check-in station with the smallest number in queue. There are several ways to model this situation. Perhaps the most obvious way is to add a TRANSFERNODE that branches passengers to the shortest queue. The distance from the **SrcPassengersArrive** to this TRANSFERNODE is 50 yards (remember that the selection weight is 70 for this path and you will have to delete the original path to the **SrvCheckIn** object).

- The TRANSFERNODE should specify the *Outbound Link Rule* as "By Link Weight"
- Set the *Selection Weight* properties of the two paths from the TRANSFERNODE (the "value" of a "true" expression is one and zero if it is "false")

(1) To **SrvCheckIn**:
```
SrvCheckIn.InputBuffer.Contents.NumberWaiting <=
        SrvCheckIn2.InputBuffer.Contents.NumberWaiting
```

(2) To **SrvCheckIn2**:
```
SrvCheckIn2.InputBuffer.Contents.NumberWaiting <
        SrvCheckIn.InputBuffer.Contents.NumberWaiting
```

Question 9: Which check-in station is chosen if there are "ties" in the size of the InputBuffer (For example, suppose both are empty)?

___

Question 10: What is the utilization of **SrvCheckIn2** and does it make sense to have both check-ins?

___

Part 4.5: Work Schedules

Suppose you decide only to staff the second inside check-in station from 10am to 6pm. This kind of concern requires a work schedule which alters the capacity of a server over time. Work schedules will be built from the *"Schedules"* component in the *"Data"* tab.

Step 1: Click on *"Work Schedule"* button under in the "Schedules" icon in the "Data" tab to add a new work schedule. By default, the StandardWeek schedule has already been defined as seen in

Step 2: Figure 4.14. Rename this schedule **SecondCheckInStation**. Change the Days to "1" – this schedule will repeat daily.[37] For each day of the work schedule, you need to specify the work pattern for that day. Therefore, you can have different patterns for different days (i.e., weekends, etc.). The **StandardDay** pattern was specified for *Day 1*. The *Start Date* property can be left as the default since we have a daily pattern.

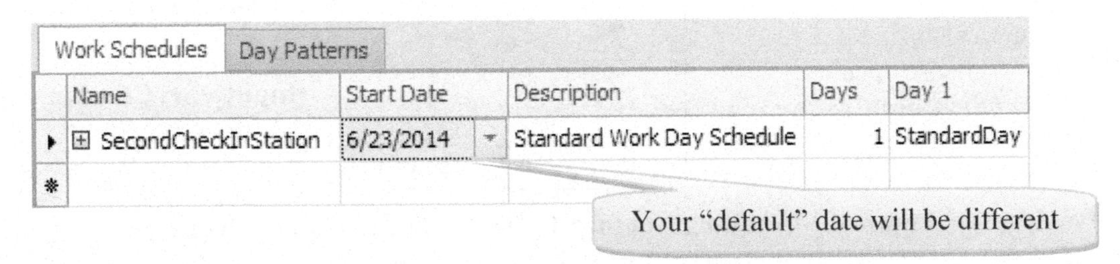

Figure 4.14: Creating SecondCheckInStation WORKSCHEDULE

Step 3: In the *"Day Patterns"* tab, you can add DAY PATTERNS by clicking the *Day Pattern* button in the *"Create"* section. SIMIO defines the **StandardDay** pattern of 8am to 5pm with an hour lunch at 12pm. Since we start the day at 10am and end the day at 6pm, we only need one work period in the pattern. Select the second work period (1pm to 5pm) and delete it. Then specify the Start Time to be 10am and the Duration to be eight hours or the *"End Time"* to be 6pm as seen in Figure 4.15. Specify a "1" for the

[37] If there are seven days in the pattern, then the days of the week will appear. In our case there is only the Day 1 pattern. SIMIO will repeat the pattern based on the number of days in the WORK SCHEDULE. For each day in the work schedule you need to specify a pattern. Also, we have the ability to specify exceptions to the day pattern or work periods for particular dates.

Value which will specify the capacity of the second inside check in to be one. If a work period has not been defined for a period in the day pattern, it is assumed to be off shift (i.e., zero capacity).

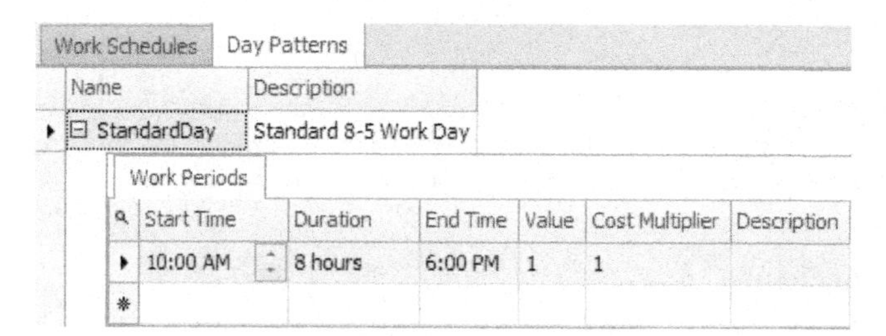

Name	Description
StandardDay	Standard 8-5 Work Day

Work Periods

Start Time	Duration	End Time	Value	Cost Multiplier	Description
10:00 AM	8 hours	6:00 PM	1	1	

Figure 4.15: Setting a Day Pattern with Different Work Periods

Step 4: Go to the "Facility" view and click on the **SrvCheckin2** object. Change the *Capacity Type* to "**WorkSchedule**" and the *Work Schedule* property to **SecondCheckInStation as in** Figure 4.16.

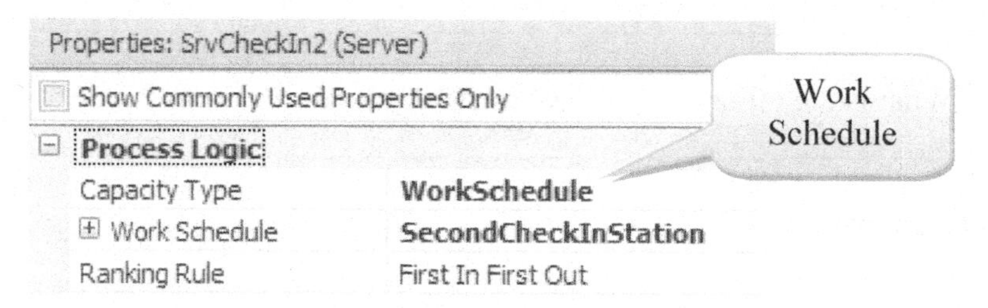

Figure 4.16: Specifying a Work Schedule

Step 5: Now we need to be sure that people who choose the inside check-in only go to the **SrvCheckIn2** station if it is open (i.e., only between hours 10 and 18). This requires another TRANSFERNODE as seen in Figure 4.17. Connect the original TRANSFERNODE to the new one via a Connector and connect the new TRANSFERNODE to the inputs of both check-in stations.

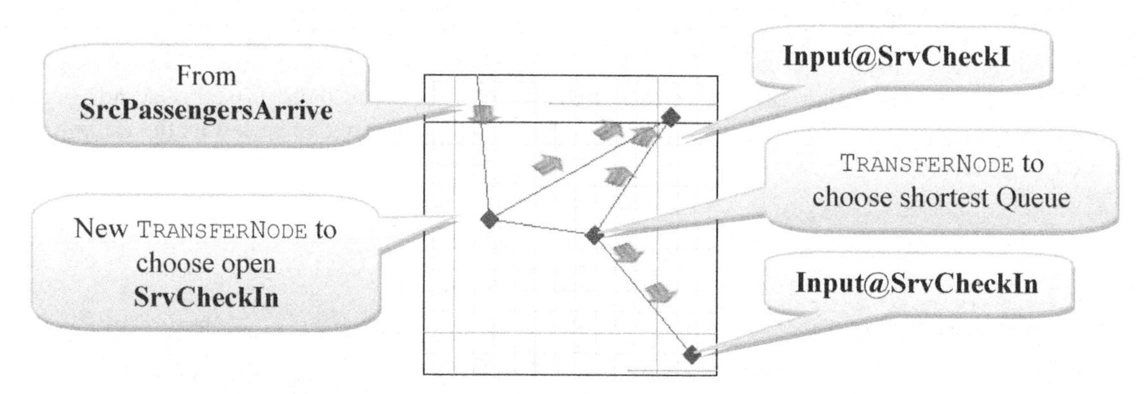

Figure 4.17: Adding Another TransferNode to Handle the Logic

- The new TRANSFERNODE should determine routing *By Link Weight* as well.
- Path to **Input@SrvCheckIn** has zero length[38] and *Selection Weight* property: `Run.TimeNow <=5 || Run.TimeNow>=13`[39]
- Path to TRANSFERNODE to choose shortest queue has zero length and *Selection Weight* property of `Run.TimeNow > 5 && Run.TimeNow <13`[40]

[38] May be easier to use a CONNECTOR instead of a PATH – CONNECTORS have zero length and take zero time to traverse.

[39] Note, SIMIO utilizes `||` as the "Or" logical operator and `&&` as the "And" logical operator. Also, this expression works only here since we are simulating just one day and 10 A.M. represents five hours into the simulation since we are starting at 5 A.M. If we were simulating more than one day, then we would need to specify `Math.Remainder(Run.TimeNow, 24)*24` which converts the running time into a number between 0 and 24.

Step 6: Note that when there is no context, the default units of time are hours. Now run the model and obtain these basic statistics.

Question 11: What is the average Check-In time for a Passenger?

____

Question 12: What is the average wait times for Curbside Check-In and Inside Check-In?

____

Question 13: What is the utilization for an employee at the Curbside Check-in?

____

Question 14: What are the utilizations for an employee at the Inside Check-in?

____

Step 7: You may have noticed that in the simulation, there are entities in the queue of the **SrvCheckIn2** that are in line at the time the station closes. These are marooned in this model. We will consider how to handle this kind of problem in a later chapter.

Question 15: What would have to change to the model every time the work schedule of the second check in server was modified?

____

Step 8: If we wanted to determine the best starting time and ending time to have the second check in station open, the link weights using the `Run.TimeNow` expression would have to be changed to match the new start and end times each time. A better approach would be to use the capacity of the server so as it goes to zero we need to close the path and open the path when the capacity goes to one. Change the link weights according and rerun the model comparing the previous results.

- Path to **Input@SrvCheckIn** *Selection Weight* property: `SrvCheckIn2.Capacity == 0`[41]
- Path to TRANSFERNODE to choose shortest queue has zero length and *Selection Weight* property of `SrvCheckIn2.Capacity > 0`

Part 4.6: Commentary

- The specification of the time-varying arrival pattern in SIMIO requires fixed units for the arrival rate, namely arrivals per hour.
- Pay special attention to the SIMIO specification of cumulative probability and its value for the `Random.Discrete` and `Random.Continuous` distributions.
- While work schedules in SIMIO offers considerable flexibility in terms of specifying exceptions etc., they are somewhat complicated to input. One has to be careful in specifying the correct times and duration. Also be careful in specifying the time to begin and end the simulation run.
- In SIMIO, when a server's capacity goes to zero while there is an entity (or entities) being processed, SIMIO chooses, by default, to "Ignore" this so that the items "in process" will be processed while at the same time the server capacity goes to zero. If another behavior is desired, then you will need to implement it using more advanced concepts, which will be described later.

---

[40] We could have modified the original link weights and avoided having to add the new paths and transfer node. Specify the expression to be (TimeNow <=5 || TimeNow >= 13) || `SrvCheckIn.InputBuffer.Contents.NumberWaiting <= SrvCheckIn2.InputBuffer.Contents.NumberWaiting` for the link weight for the path to **SrvCheckIn** and for the path link weight to **SrvCheckin2** to (Run.TimeNow >5 && Run.TimeNow < 13) && `SrvCheckIn2.InputBuffer.Contents.NumberWaiting < SrvCheckIn.InputBuffer.Contents.NumberWaiting`.

[41] Note, SIMIO utilizes a double equal sign (i.e., "==") as the logical equal.

- Note that if the length of the simulation replication exceeds the work schedule time, then the work schedule is repeated until the replication ends.

Chapter 5
Data-Based Modeling: Manufacturing Cell

Most simulation models employ substantial amounts of data in their development and execution. In SIMIO, employing a data table, as seen in the previous chapter, provides a convenient means to store and recall the data. In this chapter we will use tables more extensively by re-constructing a manufacturing cell consisting of several work stations.

A small manufacturing cell consists of three workstations through which four part types are processed. The workstations, A, B, and C, are arranged such that the parts must follow the one-way circular pattern as shown in the layout in Figure 5.1. For instance, if a part is finished at Station C, it would have to travel the distance to Station A, then the distance to Station B before it could travel to where it exits the system. We want to use a simulation of a five day week with eight hours of operation a day for a total of 40 hours.

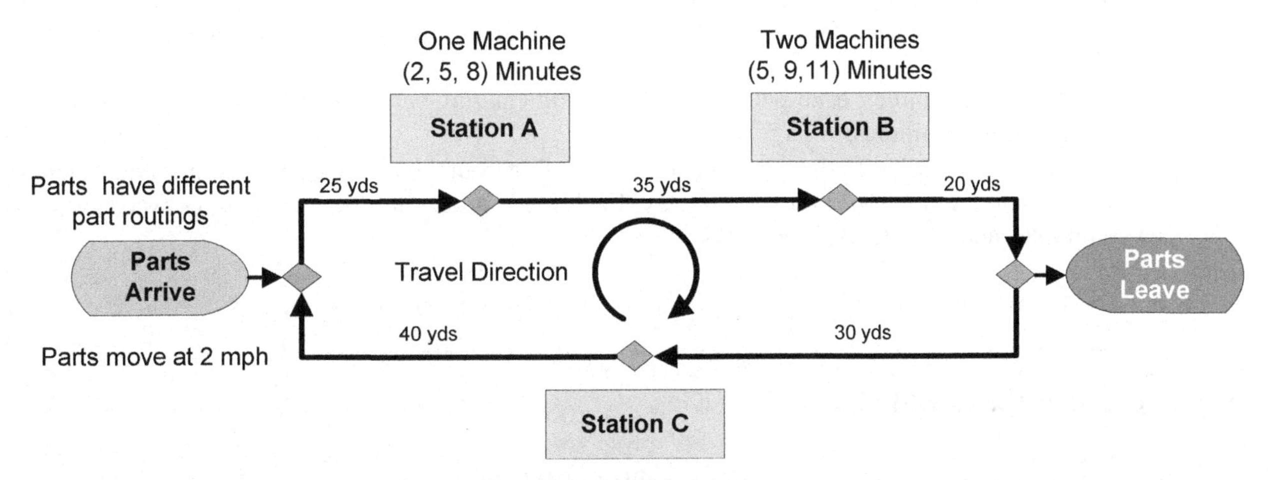

Figure 5.1: Workstation Layout

All parts arrive at the "Parts Arrive" location and leave at the "Parts Leave" location. Travel between workstations is a constant two miles per hour. Note SIMIO's default speed for entities is *meters per second.* The distances (yards) between workstations are shown in Table 5.1 and in Figure 5.1.

Table 5.1: Distance between Workstations

WorkStation Path	Distances (yds.)
Parts Arrive to Station A	25
Station A to Station B	35
Station B to Parts Leave	20
Parts Leave to Station C	30
Station C to Parts Arrive	40

Each part type arrives as follows.[42]

- Part 1's arrive randomly with an average of 15 minutes between arrivals.
- Part 2's have an interarrival time that averages17 minutes with a standard deviation of 3 minutes.

[42] For the moment, we will let you think about the distributions that these parameters may represent.

- Part 3's can arrive with an interarrival time of anywhere from 14 to 18 minutes apart.
- Part 4's arrive in batches of five exactly 1 hour and 10 minutes apart.

Question 1: What distributions will you choose to model these four arrival processes?

___

Each of the part (entity) types is sequenced (routed) across the stations differently. In fact, not all part types are routed through all machines, as seen in the sequence order given in Table 5.2.

Table 5.2: Part Sequences

Step	1	2	3
Part 1	Station A	Station C	
Part 2	Station A	Station B	Station C
Part 3	Station A	Station B	
Part 4	Station B	Station C	

Server Properties:

- Station A has one machine with a processing time ranging from two to eight minutes and a likely (modal) time of five minutes.
- Station B has two machines, each with a processing time ranging between five and eleven minutes and a likely time of nine minutes.
- Station C has two machines that operate during the first four hours of each day and one machine operating during the last four hours of the day. Each has a processing time ranging between two and eleven minutes and a mode of six minutes.

Question 2: What distributions will you choose to model the three station processing times?

___

Part 5.1: Constructing the Model

We will build the SIMIO model using four SOURCES, three SERVERS and several types of links.

Step 1: Insert four SOURCES named **SrcParts1**, **SrcParts2**, **SrcParts3**, and **SrcPart4** to create the four different part types (i.e., drag four MODELENTITIES onto the model canvas named **EntPart1**, **EntPart2**, **EntPart3**, and **EntPart4**).[43] Four SOURCES are necessary because each part has its own arrival process. Make sure to change the *Entity Type* property of the four SOURCES to produce the appropriate entity type.

Step 2: Insert three SERVERS (i.e., one for each workstation named **SrvStationA**, **SrvStationB** and **SrvStationC**) and one SINK named **SnkPartsLeave**. Be sure to name all the objects, including the entities, so it will be easy to recognize them later as seen in Figure 5.2.

Step 3: Probably the best method to simulate the circular travel pattern would be to use a BASICNODE at the entry and exit point at each station.[44] This allows the entities to follow the sequence pattern without having to go through an unneeded process. A BASICNODE is a simple node to connect two different links while a TRANSFERNODE can connect different links as well but provides the ability to select destination, path, and/or require a TRANSPORTER.

___

[43] To do this efficiently, drag one SOURCE on to the facility and name it **SrcParts**. Then copy and paste it three times which will take advantage of the SIMIO naming scheme (i.e., it will create **SrcParts1**, **SrcParts2**, and **SrcParts3**). Now, rename the first source **SrcParts4**. Repeat this method for the four entities starting with **EntPart**.

[44] A TRANFERNODE used to enter/exit from a station can disrupt the SIMIO internal sequencing of sequence steps if one is not careful.

Step 4: Now connect all the objects using CONNECTORS and PATHS. Note that because there is no time lapse or distance between the circular track and the server objects, we will use connectors. The distances between stations will be calculated using the five paths connecting each entry and exit node. Be sure that you draw the connectors and paths in the right direction – zoom in on your model to see the "arrows" as shown in Figure 5.2 (i.e., links are entering the input nodes of the SERVERS while leaving the output nodes).

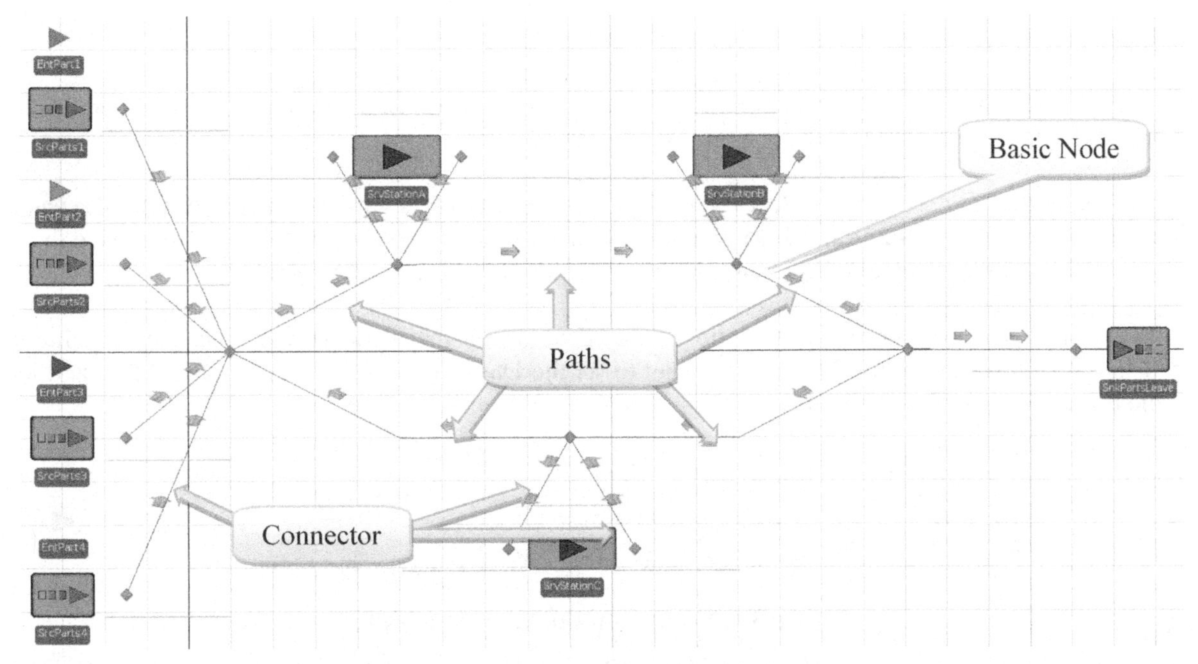

Figure 5.2: Initial Model

Step 5: Change the properties of each entity so that they now travel at the desired speed of two miles per hour.[45] Also change the color of the entities so you can distinguish between the four part types.

Step 6: Change the properties of the four source objects so that they correspond to the correct parts and have the desired interarrival times and entities per arrival. We chose an Exponential, Normal, and Uniform as the distributions of interarrival times for part types one, two, and three respectively with the parameters given. For part type four, we chose a constant interarrival time of 70 minutes and *Entities Per Arrival* property set to 5 as seen in Figure 5.3.

Properties: SrcParts4 (Source)

☐ Show Commonly Used Properties Only

⊟ **Entity Arrival Logic**	
Entity Type	**EntPart4**
Arrival Mode	Interarrival Time
⊞ Time Offset	0.0
⊟ Interarrival Time	**70**
Units	Minutes
Entities Per Arrival	**5**

Figure 5.3: Specifying Parameters for the SrcPart4

Step 7: Set the distances for each of the five paths so they correspond to the distances between the stations. You will have to set the *Drawn To Scale* property for each to "*False*" in order to do this and then specify the logical length and units of type "Yards".

[45] Select all four entities using the *Ctrl* button and specify the speed once for all entities.

Step 8: Set the processing times for all the stations using the Pert (BetaPert) distribution. In general, we prefer the `Pert` distribution to the `Triangular` distribution. It is specified like the Triangular (minimum, most likely, maximum), but has "thinner" tails, which, we believe, better represents the expected behavior of the processing time distribution. See Appendix A for more information.

Part 5.2: Setting Capacities

We will need to set the capacities for the three workstations A, B, and C.

Step 1: Set the capacities for A and B to one and two respectively. For Station C, we will need to add a work schedule to accommodate the changes in capacity.

Step 2: Go to the "*Data*" tab and select "*Schedules*" and modify the default schedule (**StandardWeek**) and the default day pattern (**StandardDay**). Rename the work schedule to **ScheduleForC**[46] and leave the *Start Date* property alone to remain the default date the simulation will start in the *Run* setup.

SIMIO will repeat the work cycle over how many days are in the work pattern. Because our simulation will start at 12am and run for 40 hours we want to set the Days in *Work Schedule* property to 1 and let the cycle repeat after the first 24 hours.

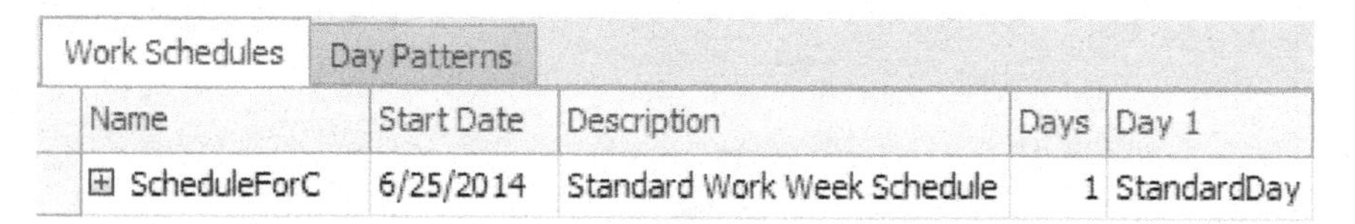

Work Schedules | Day Patterns

Name	Start Date	Description	Days	Day 1
⊞ ScheduleForC	6/25/2014	Standard Work Week Schedule	1	StandardDay

Figure 5.4: Setting up the WorkSchedule

Now select the "*Day Patterns*" tab to modify the **StandardDay** day pattern to match the one in Figure 5.5. Change the *Start Time* property of the first row to 12:00 AM which will cause the *Duration* to change to match the *End Time.* Change it back to "4 hours" and click the *Enter.* After changing the value of the capacity to two, repeat this process for the second row using one as the capacity value. Next, add four new work periods by specifying the *Start Time*, the *Duration*, and the capacity value.

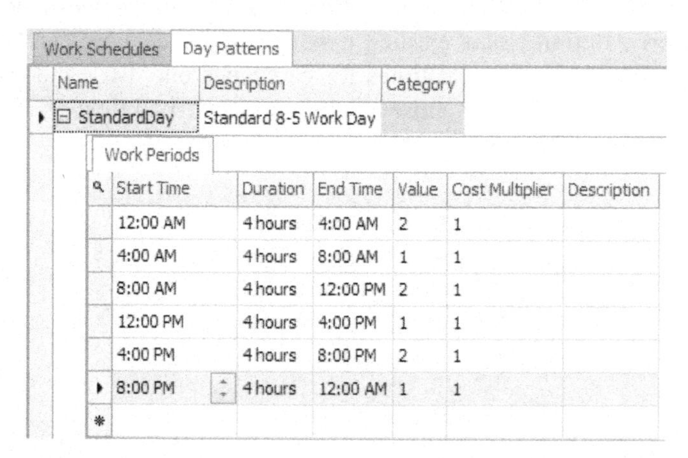

Work Schedules | Day Patterns

Name	Description	Category
⊟ StandardDay	Standard 8-5 Work Day	

Work Periods

Start Time	Duration	End Time	Value	Cost Multiplier	Description
12:00 AM	4 hours	4:00 AM	2	1	
4:00 AM	4 hours	8:00 AM	1	1	
8:00 AM	4 hours	12:00 PM	2	1	
12:00 PM	4 hours	4:00 PM	1	1	
4:00 PM	4 hours	8:00 PM	2	1	
8:00 PM	4 hours	12:00 AM	1	1	

Figure 5.5: Modifying the Day Pattern to Represent Three Eight Hour Days

Step 3: Go back to the "*Facility*" tab and set the *Capacity Type* property of Station C to "WorkSchedule" and add your work schedule, namely **ScheduleForC**.

---

[46] Double clicking the *Standard Week* name will allow you to change the name.

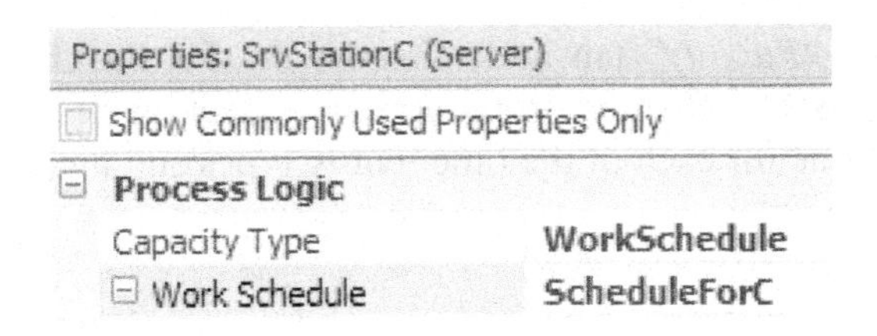

Figure 5.6: Specifying the WorkSchedule for the StationC

Question 3: For how many hours during a 24 hour day is the capacity of Station C equal to two?

____

Step 4: Save and run the model for 40 hours observing the path the different entities travel.

Question 4: How long is each part type in the system (enter to leave)?

____

Question 5: Why do you think the time in the system is large for all four part types?

____

Part 5.3: Incorporating Sequences

The way the current model is set up, entities will branch at nodes based on link weights and the direction on the link. However, we want the entities to follow a desired sequence of stations as seen in Table 5.2. We would like to be sure that entities will continue along the circular pattern until they reach the node that takes them to their particular station. To accomplish this we will use "Sequence Tables" (from the "*Data*" tab).

Step 1: Click the "*Add Sequence Table*" button in the tab and rename the new table **SequencePart1** to correspond to the first entity type.

Step 2: Now add the sequence of nodes that you want Part 1 to follow (i.e., Station A, Station C and Exit). You only need to include the visited nodes at each station as well as the exit. Once the entities sequence has been set, it will always travel the shortest route to get to the first node on the list. The correct sequence for **Part1** is given in Figure 5.7.[47]

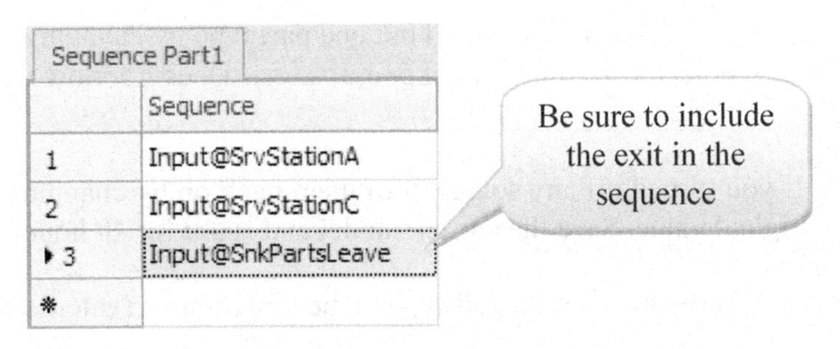

Figure 5.7: Sequence for Part1

Step 3: Next, add three more sequence tables to correspond to the remaining three entities (i.e., **SequencePart2**, **SequencePart3**, and **SequencePart4**). Set the sequences for each of the other three parts in their corresponding sequence tables as well. Make sure the last node is associated with the SINK. Note that the station's names may appear differently in your sequence if you named them differently.

[47] Note the input nodes of the associated SERVERS and SINK are specified rather than then actual object. Note the names of the nodes are in the form **Input@ObjectName**.

Step 4: Now go back to the "*Facility*" tab and click on the **EntPart1** entity. Set the *Initial Sequence* property to the "**SequencePart1**" under the "*Routing Logic*" section which you just created for "Part 1" as seen in Figure 5.7. Do the same for each of the other entity types and their associated sequences.

Routing Logic	
Initial Priority	1.0
Initial Sequence	**SequencePart1**

Figure 5.8: Specifying an Entity to follow a Particular Sequence

Step 5: For each of the seven TRANSFERNODES (output nodes at each source and server), change the *Entity Destination Type* property to "By Sequence" for each one.[48] The first time an entity passes through a transfer node where its destination is "By Sequence", the node sets the destination of the entity to the first row index of the SEQUENCE TABLE. After that, every time an entity passes through a transfer node with the "By Sequence" destination type the row index of the table is increased by one and the destination of the entity is set to the next node in the table.[49]

Table 5.3: Definition of Basic and TransferNodes

BasicNode	Simple node to support connection between links. Often used for Input nodes of objects (e.g., SINK, SERVER, etc.).
TransferNode	Supports connection between links but has the ability to select entity destination as well as request a transporter. Output nodes of objects (SOURCE, SERVER, etc.) are TRANSFERNODES

Question 6: Will passage through a BASICNODE also change the row index of the SEQUENCE TABLE?

____

Question 7: Will passage through a TRANSFERNODE always change the row index of the SEQUENCE TABLE?

____

Step 6: It will also help to make sure they follow the correct sequence. Run the simulation for 40 hours at a slow speed to be sure each of the parts follows their sequence appropriately.

Question 8: Turn off the arrivals of all but one part type by changing the *Entities Per Arrival* property of the SOURCE object to zero for all but one SOURCE. Does it follow the expected sequence?

____

Step 7: If you turned off any source, turn them back on by changing the *Entities Per Arrival* property back to the original value. Save the current model and run it for 40 hours.

Question 9: How long is each part type in the system now (enter to leave)*?*

____

[48] To set them all at once, press the *Ctrl* key while selecting all seven TRANSFERNODES and then set the *Entity Destination* to "By Sequence."

[49] If the current row index is on the last row and the entity enters a TRANSFERNODE that has "By Sequence" before reaching the current destination node, the row index will start over at one.

Part 5.4: Embellishment: New Arrival Pattern and Processing Times

We now learn that the parts arrive in one stream of arrivals (i.e., with a mean arrival rate of 6 per hour, Poisson distributed[50]), instead of four. There are still four part types but the type is random, based on the percentages specified in in Table 5.4. The number of each that arrive doesn't change (meaning Part 4 arrives 5 at a time).

Table 5.4: Part Type Percentage

Part Type	Percentages	Number to Arrive
Part 1	25%	1
Part 2	35%	1
Part 3	25%	1
Part 4	15%	5

Question 10: When do you think a single source type of model is preferred over a multiple source model?

____

Previously, the processing times at the various stations were the same regardless of the part type. Now we discover the processing times depend on the particular part type as given in Table 5.5. In this table, the sequences are the same as before, but here the processing times have been added.

Table 5.5: Entity Sequences and Processing Times (in Minutes)

Step	1	2	3
Part 1	Station A (Pert(2,5,8))	Station C (Pert(2,6,11))	
Part 2	Station A (Pert(1,3,4))	Station B (Uniform(5 to 11)	Station C (Uniform(2 to 11))
Part 3	Station A (Triangular(2,5,8))	Station B (Triangular(5,9,11))	
Part 4	Station B (Pert(5,9,11))	Station C (Triangular (2,6,11))	

Question 11: Do you think that it is realistic that the processing time for Part 1 on Station A might be different for Part2 on that same Station A?

____

Step 1: Now delete all but one of the SOURCES referring to Figure 5.9 as an example renaming the one SOURCE as **SrcParts**. We will return to specifying this source after we specify the processing times.

[50] Remember that this Poisson arrival rate is equivalent to an Exponential interarrival time with a mean of 10 minutes. SIMIO wants the interarrival time, not the arrival rate – so be prepared to make the conversion.

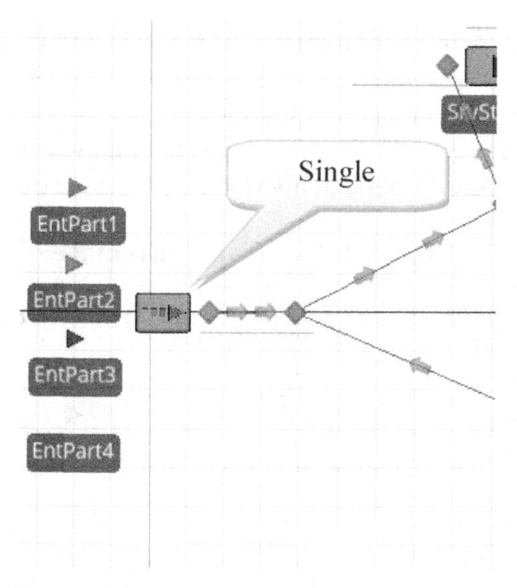

Figure 5.9: Manufacturing Cell Model

Step 2: To model the processing times which are dependent on the part type, the processing times will be added to the original sequencing tables. When the entity is assigned a destination (i.e., a row in the sequence table is selected), the processing time column can be accessed for that a row. (Note, you could use a more complicated expression for the processing time by checking to see if this is a Part A, then make the processing time X, etc. but is not very easy if there is a large number of parts).

Step 3: Go to the "*Data*" tab and click on the **SequencePart1** table which is associated with Part 1 types. Since the processing times can be any distribution (i.e., any valid expression), add a new *Expression* type column from *Standard Property* drop down named **ProcessingTimes.** Make sure the column has the appropriate units by making the *Unit Type* be "*Time*" and the *Default Units* to be "*Minutes*."

Step 4: Set the Processing times for **SrvStationA** (i.e.., the first row) to a `Random.Pert(2,5,8)`, **SrvStationC** (i.e., the second row) to a `Random.Pert(2,6,11)` and finally the **SnkPartsLeave** or third row will remain zero as seen in Figure 5.10.

Sequence Part1 | Sequence Part2 | Sequence Part3 | Sequence Par

	Sequence	ProcessingTimes (Minutes)
1	Input@SrvStationA	Random.Pert(2,5,8)
2	Input@SrvStationC	Random.Pert(2,6,11)
3	Input@SnkPartsLeave	0.0
*		

Figure 5.10: New Sequence Table for Part Type 1

Step 5: Repeat the last two steps for the remaining three part sequence tables using the appropriate processing times from Table 5.5.

Step 6: In order to have a single source produce multiple types of entities, a new data table has to be created to specify the part type entities and the part mix percentages. Add a new data table called **TableParts** with four columns with the specified types as seen in Table 5.6.

Table 5.6: Column Definitions

Column	Column Name	Column Type
1	**PartType**	*Entity*
2	**ProcessingTimes**	*Expression*
3	**PartMix**	*Integer (or Real)*
4	**NumberToArrive**	*Integer (or Real)*

Step 7: Next specify each entity as a new row with the appropriate processing times, part mix percentages and sequence table as shown in Table 5.7.

Table 5.7: Content of TableParts

Table Parts | Sequence Part1 | Sequence Part2 | Sequence Part3 | Sequence Part4

	Part Type	ProcessingTimes (Minutes)	Part Mix	Number To Arrive
1	EntPart1	SequencePart1.ProcessingTimes	25	1
2	EntPart2	SequencePart2.ProcessingTimes	35	1
3	EntPart3	SequencePart3.ProcessingTimes	25	1
4	EntPart4	SequencePart4.ProcessingTimes	15	5

Specifying `SequencePart1.ProcessingTimes` states that the processing time will come from the column associated with `ProcessingTimes` in the table `SequencePart1` for the current row.

Step 8: In order to use the dependent processing times, change each of the processing time expressions in the three stations to be **TableParts.ProcessingTime**. This logic specifies that the processing time will come from the *ProcessingTimes* column of the data table **TableParts**. As the parts move through their individual sequence table (i.e., row to row), it will retrieve the processing time for the current row. The property is now a reference property type Processing Time | **TableParts.ProcessingTime** which can only be changed back to a regular property by right clicking on the label (*Processing Time*) and resetting it.

Step 9: To generate multiple part types from the same source, the *Entity Type* needs to be changed from a specific entity type. In the SOURCE, specify the *Entity Type* to come from the **TableParts.PartType**. This will allow the *Entity Type* to come from one of the four entities specified in the table.

Step 10: Set the interarrival time of entities to be an `Exponential` with a mean of ten minutes (i.e., same as a Poisson arrival rate of six per hour). Set the *Entities Per Arrival* to **TableParts.NumberToArrive**.

Step 11: To determine which specific part type is created, the part mix percentages need to be utilized. The SOURCE has a property category called "Table Reference Assignment" which can assign a table and a row of that table to the entity.[51] There are two possible assignments: *Before Creating Entities* and *On Created Entity*. We want to do the assignment before the entity is actual created in order to create the correct entity type and the correct number to arrive. Therefore, specify the table shown in Figure 5.11.

[51] In later chapters we will utilize *Processes* and the `Set Row` step to assign a table to an entity.

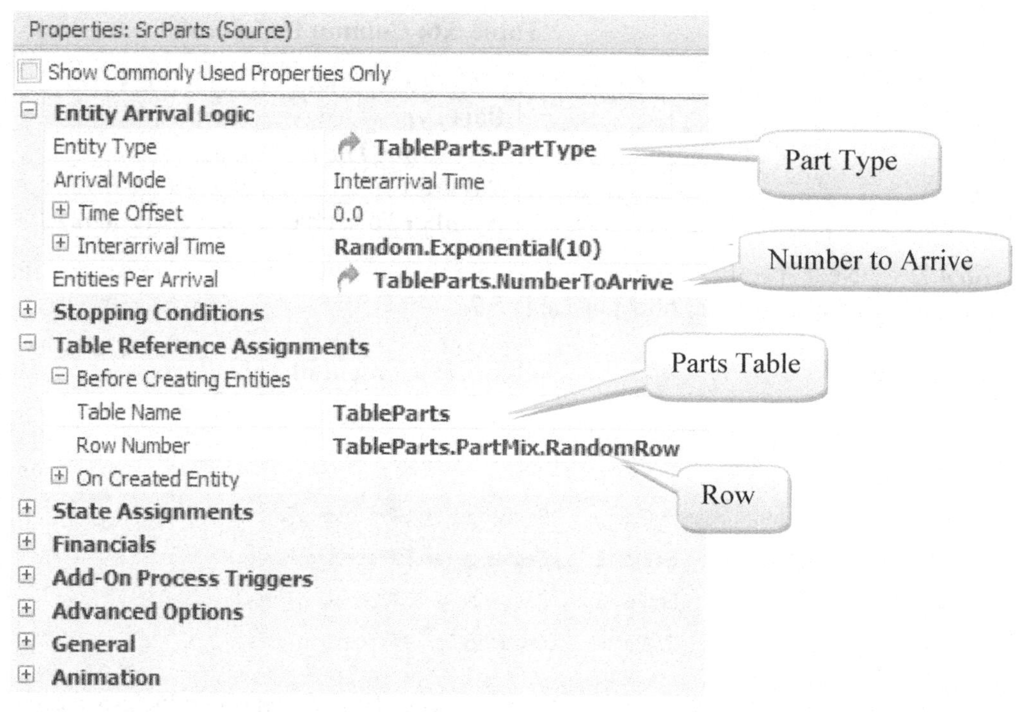

Figure 5.11: Making the Table Reference Assignment

You specify which table (**TableParts**) and which row in the table to associate the object. In this case, the row is determined randomly by using the `PartMix` column and specifying that a random row should be generated (`TableParts.PartMix.RandomRow`).[52]

Step 12: Save and run the simulation model for 40 hours at a slow speed to be sure each of the parts follows their sequence appropriately and answer the following questions.

Question 12: How long is each part type in the system (enter to leave)?

___

Question 13: What is the utilization of each of the SERVERS?

___

Question 14: What is the overall time for all part types in the system?

___

Part 5.5: Using Relational Tables

The original model utilized just four part types, however we may need to add additional parts. This would require adding a new entity, adding rows into the **TableParts** along with creating a new sequence table for each part. Each addition now requires design time to continue to update the model as well as adding sequence tables. Instead, can we drive these specifications from a spreadsheet that contains the number of parts and their sequences? Or is there an easier way to implement the creating and sequencing of parts?

One of the inventive modeling constructs in SIMIO is the use *relational data tables*. SIMIO has implemented limited database capabilities that can facilitate the modeling of relationships. Figure 5.12 shows a "relational diagram"[53] where the **TableParts** data table is related to the **SequencePart** Table via the **ID** column. **ID** is a

[52] SIMIO will normalize the numbers so the probability sums to one. Therefore, you may specify whole numbers 25, 25, and 50 rather than percentages 0.25, 0.25, and 0.5.

[53] More formally it is called an Entity Relationship Diagram, but the word "Entity" is use differently than it is used in SIMIO.

primary key (PK) in the **TableParts** which uniquely identifies each row in the table (i.e., cannot be any duplicates). The child table (i.e., **SequencePart**) inherits the primary key from the parent table (i.e., **TableParts**) which is denoted as a *foreign key* (FK). A particular part can have many rows (records) in the child table.

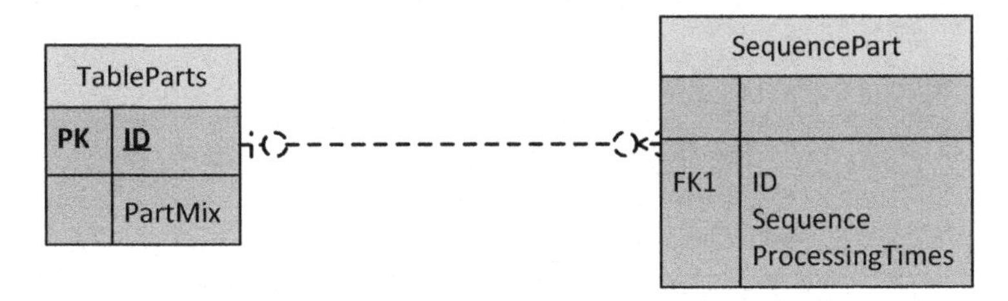

Figure 5.12: Setting up a Relation between Table Parts and the Sequence Part

Step 1: First, we will no longer need just four part types since we want to make this a more generic model. Delete all but one of the part type ENTITIES and rename the remaining entity **EntParts**. From the *Additional Symbols* section, add five additional symbols making sure to color each new entity as seen in the rotated Figure 5.13.[54]

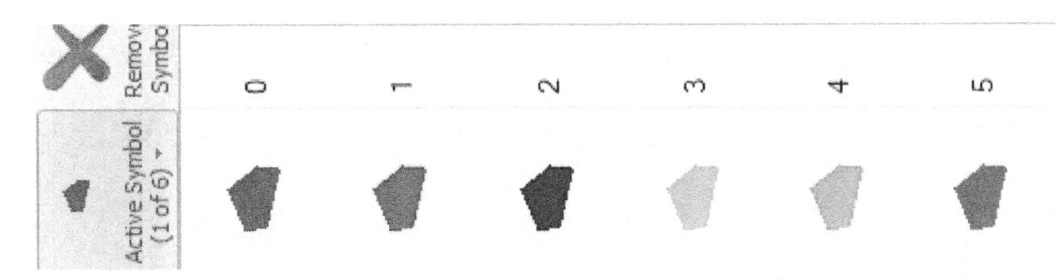

Figure 5.13: Adding Additional Symbols for the Part Entity

Step 2: For the **SrcParts**, set the *Entity Type* property to **EntParts**.[55]

Step 3: From the *Data* tab, delete all the columns of the **TableParts** except for the *Part Mix* and *Number to Arrive* columns. Insert a new *Integer Standard Property* named **ID**. From the *Edit Column* section, you can move the new column to the left and set the column as a PK using *Set Column As Key* button in the same section as seen Figure 5.14. Set the **ID** for each of the part types to a unique number (i.e., 1, 2, 3, and 4).

Table Parts | Sequence Part

		ID	Part Mix	Number To Arrive
1	⊞	1	25	1
2	⊞	2	35	1
3	⊞	3	25	1
4	⊞	4	15	5

Figure 5.14: Parent Table TableParts with the ID Primary Key

Step 4: Select the **SequencePart1** sequence table and rename it **SequencePart.** From the *Add Column* section, select the *Foreign Key* button. Set the properties for the column to specify **TableParts.ID** as the primary table key to relate column too as seen in Figure 5.15. Change the name of the column to be ID.

[54] Note the symbols start with "0" index.

[55] Right click the property and reset it back from the reference property.

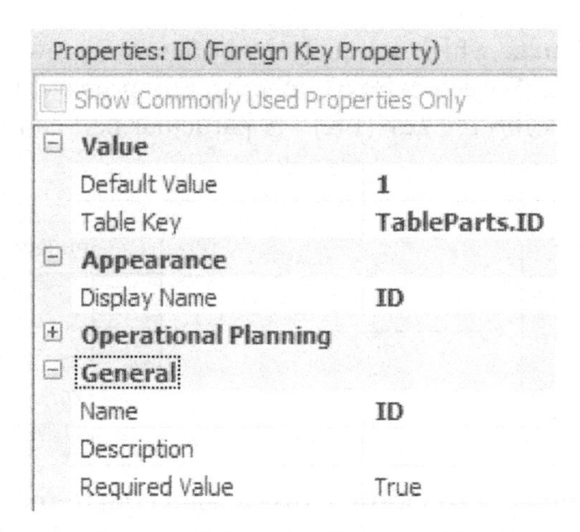

Figure 5.15: Specifying the Foreign Key Property.

Step 5: Set the **ID** for all the current rows to be one.

Step 6: Select the **SequencePart2** tab and then highlight all the rows by selecting the first row using the row selector and dragging down. Then copy all the rows and transverse back to the **SequencePart** table. Select the last row and then paste in all the values. Change the IDs of this new part to two.

Step 7: Repeat for the last two part sequence tabs setting their **IDs** to three and four respectively as seen in Figure 5.16.

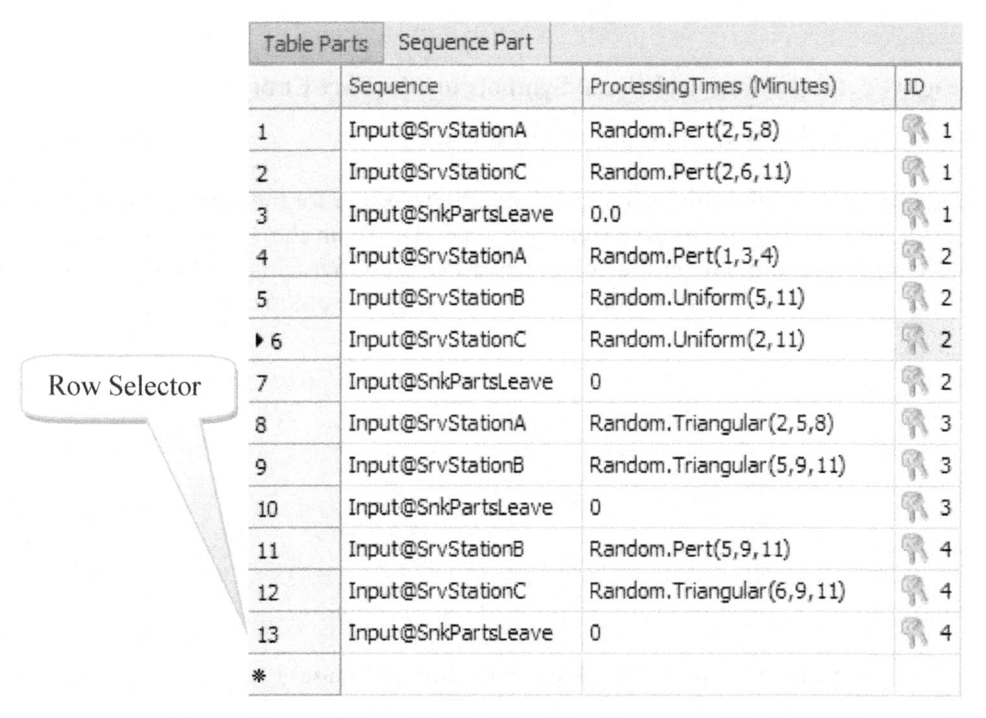

	Sequence	ProcessingTimes (Minutes)	ID
1	Input@SrvStationA	Random.Pert(2,5,8)	1
2	Input@SrvStationC	Random.Pert(2,6,11)	1
3	Input@SnkPartsLeave	0.0	1
4	Input@SrvStationA	Random.Pert(1,3,4)	2
5	Input@SrvStationB	Random.Uniform(5,11)	2
▸6	Input@SrvStationC	Random.Uniform(2,11)	2
7	Input@SnkPartsLeave	0	2
8	Input@SrvStationA	Random.Triangular(2,5,8)	3
9	Input@SrvStationB	Random.Triangular(5,9,11)	3
10	Input@SnkPartsLeave	0	3
11	Input@SrvStationB	Random.Pert(5,9,11)	4
12	Input@SrvStationC	Random.Triangular(6,9,11)	4
13	Input@SnkPartsLeave	0	4
*			

Figure 5.16: The New Sequence Parts Table

Step 8: Once the table is set up, select the **TableParts** table again and observe that each of the rows now has an expander (⊞) associated with it. Figure 5.17 shows the related records for "Part type 0." If an entity is assigned a row from the **TableParts** table, it will automatically be assigned the related records in the **SequencePart** table.

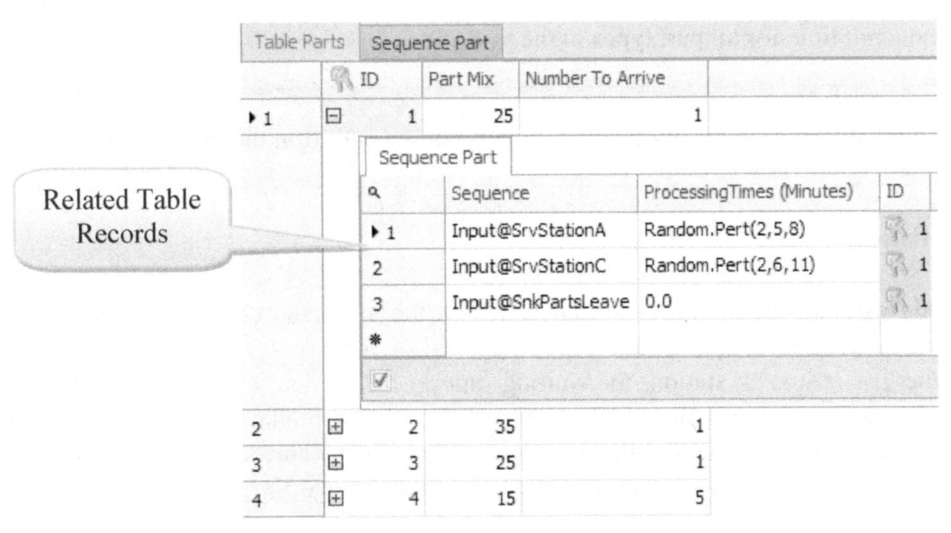

Figure 5.17: Seeing the Related Table Information

Step 9: We need to modify the entity properties to utilize the new sequence table and id for changing the picture symbol. Under the *Animation* section change the *Current Symbol Index* to be "`TableParts.ID-1`" so the color will change appropriately.[56] Also make sure the *Initial Sequence Property* is set to "***SequencePart"*** as seen in Figure 5.18. Note, the entity will only receive the related records and not the entire **SequencePart** table.

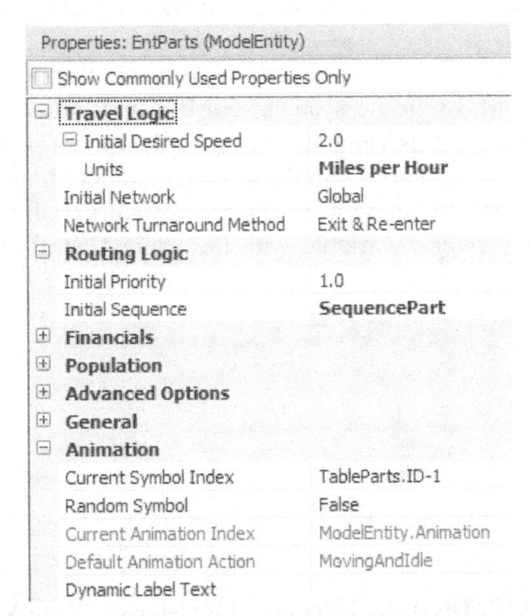

Figure 5.18: Modifying the EntParts Property to Utilize the New Tables

Step 10: For each of the three server stations, change the *Processing Time* to be the reference property ***SequencePart.ProcessingTimes*** rather than ***TableParts.ProcessingTimes*** since this column is now only in the related table.

Step 11: Save and execute the model.

Question 15: What is the utilization of each of the SERVERS now?

____

[56] If we had a lot of parts, we can utilize a status label to display the ID as the part moves through the network making this work for any number of parts. This concept will be explored in later chapters.

Question 16: What is the overall time for all part types in the system?

___

Question 17: Is there anything different in this model as compared to the one from the previous section?

___

Part 5.6: Creating Statistics

SIMIO automatically produces statistics for various objects within the simulation. For example, there are statistics on the MODELENTITY number in system and time in system or for any SERVER there is a scheduled utilization, a number in the INPUTBUFFER station, the waiting time in the station, etc. However, in spite of their comprehensive nature, that may be insufficient. For example, our latest model has only one entity type, so the MODELENTITY statistics are not separated by the part type. What if we wanted to know, on the average, how many Part 3s are in the system during simulation, and what is the average time these parts spend in the system?

The two major statistical categories are: TALLY statistics and STATE statistics. The TALLY statistic is based on observations like waiting time, time in system, cycle time, etc. The STATE statistic such as number in queue, number in inventory, etc., is a time-weighted statistic such that the time a value persists is used to compute the statistical value. While the STATE statistic may appear to be the more complicated, it is the easiest to specify in SIMIO. For example, let's add a STATE statistic to determine the average number of Part 3's in the system first and then add a TALLY statistic for the average times in the system.

Step 1: Before creating a STATE statistic, it is necessary to create a state variable the will hold the values for example the number of Part 3's in the system. SIMIO will monitor the value of the state variable through a STATE statistic and weight its value according to the time a particular value persists throughout the simulation. Since the number of Part 3's in the system is a model-based characteristic (not associated with particular entities), we need to define a model-based state variable. From the *Definitions→States* section, insert a new *Discrete, Integer* variable with the name **GStaPart3NumberInSystem** as seen in Figure 5.19

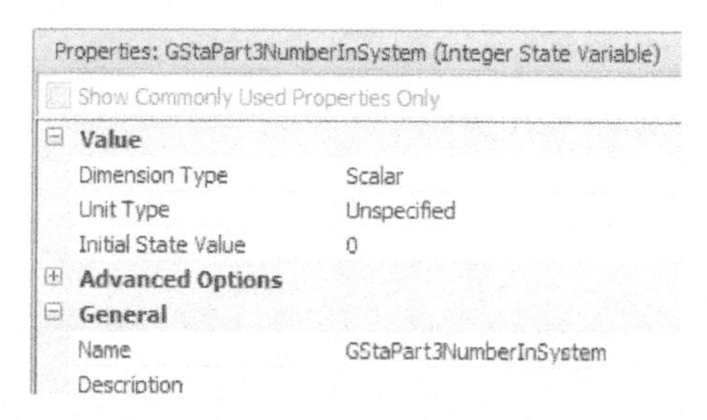

Figure 5.19: Discrete, Integer, Model-based State Variable

Step 2: The variable needs to be incremented when a Part 3 enters the system and then decremented when a Part 3 leaves the system. Let's use the *State Assignments* of the **SrcParts** for *Before Exiting* employing the *Repeating Property Editor* to increment the new state variable as shown in Figure 5.20.

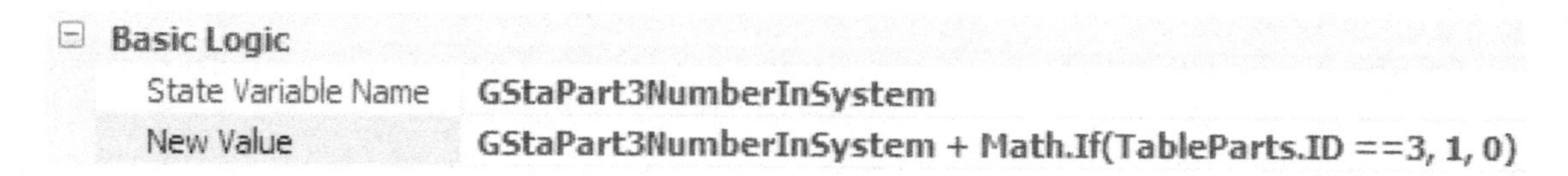

Figure 5.20: Incrementing the State Variable

Step 3: The `Math.If()` function simply tests if **TableParts.ID** has the value of "3" which refers to Part 3's then adds "1" to the state variable, otherwise the expression adds "0."

Step 4: In a similar fashion, the state variable can be decremented in the *State Assignments* of the **SnkPartsLeave** for *On Entering* as shown in Figure 5.21.

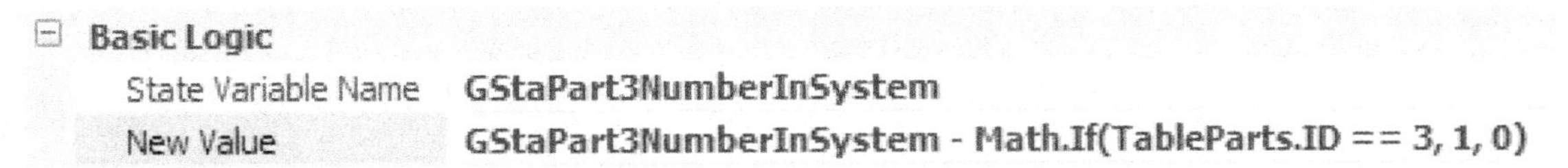

Figure 5.21: Decrementing the State Variable

Step 5: Now all you need to do is to insert STATESTATISTIC from the *Definitions→Elements* section named StateStatNumberPart3InSystem that watches the state variable as seen Figure 5.22.

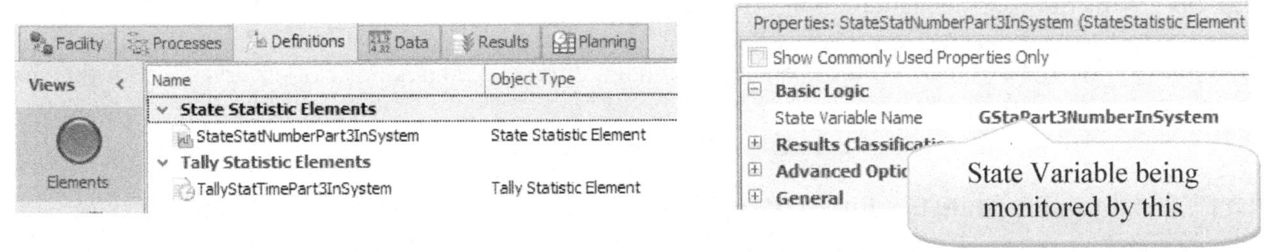

Figure 5.22: Defining the State Statistic Element

Step 6: Save your model and run your simulation for 40 hours.

Question 18: What did you get for the average and the maximum number of Part 3's in the system?

____

Step 7: Now, we would like to collect time in system for the Part 3's which is a TALLY statistic. The TALLY statistic needs to record individual observations. Tally statistics can be collected at any BASIC or TRANSFER node in their *Tally Statistics* property.[57] Unlike the STATE statistics, the TALLY statistic needs to be defined first. Using the *Definitions→Elements* section insert a Tally Statistic named **TallyStatTimePart3InSystem.** See Figure 5.23 for specifying the ***Unit Type*** property of "Time."

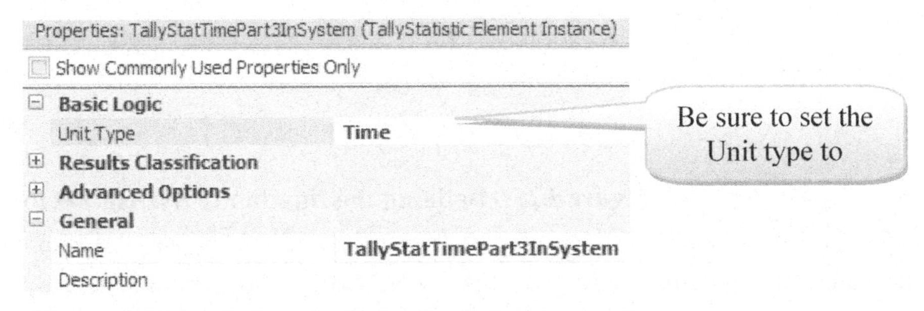

Figure 5.23: Defining the Tally Statistic Element

Step 8: Select the **Input@SnkPartsLeave** node (i.e., the one attached to the SINK). Under the ***Tally Statistics*** Section using the *On Entering* tally property, specify which observation (i.e., `ModelEntity.TimeInSystem`) to record using the *Repeating Property Editor*. Specify the collection of the tally statistic as shown in Figure 5.24.

[57] In later chapters we will use the `Tally` process steps to collect statistics anywhere.

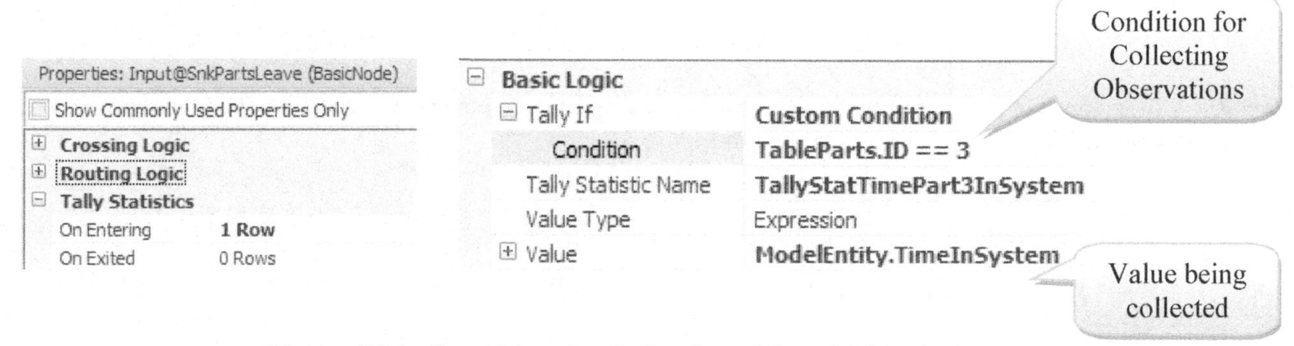

Figure 5.24: Specifying the Collection of Part 3 Time in System

Step 9: Note that a *Custom Condition* causes only the Part 3's time in system to be collected.

Step 10: Now run your simulation for the 40 hours.

Question 19: What did you get for the average, maximum, minimum, and number of observations for time in system for Part 3's?

____

Part 5.7: Obtaining Statistics for All Part Types

Since statistics like time in system and number in system would be useful for each part type, we will extend the statistics collection to all part types rather than just Part 3's.

Step 1: First, let's create the time in system statistics. Navigate to the *Definitions→States* section, and modify the definition of **GStaPart3NumberInSystem**. Rename the variable to **GStaPartNumberInSystem**. Set the ***Dimension Type*** property of the state variable to a "**Vector**" with "**4**" rows, as shown in Figure 5.25. This will create a vector of four state variables one for each part type.

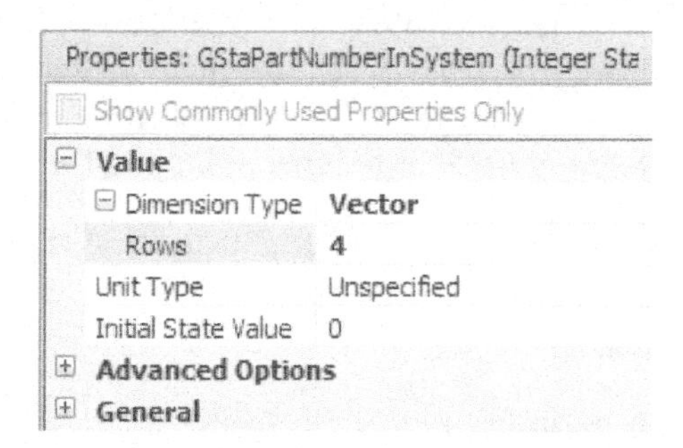

Figure 5.25: Defining the Number in System Vector

Step 2: Next change the state assignments at **SrcParts** and **SnkPartsLeave** to increment and decrement the number in system for each part type respectively as shown in Figure 5.26. Note the use of the row (`TableParts.ID`) in the vector.[58]

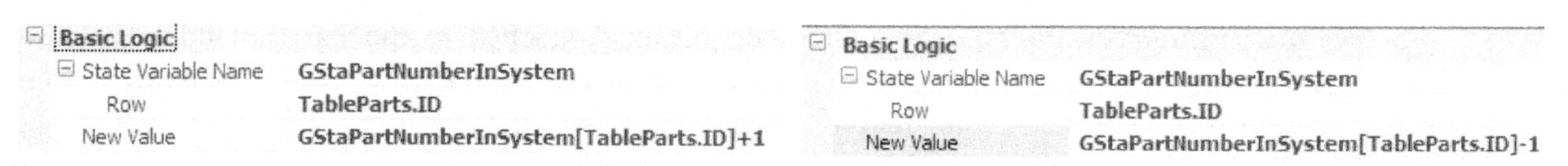

Figure 5.26: Incrementing and Decrementing the Number in System

[58] The row index of the first row for Vectors, Matrices, and Data Tables is one while symbol indexes start at zero.

Step 3: You can now define state statistics for all the parts. Under *Definitions→States*, create three more state statistics named **StateStatNumberPart1InSystem**, **StateStatNumberPart2InSystem**, and **StateStatNumberPart4InSystem**. You need to specify the state variable for each of the statistics using the previously defined vector state variable. This specification is the same for each statistic, except for the row. The specification for Part 2's is shown in Figure 5.27. Modify all the four state statistics.

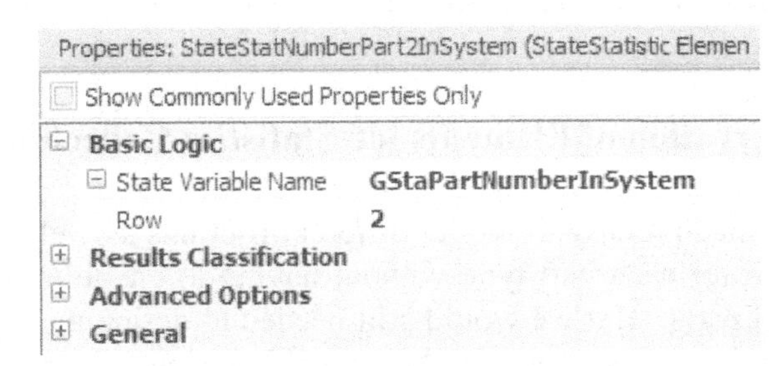

Figure 5.27: Specification for StateStatNumberPart2InSystem

Step 4: Now define the TALLY statistics for the other three part typess (Parts 1, 2, and 4), just as was done for **TallyStateTimePart3InSystem** (see Figure 5.23 making sure the ***Unit Type*** property is equal to "**Time**".

Step 5: Navigate to the *Data→Tables* section and select the **TableParts** DATA TABLE. Insert a new *Tally Statistic Element Reference* property named **TallyStatTimeinSystem,** which will allow us to specify a tally statistic for each part. Therefore each entity (i.e., part type) will know which statistic to update. There is little need to insert State statistics into the data table since we don't reference them within the model – only the state variable that the state statistics monitor. Fill in the tables as shown in Figure 5.28.

Table Parts | Sequence Part

		ID	Part Mix	Number To Arrive	Tally Stat Time In System
1	⊞	1	25	1	TallyStatTimePart1InSystem
2	⊞	2	35	1	TallyStatTimePart2InSystem
3	⊞	3	25	1	TallyStatTimePart3InSystem
4	⊞	4	15	5	TallyStatTimePart4InSystem

Figure 5.28: Data Table Including Tally Statistics

Step 6: Once the Table row reference has been assigned, each MODELENTITY (i.e., part) will have an associated TALLY statistic which will eliminate using complicated deciding logic to determine which TALLY statistic to update. Select the **Input@SnkPartLeave** BASIC node in front of **SnkPartLeave** and under the *Tally Statistics* entry click on the *On Exited.*[59] Use the *Tally Statistic Name* will be based on the row that has been assigned to the entity (i.e., `TableParts.TallyStatTimeinSystem`). The observation *Value* will be the model entity time in system value as seen in Figure 5.29 to utilize the entity's tally statistic.

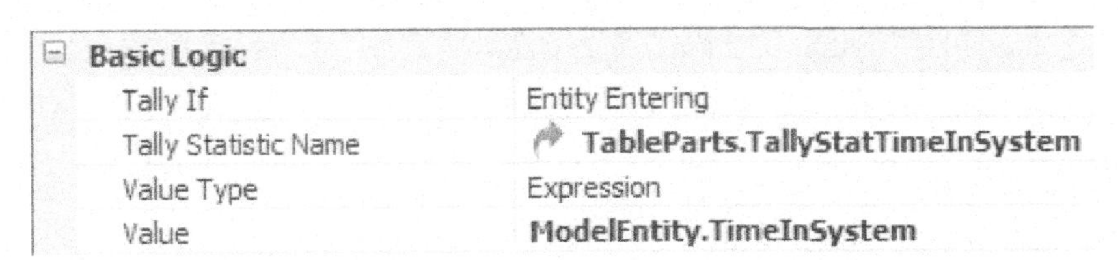

Figure 5.29: Time in System Tally Statistics

[59] The *On Entering* could be used as well.

Step 7: Now re-run the simulation and examine the results.

Question 20: Do the tally statistics on time in system for each part type seem correct?

____

Question 21: Do the state statistics for number in system for each part type seem correct?

____

Part 5.8: Automating the Creation of Elements for Statistics Collection

Adding intelligence to entities made it easy to model different routings as well as to calculate time in system and number in system statistics for each part type without having to create multiple sink and source objects corresponding to the individual parts. But we would still needed to define the associated statistic objects and state variables. Such an approach becomes very cumbersome if we want to add many new part types.

An advantage of using related tables is that they may be linked (bound) directly to a spreadsheet containing many part types and sequences. Nevertheless, we may want statistics for time in system (i.e., cycle times) and number in system for each of the different part types. However, this situation will require a TALLY and STATE statistic for each part type. Having to define each statistic could negate the advantages of using related tables as individual tally and state statistic elements need to be created and then the correct statistic must be collected for the particular part. However, SIMIO allows for the automatic creation of state variables and statistic elements.

Step 1: If you are using the model from the previous section, you should delete all the TALLY *and* STATE STATISTIC *elements*, so these can be created automatically.

Step 2: Since the number of part types might change, the state variable **GStaPartNumberInSystem** will need to be modified to match the number of parts in the table each time. In our case, we will need to set it equal to maximum possible number of part types in the system (e.g., 10).[60] The size of the state variable can be tied to the number of rows and columns of a data tables. From the *Definitions→States* section change the *Rows* property from "4" to "10".

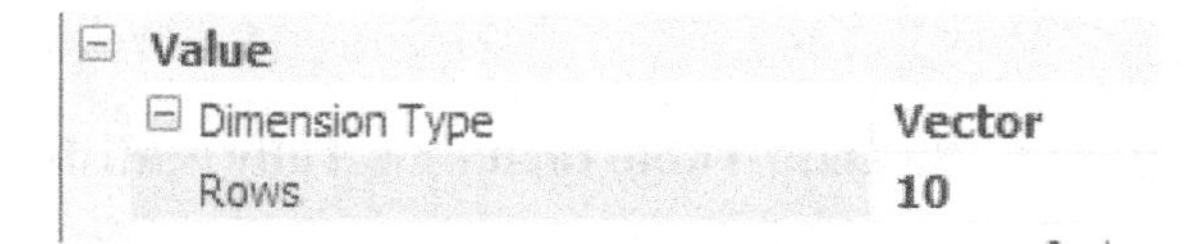

Figure 5.30: Specifying the Size of the State Variable to Match the Table

Step 3: Navigate to the *Data→Tables* section and select the **TableParts** DATA TABLE. Notice *Statistic Element Reference* property named **TallyStatTimeinSystem. Remove the tally statistics from that column, so it looks like** Figure 5.31. **Note that the tally statistics references are missing.**

Table Parts | Sequence Part

	ID	Part Mix	Number To Arrive	Tally Stat Time In System
1	0	25	1	
2	1	35	1	
3	2	25	1	
4	3	15	5	

Figure 5.31: Adding a Tally Stat Reference Column

[60] "Matrix from Table" is another *Dimension Type* for state variables which will create a two dimensional matrix based on the number of rows and numeric columns of a table. Dynamically. However, it initializes the variable based on the table values.

Step 4: Once one specifies entries (i.e., names), the elements will be created. The element's initial properties can be manually specified (e.g., *Category, Data Item, Unit Type*, etc.) from the *Definitions→Elements* section. However, since the number of part types can be dynamic we do not want to have to specify these properties each time, which would defeat the benefit of automatic specification of the statistical elements. Therefore, the elements that are created can also have their properties initialized by specifying another column in the table as their values. For our problem, the category and data item classification of the TALLY statistics can be specified as we want it to be the same as other time in system statistics (i.e., "FlowTime" and "TimeInSystem"). Therefore, insert two *Standard String* property columns named **Category** and **DataItem** as seen Figure 5.32.

Table Parts | Sequence Part

	ID	Part Mix	Number To Arrive	Tally Stat Time In System	Category	Data Item	Unit Type	State Stat Num In System
1	⊞ 1	25	1		FlowTime	TimeInSystem	Time	
2	⊞ 2	35	1		FlowTime	TimeInSystem	Time	
3	⊞ 3	25	1		FlowTime	TimeInSystem	Time	
4	⊞ 4	15	5		FlowTime	TimeInSystem	Time	

Figure 5.32: Adding Properties to the Statistical Elements

Step 5: For the *Unit Type* property of the TALLY statistic, insert a *Standard Enumeration* property column named **UnitType** into **TableParts**, as was done in Figure 5.32. An enumeration is a property whose value is specified from a list of potential values. In this case unit types can be "Unspecified", "Time", "Length", etc. For the *Enumeration* column property, specify the *Enum Type* to be "UnitType" to pull its values from unit type enumeration list and the *Default Value* should be "Time" as seen in Figure 5.33.

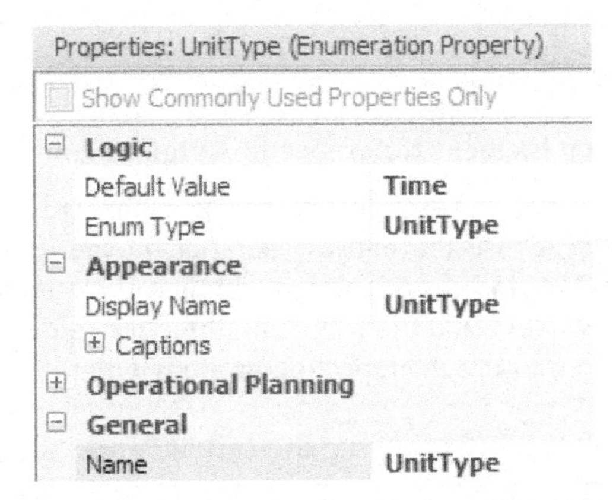

Figure 5.33: Setting up the Enumeration Property Column

Step 6: Next, insert a *State Statistic Element Reference* property named **StateStatNuminSystem** as seen in Figure 5.32.

Step 7: Rather than creating each TALLY statistic to specify as entry values, SIMIO has the ability to automatically create elements (TALLY and STATE statistics, MATERIALS, etc.) specified from a table column. Therefore, each part row can have a specified TALLY statistic (i.e., name) that will automatically be created and used to keep track of its on time in system statistics. By default the *Element Reference* property column will be of a "Reference" type meaning the element would need to be already created in order to be specified as an entry in the table.

- Therefore, change the *Reference Type* property to "Create" which is explained in Table 5.8 and seen in Figure 5.34.
- Set the *Auto-set Table Row Reference property* to "True" as well.

Table 5.8: Specifying Properties to Automatically Create the Element

Property	**Description and Value**
Reference Type	"*Reference*" - It allows you to simply specify an element that is already been defined under the "Definitions" tab.
	"*Create*" - Entries in this column will now automatically create a new element of the specified type with the name of the entry. Note these elements will appear in the *Definitions→Elements* but they cannot be deleted from here. Any changes in names in either location (i.e., TABLE or *Definitions→Elements*) will be reflected in both.
Initial Property Values	This can be used to initialize the element based on values in other columns of the table.
Auto-set Table Row Reference	If "True", then the element that is pointed to by each row will automatically be given a table reference set to that row. This has to be case if your creating and initializing the element based on other columns

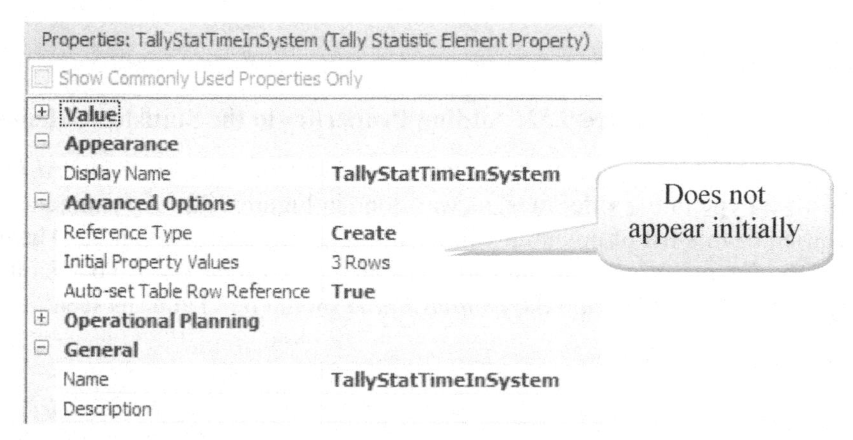

Figure 5.34: Specifying Element Reference to Automatically Create the Element

Step 8: After specifying the "Create", the statistical elements be created but they will not have their properties referenced. In order to use the entries specified in these three new columns as the initial properties, select the **TallyStatTimeinSystem** column and then click the *Repeating Property* button of the *Initial Property Values* property. Next, add the following three properties and values as specified in Table 5.9.[61] To specify the initial properties use the repeating property window as shown in Figure 5.35.

Table 5.9: Specifying the Initial Properties

Property Name	**Value**
Unit Type	`TableParts.UnitType`
Category	`TableParts.Category`
DataItem	`TableParts.DataItem`

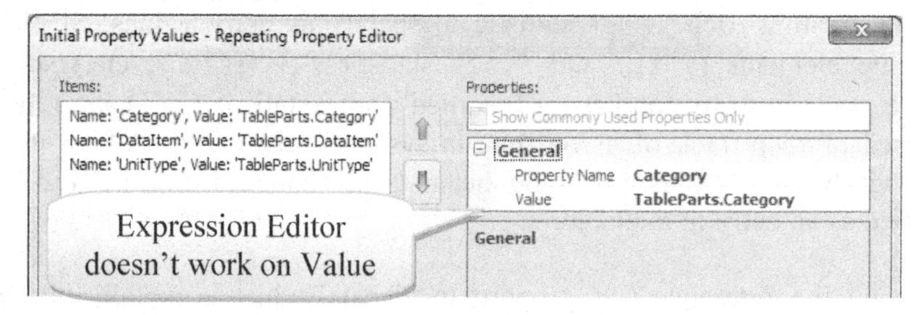

Figure 5.35: Specifying the Initialization Properties of the TallyStat Reference Column

[61] You will need to type in the values directly as this is not an expression editor.

Step 9: Select the **StateStatNumInSystem** and specify that this column will be created as well as seen in Figure 5.36.

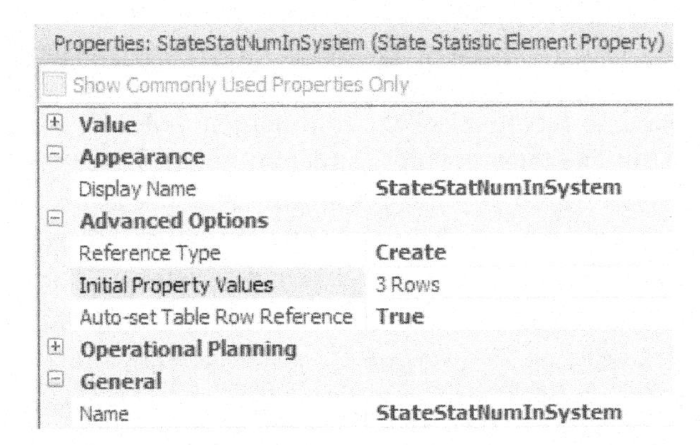

Figure 5.36: Having the State Statistics to be Automatically Created

Step 10: Next we need to initialize the new state statistics in a similar fashion as the Tally statistic column. However, this time we will not setup additional columns to specify the parameters. Set the "Category" and "DataItem" properties to string values and the "StateVariableName" property will use the `TableParts.ID` to specify which state variable the statistic should monitor as seen in Table 5.10 and Figure 5.37.

Table 5.10: Specifying the Initial Properties

Property Name	Value
StateVariableName	`GStaPartNumberInSystem[TableParts.Id]`
Category	`Content`
DataItem	`NumberInSystem`

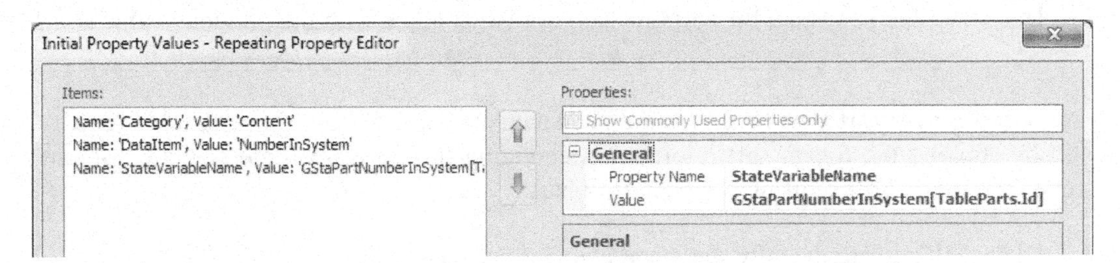

Figure 5.37: Initial Parameters of the State Statistic Column

Step 11: Add in the specifications for the properties of the four part types as shown in Figure 5.38.

Table Parts | Sequence Part

	ID	Part Mix	Number To Arrive	Tally Stat Time In System	Category	Data Item	Unit Type
1	0	25	1	⊗	FlowTime	TimeInSystem	Time
2	1	35	1	⊗	FlowTime	TimeInSystem	Time
3	2	25	1	⊗	FlowTime	TimeInSystem	Time
4	3	15	5	⊗	FlowTime	TimeInSystem	Time

Figure 5.38: Specifying the Properties of the Statistical Elements

Step 12: You may now want to reset the *Reference Type* property to back to "Reference". Then when you set the *Reference Type* property to back to "Create", the statistical elements will be automatically be created as shown in *Definition→Tally Statistical Elements (Auto-Created)* section. Note how each statistic takes its *UnitType*, *Category*, and *DataItem* from the DATA TABLE.

Step 13: Now for every new row that is added, a new TALLY and STATE statistic will be created that is initialized from the entry values of the *Category*, *DataItem* and *UnitType* columns. For any rows that existed before the two columns were created, the values will not be initialized.[62] We can cut and paste values back and forth from SIMIO tables and Microsoft Excel™ spreadsheets. Select the entire table by highlighting the row selectors and then cut the rows (*Ctrl-x*). Next the paste them into a Microsoft Excel™ spreadsheet so you don't have to retype all of the information. Modify the names of the tally and state statistics, categories, data items and the unit types as seen in Figure 5.39.

	A	B	C	D	E	F	G	H
1	1	25	1	TallyStatP1	FlowTime	TimeInSystem	Time	StateStatP1NumInSystem
2	2	35	1	TallyStatP2	FlowTime	TimeInSystem	Time	StateStatP2NumInSystem
3	3	25	1	TallyStatP3	FlowTime	TimeInSystem	Time	StateStatP3NumInSystem
4	4	15	5	TallyStatP4	FlowTime	TimeInSystem	Time	StateStatP4NumInSystem

Figure 5.39: Using Excel to Modify the Values

Step 14: Copy the values in Excel and the select the empty row in the SIMIO data table by clicking on the row selector and then paste the values into the table which should now look like the one in Figure 5.31.

Step 15: Save and run the model for 40 hours.

Question 22: What is the average time in the system and number in system for each of the parts?

____

Question 23: Does the time in system for Part3 in TallyStatistic3 agree with the previous value computed?

____

Step 16: Now add a fifth part type by creating a sequence and update only the data table (don't create any elements) by copying the fourth part information changing the tally and state statistic column values as well as copy the fourth part sequences for the fifth part.

Question 24: Did SIMIO create the tally statistical element automatically?

____

Question 25: When you execute the model, did the flow time and time in system statistics for the fifth part type appear with the proper unit type, category, and data item?

____

Part 5.9: Commentary

- The *Start Date* property of the Work Schedule intuitively would be when the schedule will start. So if one specified a future date then nothing would happen until that date. That is not the case. The *Start Date* property represents the particular day the first day of the pattern will starts. If this date is in the future compared the start of the simulation, it will repeat backwards to the current date based on the pattern. For example, if we have a three day pattern named Day1, Day2, and Day3 and the work schedule start date is set to 09/17/15, schedule will follow the Day1 pattern for this date. If the simulation starts on 09/13/15, then the pattern for 09/13 is Day3, 09/14 is Day1,

[62] We can either modify the properties our self or we can delete the rows and recreate them. You can select the AutoCreate dropdown in each column value.

09/15 is Day2, 09/16 Day3 which makes Day1 be on 09/17 and Day2 on 09/18, etc. In order to have a work schedule not start until a particular day, exceptions to the work day have to be employed.

- Another way to look at SIMIO properties and states is that property values are established for each object in the Facility window at the beginning of the simulation execution and cannot be changed during "run-time", whereas the value of states are also established at initialization but the can be changed during "run-time". Objects that are created during run-time, such as ENTITY, have their properties established during an instance of run-time which is during the initialization of that instance, but they otherwise cannot be changed during the simulation execution.
- The use of relational tables offers a tremendous advantage over other simulation languages in being able to assign one table and then automatically be assigned all records of related tables. We will explore this feature more in later chapters. In later chapters, we will demonstrate how you can bind data tables to Excel spreadsheets which could facilitate the automatic creation of new parts and sequences.

Chapter 6
Assembly and Packaging: Memory Chip Boards

A common simulation environment in production is assembly and packaging. What makes these operations challenging is that entities are being combined together. The way they are combined can be complicated. Sometimes the assembly is temporary and the assembly (called a batch) must be separated later. These considerations cause us to have interest in the COMBINER and SEPARATOR object.

Part 6.1: Memory Board Assembly and Packing

Memory boards have memory chips inserted into them at an assembly operation, whose capacity is four. Each memory board requires four chips. Memory boards as well as individual memory chips arrive randomly, so the assembler waits for four chips of the same type and one board to be available before doing the assembly. After the board is assembled it is sent to one of three packing stations, each of which has one worker. Boards are sent to the packing station with the smallest queue. Figure 6.1 depicts the memory board assembly and packing process.

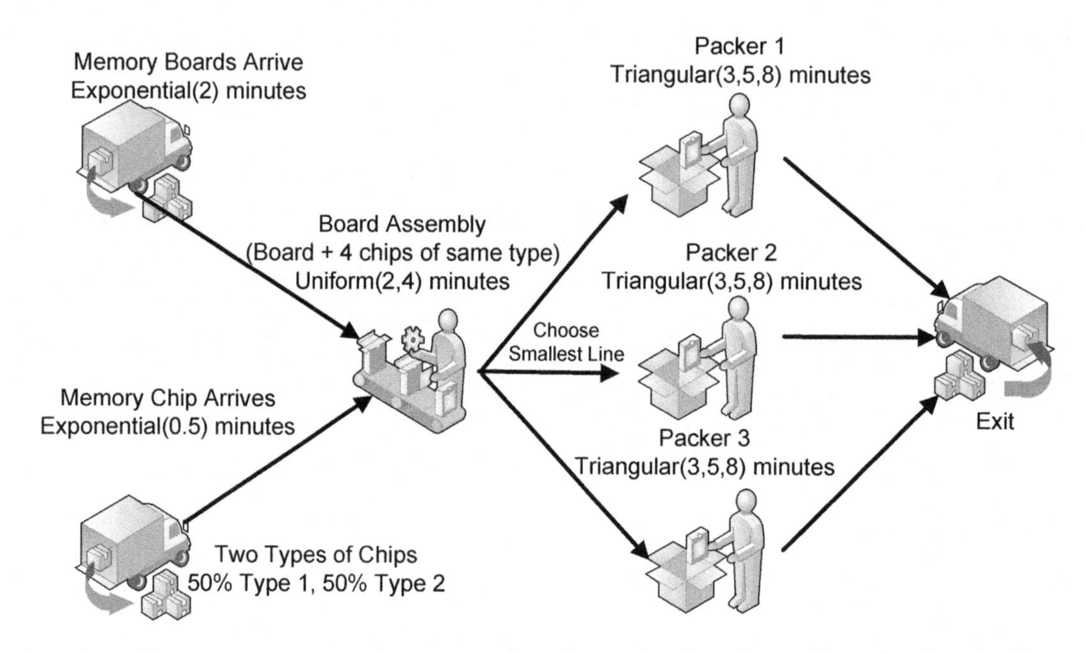

Figure 6.1: Memory Board Assembly and Packing

The basic numerical characteristics of the assembly problem are found in Table 6.1.

Table 6.1: Numerical Distributions of the Memory Board Assembly

	From	To	Travel time (minutes)
Travel Times	Memory board arrivals	Board Assembly	4 minutes
	Memory chip arrivals	Board Assembly	5 minutes
	Board Assembly	Packer Station 1	Pert(10,12,15) minutes
	Board Assembly	Packer Station 2	Pert(5,7,10) minutes
	Board Assembly	Packer Station 3	Pert(4,5,7) minutes
	Any Packer Station	Exit	3.5 minutes

Arrival Information	Interarrival time for Memory Boards	Exponential(2) minutes
	Interarrival time for Memory chips	Exponential(.5) minutes
	50% of memory chips are of type 1 and 50% are of type 2	Discrete
Processing Times	Board Assembly processing time	Uniform(2,4) minutes
	Packing time for all Packers	Triangular(3,5,8) minutes

Step 1: Create a new model that has two SOURCES named **SRCMEMORYBOARD** and **SRCMEMORYCHIP** (i.e., one for boards and one for memory chips). Add two MODELENTITIES into the model, one for boards named **EntMemoryBoard** and one for memory chips named **EntMemoryChip**.[63] Make sure the *Entity Type* property of the two sources will create the appropriate entity type.

Step 2: Give the entity type associated with boards a different symbol as seen in Figure 6.2. You can use one of the stock SIMIO symbols or download a symbol from Trimble 3-D warehouse like Chapter 2. You can also make your own symbols by clicking the "*Create New Symbol*" menu item from the *New Symbol* dropdown in the "*Create*" section of the "*Project Home*" tab. In the properties window, name the symbol **SymMemoryBoard**. When creating a new symbol, you should give the symbol "height" by specifying a value (in meters). In the case shown in Figure 6.2, a rectangle was created with a line width of 0.1 meters and a height of 0.25 meters[64]. The rectangle was given a fill color of grey with a line color of green. One should not be worried too much about the exact specifications since the symbol can be adjusted manually.

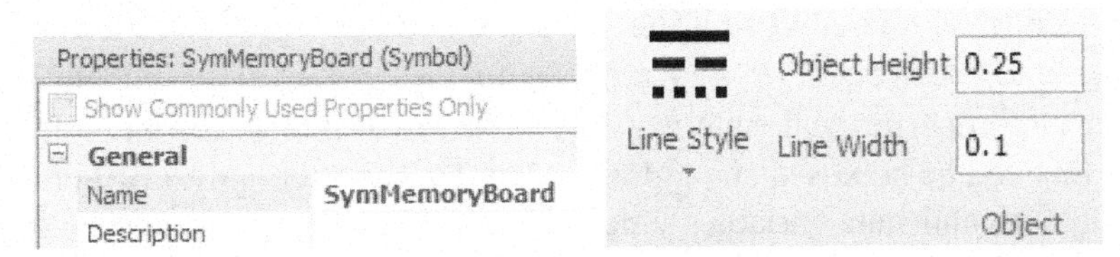

Figure 6.2: Changing the Name of the Symbol and Specifying a Height and Line Width

Step 3: Select the "*Model*" in the [*Navigation Panel*] to return back to simulation model. Select the **EntMemoryBoard** model entity and then click on the **SymMemoryBoard** symbol in the *SYMBOLS→PROJECT SYMBOLS* section to utilize our new symbol.

Step 4: For the memory chips, add an additional symbol[65] since we have two different types of memory chips. You should color the second symbol a different color (e.g., red) to distinguish them in the system.[66]

Step 5: Insert a COMBINER named **CmbBoardAssembly** into the model for the assembly operation as shown in Figure 6.3 which will batch/combine a parent entity object with a member entity object. Set the

[63] To add entities, drag MODELENTITY from the [*Project Library*] panel.

[64] Note memory boards are not ¼ meter thick but were made this thick to show a depth in 3-D view.

[65] Select the **EntMemoryChip** MODELENTITY and the click *Symbols→Additional Symbols→Add Additional Symbol* button.

[66] Select the 3-D view and color the sides of the new symbol as well.

initial capacity of the **CmbBoardAssembly** to four and the processing time to Uniform(2,4) minutes. The combiner has two input nodes (i.e., a PARENTINPUT and a MEMBERINPUT). The parent object is the board while the member object is the memory chip. Connect the two SOURCES via TIME PATHS to the appropriate nodes using the time specified in Table 6.1.

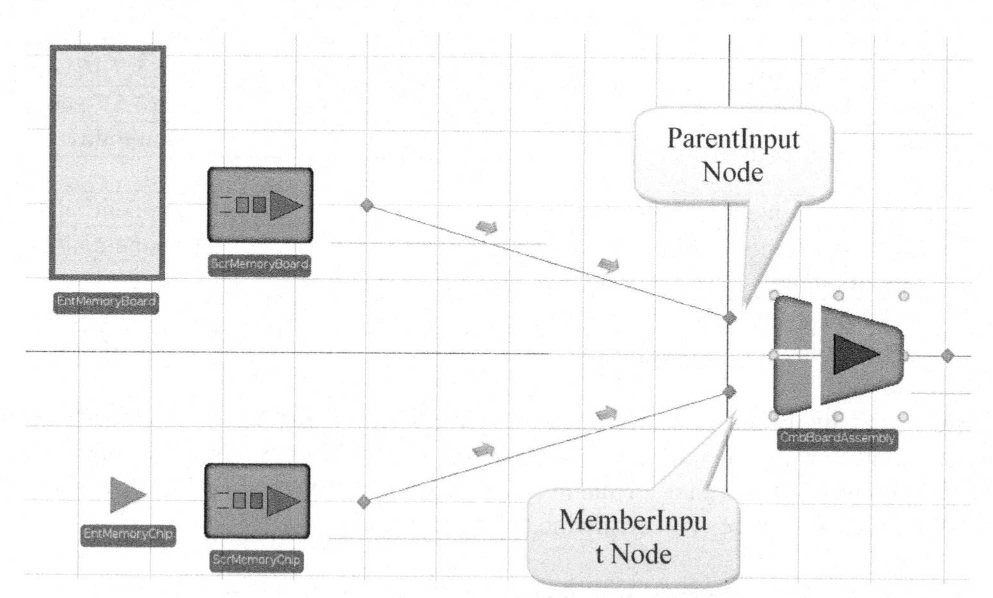

Figure 6.3: Initial Memory Board Model

Step 6: Objects have characteristics that describe and /or define their behavior or attributes. Properties and state variables are two types of characteristics that are very similar and allow the user to specify certain characteristics of the object (e.g., processing time for a SERVER object). Properties are characteristics that are set (initialized) when the object is created. For most objects, the initialization occurs at the beginning of the simulation and, therefore, cannot change during the run of the simulation. State variables are dynamic characteristics which allow their values to change during the simulation run and should be used when their values need to be altered during the simulation. There are two object definitions that we can modify (i.e., MODEL and the MODELENTITY) as seen in the [*Navigation*] panel. Both the MODEL and MODELENTITY can have properties and state variables. However the MODEL has only one set of properties and state variables that can be accessed (i.e., global scope) while each individual entity that is created will contain its own properties and state variables independent of the other entity instances.[67]

Step 7: To make the state variables clear when they are used in various places in a model, we prefix a MODELENTITY state variable with **ESta...** and a MODEL global state variable with **GSta...**.

Step 8: Recall a board requires four identical memory chips. Therefore, the COMBINER should batch four members of the same type. Let's add an entity-based state variable to distinguish between the two types of memory. Select the MODELENTITY in the [*Navigation*] panel and then choose the "*States*" section from the "*Definitions*" tab. Since there are two distinct chip types, insert a new "*Integer Discrete State*" variable named **EStaChipType** as seen in Figure 6.4.

[67] Both the MODEL and MODELENTITY can have a state variable named **Color**. The only difference is how they are accessed. Specifying `Color` will access the variable of the MODEL while `ModelEntity.Color` will access the entity's variable.

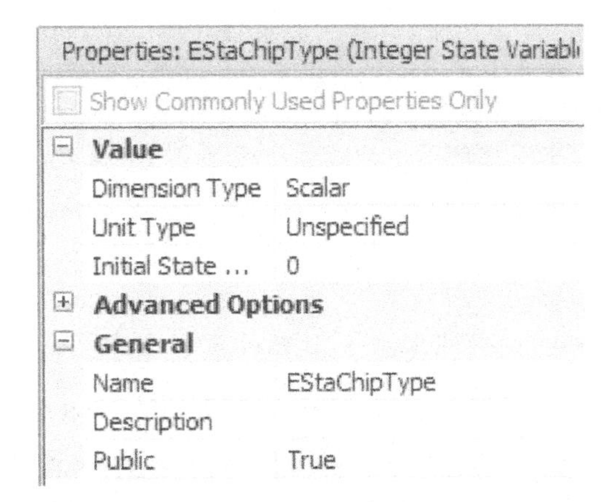

Figure 6.4: Specifying the Discrete Integer State Variable

Step 9: Returning to the "*Facility*" tab of the **Model**, click on the **SrcMemoryChip** source. To assign the new **EStaChipType** variable, a *Before Exiting State Assignment* will be specified which makes the assignment right before the entity leaves the source. Click on the box to the right of *Before Exiting State Assignments* which brings up the "Repeating Property Editor" and add the following properties.

- Since it is equally likely the chip will be type one or type two, a random discrete distribution will be used to assign the **EStaChipType** value a one or a two.

State Variable Name: `ModelEntity.EStaChipType`
New Value: `Random.Discrete(1,0.5, 2, 1.0)`[68]

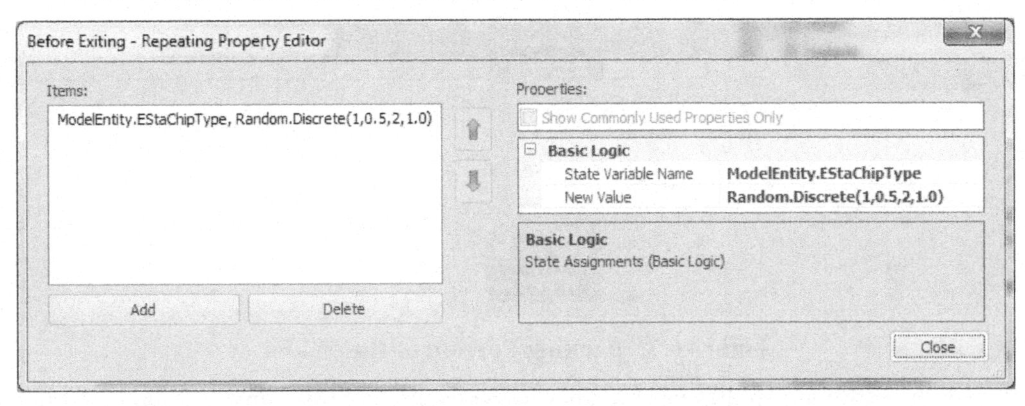

Figure 6.5: Setting the Entity's ChipType in a State Assignment

Step 10: Since the chip entities can be distinguished, batch four members (i.e., chips) at the COMBINER by specifying **Match Members** as the *Matching Rule* with `ModelEntity.EStaChipType` as the *Member Match Expression* property as seen in Figure 6.6.[69]

[68] The `Discrete` distribution must use the appropriate cumulative probabilities.

[69] Combiners can match any entity, match members or match certain members and parents based on a criterion.

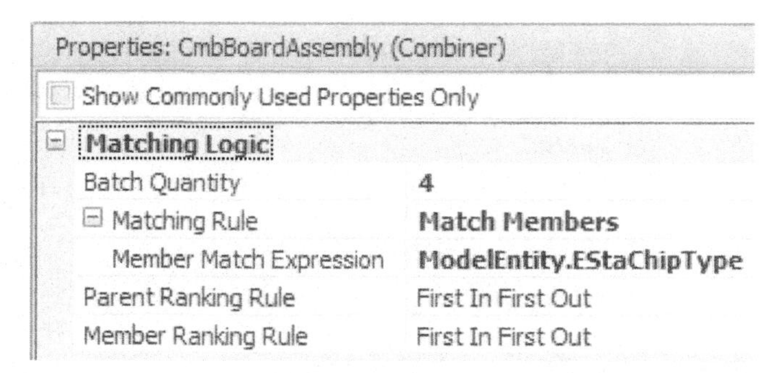

Figure 6.6: Specifying the Matching Logic for a Combiner

Step 11: Insert three SERVERS, (i.e., **SrvPacker1**, **SrvPacker2**, and **SrvPacker3**) one for each packer with the appropriate processing times (see Table 6.1) and one SINK named **SnkExit** (see Figure 6.7). Utilize TIMEPATHS to connect the assembly to the packers and the packers to the exit. Specify the appropriate travel times specified in Table 6.1 for the time paths between the assembly and the packers, but do not specify the times between the packers and the exit quite yet.

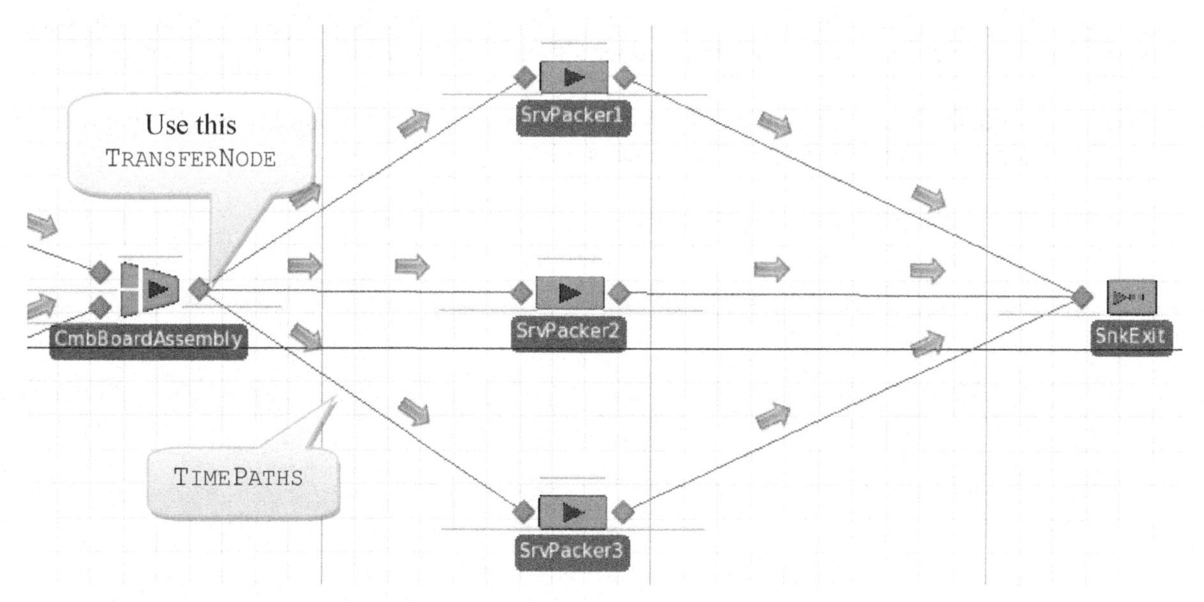

Figure 6.7: Packing Portion of the Model

Step 12: Recall, the board assemblies need to choose the packer with the smallest number assigned to it. At the output TRANSFERNODE associated with the COMBINER, the branching decision needs to be specified using the *Entity Destination Type* property specification in the TRANSFERNODE.

Step 13: Once the board finishes assembly in the COMBINER it can select one of the three destinations: **Input@SrvPacker1**, **Input@SrvPacker2**, or **Input@SrvPacker3** depending on the number of already boards assigned/designated to each packer. Create a new Node LIST (see *Definitions→Lists* tab) named **LstPackers**, beginning with **Input@SrvPacker1** as seen in Figure 6.8. Lists are necessary whenever you need to choose from a set of objects (e.g., nodes, resources, etc.).

Figure 6.8: Node List for Selecting the Best Packer

Step 14: Now at the TRANSFERNODE, specify that entities should transfer to the destination according to the following properties as seen in Table 6.2. Once you specify the *Selection Goal* property, you need to expand it to access the *Selection Expression*. We need an expression that can be evaluated for the "shortest queue". However, do we just mean the number in the respective queues? Probably not. How about including the number of entities being processed in the SERVERS as well as the number of entities traveling to the SERVER.

Table 6.2: Properties for Selecting the Shortest Line

Property	**Value**
Entity Destination Type:	**Select from List**
Node List Name:	**LstPackers**
Selection Goal:	***Smallest Value***
Selection Expression:	`Candidate.Node.NumberTravelers.RoutingIn + Candidate.Server.InputBuffer.Contents.NumberWaiting + Candidate.Server.Processing.Contents.NumberWaiting`
Selection Condition:	**Left Blank**[70]

The CANDIDATE designation is a "wildcard" reference that it takes on the identity of the object it is referencing. Here it refers to the node to which the entity is being routed and the particular INPUTBUFFER and PROCESSING queue of the SERVER at the end of the path. In this example, the word CANDIDATE is replaced by each node in the list and the expression is evaluated. In a later chapter, ASSOCIATEDSTATIONS functions will be used to simplify this expression. Leave the *Selection Condition blank and leave the Blocked Destination Rule as the default value.*

Question 1: How does the *Selection Expression* cause the alternative destinations for the entity to change, relative to each destination?

____

Step 15: Since the travel time to exit is the same from all packers and we don't expect to change it during the simulation, let's specify the travel from the packers to the exit using a "Model Property" via the "*Definitions*" tab. Insert an *Expression* property from the "Standard Property" drop-down named **TimeToExit** and give it a *Default Value* of 3.5 with *Unit Type* of **Time** with *Default Units* as **Minutes**. Since this is an "Expression" property then one could specify an expression for the time to exit (for example, `Random.Uniform(2,10)`) while a numeric property only allows constant (i.e., 3.5).

Step 16: Now you can specify **TimeToExit** as a referenced property for the *Travel Time* on each of the time paths to the exit[71]. Now if you want to change the travel time on these three paths at the same time,

[70] It is very easy to specify the *Selection Goal* property as the *Selection condition*. The *selection condition* property is an expression that has to evaluate to "True" before the item can be selected from the list (e.g., SERVER has not failed) as a possible choice.

[71] Select all three time paths and then right click on the *Travel Time* property and select the new reference property **TimeToExit**.

only the new property has to be changed or you want to experiment on the impact of this time. You can access the model properties by right clicking on the MODEL in the [*Navigation*] and then expanding the "*Controls*" category to access our new property as seen in Figure 6.9.[72]

Step 17: The MODEL properties become controls to the simulation. These can be changed at the beginning of any run. Furthermore they become changeable scenario properties in experiments as we saw earlier in Figure 6.9.

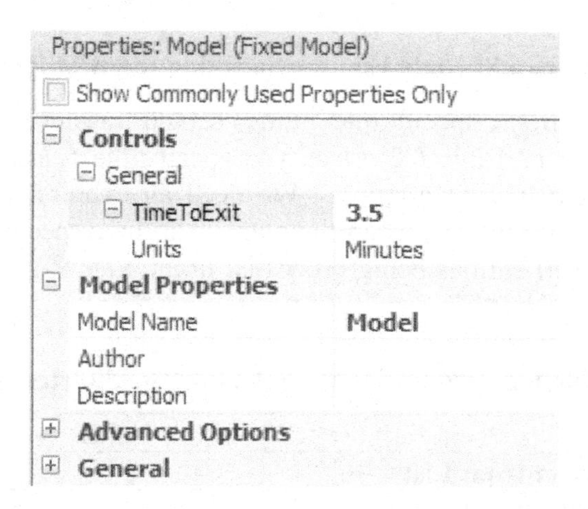

Figure 6.9: Changing the Properties of the Model

Step 18: Save and run the model for 10 hours answering the following questions.

Question 2: How long do the entities stay in the system (enter to leave)?

____

Question 3: What is the utilization of the **CmbBoardAssembly**?

____

Question 4: What is the utilization of each of the packers?

____

Part 6.2: Making the Animation Reveal More Information

Let's first fix the memory board animation so the "chips" look attached to the board.

Step 1: Select the **EntMemoryBoard** MODELENTITY and insert the BATCHMEMBERS queue from the "*Attached Animation*" in the "*Draw Queue*" drop down. Position it on top of the **EntMemoryBoard** as seen in Figure 6.10.

[72] Note that if a default value is not given, the simulation model may not run until you have set the property because it will be Null.

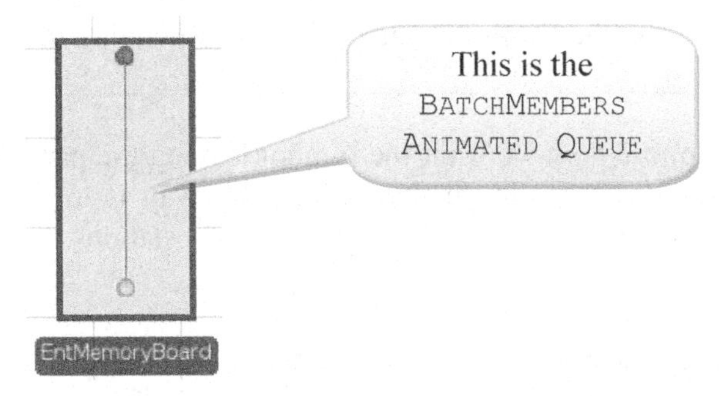

Figure 6.10: Added Batch Queue

To put an object on top of another you need to select the object and hold down the *Shift* key to stack the queue on top of the memory board symbol. It's probably best to do this while switching back and forth from 2-D to 3-D.[73]

Step 2: Now let's use the color of the memory chips symbol to designate the two different chip types. To do this, click on the MEMORY model entity and click on "*Add Additional Symbol*" in the "Additional Symbols" section of the" Symbols" tab. You can color one symbol red and the other green by simply selecting one of the "Active Symbol" and changing its color. Please note that these two symbols are numbered zero and one (**not one and two**).

Step 3: Next, we need to "assign" the colors to the different chip types. Select the **SrcMemoryArrivals** SOURCE. Click on the box to the right of *Before Exiting State Assignments*, since we want to assign the color before the entity leaves the source. This brings up the "Repeating Property Editor". Add an additional property with the following characteristics.

State Variable Name: `ModelEntity.Picture`
New Value: `ModelEntity.EStaChipType-1`[74]

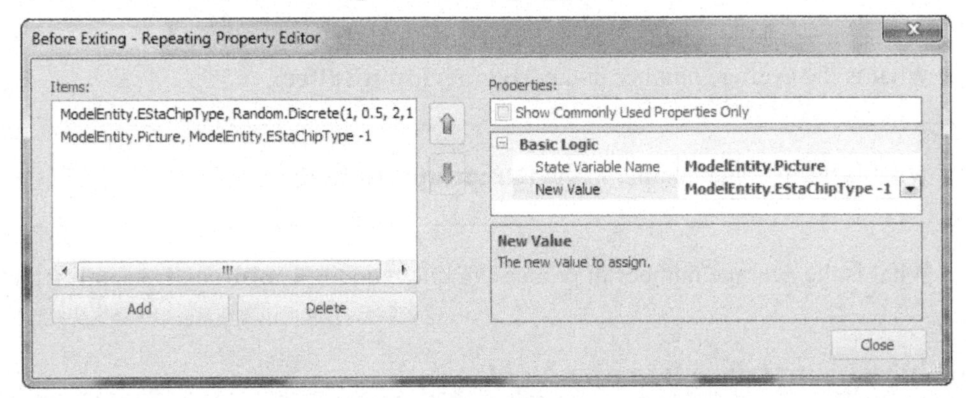

Figure 6.11: Changing the Chip Entity to have a Different Color

A "state variable", unlike a property, can be changed throughout the simulation and here the picture is a SIMIO defined characteristic of the entity (in some simulation languages called an attribute of the entity).

Question 5: Why do we subtract one from the **EStaChipType** property of the entity?

____

Step 4: Run the model to be sure the animation behaves as expected.

[73] Remember that hitting the "h" key in the model window brings up instructions for moving around in 2D and 3D and also in manipulating objects. Hitting the "h" key again removes the instructions.

[74] Recall the picture symbols start at zero which is why we need to subtract one.

Question 6: Do the entities change color corresponding to their chip type?

____

Step 5: Suppose now we want the symbol leaving the packing station to represent a "package". There are several ways one might do this. One easy way is to use the *Before Exiting State Assignments* at the packing stations. Now at one of the Packing stations, bring up the *Repeating Property Editor* and this time add:

State Variable Name: `ModelEntity.Size.Height`
New Value: `1`

Step 6: You may need to experiment with the value. The idea is to change the memory board's "height" so that it encases the "chips" so they cannot be seen (i.e., hides the batch members inside the board).

Step 7: Run the model and adjust the "height" so it shows a "package" leaving the packer. You will need to add this change in "height" at each packer station, when the entity exits.

Step 8: Look at the model in 3-D and notice the changes.

Step 9: Save the model as Chapter 6.2.spfx and run the model for 10 hours.

Question 7: How long are the entities in the system (enter to leave)?

____

Question 8: What is the utilization of the **CmbBoardAssembly**?

____

Question 9: What is the utilization of each of the packers?

____

Question 10: What is the average number in the **MemberInputBuffer**?

____

Question 11: What is the average number in the **ParentInputBuffer**?

____

Question 12: What is the average number in processing queue in the **CmbBoardAssembly** queue?

____

Part 6.3: Embellishment: Other Resource Needs

Sometimes capacity is not just limited at objects like SERVERS and COMBINERS. The limitation may not even be fixed at a particular location. For example, suppose the actual board assembly requires a particular fixture in order to assemble the chips to the board. Furthermore that fixture is used to transfer the assembly to packing. For each assembly, the fixture must be first obtained from the packing operation, which takes about three minutes which we will relax the assumption. Let's assume there are ten identical fixtures available for assembly-packing and once they are released at the pacing they can be used immediately at the assembly.

Step 1: First, drag and drop a RESOURCE object[75] onto the modeling canvas, maybe just below the COMBINER as shown in Figure 6.12. Name the resource **ResFixture** and specify an *Initial Capacity* of 10 (we won't be changing its capacity in this case).[76]

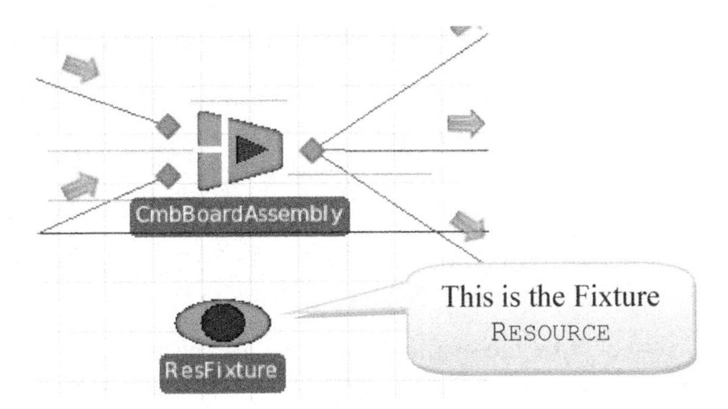

Figure 6.12: Adding the Fixture Resource

Next, logic needs to be specified on how this RESOURCE is to be utilized. The fixture will be "seized" just before assembly operation begins and then released just after packing operation finishes.

Step 2: For the **CmbBoardAssembly** COMBINER, select the *Secondary Resources* and then the *Other Resource Seizes* and then the *Before Processing* as seen in Figure 6.13, Click the ellipses box to bring up the *Repeating Property Editor* for seizing resources and request one unit of the **ResFixture** resource capacity to be seized. We won't worry about which fixture is seized, since they are identical.

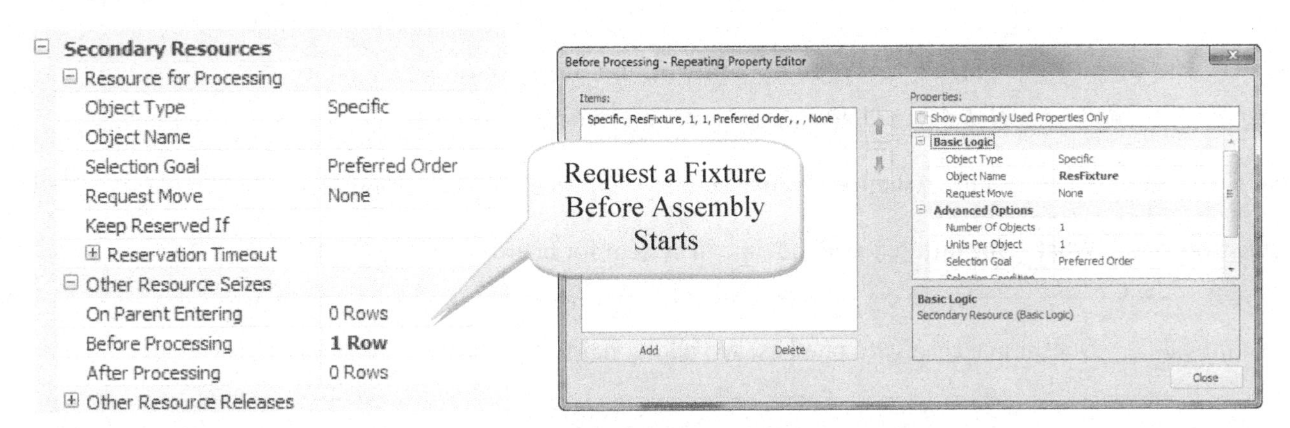

Figure 6.13: Seizing One Unit of Resource Capacity

Step 3: When the package has been finished, the fixture needs to be released for the next assembly. This process has to be done for each of the three packers. For each packer, select the *Secondary Resources* and then the *Other Resource Releases* and then the *After Processing*. Click the ellipses box to bring up the *Repeating Property Editor* for releasing resources. As shown in Figure 6.14, one unit of the **ResFixture** resource capacity will be released.

---

[75] Later chapters will go in more detail on the RESOURCE object.

[76] Note, a RESOURCE is fixed object and cannot travel throughout the network. Other types of resources (i.e., WORKER, VEHICLE, and ENTITY) can be dynamic and move through out a network.

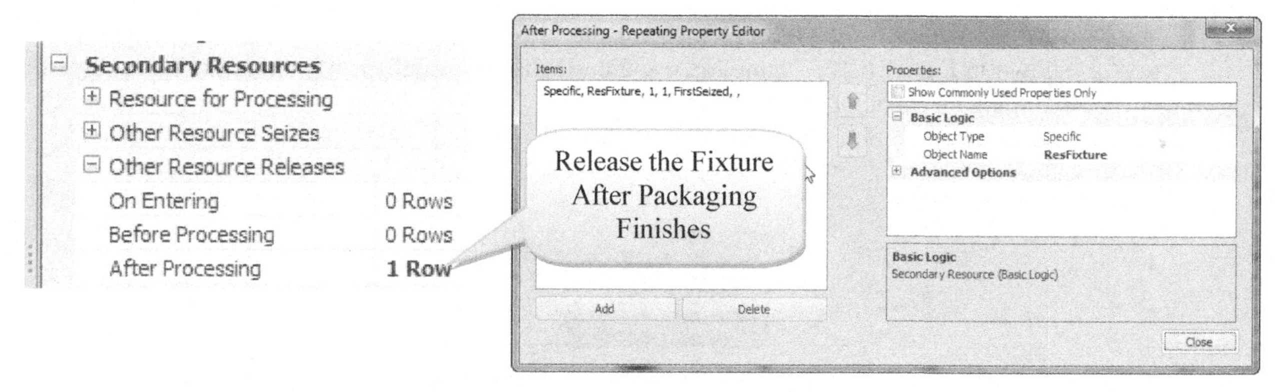

Figure 6.14: Releasing the Fixture

Step 4: In order to "see" how the available capacity of the resource changes during the simulation, add a *Status Label* from the *Animation* section in the ribbon. Make sure no objects are selected.[77] When the status label is selected, the cursor can be used to draw a box onto the modeling canvas as shown in Figure 6.15. Note that the expression gives the remaining capacity of **ResFixture**.

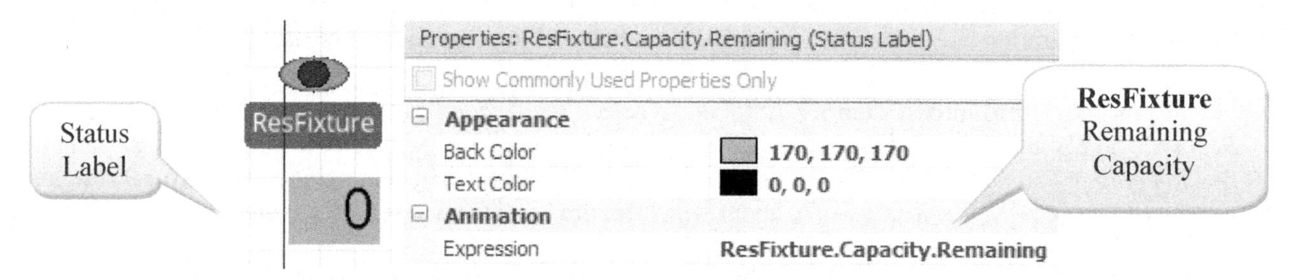

Figure 6.15: Adding a Status Label for Remaining Capacity

Step 5: Run the model and record the time in system for boards and chips.

Step 6: Make the capacity of the **ResFixture** equal to "20" and re-run the model.

Question 13: What is the difference in the time in system for boards and chips for 10 versus 20 fixtures?

____

Question 14: Does it make sense to purchase ten additional fixtures?

____

Part 6.4: Changing Processing Time as a Function of the Size of the Queue

Suppose you find out that **SrvPacker1** changes its processing time depending on the number currently waiting (i.e., size of the queue). Specifically, the packer is only 80% as efficient when there is zero in the input buffer queue, 90% when there is one, and 100% for all others.

Step 1: You need an "efficiency" table now to determine the processing time. This table can be modeled as a "Discrete Lookup" table in SIMIO (i.e., *f(x)*). From the "*Data*" tab, add a "*Lookup Table*" whose values are 0.8 for 0, 0.9 for 1, 1 for 2, and 1 for 100 as shown in Figure 6.16. Name the table **LookupEfficiency**.

[77] If we had selected the **ResFixture** first before inserting the status label, the label would be attached to the resource and therefore the expression would be just `Capacity.Remaining`. If we moved the Resource in the model the status label would move as well.

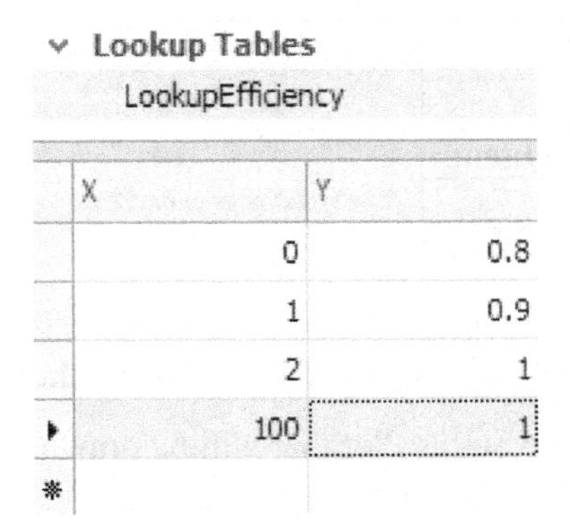

Lookup Tables

LookupEfficiency

X	Y
0	0.8
1	0.9
2	1
100	1

Figure 6.16: Lookup Table for Efficiency Calculation

Remember the table gets interpolated for values in between the discrete points. To access the efficiency based on the number waiting in the queue at the SrvPacker1, we will use the following expression: `LookupEfficiency[SrvPacker1.InputBuffer.Contents.NumberWaiting]`. Notice, to access rows into (i.e., indexing) into tables and arrays, SIMIO uses the [] notation rather than ().

Step 2: We are planning on using the expression more than once in our model. Since the expression is very long and complicated, we will utilize SIMIO's function expression creator to alleviate mistakes.[78] From the "*Definitions*" tab under the *Functions (f(x))* section, insert a new FUNCTION named **FuncEfficiency**. Set the *Expression* property to the appropriate value as seen in Figure 6.17.

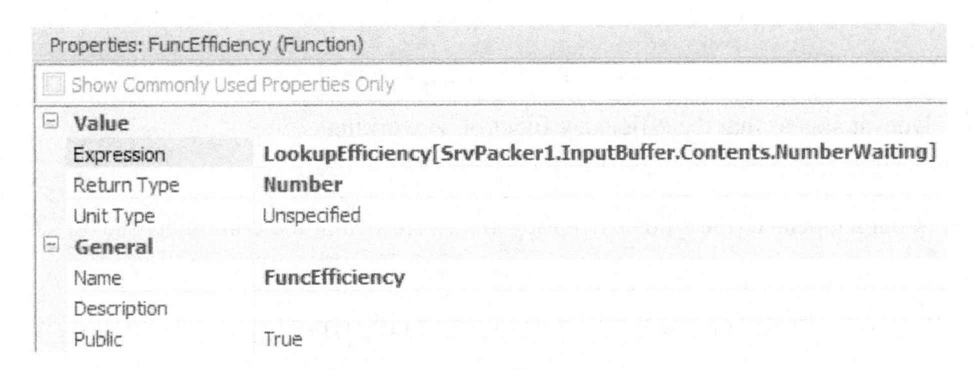

Figure 6.17: Specifying a SIMIO Function Expression

Step 3: Now in the *Processing Time* property for the **SrvPacker1**, change the expression such that the Triangular distribution is divided by the efficiency function to get the actual processing time. The new expression is defined as the following.

```
Random.Triangular(3,5,8)/FuncEfficiency
```

Question 15: Why do we divide the original processing time by the efficiency?

____

Step 4: We want to be sure our "efficiency" function is working correctly. To do that, we will use a SIMIO "Status Plot" which you will find under the "*Animation*" tab in the ribbon. Use the following specifications for the plot which are in the "*Labels*" and "*Time Range*" section of the "*Appearance*" tab.

[78] If we were to change the expression, it will only need to be modified in one place rather than in every location we used the direct expression.

Title:	**Efficiency at Packer 1**
X Axis:	**Time**
Y Axis:	**Number in Packer 1 Queue**
Text Scale:	**1**
Time Range:	**10 Minutes**

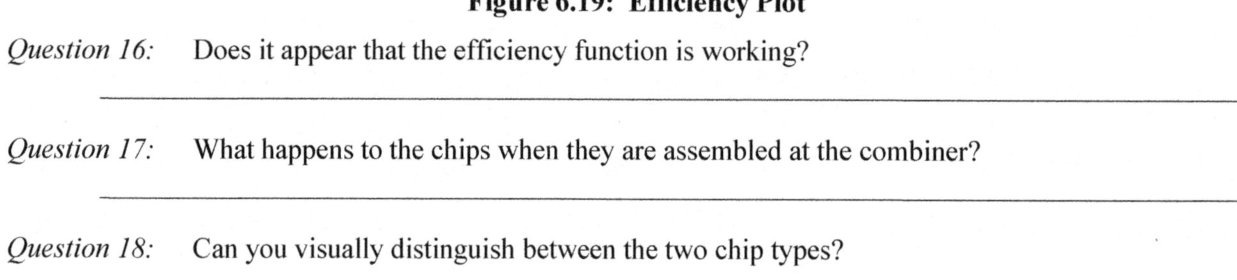

Figure 6.18: Setting the Properties of the Status Plot

Also, add the *Expression* property of the Status Plot that will be drawn which is `FuncEfficiency` where the plot is given in Figure 6.19.

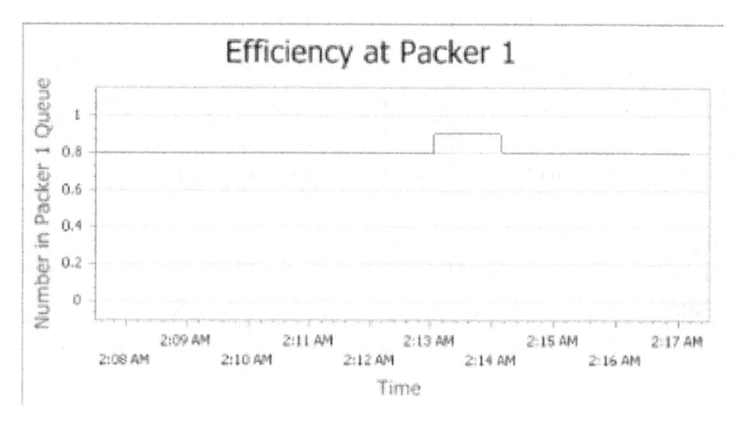

Figure 6.19: Efficiency Plot

Question 16: Does it appear that the efficiency function is working?

___

Question 17: What happens to the chips when they are assembled at the combiner?

___

Question 18: Can you visually distinguish between the two chip types?

___

Part 6.5: Creating Statistics

Here the SIMIO "automatic" statistics collection process may not provide the statistics that interest you. For example, we might want the time the boards and chips get from entry to exit or we might want to know what the average efficiency is for **SrvPacker1**. Time in system or time in a "sub-system" is an example of "observation-based" statistics. Recall this type of statistic is referred to in SIMIO as a "Tally Statistic". In the case of the efficiency, its value is changing (discretely) with respect to time and it is an example of what is generally referred to as a "time-persistent" or "time-weighted" statistic. In SIMIO this type of statistic is a "State Statistic" and since efficiency changes instantaneously (discretely) it is referred to as a "Discrete State Statistic".[79]

Step 1: Consider first the "state" statistic "efficiency". To collect a statistic on this characteristic, we need to define a model-based (global) state variable for efficiency. From the "*Definitions*" tab, select "*States*" and in the "*Discrete*" section click on "*Real*" to define a new state variable named **GStaPacker1Efficiency**.

---

[79] Don't confuse a "state variable" with a "state statistic".

Question 19: Notice there are seven general discrete types of "state" variables, but only the "*Discrete→Real*" one is relevant here. What is there about the efficiency change that is "discrete" and "real"?

Question 20: What are some of the other "value types" within the "*Discrete*" variable type?

Step 2: To ensure that **StaPacker1Efficiency** is correctly monitored for its value by SIMIO, let's modify the "State Assignments" in the **SrvPacker1** SERVER. In particular, we will make the assignment *On Entering* the server and use the following assignments.

State Variable: `GStaPacker1Efficiency`
New Value: `FuncEfficiency`

Step 3: Finally (this is the last step in collecting a state statistic), we need to define a "State Statistic". Defining new statistics can be done from the "*Definitions*" tab, but in the "*Elements*" section. Elements represent special objects that add modeling flexibility. Let's call the new statistic the **StateStatPacker1Utilization** and use **GStaPacker1Efficiency** as the *State Variable Name* as seen in Figure 6.20.

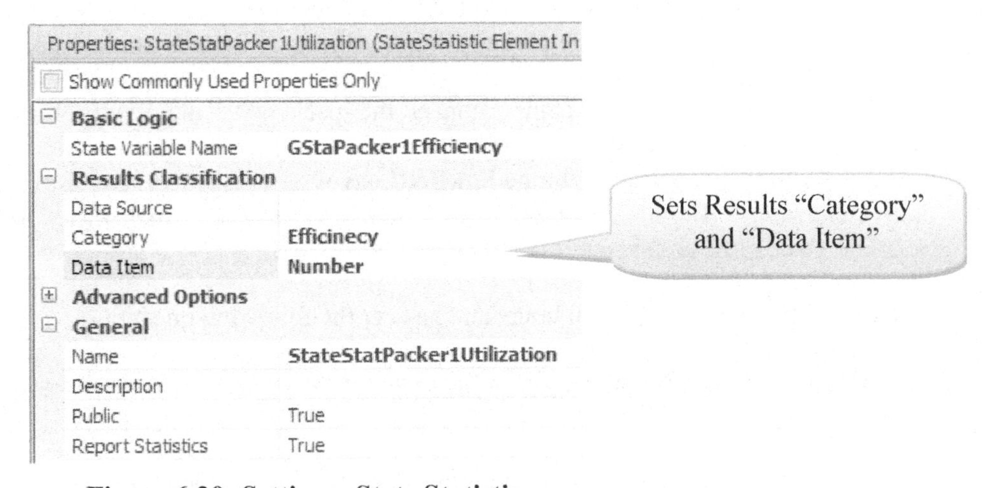

Figure 6.20: Setting a State Statistic

Step 4: Save and run the simulation model for ten hours. Filter the results of the "Object Type" to "Model" to see the new statistic.

Question 21: What did you get for the average **StateStatPacker1Utilization**?

Step 5: Next let's obtain the "time in system" statistics for the boards and chips to go from entry to exit. There are several ways to do this. Although you might think you need to give some special instructions to SIMIO, this is a case where we can take advantage of the way SIMIO provides statistics. Recall that we automatically get "time in system" statistics on the entities exiting through a sink. We will exploit this modeling approach.

Step 6: To obtain the statistics on the boards and chips, we can separate them and send them through separate exits (i.e., SINKS) – and do all this without changing the fundamental statistics. Our approach is shown below in Figure 6.21.

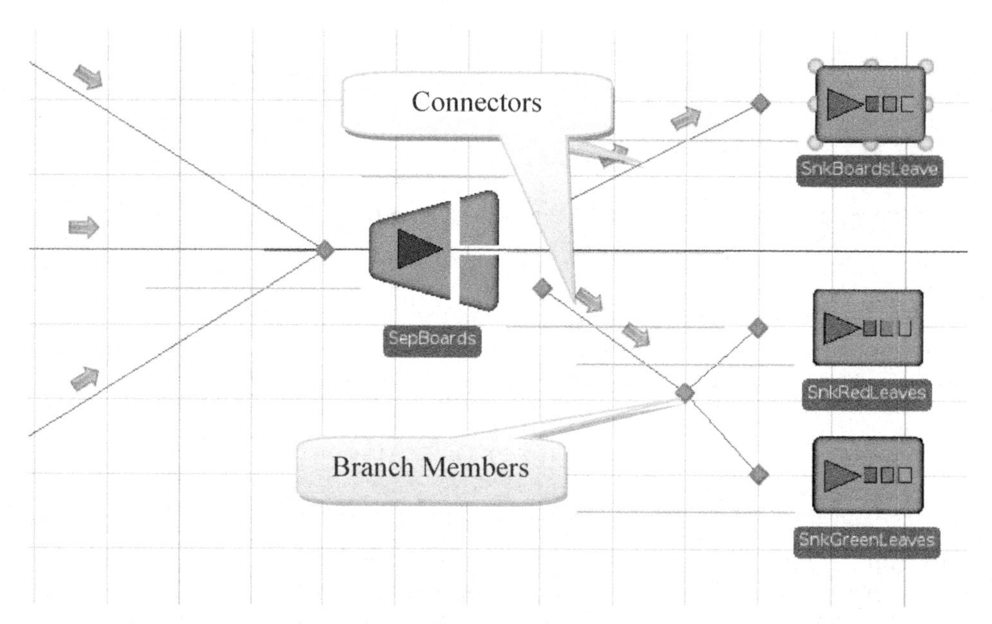

Figure 6.21: Separating Entities

The approach will substitute a SEPARATOR (from the [*Standard Library*]) for the original SINK and then branches the "parent" and "members" output to the appropriate SINK. The members get further branched to separate SINKS. CONNECTORS are used to connect objects so there is no time taken.

Step 7: Again, using time paths connect the packers to the input of the SEPARATOR specifying the ***TimetoExit*** property as the travel time. The branching of members uses "Link Weight." The *Selection Weight* property for the link connecting to the **SnkRedLeaves** SINK is `ModelEntity.EStaChipType==2` while the *Selection Weight* property for the connector to the **SnkGreenLeaves** SINK is `ModelEntity.EStaChipType==1`.

Step 8: Run the model for 10 hours and answer the following questions.

Question 22: How long are boards in the system from the time they enter to when they leave?

____

Question 23: How long are "red" chips and "green" chips in the system (enter to leave)?

____

Question 24: What is the average utilization of Packer 1 over the 10 hours?

____

Part 6.6: Commentary

- Good animation can add to modeling insight as well as providing a communication vehicle

Chapter 7
Using SIMIO Processes

Most simulation languages hide the details of executing the processes associated with an event. Generally, SIMIO has taken care of performing these steps and we have been somewhat ignorant of what specifically happens unless you have looked at the trace. Also, it is usually difficult to look "under the covers" in most simulation languages as well as have the ability to add your own features to the simulation model.

An innovative feature of SIMIO is the capability to modify an object's behavior through the use of SIMIO "Processes." In essence you can determine some new things to happen based on current conditions. SIMIO processes offer a wide range of modeling opportunities, which will be employed in this and later chapters. Processes are very flexible and can model many varied behaviors, well beyond those that are intrinsic to the object (such as a SERVER). Processes allow you to extend and expand the modeling capability of native SIMIO. There are two basic considerations when using and developing processes.

1. "When" is the process invoked or executed?
2. "How" do you write a process to produce the desired behavior?

By now you realize that a simulation executes by moving from event to event. SIMIO has defined certain events from which you can respond with a process changing the models behavior. Processes are a sequence of simulation "steps" typically invoked by actions of objects within the simulation or by events whose response you write. Writing a process is similar to creating a stylized programming flowchart, except that the components of the flowchart are logical simulation "steps" such as *Assign*, *Decide*, *Tally*, *Transfer*, *Seize*, *Delay*, *Release*, and so forth. The actual execution of the process is performed by a "token." The TOKEN is a SIMIO object that can have state variables but it is not the same as an entity. The tokens execute the steps of a process on behalf of an object, typically the object whose behavior is being modified by the process or where the process resides (i.e., PARENT OBJECT) or the object causing the modification (i.e., ASSOCIATED OBJECT). The relationship among the token, objects, and process is shown in Figure 7.1.

Part 7.1: The Add-On Process

We will initially focus on the SIMIO "Add-on Process" triggers. These are events triggered by actions of objects but mostly by entities to change the behavior or cause an action for an object that is needed for your particular model. In the case of an Add-on process, the PARENT OBJECT is primarily the main MODEL since the Add-on processes are created within the main **Model** and are viewed and edited in the "*Processes*" tab of the MODEL. Often, the ASSOCIATED OBJECT is the MODELENTITY that triggers the particular process to be executed.

Figure 7.1 shows an example of the "*Entered*" Add-on process trigger. When **Entity30** enters the **SrvPacker1** server, the **SrvPacker1_Entered** process is triggered. At this point, a TOKEN associated with **Entity30** is created at the *Begin* endpoint and travels step to step in zero time (e.g., the TOKEN is currently at the begging getting ready to execute the *Assign* step) until it reaches the *End* endpoint at which time it is destroyed. In this example, the ASSOCIATED OBJECT is **Entity30** and the PARENT OBJECT is the main Model. The TOKEN is executing the steps on behalf of **Entity30**.[80]

[80] You can have more than one TOKEN associated with the same MODELENTITY which will be seen in later chapters.

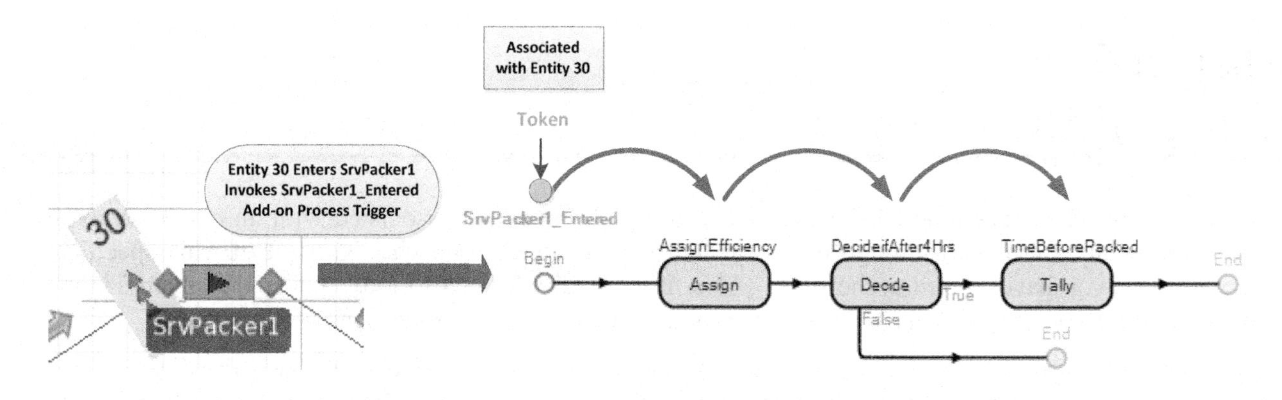

Figure 7.1: Example Explaining Entity Triggering an Add-on Process

Part 7.2: The Add-On Process Triggers: Illustrating "Assign" Step

Almost all SIMIO objects have "*Add-On Process Triggers.*" These triggers cause external processes to be invoked as part of other processes. So as a part of the execution of the object, arbitrary "processes" of your design can be also executed which helps change/expand the underlying object's behavior (i.e., SERVER) without having to look inside the object which will be explored in a later chapter.

In Chapter 6, the characteristics of the entities were changed. In one case, the entity types (i.e., the memory chips) were colored and in the second case the memory board entity's height was changed. In both cases, we used "state variables" whose values were changed using the "*State Assignments*" properties of the SOURCE and SERVER objects. You will see that there are only a few opportunities within an object to make state assignments, while there are many more Add-On Process Triggers. Furthermore, the "*State Assignments*" can only assign state variable values, whereas a process can incorporate more logic than just simple assignments.

Step 1: Let's see how to replicate the "*State Assignments*" using Add-On Process Triggers. Bring up the model from the prior chapter. First, remove the "*State Assignment*" from the **SrcMemoryChip** SOURCE by deleting them within the *Repeating Property Editor*.

Step 2: Make sure the **SrcMemoryChip** is selected and click on the "*Add-On Process Triggers*" menu to expand it.

Question 1: What *Add-On Process Triggers* are available in the SOURCE object?

____

Question 2: When "*State Assignments*" are performed on entities made in the SOURCE object?

____

Step 3: Choose the "*Exited*" Add-On Process trigger. Double clicking on this trigger automatically creates a new process named **SrcMemoryChip_Exited** and then selects the "*Processes*" tab. You will see a "flowchart" with a "`Begin`" and an "*End*" as seen in Figure 7.2. By clicking on one of the process *Steps*, you can insert it into the flowchart, which will adjust itself automatically.

Question 3: Name a few of the steps available that you might use in creating a process?

____

Question 4: Why did you choose these?

____

Step 4: Insert an `Assign` step as shown in Figure 7.2.

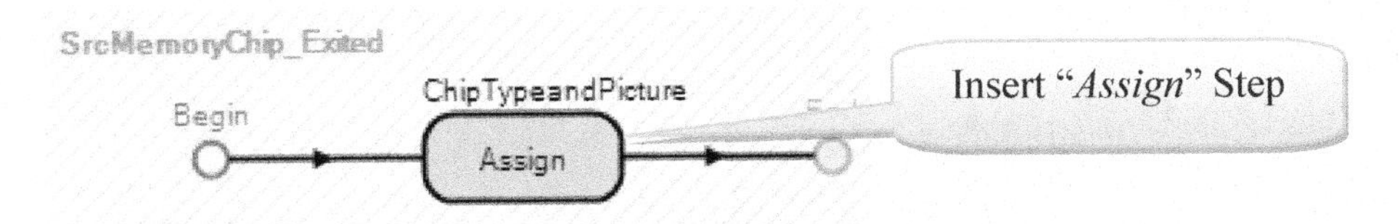

Figure 7.2: Inserting the Assign Step

Step 5: Under the properties window you will see a place to put the *State Variable Name* and its *New Value*. You can only assign one state variable. However, you can also invoke the *Repeating property Editor* by clicking beside the "*Assignments (More)*." Add the following two assignments using the latter approach as seen in Figure 7.3.[81] It should look very familiar as the "State Assignments" utilize an `Assign` step internally.[82]

State Variable Name: `ModelEntity.EStaChipType`
New Value: `Random.Discrete(1, 0.5, 2, 1.0)`

State Variable Name: `ModelEntity.Picture`
New Value: `ModelEntity.EStaChipType - 1`

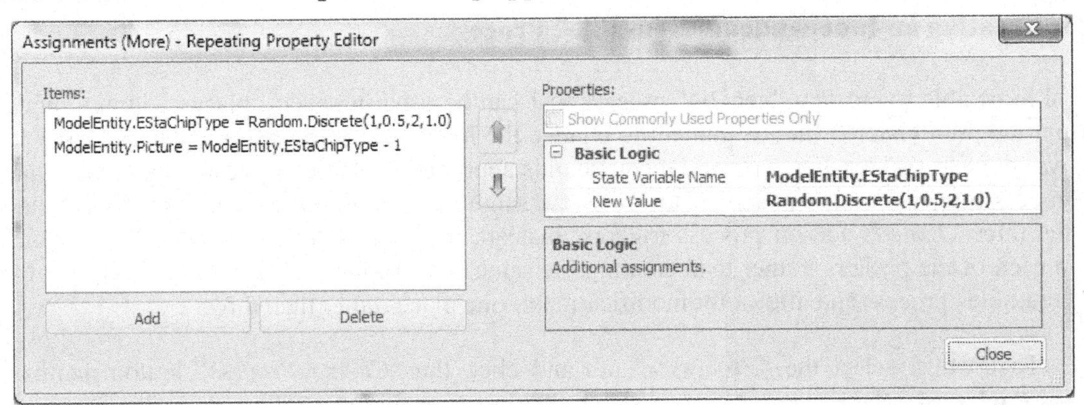

Figure 7.3: Repeating Property Editor for Assignment Process Step

Step 6: Recall from the previous chapter, we calculated statistics on the efficiency of packer one. The efficiency was used to change the processing time of the packer based on the number in the queue. The state variable was updated when the entity entered the packer including the current entity rather when the entity entered processing where the efficiency value is used. The reason for not calculating the statistic correctly is there are only two default places for assignments (i.e., entering or exiting the server) but processes allow more flexibility. To fix the issue, delete the state assignment from the "*On Entering*" at **SrvPacker1**. Create a "*Processing*" Add-On Process Trigger which will calculate the efficiency right before the processing starts. Insert an `Assign` process step that does the following – see Figure 7.4.[83]

State Variable Name: `GStaPacker1Efficiency`
New Value: `FuncEfficiency`

---

[81] When performing more than one assignment, it is recommended you perform all of them in the *Assignments (More)* to avoid issues of duplication, etc. The single assignment is performed before the *Assignments (More)*.

[82] In the later chapters on sub-classing we will explore how SIMIO does the assignments when we take apart some of the basic objects.

[83] Utilize the single assignment for this exercise.

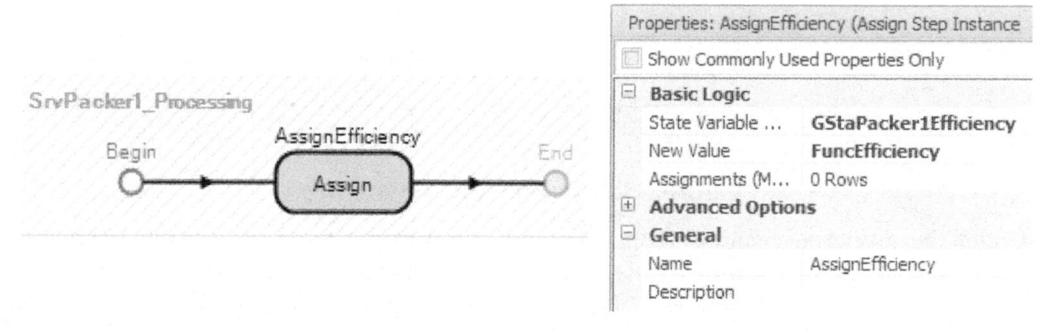

Figure 7.4: Recreating State Efficiency Assignments

Step 7: Save the model and run this model to make sure it replicates the behavior from the previous chapter.

`Question 5:` What was the average efficiency and was it much different than the previous chapter?

Part 7.3: Creating an Independent "Reusable" Process

It is useful to be able to create a "generic" process that can be used in several places within a model. In the model from the prior chapter, the assignment to change the height of the board to make the entity look like a package was done at each packer station. Of course, this approach could be duplicated by creating an "*Exited*" add-on process trigger for each packer. However, the same exact logic of assigning the height has to go in each of the three "*Exited*" add-on process triggers. Instead, let's create a "common" process which we can refer to in each of the packers, rather than creating the same process for each packer. This type of process is called a "reusable" process and allows for modification in one place rather than three.

Step 1: To do this, select the "*Processes*" tab and click the "*Create Process*" button naming the new process **Packaged**. Next add an `Assign` step to its flowchart with the following assignment.

State Variable Name: `ModelEntity.Size.Height`
New Value: `1 (Meters)`

Step 2: Delete the *Before Exiting* "*State Assignments*" at each of the packer stations.

Step 3: For each of the packer object's "*Exited" Add-On Process Trigger*, instead of creating a new one; just select the **Packaged** process from the drop-down list. Again select all three packers to change them all at once as seen in Figure 7.5.

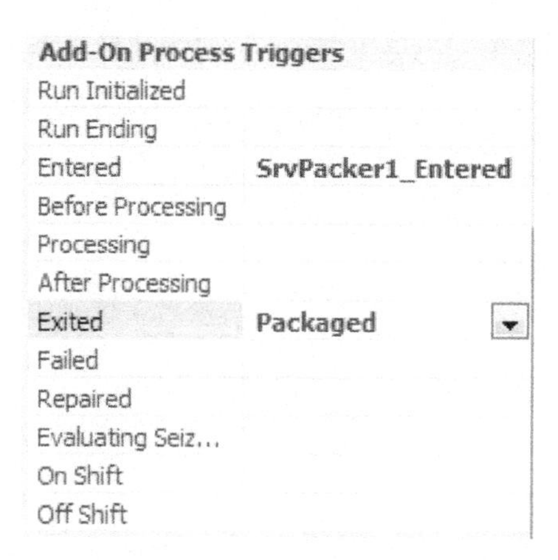

Figure 7.5: Utilizing a Common Process in all Three Packers

Question 6: What is the advantage of have one process and referring to it at each packing operation over having duplicated processes at each packing station?

____

Step 4: Run this model and make sure it replicates the previous behavior.

Part 7.4: Collecting Tally Statistics

Recall, observation-based statistics are referred to as TALLY statistics in SIMIO. These statistics are gathered as observations as the simulation executes. Examples would be queuing times, flow times, cycle times, and time in system. For the most part, these are statistics whose observations are intervals of time. Previously, TALLY statistics collection was limited to BASICNODES or TRANSFERNODES or the default results. Now we want to broaden greatly the possible collection points, namely in any PROCESS.

Step 1: Let's consider collecting TALLY statistics on the time boards and chips spend in the system from arrival time to the time they start to be packed at Packer1 only. To accomplish this addition, we need to first define a new "*Element.*"[84] (When working with TALLY statistics, we define them before we determine their method of collection.) Under the *Definitions→Elements* section, click on "*Tally Statistic*" in the ribbon to insert a new statistic named **TallySystemTimePacker1** as seen in Figure 7.6.

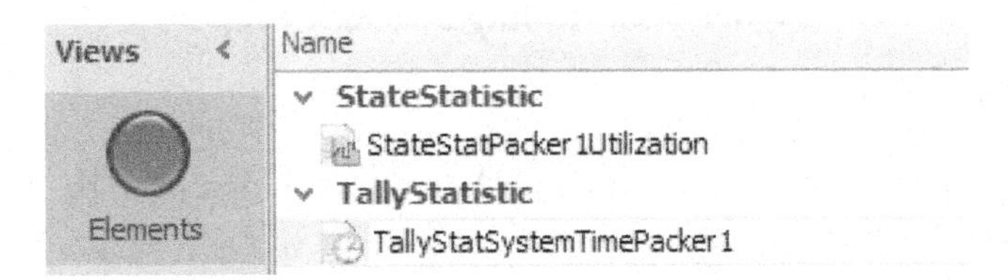

Figure 7.6: Specifying a new Tally Statistic in the "*Elements*" Section

Step 2: Be sure to make the "*Unit Type*" for the statistic "*Time*".

Step 3: Next, we need to define the collection process (i.e., logging each observation). This done by inserting a `Tally` step into the **SrvPacker1_Entered** add-on process trigger as seen in Figure 7.7.

TallyStatistic Name: **TallyStatSystemTimePacker1**
Value Type: Expression
Value: `Run.TimeNow-ModelEntity.TimeCreated`[85]

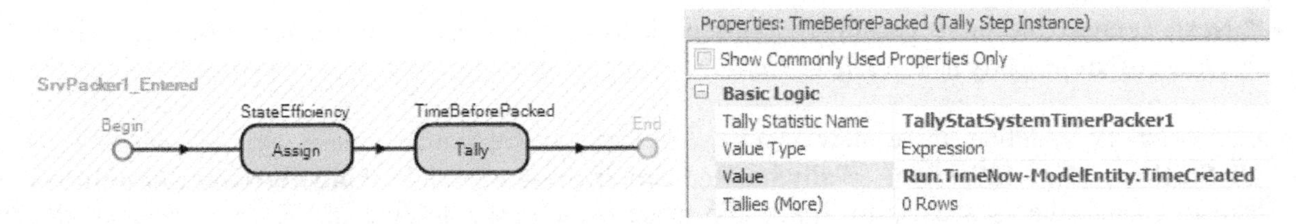

Figure 7.7: Using a `Tally` Step to Track Time before Packing Station 1

Question 7: Could the *Value* expression simply be `ModelEntity.TimeInSystem`?

____

Step 4: Save and run the model observing the results.

[84] Recall, *Elements* are behaviors of objects that can change state over time.

[85] `Run.TimeNow` is the current time of the simulation and `TimeCreated` function returns the time the entity was created.

Question 8: What did you get for the arrival time for boards and chips from the time they arrive to the time they start to be packed at Packer1?

____

Step 5: Suppose now we want only the time in system for assemblies as they start to be packed at **SrvPacker1** after the first four hours of the simulation? We need to modify our process so only observations of boards are collected after the first four hours.

Step 6: Modify our "***SrvPacker1_Entered***" add-on process by adding a `Decide` step before the `Tally` as shown in Figure 7.8. The `Decide` step is similar to an If statement in Excel where when the condition is true the process (i.e., TOKEN) will continue via the "True" branch otherwise it will follow the "False" branch. The *Decide Type* property should be "ConditionBased" with the *Expression* property set to `Run.TimeNow > 4`.

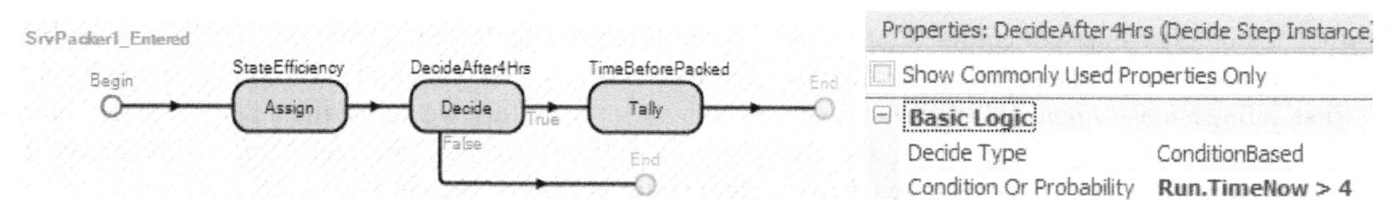

Figure 7.8: Adding a Decide Step

Step 7: Save and run the model.

Question 9: What did you get for the arrival time for boards and chips from the time they arrive to the time they start to be packed at **SrvPacker1**?

____

Step 8: It is also possible to use process steps to replace the modeling changes we made to the original model in order to produce time in system statistics for each chip type (i.e., red and green) and board. The previous approach placed additional objects (i.e., SINKS) in the model, which were not be part of the real system and might confuse management (or anyone else) when looking at the model.

- First, insert the original **SnkExit** (SINK).
- Delete all the additions (i.e., SEPARATOR and the three SINKS) when the original **SnkExit** SINK was replaced.[86]
- Next, connect the three packing stations back to the SINK as before, making sure to specify the **TimeToExit** property as the *Travel Time*.

Step 9: Next let's define three new TALLY STATISTICS (remember these are "*Elements*") named **TallyStatBoardTimeInSystem**, **TallyStatRedTimeInSystem**, and **TallyStatGreenTimeInSystem.** Make to change the *Unit Type* to ***Time***. To make these easier to find in the output, under "Results Classification", set the "*Category*" to **FlowTime**, and the "*Data Item*" to **TimeInSystem** as seen in Figure 7.9.

[86] When you delete a node, all the connections to it disappear. You can quickly connect nodes by holding the *Shift* and *Ctrl* key down, and clicking on the origin node and dragging a connector to the destination. You will need to select the type of connection (here we used TIMEPATHS whose *TravelTime* is a reference "**TimeToExit**".

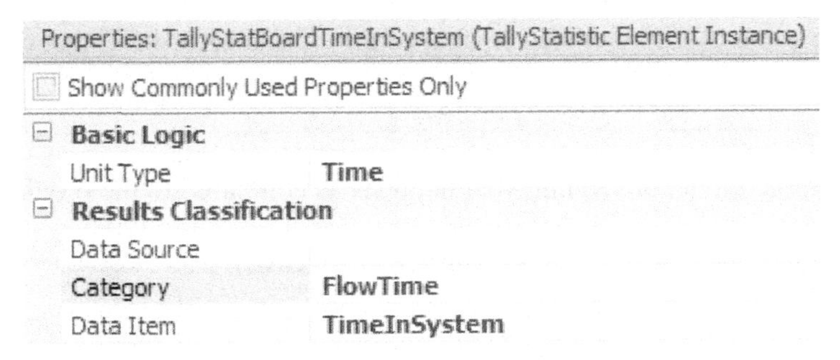

Figure 7.9: Specifying Time in System Tally Stats

Step 10: We will now create an *Add-On Process* in the SINK using the "*Destroying Entity*" add-on process trigger. This process is called **SnkExit_DestroyingEntity** and its process is shown in Figure 7.10.[87]

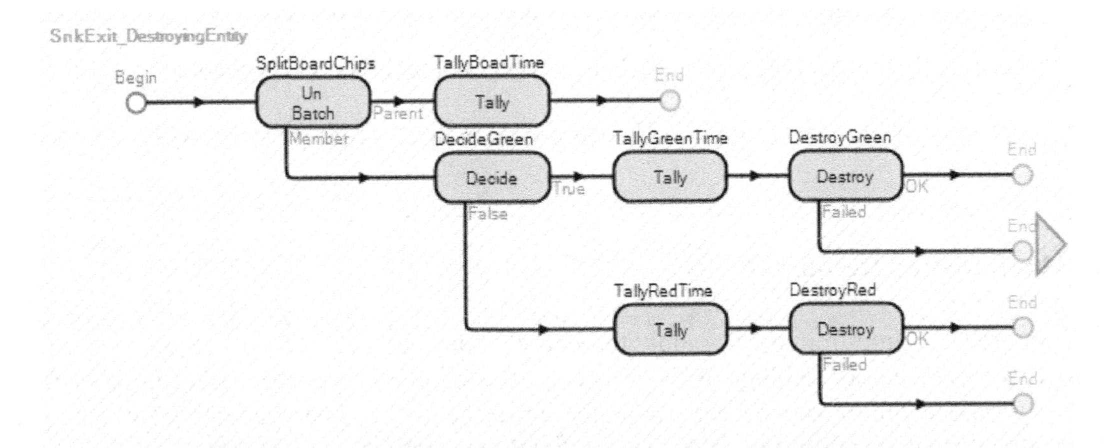

Figure 7.10: Destroying Entity Process (Tallying Statistics)

Note, the `UnBatch` step comes from the "All Steps (A-Z)" category in the steps panel. The `UnBatch` works much like the SEPARATOR object except it breaks up the parent and the members separately in the batch and sends them along the correct branches of the step.[88] Here two tokens are active on behalf of the same associated object: one TOKEN executes on behalf of the PARENT and the second TOKEN executes on behalf of the MEMBER[89].

- The `Decide` step is conditional, based on the condition: `ModelEntity.GStaChipType == 1`
- The `Tally` steps simply record `ModelEntity.TimeInSystem` at each step for each of the three appropriate TALLY statistics.
- The `Destroy` steps are needed to eliminate the "unbatched" members. The Parent will be destroyed at the sink, but the members are unattached and will remain in the model[90] unless removed.

Question 10: What are the advantages and disadvantages of this approach versus the original model where the symbolic model was changed to gather the different statistics?

---

[87] Double clicking on the *Exited* Add-on process trigger property will automatically create the process as well as take you to the "*Processes*" tab.

[88] Internally the SEPARATOR utilizes an `UnBatch` step to separate the parent and member entities and then sends them out the appropriate output nodes.

[89] In this case the second token will cause four tallies – one for each of the four chips. Furthermore, the execution of the "member" occurs before the execution of the "parent" from an `Unbatch`.

[90] The members are unbatched into "FreeSpace", which designates nowhere in particular, but they take up computer memory.

Question 11: What are some of the other steps that appear in "*All Steps (A-Z)*" that don't appear in the "*Common Steps*"?

____

Step 11: Save and rerun the model observing what happens as compared to the previous model.

Question 12: How long are the boards in the system (enter to leave)?

____

Question 13: How long are the "red" chips in the system (enter to leave)?

____

Question 14: How long are the "green" chips in the system (enter to leave)?

____

Step 12: The results should be comparable to what you obtained in the previous chapter.

Question 15: Since using processes duplicated what we did with the original model, which approach is the easiest for you? [91]

____

Part 7.5: Manipulating Resources

Previously, we used a fixture from the assembly through the packing. A RESOURCE object with a capacity of ten was used to model the availability of the fixtures. We employed the SERVER's built in "*Secondary Resources*" to seize one unit of resource capacity before processing at the board assembly. We did not release that capacity until the entity had finished processing at the packer station.

Unfortunately, if additional resources are needed at a SERVER or COMBINER, the places where capacity may be "seized" or "released" is limited. Furthermore, all that can be done at those locations is a "seize" or "release" (seize and release if used for only for "processing"). Recall, before the fixture can be used it has to be retrieved from the packaging area prior to processing and then torn-down after packaging (i.e., packaging separated from the memory board). One approach would be to add server objects for these activities and place them into the model. These additional objects again begin to clutter the model with substantial objects that perform only simple functions which processes can handle as well as model these situations easier.

Whenever we are modeling the use of capacity within an object, there are three steps that are often employed (i.e., `Seize`, `Delay`, and `Release`). Using these steps you "seize", "delay" for some time, and then "release" capacity. When you seize capacity it is important to know how many units of a resource is being seized. It is possible to seize units of multiple resources at the same time.

Now, we want to delay a Uniform(1,3) minutes after seizing the resource to setup and retrieve it and then delay a Uniform(2,4) minutes for a tear down separation process before releasing the resource. Before we simply choose one of the Add-on processes, let's consider the choices related to the processing as seen in Table 7.1.

[91] People often try to avoid using processes at first, especially if they can model without them (this is part of the "different thinking" required in SIMO). However you will find that the time invested in learning to use processes will pay off handsomely as you encounter more complex systems. The remaining chapters will illustrate why processes are so essential.

Table 7.1: Few of the SERVER' S Add-on Process Triggers

Add-on Process Trigger	**Description**
Before Processing	Occurs when an entity has been allocated SERVER capacity, but before entering (i.e., or ending transfer into) the processing station.
Processing	Occurs when an entity has been allocated SERVER capacity, ended transfer into the SERVER' S processing station, and is now about to start the processing time.
After Processing	Occurs when an entity has completed the processing time and is about to attempt its exit from the SERVER' S processing station.
Exited	Occurs when an entity has exited the SERVER object

We will (somewhat arbitrarily) choose *Before Processing* for the place to seize the fixture and *After Processing* for when it should be released. Note that the delay times setting up and tearing down the fixture therefore contributes to the time being processed at the objects.

Step 1: First, remove from the **CmbBoardAssembly** the "*Secondary **Resources/Other Resource Seizes/Before Processing***" the seize of one unit of capacity of **ResFixture** by deleting the entry in the *Repeating Property Editor*.

Step 2: Next, remove from each of the Packing stations the "*Secondary Resources/Other Resource Releases/after Processing*" the release of one unit of capacity of **ResFixture** by deleting the entry in the *Repeating Property Editor*.

Step 3: For the **CmbBoardAssembly** COMBINER, create the "*Before Processing*" add-on process trigger named **CmbBoardAssembly_BeforeProcessing**. This trigger is executed right before the COMBINER is ready to start processing on the assembly operation. The `Seize` step will be used to specify that one unit of capacity of **ResFixture** is needed at processing as shown in Figure 7.11. Once a fixture has been seized, a Uniform(1,3) minute `Delay` will be the time to setup and retrieve the fixture.

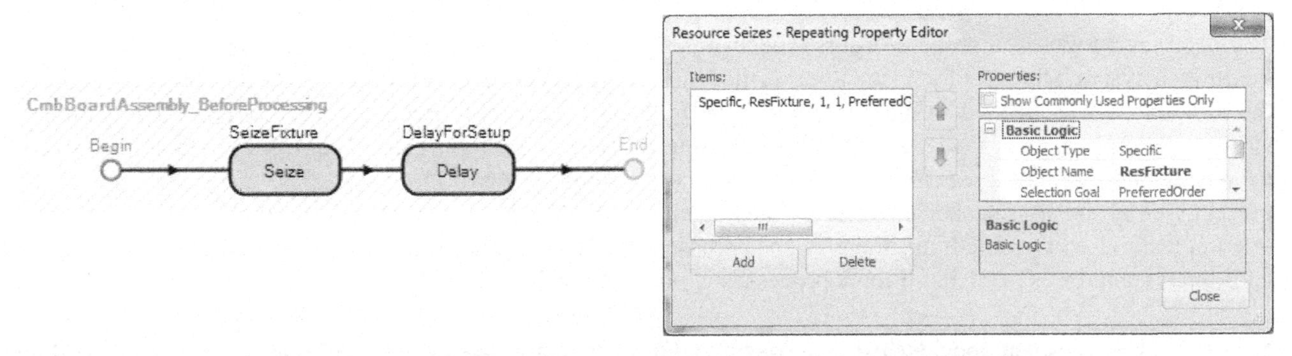

Figure 7.11: Seizing One Unit of Fixture Capacity and Delaying

Step 4: When the package has been finished, the fixture needs to be removed and released for the next assembly. This process has to be done for each of the three packers, so it will be easier to create a reusable process which is invoked when the board exits the packing stations. From the "*Processes*" tab click the "*Create Process*" button naming the new process **RemoveFixture**. The `Delay` step and the `Release` step are shown in Figure 7.12. The delay will take a Uniform(2,4) minutes to tear-down the fixture before the fixture is released.

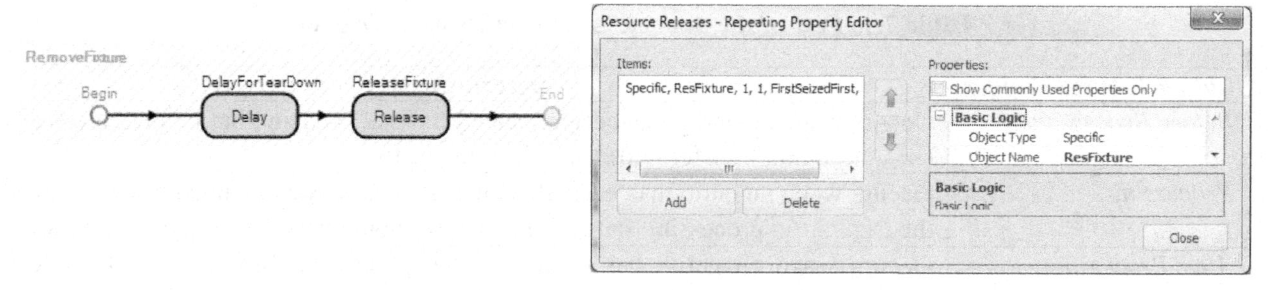

Figure 7.12: Tear-Down and Release the Fixture

Step 5: Now select the *After Processing* Add-On trigger of each of the Packers and select the **RemoveFixture** from the drop down list.

Step 6: Run the model for 8 hours.

Question 16: After looking at the results, what concerns do you have?

____

Question 17: What changes to the system could we suggest to make it more reasonable?

____

Part 7.6: Tokenized Processes

The previous section showed how the *exact* same process could be reused in multiple locations. However, if the process was not completely identical (i.e., the tear down time depended on which packer). A process may be given arguments through the tokens that execute them. These execute in a fashion similar to arguments in a programming language subroutine or subprogram. They parameterize the process or what we call a "tokenized" process since it requires custom tokens.

The need for a tokenized process occurs whenever you have several almost identical processes. For example, the time in system statistics for assemblies as they start to be packed at **SrvPacker1** after the first four hours of the simulation was needed. The "solution" was to create a process specialized to computing the time in system to that packer. If we wanted that same time in system for **SrvPacker2** and **SrvPacker3**, we would need to create two new processes, one for each packer, along with two more statistics. To do this we would probably copy the first packer process for the second two and edit them accordingly. Of course every time, one copies and edits something, there is an increased chance of making an error. Also, if later a change or additional behavior is needed, all three processes would need to be updated.

Step 1: Let's instead consider using a *Tokenized Process* that passes-in the particular statistic to record the operation. First, define new time in system (TALLY) statistical elements for Packer2 and Packer3 as seen in Figure 7.13 and do not forget to change the *Unit Type* property to "*Time.*"

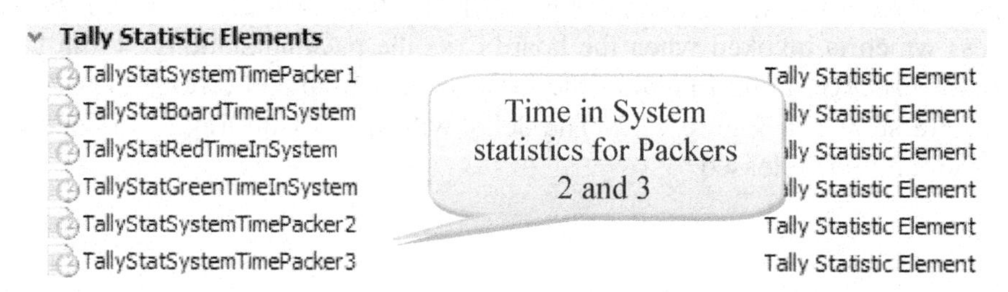

Figure 7.13: Tally Statistics for Packers 2 and 3

Step 2: Recall a TOKEN is created in order to execute the steps of a process. Therefore, parameters are passed to the process via a custom TOKEN. From the *Definitions→Token* section, add a custom TOKEN named **TknPacker**. This TOKEN needs an additional "Element Reference" state variables[92] named **TStaTallyStat**, to reference the correct statistic that will be used to store a parameter as seen in Figure 7.14.

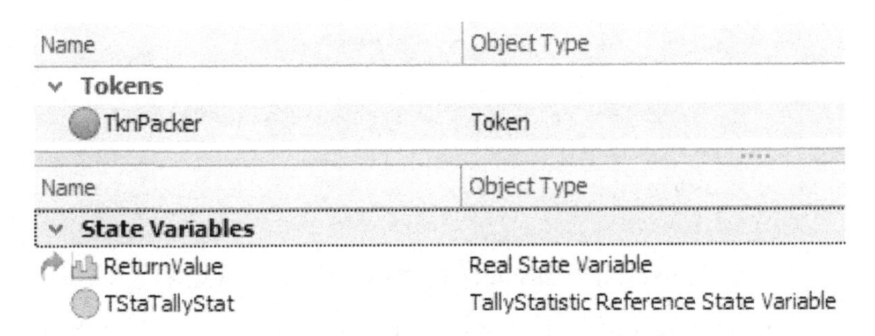

Figure 7.14: TknPacker State Variables

Step 3: Rename the **SrvPacker1_Entered** add-on process to **SrvPacker_Entered** and specify its properties as seen in Figure 7.15. Change the type of TOKEN that will execute the process steps to be **TknPacker**. Then specify the *Input Arguments* property with the one row (input arguments) as defined in Figure 7.16which specifies what the input parameter will be named and how it is linked to the customized token.[93]

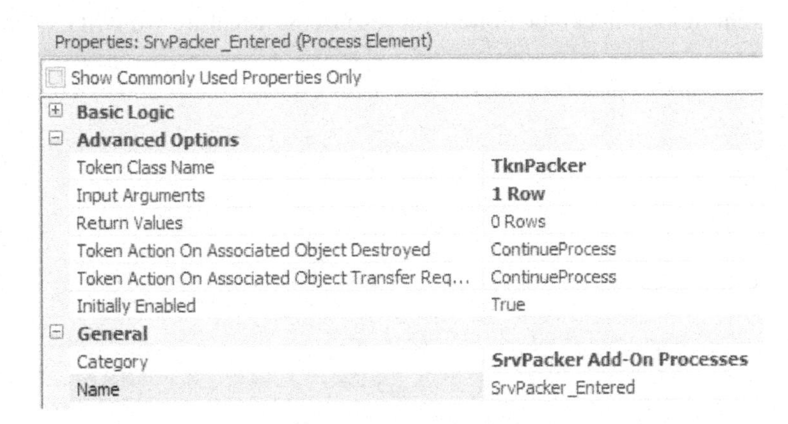

Figure 7.15: Properties of the SrvPacker Process

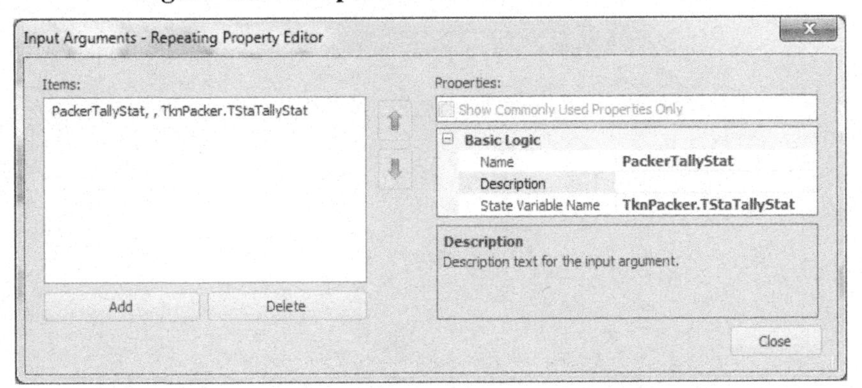

Figure 7.16: Input Arguments for the Process

[92] Tokens always have the state variable `ReturnValue`.

[93] Notice, you can also specify the process return values which might be useful if one process calls another process.

Step 4: Finally at each packer invoke the **SrvPacker_Entered** by specifying its own argument. Figure 7.17 shows the example for Packer 2.

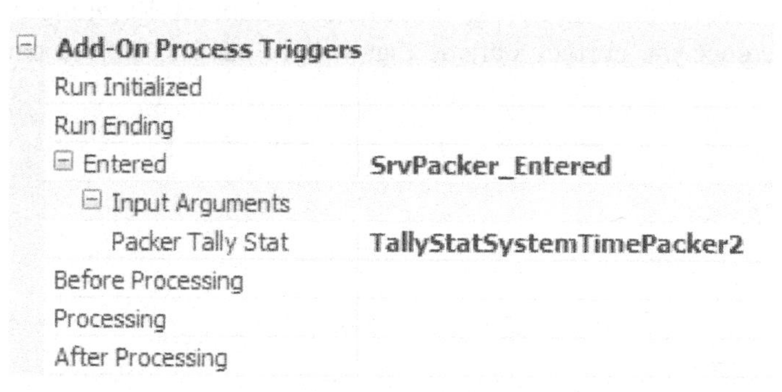

Figure 7.17: Invoke the Tokenized Process

Step 5: Execute the model and note the tally time in system statistics for each of the three packers.

Part 7.7: Commentary

- The process capability in SIMIO is one of its standout features, allowing for extension and customization in powerful and unique ways. To use it requires a potential change in thinking towards simulation modeling.
- A lot of different modeling can be done with resources, a feature we will explore more in later chapters.

Chapter 8
Working with Flow and Capacity: The DMV

Almost every object in the SIMIO standard library has capacity, which means that their capacity may be seized and released, typically within the flow of entities - but the SERVER and the RESOURCE objects are used the most often. As seen in earlier examples, the SERVER is often the basic flow component of many models. When the RESOURCE has been employed, it is often a secondary resource to a SERVER. But their relationship can be convoluted by the need to consider systems with multiple capacities in the context of complicated entity flow.

The SERVER object is composed of three stations (locations) where entities are held (i.e., INPUTBUFFER, PROCESSING, and OUTPUTBUFFER). They are illustrated with "contents" queues, each having their own capacity (see Figure 8.1). A BASICNODE is used as input to the SERVER while a TRANSFERNODE for output.

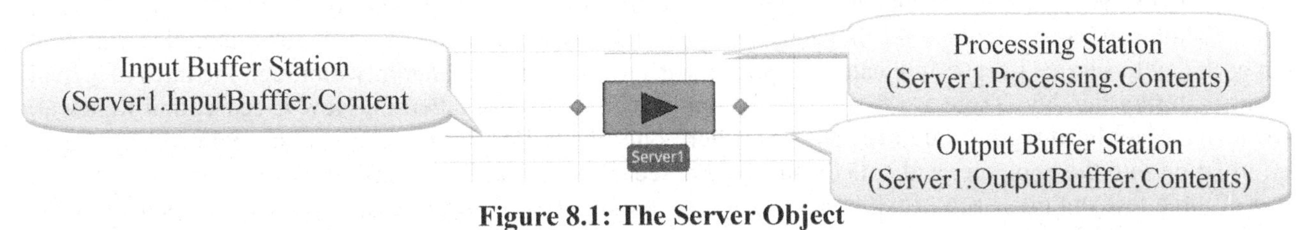

Figure 8.1: The Server Object

The capacity of the SERVER can be a fixed value or have its capacity controlled by a *WorkSchedule*. Recall that capacity simply refers to the number of entities that can be simultaneously served by the object. When no capacity of the processing station is available and an INPUTBUFFER exits (i.e., Input Buffer capacity is greater than zero), then the entities wait in the INPUTBUFFER. The *Processing Time* property is the time required for the server to process an entity in the processing station and is a required property of the server.[94] The SERVER can occupy one of five states as explained in Figure 8.2 and SIMIO provides additional symbols to automatically animate the current state of the SERVER. The state applies to all its capacity (i.e., if one unit of its capacity is allocated, the server is considered processing) and statistics are collected on the amount of time spent in each state.

Server State	Meaning	Additional Symbols
Starved	Currently idle since no unit of capacity has been allocated.	Active Symbol (1 of 5) · Remove Symbol
Processing	At least one unit of capacity has been allocated to process an entity.	Starved (0) · Processing (1) · Server states have a color, a name, and a number
Blocked	The current entity has finished processing but cannot proceed into the network	Blocked (2)
Failed	The server has failed because of some condition	Failed (3)
OffShift	Currently the capacity of the Server is zero meaning it is unavailable.	OffShift (4)

Figure 8.2: Active Symbol Definitions

[94] The *Process Type* property can be changed to handle more complicate sequences rather just a "Specific Time" and will be explored in a later chapter.

A RESOURCE is a much simpler object than the SERVER but shares a lot of common properties including capacity specification but is not composed of stations. Therefore has no place for entities to wait or be served and there is no processing time specification used to control when the resource is released. So a resource must have its capacity seized and released in a SIMIO process externally, as seen previously.

Flow and capacity are often the primary concern in many different types of systems (i.e., production, manufacturing, service, healthcare, etc.). Capacity maybe used to restrict flow but cannot be arbitrarily expanded. So finding ways to expand capacity and extend flow are significant challenges where simulation offers one way to experiment with different configurations of capacity and different flow mechanisms.

Part 8.1: The DMV Office

A Department of Motor Vehicle (DMV) often operates multiple offices of various sites across a state. If a simulation model can be created for one site, then perhaps it can be "reused," with some minor changes, for other sites (and thus the cost of model development is spread over several installations). A typical DMV office tests driver license applicants and supplies driver licenses to those passing the tests. A typical office size is being used as a reference for the analysis. The office typically opens up at 8:30am and closes at 4:00 pm each day to service license applicants. Applicants arrive more or less randomly although the rates of arrival changes during the day. There are three types of applicants for licenses: Permit, New License, and Renewal. All arriving applicants go through registration where one of the DMV clerks enters an applicant's personal information into the computer and collects the license fee. Next all applicants take a written and visual test on one of three test stations, each of which are staffed by its own DMV Tester. Not everyone passes these tests. Those getting a "New License" must also pass a driving test accompanied by a DMV officer where some fail the driving test. People who fail any (written/visual and driving) of the tests leave the office and return another day. Those who pass all their tests have their picture taken and their license processed and printed. The DMV is interested in the service of the office to applicants and the use of DMV resources. Consequently flow times of applicants are a major concern. Since queuing times are a part of flow time, it is important. If congestion in the office becomes high, then it also is a concern. Finally, since these are public facilities, the cost of the facilities, especially the use of DMV resources must be balanced against the service to applicants.

Developing a Base Model

In any modeling activity, try to start with the simplest model you can develop quickly. Make whatever assumptions you need to get a model "up and running" right away. It's alright if your assumptions don't correspond to reality. We want something working, because we want to extend and evolve our model, rather than creating a final model at the beginning. Most people new to simulation modeling want their model to be close to the final one, even though it is their initial attempt. What they discover is that their model has all kinds of complexities that they have trouble unraveling when they encounter an error. If they started with a simple (maybe completely unrealistic) model that is working, they can embellish it incrementally and adapt it into a final working model with far less difficulty than trying to create the final model at the beginning. Finally, it is not possible to build a model that completely represents the real system in every detail. If that is your plan, then you are better off experimenting with the real system. Remember a model is only an approximation of the real system. How detailed your model is depends on what you want the model to reveal. In general, your model needs only to be as detailed as needed to obtain the performance measures of interest. Any more detail is unnecessary.[95]

A second recommendation is not to worry about the input right away. The critical concern is to get a model that appears most closely to match the system, so you can obtain the needed performance measures reliably. As you are building your model, it will become clear what input data is needed. Let the model inform your data collection, so you don't spend a lot of time collecting the unimportant or wrong data.

---

[95] A problem often arises when a client/boss sees the animated model not conforming exactly like the real system. Some people (naively) equate the "quality" of the model with the quality of the animation. Sometimes you are forced into animation details that have little to do with performance measures.

Step 1: Draw a flowchart of the service process which can come from a "value-stream map".

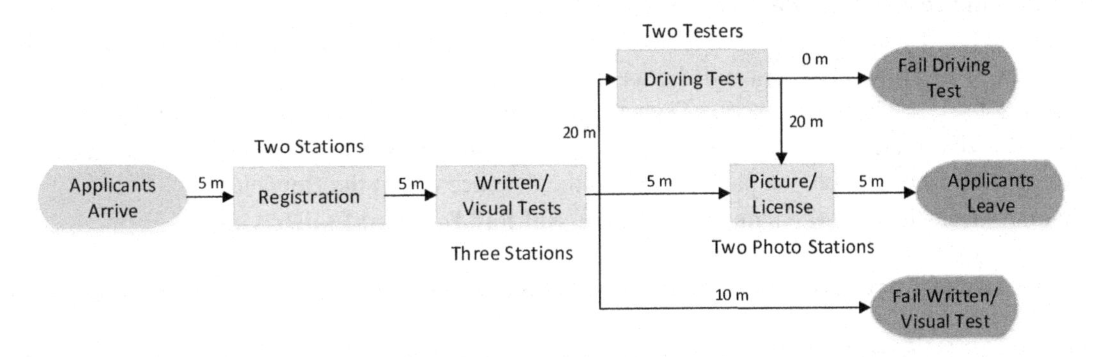

Figure 8.3: DMV Flowchart

Step 2: Since our flowchart in Figure 8.3 corresponds easily to the basic SIMIO objects, it can be used as a basis for the initial model. Create a model with the appropriate objects as seen in Figure 8.4.

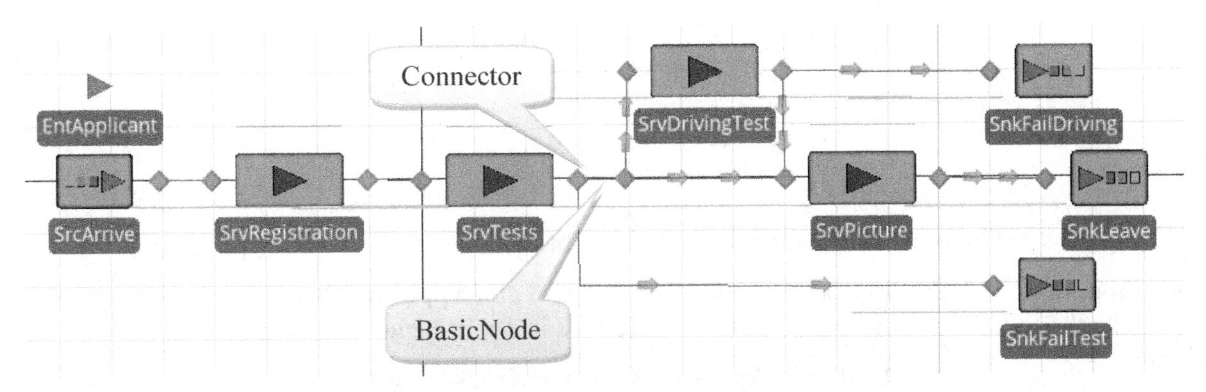

Figure 8.4: Initial DMV Model

- Insert a new ModelEntity named **EntApplicants** with the *Initial Desired Speed* property set to "**1**" meter per second. The applicants arrive randomly, but in a time-varying arrival process Based on information, general estimates of the arrival rates are shown in Table 8.1. From the *Data* tab, insert a new Rate Table named **ArrivalRateTable** that has 15 intervals with an interval size of 30 minutes.

Table 8.1: Hourly Arrival Rates for Applicants

Time	Hourly Arrival Rate
8:30-9:00am	15
9:00-9:30am	10
9:30-10am	8
10:00-10:30am	12
10:30-11:00am	15
11:00-11:30 am	18
11:30-Noon	15
Noon-12:30pm	12
12:30-1:00pm	10
1:00-1:30pm	8
1:30-2:00pm	6
2:00-2:30pm	7
2:30-3:00pm	5
3:00-3:30 pm	5
3:30-4:00pm	5

- Insert a new Source named **SrcArrive** that uses the new arrival table **ArrivalRateTable**.

- The four basic activities will be modeled using SERVERS named **SrvRegistration**, **SrvTests**, **SrvDrivingTest** and **SrvPicture** as seen in Figure 8.4.
- From Figure 8.4, notice the BASICNODE named **BNodePass** that was inserted after **SrvTests** which is not part of the flow chart but will be used to help branch applicants who fail the written or eye test. Use a CONNECTOR between the **SrcTests** SERVER and the BASICNODE so it doesn't take any time. Use a 20 and a five meter path between the BASICNODE and **SrvDrivingTest** and **SrvPicture** respectively. Connect all other objects via paths based on the distances from the flow chart.
- Finally, insert three SINKS named **SnkLeave, SnkFailTest, SnkFailDriving**.[96]

Step 3: **Recall there are three types of applicants (i.e., permits, new licenses, and renewal licenses).**

- In order to keep track of time in system for each type, insert a TALLY Statistic as seen in Figure 8.5. Make sure your TALLY statistics have **Time** as the *Unit Type*.

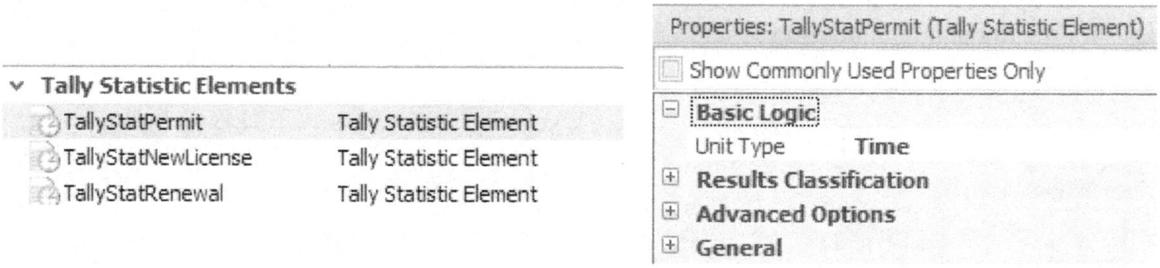

Tally Statistic Elements	
TallyStatPermit	Tally Statistic Element
TallyStatNewLicense	Tally Statistic Element
TallyStatRenewal	Tally Statistic Element

Properties: TallyStatPermit (Tally Statistic Element)
Show Commonly Used Properties Only
Basic Logic
Unit Type Time
Results Classification
Advanced Options
General

Figure 8.5: Creating Tally Statistics to Track Time in System for Each Type of Applicant

- From the *Data* tab, insert a new data table named **TableApplicant** that has six different properties with the following particular characteristics defined in Table 8.2 and seen in Figure 8.6.

Table 8.2: Applicant Characteristics Data Table

Property Name	Property Type	Description
Name	String	The name of applicant type
ApplicantType	Integer	Will be used to set the entity's animation picture
ApplicantMix	Real	The percentage of time this applicant type arrives
RegistrationTime	Expression (*Unit Type* set to "Time" with minutes as the *Default Units*).	The registration time varies by the applicant type
PassTests	Real	Probability of passing varies by applicant type
PassDrivingTest	Real	Probability of passing the driving test
TallyStatTimeInSystem	Tally Statistics Element	Used to track each type's time in system

Table Applicant

	Type	Applicant Type ID	Applicant Mix	RegistrationTime (Minutes)	Pass Tests	Pass Driving Test	Tally Stat TIme In System
1	Permit	0	25	Random.Pert(6,8,10)	90	0	TallyStatPermit
2	NewLicense	1	25	Random.Pert(8,10,14)	95	90	TallyStatNewLicense
▸3	Renewal	2	50	Random.Pert(5,7,8)	98	0	TallyStatRenewal

Figure 8.6: Table Values for the Three Applicant Types.

- Select the **EntApplicant** and add two additional symbols coloring them red and blue respectively.[97]

[96] Note, different SINKS will be used to separate the statistics for those that fail test versus not. From an animation standpoint, one can put the three SINKS on top of one another to demonstrate the applicants leaving the same door so it visually looks correct.

[97] Switch to 3-D mode to color the entire entity widget.

- Next, as **EntApplicants** are created by the **SrcArrive**, assign the applicant type and associated picture using the *Created Entity Add-on Process Trigger as seen* in Figure 8.7.[98]

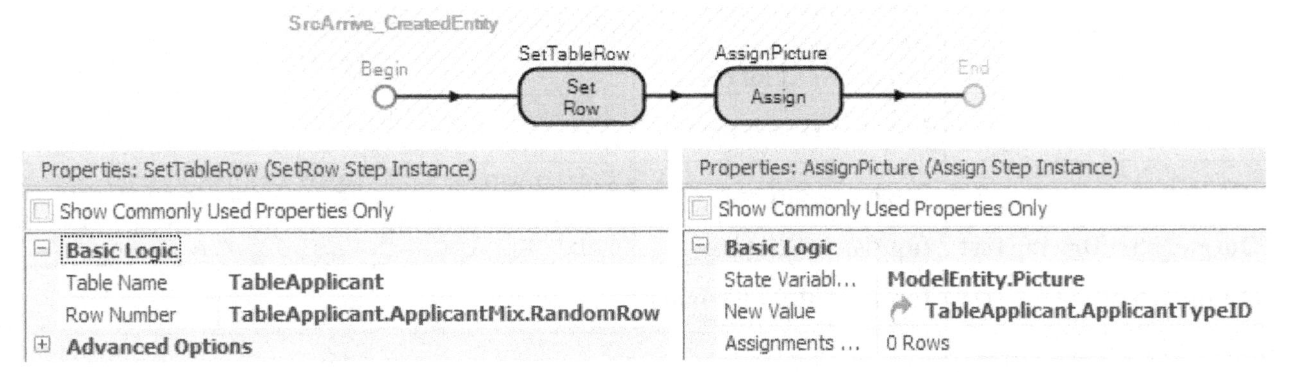

Figure 8.7: Assigning the Applicant Type and Picture

- Once applicants leave the system, tally the time in system through the *Entered Add-on Process Trigger* of the **SnkLeave** as shown in Figure 8.8.

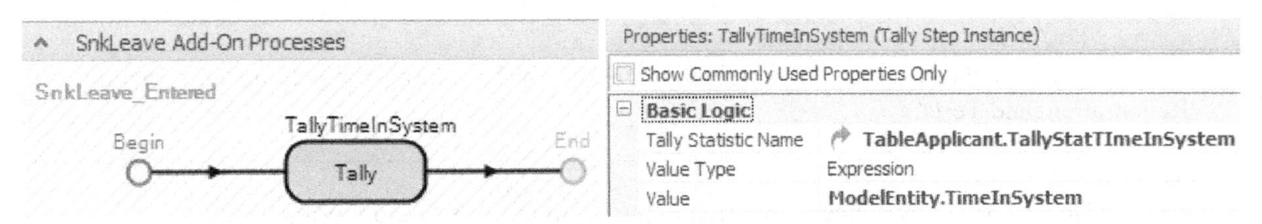

Figure 8.8: Tally Time in System

Question 1: Will these Tally statistics by applicant type include the time in system for those applicants that fail any of the tests (why or why not)?

____

Step 4: Use Table 8.3 to set the processing times and capacities for each of the four servers.

Table 8.3: Processing Time Property Values

SERVER	Processing Time Property Value	Initial Capacity
SrvRegistration	`TableApplicant.RegistationTime`	2
SrvTests	`Pert(10,14,20)` minutes	3
SrvDrivingTest	`Pert(15,25,40)` minutes	2
SrvPicture	`Pert(1,2,3)` minutes	2

Question 2: Why would the Pert distribution be preferred over the Triangular for *Processing Time*?

____

Step 5: If you save and run the model, the applicants will choose the various branches equally likely since the *Selection Weight* properties are all set to one. Recall only new licenses go through the driving test and the passing rate of the driving and written/eye tests are stored in the table. Change the *Selection Weight* properties for the paths so they account for how the applicant types should branch as seen in Table 8.4.

[98] We have a strong preference for processes over the specifications in objects (e.g., the table, state, tally properties), since processes give use the most flexibility in specifying and re-specifying these additions to the model.

Table 8.4: Branching Conditions

From Node	To Node	Selection Weight Property
Output@SrvTests	BASICNODE named **BNodePass**	`TableApplicant.PassTests`
Output@SrvTests	**Input@SnkFailTests**	`100 - TableApplicant.PassTests`
BNodePass	**Input@SrvDrivingTest**	`TableApplicant.ApplicantTypeID == 1`
BNodePass	**Input@SrvPicture**	`TableApplicant.ApplicantTypeID != 1`
Output@SrvDrivingTest	**Input@SrvPicture**	`TableApplicant.PassDrivingTest`
Output@SrvDrivingTest	**Input@SnkFailDriving**	`100 - TableApplicant.PassDrivingTest`

Step 6: Save and run the model, observing it for a while before fast-forwarding to complete one replication.

Question 3: What are the average times in system for New License, Permit, and Renewal applicants?

____

Question 4: What are the observed "ScheduledUtilization" of the stations for Driving Test, Picture, Registration, and Tests?

____

Question 5: Do you have any concerns relative to the behavior of the model or the results?

____

Question 6: Why do we need to be careful about drawing conclusions from a simulation composed of only one replication?

____

Question 7: How many units of capacity need to be unavailable for the resource to occupy the "Offshift(4)" state?

____

Question 8: What bothers you about the model realism, as seen in its animation, based on your experience in a DMV office?

____

Part 8.2: Using Resources with Servers

After building the initial model, it was discovered that the people who register applicants also take the pictures as well as hand out the licenses. These DMV clerks work in the same physical area and share the workload, meaning there are four clerks to perform both functions. However, when we ask about the working environment, we find that only three applicants can be processed at once at the registration and that only two people can have their picture taken at the same time.

Step 1: Change the capacity of the **SrvRegistration** to three, meaning that no more than three entities can be processed at the same time at this server. The **SrvPicture's** capacity will remain two.

Step 2: Insert a RESOURCE object into the model named **ResClerks** with an initial capacity of four which corresponds to the four clerks. A resource object can have its capacity "used" at several servers within a model while server capacity is limited to a specific server.

Question 9: Even though there are four clerks, why can only three be used at the registration process at the same time?

___

Step 3: Next, at both the **SrvRegistration** and **SrvPicture** servers, use their *Secondary Resources*[99] section for "*Resource for Processing*" and specify the use of **ResClerks**[100] as seen in Figure 8.9.

Secondary Resources	
Resource for Processing	
Object Type	Specific
Object Name	**ResClerks**
Selection Goal	Preferred Order
Request Move	None

Figure 8.9: Specifying the use of Clerks at the Server

Step 4: Save and run the model.

Question 10: Does the model now behave as expected?

___

Question 11: Did you notice that the RESOURCE shows its "Busy (1)" state whenever any of its capacity is being used similar to the Processing state of the server?

___

If a server employs resources for service, then the capacity to serve entities may be limited by either the resource capacity availability, the server capacity availability, or both. If only resources limit the capacity, then the server capacity should be made arbitrarily large (i.e., infinity) so as to not constrain the system.

Part 8.3: Handling Failures

Suppose we now discover that the reception area is interrupted by phone calls. These phone calls arrive randomly (i.e., Exponentially distributed) on average every 30 minutes and they take anywhere from 1 to 4 minutes, with 2 minutes being typical.

Both SERVERS and RESOURCES can experience failures as can WORKSTATIONS, COMBINERS, SEPARATORS, VEHICLES, and WORKERS. The failure options are found under the object's "Reliability Logic" property section. In SIMIO, the failure concept is intended to be used for any kind of downtime, exception, or (non-modeled) interruption[101]. There are four different types of failures that can be specified, two of which are based on "counts" while the others are based on time as described in Table 8.5.

___

[99] It is convenient to use the *Secondary Resources* when you don't need the flexibility of the add-on processes. Resources for Processing does both the seizing and releasing of the resources.

[100] By default the capacity requested will only be one. You have more flexibility with the `Seize` step.

[101] An `Interrupt` step can be used to model interrupts.

Table 8.5: Different Types of Failure Types

Failure Type	Description
No Failures	No failure or exception will occur.
Calendar Time Based	The user specifies the time between exceptions or failures. Often used model general reliability or maintenance programs.
Event Count Based	The user specifies that an exception or failure will occur after a particular number of events has happened (e.g., the number of tool changes or certain number of arrivals).
Processing Count Based	The user specifies that an exception or failure will occur after a certain number of operations has occurred the SERVER.
Processing Time Based	The user specifies that an exception or failure will occur after the object has operated a certain amount of time (i.e., after a 100 hours of operation the machine has to go through maintenance).

When an object is interrupted, its entire capacity is interrupted – its state is changed to "Failed(3)". So do we interrupt the service at the server or do we interrupt the resource?

Step 1: To model the phone call interruption, one can use a "Calendar Time Based" type failure. Select the **SrvRegistration** SERVER object and specify the "reliability logic" as shown in Figure 8.10, using the *Uptime Between Failure* and the *Time to Repair* expressions corresponding to the phone calls.

Reliability Logic	
Failure Type	Calendar Time Based
Uptime Between Failures	Random.Exponential(30)
Units	Minutes
Time To Repair	Random.Pert(1,2,4)
Units	Minutes

Figure 8.10: Specifying the Failure

Step 2: Implement this reliability logic in the **SrvRegistration**. Run the model and observe its behavior.

Question 12: Did you see the interruptions, displayed by the sever being colored red?

____

Question 13: If you simulate in fast-forward the simulation ends at 7.5 hours, how many phone calls were handled at the registration desk and what percent of time was the registration desk failed?

____

Question 14: Did the failures of the server cause the secondary resource to fail? What was its state while the server is failed?

____

Question 15: Did the interruptions in the server affect the "TimeProcessing" in the registration?

____

Step 3: In reality, the clerks are the ones that need to answer the phone calls. Take the failure off the server and put it on the resource **ResClerks**. Run the model and observe its behavior.

Question 16: Did you see the interruptions, displayed by the resource being colored red?

____

Step 4: If you simulate in fast-forward the simulation ends in 7.5 hours. Examine the resource statistics for **ResClerks** and you will notice a "TimeFailed" and a "TimeFailedBusy" state statistics. The "Failed(3)" state occurs when the resource is not busy and the "FailedBusy(5)" state occurs when it is busy (at least one unit of capacity is being used). Notice that the colors for both states are the same, so you can't tell in the animation between the two failed states.

Question 17: From the statistics on the resource, how many phone calls were handled by the resource when it was idle and how many phone calls were handled when the resource is busy?

Question 18: Did the interruptions in the resource affect the "TimeProcessing" in the registration?

Question 19: Why do the interruptions of resource cause an increase in the time being processed?

Step 5: Unfortunately, failing the entire server or resource is probably not our best option for modeling the handling of phone calls. Since the phone calls are answered individually, we need to spread them over the clerks. One possibility would be to create entities to represent the phone calls and have them interrupt existing service. However, for our present purposes, let's use the pre-defined reliability logic. However each clerk needs to handle phone calls individually which means we need to have individual resources. Replace our **ResClerks** with four separate RESOURCE objects, each with capacity one (see Figure 8.11). Since phone calls occur about 30 minutes apart (on the average), we will specify the time between phone calls to be 120 minutes for each clerk, using the previous time to repair.[102]

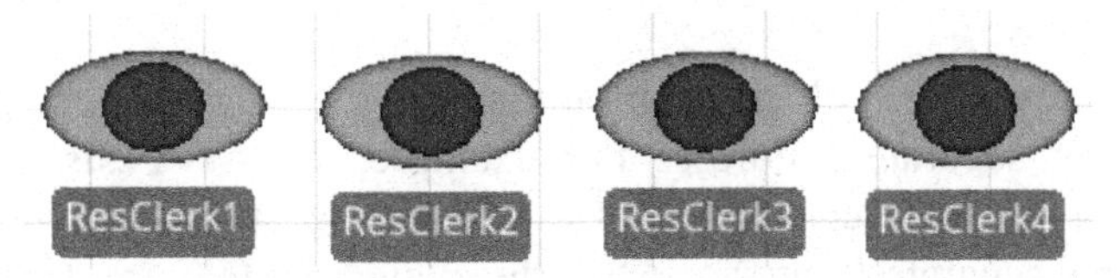

Figure 8.11: Individual Clerks

Step 6: Unfortunately, secondary resources[103] can only seize a single specific object (i.e., the **ResClerks** which had a capacity of four) but now there are four separate RESOURCES that need to be seized. From the "*Definitions*" tab, create an OBJECT LIST called **ListClerks** that contains the four clerk Resources as seen in Figure 8.12. The list will allow us to select one of the clerks based on some criteria.

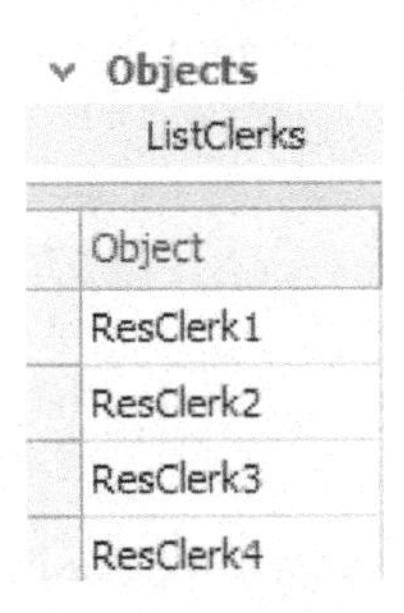

Figure 8.12: Creating an Object List of Resources

Step 7: At both the **SrvRegistration** and **SrvPicture** SERVERS, under the *Secondary Resources* section for *Resource for Processing* choose the "Select From List" option as the *Object Type* property. Specify the **ListClerks** as the *Object List* Name property value as seen in Figure 8.13. Using "Random" as the *Selection Goal* forces SIMIO to uniformly use the clerks at the servers.

---

[102] You can modify the existing resource object so it has capacity one and its *Uptime Between Failures* as Random.Exponential(120) minutes. Then copy the object three times and name them accordingly.

[103] The *Seize* process step has the same issue of being able to seize from a single object.

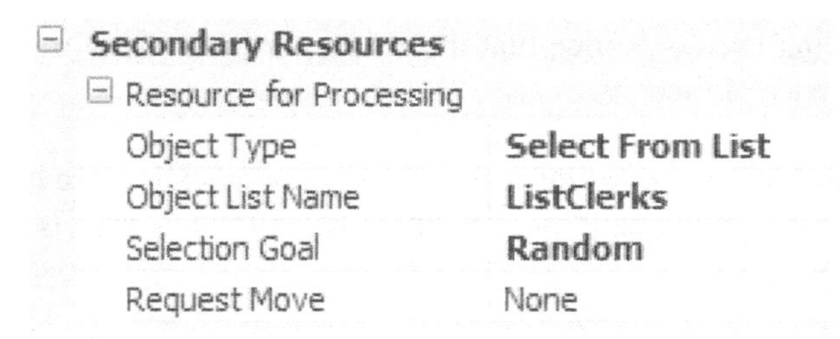

Figure 8.13: Select a Clerk

Step 8: Save and run the model, both looking at the animation and finally looking at the statistics at the end of 7.5 hours. Now four resources are being modeled, so there are individual statistics for each.

Question 20: What is the ScheduledUtilization for each of the clerks?

____

Question 21: How many phone calls did each of the clerks process while they were idle?

____

Question 22: How many phone calls did each clerk process while they were busy?

____

Part 8.4: Server Configuration Alternatives

Sometimes there are alternative desirable configurations for the servers in a model, often motivated by animation needs or handling separate failures. For example, the **SrvTests** object in the current model represents the processing of up to three applicants at the testing stations. Perhaps it would be better to represent this service as three separate SERVERS. Also, the evaluators that are doing the test take a break after processing between eight and twelve applicants, so separate SERVERS will be needed.

Step 1: The evaluators that assess the written and eye tests take a break after processing between eight and twelve applicants. The reliability logic can be used to handle this type of break. Set the *Failure Type* property to "Processing Count Based" to count the number of applicants where *the Count Between Failures* set to a uniform distribution between eight and twelve with the time of the break taking between two and six minutes but generally they take four minutes as seen in Figure 8.14.

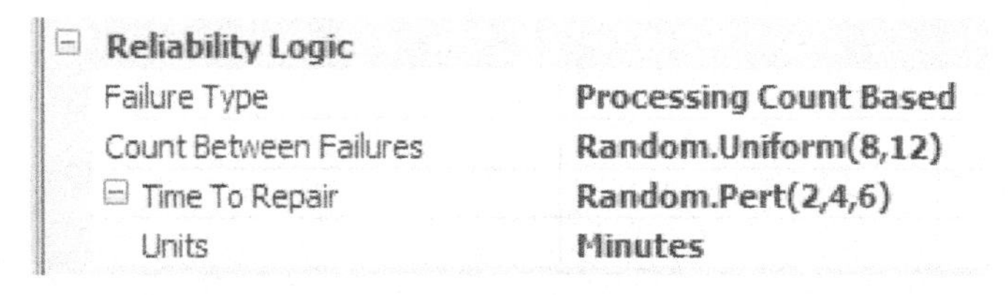

Figure 8.14: Taking a Break Based on the Number Applicants Processed

Step 2: Since the evaluators doing the test will not break at the same time, we need to have individual test stations. Change the name of **SrvTests** to **SrvTest2** and the capacity back to "**1.**" Copy it two times creating a total of three SERVER objects as shown in Figure 8.15. Connect the test stations with the registration using five meter PATHS. We will assume that the branching to each testing station is equally likely (i.e., equal *Selection Weights*). Connect the other two tests with CONNECTORS to the **Ouput@ SrvTest2** which allows us to use the same branching logic to driving test, picture taking, or failure.

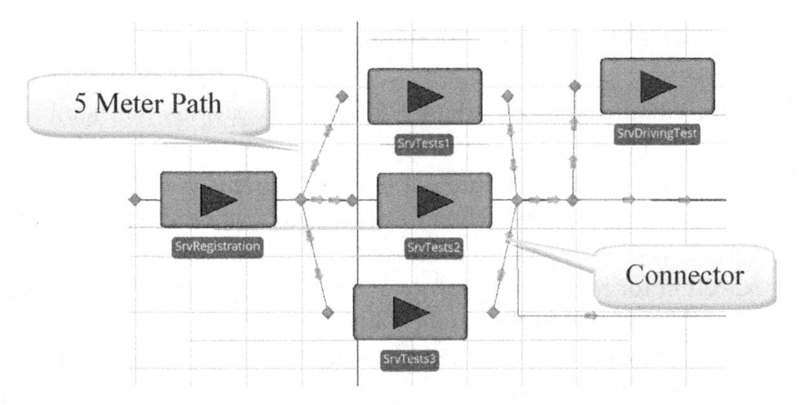

Figure 8.15: Separate Testing Stations

Step 3: Now save and run this model observing the animation.

Question 23: How do the applicants choose which of the three testing stations to enter?

Question 24: Is there anything about the behavior of this model that causes you concern?

Step 4: Typically there is one waiting station after registration where they wait for the next available testing station. To prevent the waiting of applicants at the testing stations, first set the input buffer capacity of each of testing servers to zero under the *Buffer Capacities* section of the SERVER object.

Step 5: Save and rerun the model, observing the animation.

Question 25: Now, how do the applicants wait at the testing stations? Is that a desired behavior?

The applicants now wait at the input node of the testing stations, unable to enter the SERVER's station since the input buffer capacity is zero. Furthermore the entities seem to make a decision about which path to take independent of the traffic on that path, the number at the associated test station, or if the tester is on break.

Step 6: To handle those issues, the branching process needs to be modified to choose an appropriate testing station. From the "*Definitions*" tab, create a NODE LIST called **ListTesters** that contains the input node (i.e., **Input@**) for each of the three testers as seen in Figure 8.16. Having a list will allow SIMIO to select one of the input nodes using some criterion.

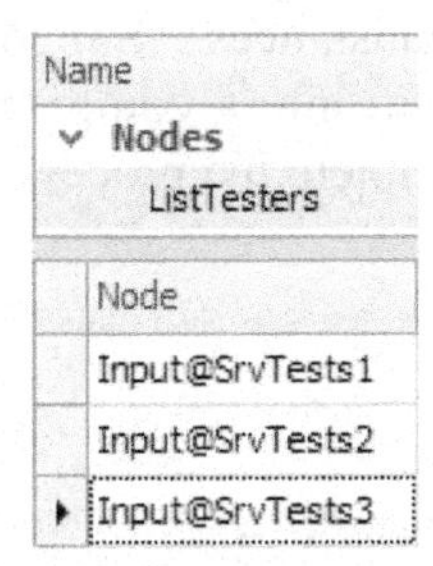

Figure 8.16: Creating a Node List of Testers

Step 7: Select the **Output@SrvRegistration** transfer node that connects the registration to the servers, change the *Entity Destination Type* property to "Select From List" and in the *Node List Name* property select the **ListTesters** list that was just created.

- Set the *Selection Goal* property to be the **Smallest Value** with the *Selection Expression*[104] equal to `Candidate.Node.AssociatedStationload`[105] which chooses the server with the shortest line by determining the number that are routing to the SERVER plus the number in the `Contents` Queues of the `InputBuffer`[106] and `Processing`[107] stations as seen in Figure 8.17.
- If a tester is taking a break, applicants should not be routed to this person while they are on the break. To avoid selecting these nodes, the *Select Condition* property can be used to include only certain items from a list. This expression property has to evaluate to "True" before the CANDIDATE NODE object will be considered for selection for routing. Therefore, we will set the property to select only those servers which have not failed before choosing the smallest line
- The `Failure.Active` function will be "False" if the server is currently not in a failure mode. If the server has not failed, then allow it be a candidate to be selected using the expression `!Candidate.Server.Failure.Active` as seen in Figure 8.17.[108]

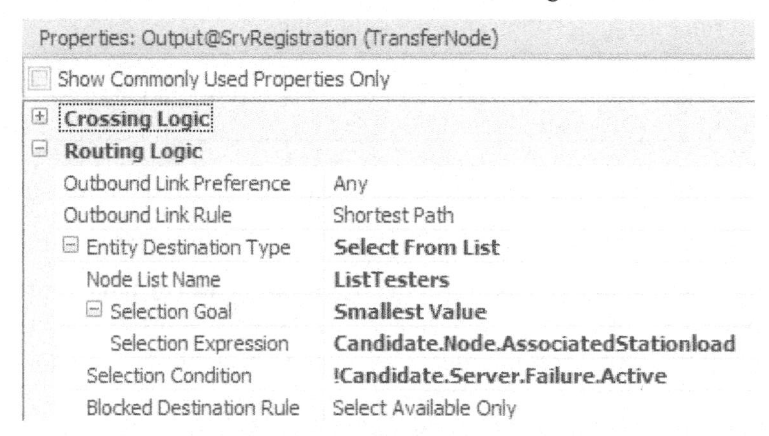

Figure 8.17: Choosing the Shortest Line

Step 8: Save and run the model observing what happens to the entities that leave registration. You may need to increase the time of the breaks to make sure the selection condition is working.

Question 26: Now does the model appear to be behaving as you want?

____

Part 8.5: Restricting Entity to Flow to Parallel Servers

Now applicants are forced to wait in the registration output node until a tester is available (i.e., the SERVER is empty) owing to the traveler capacity paths being one and the input buffer of the servers are zero. In a later chapter we will present a more complete explanation of various blocking methods. Sometimes it is necessary to block flow under somewhat complex circumstances. Suppose that the testers, who give the visual and written exam must enter the results of the exam into a computer. The applicant is allowed to return to the waiting area, but the tester remains busy to complete the process. Obviously, the tester is unavailable to other applicants until the results are entered.

---

[104] Do not confuse the *Selection Expression* property with the *Selection Condition* which is used to include/filter candidate nodes.

[105] Note the CANDIDATE object has to be used to specify a wildcard like '*'. SIMIO will replace the CANDIDATE object with the appropriate object from the list when determining the smallest value. Also, if there is a tie between nodes they will be selected in the order they were placed into the list.

[106] The `Contents` queue of the `InputBuffer Station` is used to hold entities waiting to be processed by the SERVER.

[107] The `Contents` queue of the `Processing Station` represents the location of the entities that are currently being processed by the SERVER.

[108] The ! or Not operator can be used to take the complement of a variable (i.e., ! False would equal True).

Step 1: To keep track of the status of the tester, we will utilize a STATE VARIABLE associated with each tester. From the *Definitions→States* section, insert a new discrete String state variable named **GStaTesterStatus**. Since there are three rooms, make the *Dimension Type* property be a **Vector** and set the number of *Rows* too three with an *Initial State Value* of Available[109] as seen in Figure 8.18.[110]

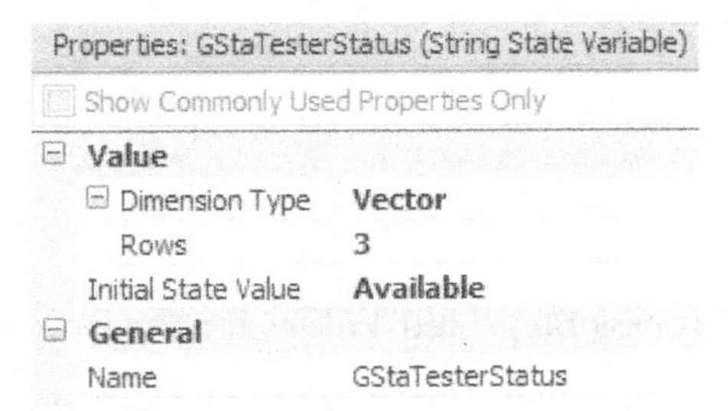

Figure 8.18: Setting a Vector State Variable

Step 2: From the *Animation* tab, insert three status labels beside each of the tester SERVERS with an expression associated with the **GStaTesterStatus** (e.g., `GStaTesterStatus[1]` for the first room as seen in Figure 8.19). Specify all three expressions accordingly. Vectors start with an index of one and each row is specified in brackets.

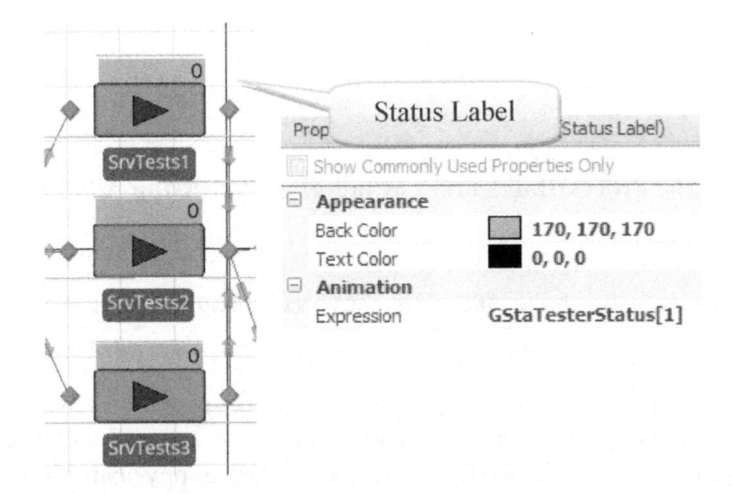

Figure 8.19: Adding Status Labels

Step 3: Once an applicant has been seen by one of the tester, we will need to set the status of the tester to "DataEntry", delay for the five to ten minutes, and then set the status to "Available". We will model this via a specialized process for each room. Navigate to the "*Processes*" Tab, from the *Process* section, use the *Create Process* button to create a new process named **ProcessDataEntry,** as seen in Figure 8.20. Note, this is a case where *State Assignments* cannot be used to model the scenario.

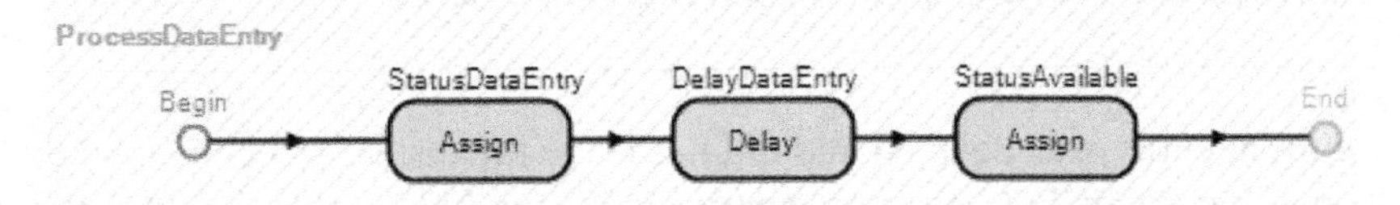

Figure 8.20: Process for Data Entry and Changing the Status of the Tester

- Insert an `Assign` step that will set the first room's status (i.e., `GStaTesterStatus[1]`) to "DataEntry".

---

109 Here "Available" is a string constant (arbitrarily chosen), not a numerical value.

110 Note we could have created three individual state variables (i.e., **GStaTesterStatus1**, **GStaTesterStatus2**, and **GStaTesterStatus3**) instead of the one variable that is a vector of size three.

- Insert a `Delay` step that with a *Delay Time* expression set to `Random.Uniform(3,6).` For a moment we will leave the *Units* set to hours which allow us to see any potential problems.
- Copy the first `Assign` step and change the *Value* to "Available".

Figure 8.21: Process Steps and Values for the ProcessDataEntry

Step 4: The new process needs to be invoked after the applicant has finished processing. In the Facility window, select **SrvTests1** and specify the **ProcessDataEntry1** as the *After Processing* add-on process trigger which is invoked immediately after the entity has finished processing.

Add-On Process Triggers
Run Initialized
Run Ending
Entered
Before Processing
Processing
After Processing ProcessDataEntry1
Exited

Figure 8.22: Using the ProcessDataEntry1 as the *After Processing* Add-On Process Trigger

Step 5: Save and run the model.

Question 27: What happens to the applicant that finishes service and then causes the data entry to happen?

____

Step 6: You should have noticed that the entity does not leave the tester until after the data entry completes which is not what we wanted. The reason this occurs is the MODELENTITY is delayed and not just the tester. The *Exited* add-on process trigger is invoked after the entity has physically left the SERVER. Instead of using the *After Processing*, specify the **ProcessDataEntry** process as the *Exited* add-on process trigger.

Step 7: Save and run the model.

Question 28: What happens to the applicant that finishes service now and causes the data entry to happen?

____

Question 29: What happens to the next applicant waiting for this tester?

____

In the first case, the current entity was delayed and did not leave the tester. When using the *Exited* trigger, the current applicant leaves the room but the next applicant is not blocked, will enter the tester, and begin processing. Also, the utilization of the tester is not correct because they are idle when performing the data entry.

Step 8: Therefore, we need to seize the tester to prevent it from being freed when the applicant leaves the room and release the server once the data entry is finished. Insert a `Seize` and `Release` steps to the **ProcessDataEntry** process as shown Figure 8.23.

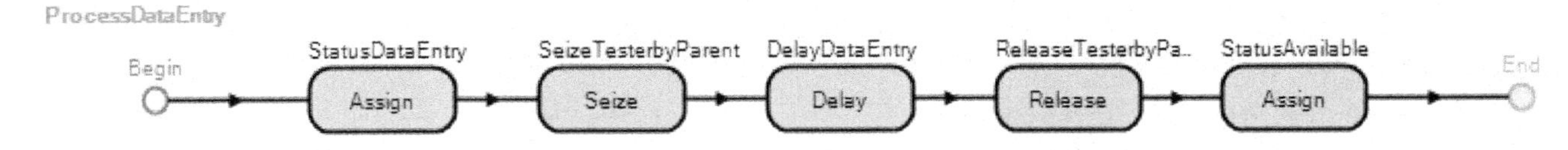

Figure 8.23: Adding a Seize and Release of Tester1

- Because the entity has left the server, it cannot be used to seize the tester. Therefore, we will use the *ParentObject* which is the MODEL in this scenario as the *Owner Type* for both the `Seize` and `Release` process steps as seen in Figure 8.24.

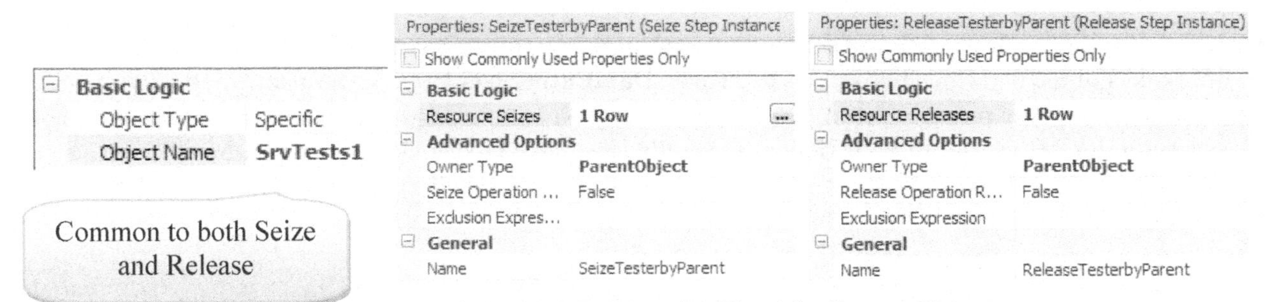

Figure 8.24: Seize and Release SrvTest1 by ParentObject

Step 9: Save and run the model observing whether the **SrvTest1** server does not accept entities now.

Question 30: Does the SERVER accept entities while doing data entry?

Step 10: The entities are still allowed to enter the path to the server even though the server is busy. Therefore, we need to block the server or the path using a different mechanism.[111] Recall, the *Selection Condition* property of the *Routing Logic* section has to evaluate to "True" in order for that particular node to be selected. Currently the condition will not select testers that have failed. Therefore, modify the condition to include not choosing servers that are currently processing (i.e., resource state equal to one) as seen in Figure 8.25.

```
!Candidate.Server.Failure.Active && Candidate.Server.ResourceState != 1
```
[112]

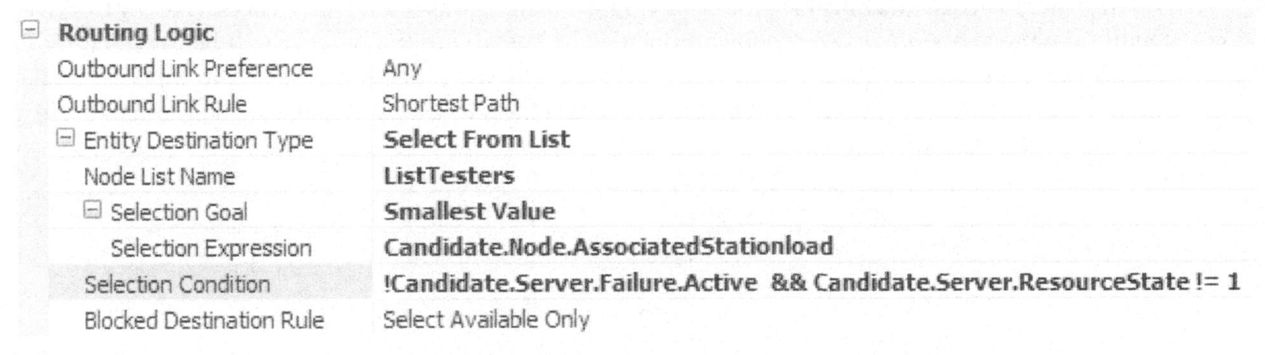

Figure 8.25: Restrict Testers that are Busy from Being Selected

Step 11: Save and run to make sure that the SERVER does not accept entities now.

---

[111] One could shut down the path by changing the capacity or the direction to None or potentially cause the TransferNode Current Travel Capacity state variable to go to zero.

[112] Note, the && is for logical "And" such that we only select servers that have not failed and currently not processing someone.

`Question 31:` So if the server's resource state is one (i.e., "Processing) when doing data entry, how does this expression insure that only idle servers are eligible for selection?

____

Question 32: Does the SERVER **SrvTests1** accept entities while doing data entry now?

____

Step 12: Since the model is working properly now, change the *Units* of time to "Minutes" for the `Delay` step of **ProcessDataEntry**.

Step 13: At this point, we could select the **ProcessDataEntry** process and copy it two times. Then change the names of the two copied processes to **ProcessDataEntry2** and **ProcessDataEntry3,** the two `Assign` steps (i.e., the *Row* property set to two or three), the `Seize` step, and the `Release` step to specify the correct tester. However, to do this more generally (i.e., have the same process used by all three testers), we will use a "Tokenized Process." This approach will be especially useful if additional testing stations are added later.

- From the *Definitions* tab, define a new TOKEN named **TknTester.** Add an INTEGER STATE variable name **TStaWhichRowNumber** and from the *Object Reference* section add an OBJECT REFERENCE STATE variable named **TStaWhichTester** as seen in Figure 8.26.

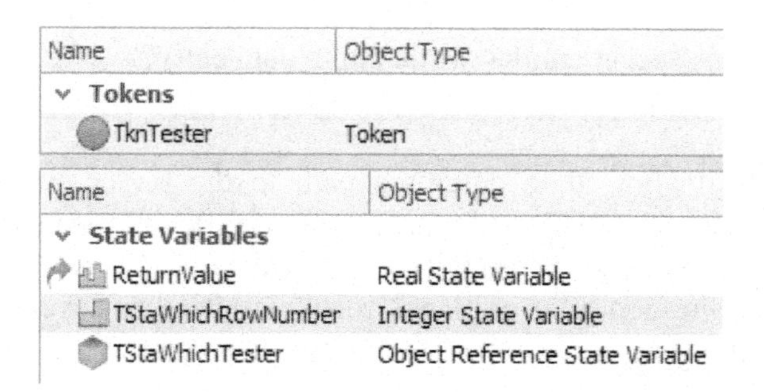

Figure 8.26: New Token TnkTester to be used in a Tokenized Process

- Modify the **ProcessDataEntry1** process by first changing its name to **ProcessDataEntry**. Next modify the process's *Advanced Options* properties by specifying the process will be executed by the **TknTester** TOKEN as shown in Figure 8.27. Next, specify the process will take two input arguments (i.e., WhichRowNumber linked to **TknTester.TStaWhichRowNumber** and WhichTester linked to **TnkTexter.TStaWhichTester**).

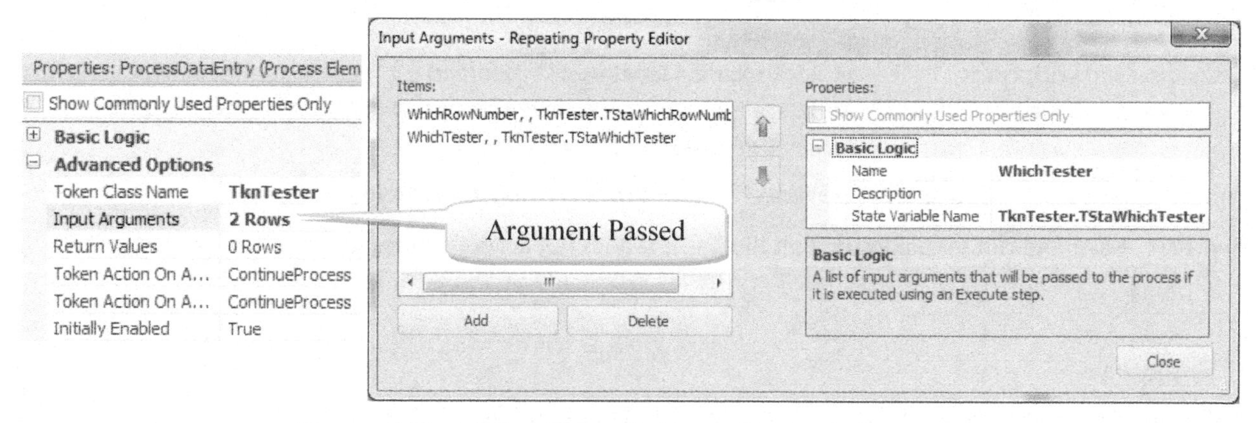

Figure 8.27: Adding Token Information to the Process

Step 14: Next change the two `Assign` steps within the process to employ the **TStaWhichRowNumber** as the *Row* property to update the **GStaTesterStatus** passed to the TOKEN **TknTester** as seen in Figure 8.28.

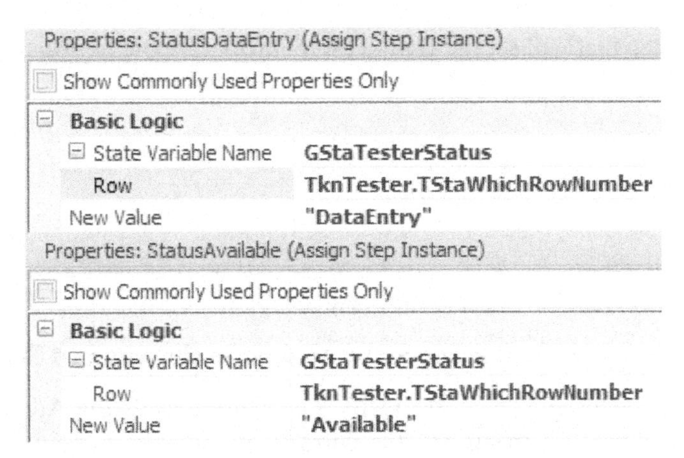

Figure 8.28: Use the Token Argument

Step 15: Update the `Seize` and `Release` steps to use **TStaWhichTester** as the specific *Object Name*.

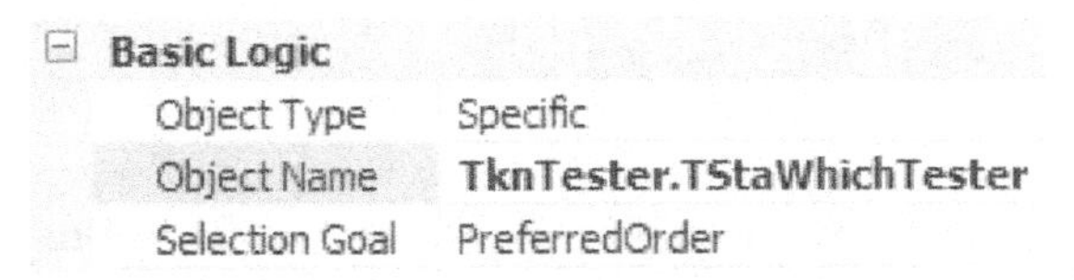

Figure 8.29: Using the TknTester to Specify Which Server to Seize and Release

Step 16: From the facility window, select **SrvTests1** and set the *Exited* add-on process trigger as seen in Figure 8.33 to pass in the correct row number and tester. Repeat for the other two testers setting the row number and which tester to the correct values (i.e., "2" and "3" and **SrvTests2** and **SrvTests3** respectively).

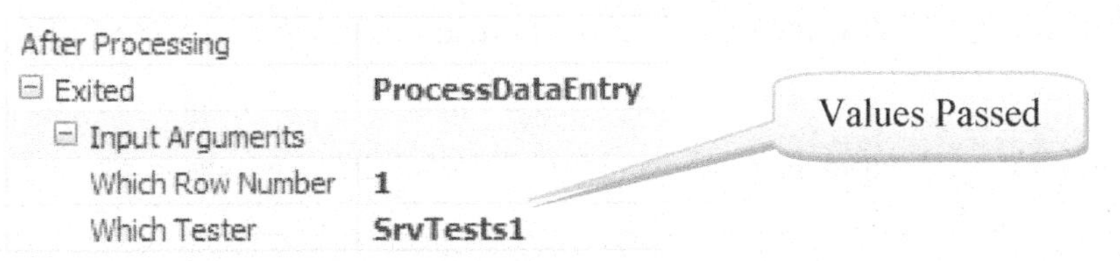

Figure 8.30: Invoking the Exited Process with its Argument

Step 17: The status labels currently change from "Available" to "Data Entry" but remain "Available" while the SERVER is processing an applicant. To fix this issue, we need to update the status to busy when the tester is servicing an applicant.

- Since it will be similar for all three testers, add another "Tokenized Process" named **ChangeStatustoBusy** as seen in Figure 8.31. Next use the Token **TknTester** as the *Token Class Name* property and insert one input argument (i.e., the row number).

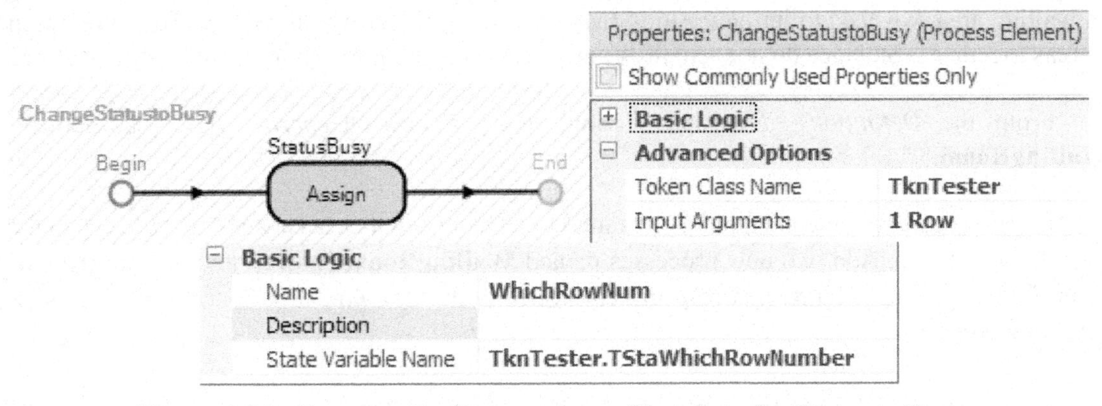

Figure 8.31: Creating the Process that will update the Status to Busy

- Insert an `Assign` step that will set the **GStaTesterStatus** variable to "Busy" as seen in Figure 8.32.

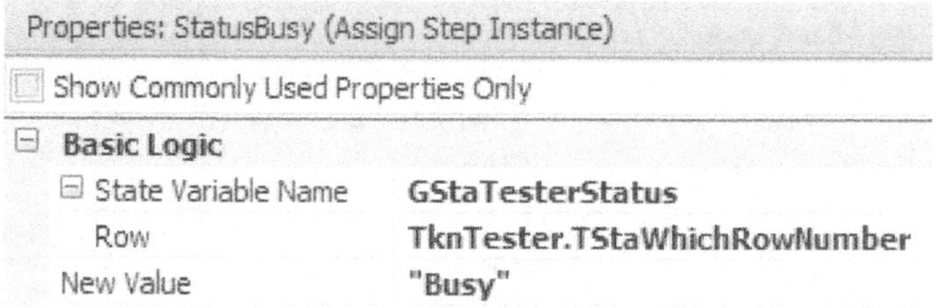

Properties: StatusBusy (Assign Step Instance)

Show Commonly Used Properties Only

Basic Logic	
State Variable Name	GStaTesterStatus
Row	TknTester.TStaWhichRowNumber
New Value	"Busy"

Figure 8.32: Setting the Status to Busy in the `Assign` Step

Step 18: For each of the three tester SERVERS, specify the *Before Processing* Add-on Process triggers to use the **ChangeStatustoBusy** process specifying the correct row number as seen in Figure 8.33 for the first tester.

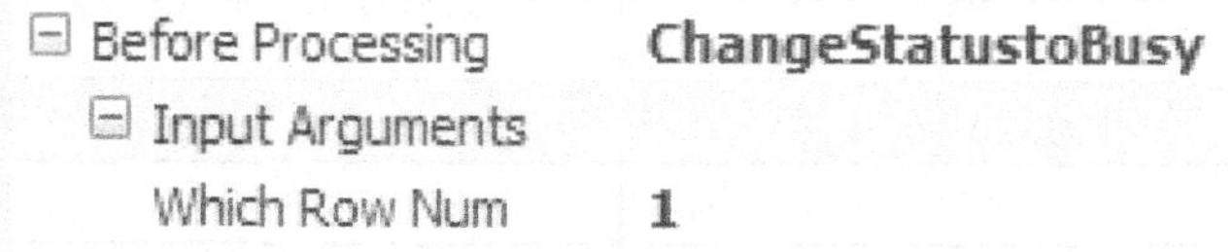

Before Processing	ChangeStatustoBusy
Input Arguments	
Which Row Num	1

Figure 8.33: Changing the Room Status to Busy

Step 19: Save and run the model making sure the status labels turn to "Busy" and that the model is working properly.

Question 33: What is the average utilization of each server?

____

Question 34: What is the total time an applicant has to wait for a tester?

____

Question 35: What is the total cycle time for each license type?

____

Question 36: Are there any drawbacks of this new model?

____

Part 8.6: The Waiting Room Size

In the real system people wait for services in a common waiting area, except for those waiting for registration which queue in front of the registration. This common area consists of waiting for the tester, waiting for the driving test, and waiting for the picture/license. Looking at each waiting area separately ignores the fact that contributions to the waiting area changes throughout the day. To visually see what is happening in the common waiting area, we need to introduce the SIMIO STORAGE object which will be used to represent all the waiting areas together. (Storages do not actually contain entities but entities hold membership in storages.)

Step 1: From the *Definitions→Elements* section, insert a new "*General*" STORAGE element named **StoWaitingRoom**.

Step 2: Entries (i.e., entity memberships) are added to STORAGE elements via an `Insert` step and removed via the `Remove` step. Add two new processes named **WaitingRoomEntered** and **WaitingRoomExite** as seen in Figure 8.34 and Figure 8.35 respectively. Specify the `StoWaitingRoom.Queue` as the *Queue State Name* property.

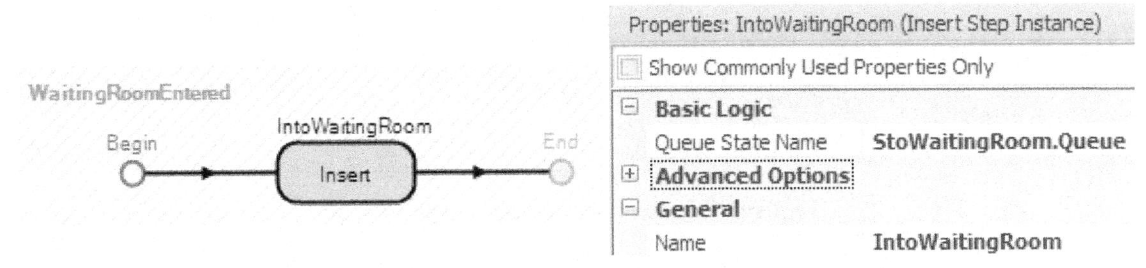

Figure 8.34: Inserting into the WaitingRoom

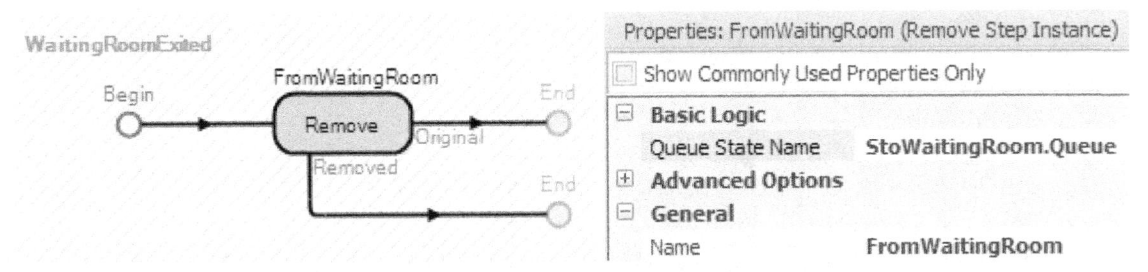

Figure 8.35: Removing from the WaitingRoom

Step 3: People enter the common waiting room when they leave the registration desk or when they are waiting at the driving test or the picture area.

- For the *After Processing* add-on process trigger of the **SrvRegistration** specify the **WaitingRoomEntered** process.
- Select both the **SrvDrivingTest** and **SrvPicture** and then specify the **WaitingRoomEntered** process as the *Entered* add-on process trigger.

Step 4: People leave the waiting room when they enter service at the testers, driving test or picture area.

- Select both the **SrvDrivingTest** and **SrvPicture** and then specify the **WaitingRoomExited** process as the *Before Processing* add-on process trigger.
- For the three testers (i.e., **SrvTests1**, **SrvTests2**, and **SrvTests3**), we will need to modify the **ChangeStatustoBusy** process since it is already specified as the *Before Processing* add-on process trigger. Insert an `Execute` step that with the *Process Name* specified as **WaitingRoomExited.**

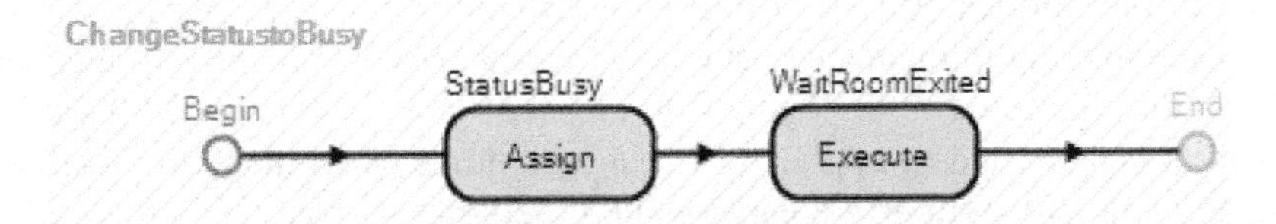

Figure 8.36: Using Execute Step to Cause Applicants to Leave the Common Area

Step 5: To "see" the waiting room, insert a ***DetachedQueue*** from the ***Animation*** tab. The ***Queue State*** property for the queue is the `StoWaitingRoom.Queue`. Draw a straight line for the queue.

Step 6: Most waiting rooms have people along the edge of a room rather than a straight line.[113]

- Select the animated **WaitingRoom** queue which by default is the green straight line.
- Change the *Alignment* to *Oriented Point* and then select the *Keep in Place* option since entities once they sit will not keep changing seats along the queue like a checkout line does as seen in Figure 8.37.
- From the *Vertices* section, add two vertices by selecting the *Add Vertex* button and dragging the green vertex (◯) onto the queue.

---

[113] From an animation standpoint, we would delete the input buffer queues of the testing stations, driving test, and picture area.

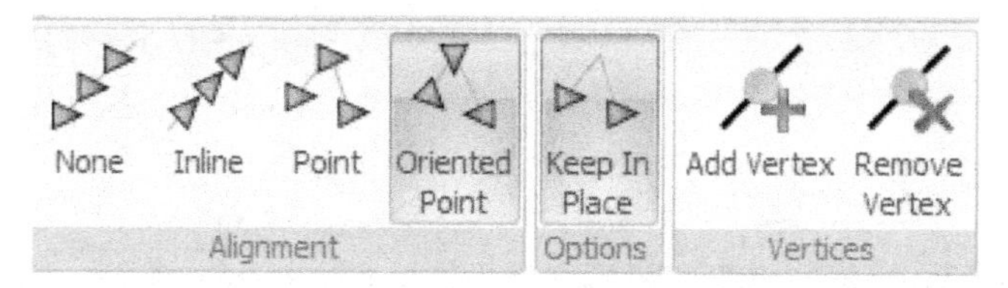

Figure 8.37: Adding an Animated Queue

- Form a U-shaped queue by moving the two new vertex points that were just added to the appropriate spots. Then continue adding vertexes and then moving the points inward to form the waiting room queue as seen in Figure 8.38.

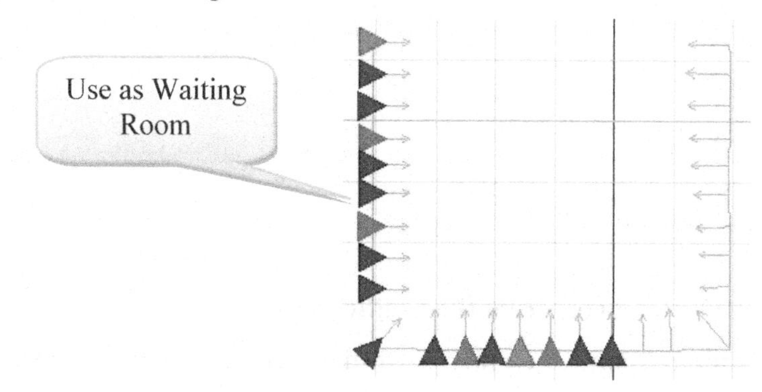

Figure 8.38: Adding an Animated Queue in a Different Shape

Step 7: Save and Run the model.

Question 37: What is the average number and the maximum number in the waiting room (storage)?

____

Question 38: How much time, on the average, is spent in that waiting area?

____

Part 8.7: Using Appointment Schedules

In response to applicant waiting concerns, the office has decided to offer a few scheduled appointments to provide better service to the applicants. The office will still allow most people to "walk-in" but will give priority to those who have an appointment. In particular, the office will schedule applicants for the following nine appointment times during the day: 8:30 am, 9:00 am, 9:30 am, 10:00 am, 2:00 pm, 2:30 pm, 3:00 pm, and 3:30 pm. The appointments schedule applicants during the times when there are lower arrival rates. Some applicants who have appointments will arrive early while others may arrive late which are called "early/late" arrivals. Also, it is expected that some may skip their appointment all together (i.e., "no-shows").

You need to realize that a "scheduled arrival process" is fundamentally different from the usual "random" arrival process in that the random arrival process has random interarrival times while the scheduled arrival process has known interarrival times. Randomness in the scheduled arrival process occurs in the early/late arrivals and the no-shows. In many "service industries" scheduled arrival processes are more prevalent than random arrival processes.

Step 1: The scheduled appointment times are specified via a data table. From the *Data* tab, add a new data table named **TableArrivals**. Insert a *Date Time* column named **ScheduledTime** from the *Standard Property* drop down, as seen in Figure 8.39. Next specify the scheduled appointments which indicate just the time of arrival.[114] At this point, these appointments are absolute arrival times (i.e., if the simulation

[114] Note, you can build the table in Excel as it can be easier and then copy the table into SIMIO.

was to start on 9/5/2015 instead of 9/6/2015 no arrivals would occur until the next day or if the simulation start date was 9/7/2015 then no arrivals would occur). See the *Commentary* section at the end of the chapter on specifying a table with relative time offsets.

Table Applicant | Table Arrivals

	Scheduled TIme
1	9/6/2015 8:30:00 AM
2	9/6/2015 9:00:00 AM
3	9/6/2015 9:30:00 AM
4	9/6/2015 10:00:00 AM
5	9/6/2015 2:00:00 PM
6	9/6/2015 2:30:00 PM
7	9/6/2015 3:00:00 PM
8	9/6/2015 3:30:00 PM

Figure 8.39: Specifying the Appointments

Step 2: Since we are assuming one person will arrive at the scheduled appointments, we will reduce the arrival rate table by one for the first and last four time periods so comparisons can be made as seen in the partial table.

Table 8.6: Hourly Arrival Rates for Applicants

Time	Hourly Arrival Rate
8:30-9:00am	14
9:00-9:30am	9
9:30-10:00am	7
10:00-10:30am	11
Middle Times Unaffected	
2:00-2:30pm	6
2:30-3:00pm	4
3:00-3:30 pm	4
3:30-4:00pm	4

Step 3: From the *Facility* tab, insert a new SOURCE object named **SrcScheduledArrivals** as seen in Figure 8.40. Next, connect it to the output node of the original SOURCE via a CONNECTOR which allows arrivals to originate from one point and travel along the same five meter path to the reception area.

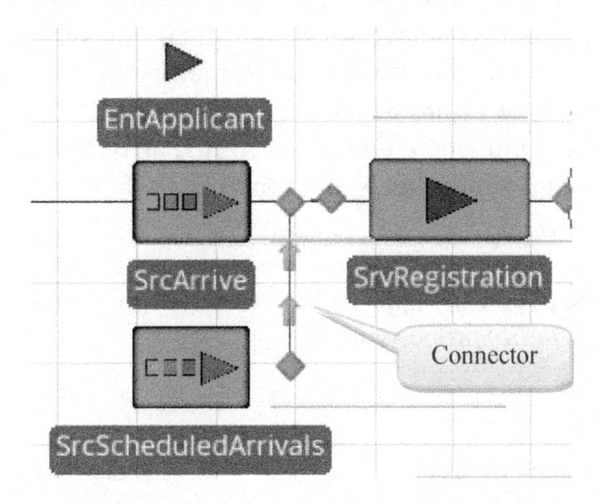

Figure 8.40: Adding an additional Source to Handle the Scheduled Appointments

Step 4: To utilize appointment schedules, change the *Arrival Mode* to "**Arrival Table**" and specify the *Arrival Time Property* to be the column **TableArrivals.ScheduledTime** as seen in Figure 8.41.

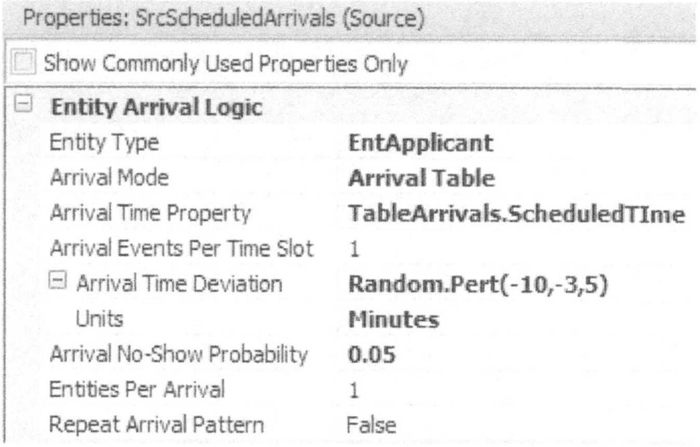

Properties: SrcScheduledArrivals (Source)

Show Commonly Used Properties Only

Entity Arrival Logic	
Entity Type	**EntApplicant**
Arrival Mode	**Arrival Table**
Arrival Time Property	**TableArrivals.ScheduledTIme**
Arrival Events Per Time Slot	1
Arrival Time Deviation	**Random.Pert(-10,-3,5)**
Units	**Minutes**
Arrival No-Show Probability	**0.05**
Entities Per Arrival	1
Repeat Arrival Pattern	False

Figure 8.41: Specifying the Appointment Scheduling in the SOURCE

Step 5: The *Arrival Time Deviation* property is used to model the deviation from the appointed scheduled time specified by the **TableArrivals.ScheduledTime**. In this example, we are stating that applicants can arrive as early as ten minutes or as late as five minutes but most of them will arrive three minutes early, using a Pert distribution[115]. See Table 8.7 for an explanation of all the appointment schedule properties.

Table 8.7: Appointment Scheduled Properties

Appointment Scheduled Property	Description
Arrival Time Property	The numeric table property that specifies the list of scheduled arrival times which can be a Date Time or any numeric or expression property.
Arrival Events Per Time Slot	This is the expected number of arrival events that will occur at each arrival time. You can think of this of how many batches will arrive where the Entities Per Arrival will determine the number of Entities per batch that will arrive.
Arrival Time Deviation	Specifies the deviation from the scheduled time which affects each batch differently. Therefore, two Arrival Events Per Time Slot and 1 Entities Per Arrival will be different than one Arrival Events Per Time Slot and 2 Entities Per Arrival when a deviation or no-show probability is specified as in the latter case the two entities will always arrive at the same time.
Arrival No-Show Probability	Specify the probability of a no-show occurring which is applied to a batch arrival.
Entities Per Arrival	The number of entities to create for each batch.
Repeat Arrival Pattern	When you reach the end of the table, determines whether or not to repeat the pattern. If you have specified actual times these are converted to deviations or time offsets. For this example, the first arrival is at 12 PM with the simulation starting at 11 am which means there is a 1 hour time offset till the first arrival. So at 4 pm when the last scheduled arrival occurs, the simulation will start back up with arrivals at 5 pm to have the same offset if this property is set to "True."

Step 6: In this case we assume 5% of the applicants are "no-shows" and skip their appointment. Therefore, set the *Arrival No-Show Probability* property to 0.05.

[115] Negative times mean the entity is early for their appointment, whereas positive number implies they are late.

Step 7: In the **SrcScheduledArrivals** source reference the process **SrcArrive_CreatedEntity** in the *Created Entity* add-on process trigger. Also, in order to distinguish the scheduled arrivals assign the `ModelEntity.Priority` a value "2" in the *Before Exiting* of the *"State Assignments"* section. Recall the default *Priority* state variable has a value of 1.[116] Refer to Figure 8.42.

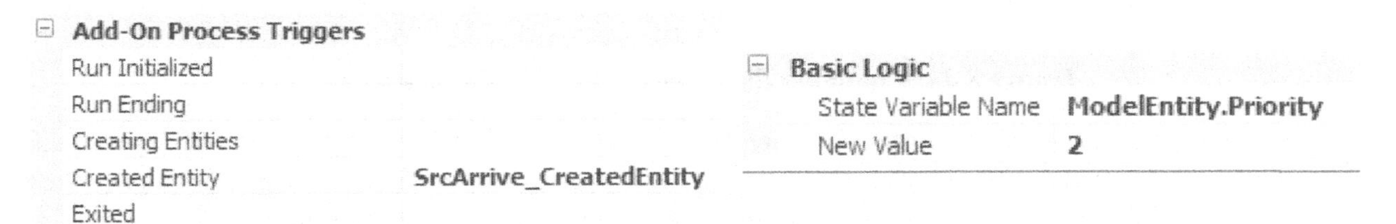

Figure 8.42: Specifying Properties of the SrcScheduledArrivals Source

Step 8: We would like to keep track of time in the system according to whether the applicant had a scheduled appointment or they walked-in. From the *Definitions→Elements* section, add two TALLY STATISTICS named **TallyStatSchedTimeInSystem** and **TallyStatWalkInTimeInSystem** with the *Unit Type* property set to "Time".

Step 9: To obtain the time in system statistics for the two types of arrivals, each needs to be tallied. One way to obtain the tallies is to use the *Tally Statistics* section of the **Input@SnkLeave** specifying ***On Exited*** as shown in Figure 8.43.

Basic Logic	
Tally If	Custom Condition
Condition	ModelEntity.Priority == 1
Tally Statistic Name	TallyStatWalkInTimeInSystem
Value Type	Expression
Value	ModelEntity.TimeInSystem

Basic Logic	
Tally If	Custom Condition
Condition	ModelEntity.Priority == 2
Tally Statistic Name	TallyStatSchedTimeinSystem
Value Type	Expression
Value	ModelEntity.TimeInSystem

Figure 8.43: Collecting Time in System by Arrival Type

Step 10: Select the experiment and add a new response for each of the TALLY STATISTICS named **WalkInTimeInSystem** with an *Expression* set to `TallyStatWalkInTimeInSystem.Average` and **ScheduledTimeInSystem** with `TallyStatSchedTimeInSystem.Average`. Both should have a *Unit Type* as **Time** with *Display Units* as **Minutes**.

Step 11: Save the current model. Reset the experiment and run for 100 replications.

Question 39: What is the average time in the system and half width for each type?

___

Question 40: Is there any statistical difference in the times in the system for the applicant types?

___

Step 12: One of the potential issues is that scheduled patients are not processed differently in the model. Applicants who have scheduled appointments should have a higher priority than walk-in applicants.

- Therefore, select **SrvRegistration**, **SrvDrivingTest**, and **SrvPicture** SERVERS in the model and change their *Ranking Rule* property so that scheduled appointments have a larger priority as seen in Figure 8.44.

[116] The `Priority` assignment could be done within the process **SrcArrive_CreatedEntity.**

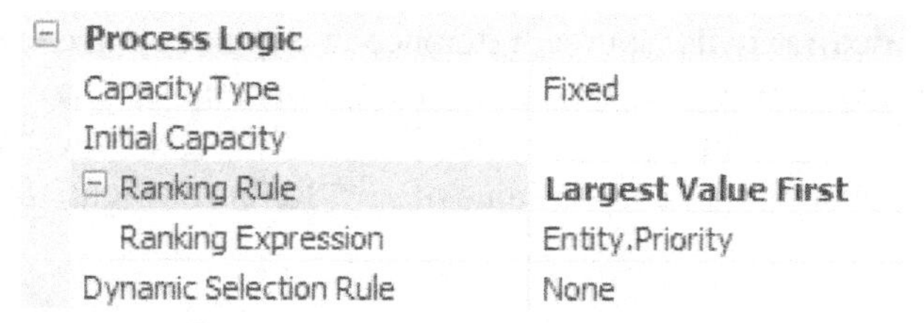

Figure 8.44: Changing the Ranking Rule in the SERVERS

- For the testing SERVERS, the expression has to be a little more complicated because both entities and the MODEL seize capacity of server. We need to give priority to the MODEL for the data entry process using the following expression seen in Figure 8.45. The `Is.Entity` function will return true if an applicant is requesting service and false otherwise.

Process Logic	
Capacity Type	Fixed
Initial Capacity	1
Ranking Rule	**Largest Value First**
Ranking Expression	**Math.If(Is.Entity, Entity.Priority, 3)**

Figure 8.45: Setting Priority Based on Entity and Model

Step 13: Save the model and rerun the experiment.

Question 41: What is the average time in the system and half width for each arrival type?

____

Question 42: Is there an improvement for scheduled appointments?

____

Step 14: There does not seem to be any difference in time when we prioritized the applicants at each of the servers. To see if the model is working as predicted, let's add a dynamic label to the entities to display the priority value. Select the **EntApplication** MODELENTITY and set the *Dynamic Label Text* property under the *Animation* section to display the priority as seen in Figure 8.46.

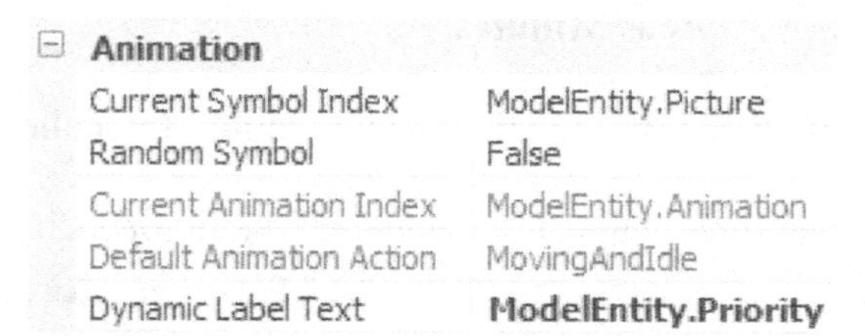

Figure 8.46: Adding a Dynamic Label that Moves with the Entity

Step 15: Run the model and observe the output buffer of the **SrvRegistration** around the 9:30 am time frame.

Question 43: Are all the scheduled applicants with priority two taken ahead of priority ones?

____

Step 16: The problem is there are three servers which have their own separate sorted queues when we really want one queue feeding those three servers to be sorted. The issue doesn't happen at the other areas as the entities are all in the same queue and sorted properly. To fix the issue, insert a RESOURCE named **ResTests** with a capacity of three and *Ranking Rule* set accordingly to act as this single queue. Also, select the **ResTests** and add the ALLOCATION QUEUE from the *Draw Queue* in the *Attached Animation* section.

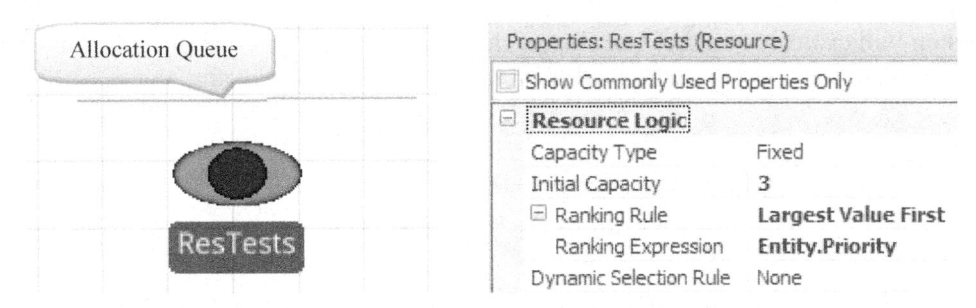

Figure 8.47: Adding a Resource to Act as Single Queue Feeding Three Servers

Step 17: Next, we need to seize the new **ResTests** once the applicants enter the output node of the **SrvRegistration** as they try to get service at testing. Select the output node of **SrvRegistration** (i.e., **Output@SrvRegistration**) and add an *Entered* Add-On Process Trigger as seen in Figure 8.48. The **ResTests** will determine which applicant to serve next and allow the applicant to move on to the testers.

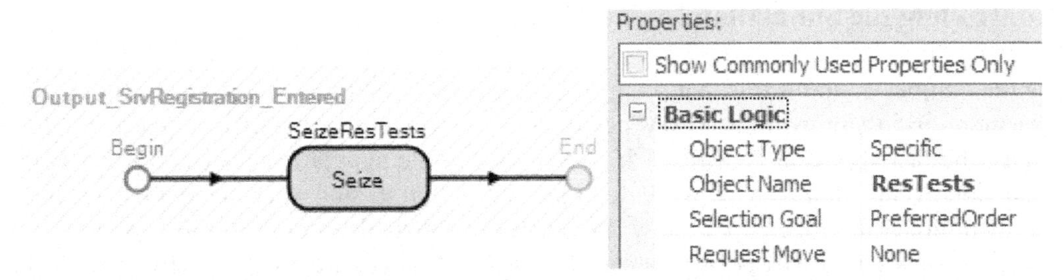

Figure 8.48: Seizing the ResTests in the Output Node of the SrvRegistration

Step 18: Once the entity finishes the testing, we need to release the **RestTests**. Insert a `Release` step in the **ProcessDataEntry** right after the status is updated as seen in Figure 8.49.

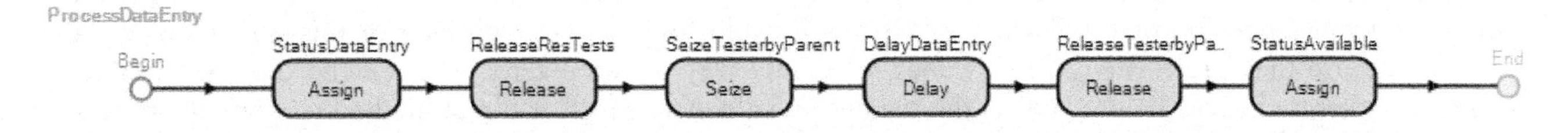

Figure 8.49: Releasing the ResTests **After Finishing Testing**

Step 19: Save and run the model observing whether or not the applicants are being ranked correctly now.

Step 20: Next, rerun the experiment.

Question 44: What is the average time in the system and half width for each arrival type?

___

Question 45: Is there an improvement for scheduled appointments?

___

Step 21: The applicants are now being processed correctly now. However, on further reflection, if a walk-in applicant has been in the system for more than 30 minutes then they should be given greater consideration/priority. Because the time in system changes while the applicant is waiting in the input buffer, the static *Ranking Rule* property cannot be used as it orders the queue as the entities enter the queue and is not reordered later. On the other hand, the *Dynamic Selection Rule* is evaluated each time the SERVER capacity becomes available in order to choose the next applicant. Select the **SrvRegistration**, **SrvDrivingTest**, and **SrvPicture** SERVER objects and specify the "**Largest Value First**" as the rule as seen in Figure 8.50 with the *Value Expression* equal to the following.

```
(Candidate.Entity.Priority==2)||(Candidate.Entity.TimeInSystem>MaxWaitTIme)
```

The expression will evaluate to either one (i.e., if the entity is a scheduled applicant or the entity has been in the system ¾ hour) or zero otherwise.[117] Repeat for the **ResTests** RESOURCE object.

Ranking Rule — First In First Out
⊟ Dynamic Selection Rule — **Largest Value First**
Value Expression — **(Candidate.Entity.Priority==2) || (Candidate.Entity.TimeInSystem>0.5)**
Filter Expression

Figure 8.50: Specifying the Selection Rule to Prioritize Scheduled Applicants and Walk-ins

Step 22: Save the model and rerun the experiment.

Question 46: What is the average time in the system and half width for each arrival type?

Part 8.8: Controlling the Simulation Replication Length

Of concern is the number of applicants that seem to be in the system at the end of the 7.5 hours that the DMV is open. After more discussion with the DMV personnel, you realize that the office closes at 4:30 pm, but the people stay until the last applicant is served. It is easy to run the simulation for exactly 7.5 hours using the "*Run Setup*" section of the "*Run*" tab. But we need the simulation to stop after the last person is served. This concern raises two modeling issues. First, how do we stop the arrivals after 4:30 pm and second, how do we stop the simulation run after the last person has left the system? There are a variety of ways to stop the arrival of applicants after 7.5 hours. For example, we could transfer them to a sink if the current time is after 4:30 pm (i.e., using selection weights on the paths). Another way would be to create a MONITOR that would cause a process to alter the interarrival time or the entities per arrival.

Note, the current **ArrivalRateTable** only has 15 time periods representing the 7.5 hours. If we extend the simulation run length beyond 7.5 hours, SIMIO will cycle back to the beginning of the table repeating the values because they are relative offset values. The only way to force zeros is to extend the arrival table to contain 24 items (i.e., 12 hours) and set the remaining values to zero. A direct method to stop the arrivals is to use the "*Stopping Conditions*" within the SOURCE, which turns off arrivals based on either specifying a maximum number of arrivals[118], some maximum time, or some stoppage event.

Step 1: Select the **SrcArrive** SOURCE, specify the *Maximum Time* property a value of 7.5 under the *Stopping Conditions* section.

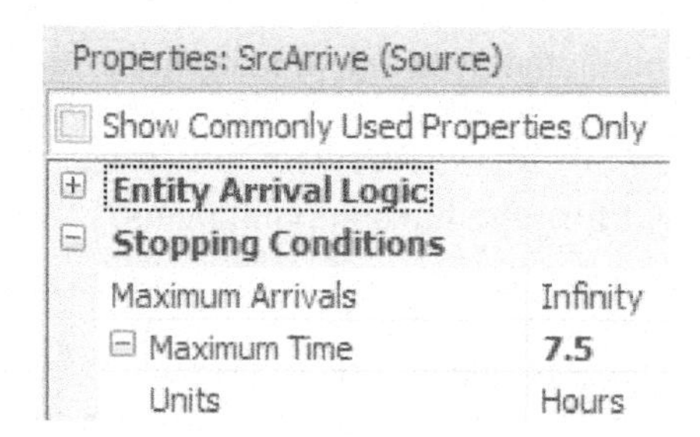

Figure 8.51: Stopping Arrivals after 4:30

Step 2: Set the run length to 12 hours and then run model. You will notice that the applicants stop arriving at 4:30 pm but the simulation continues displaying services (i.e., failures). A simulation run can be

---

[117] If there is a tie (i.e., multiple entities have the same value), they are ordered based on First in First out rule.

[118] Note that this is the number of total "arrival events" and not the total number of entities arriving. One arrival event may yield more than one entity arriving.

stopped using the `End Run` step. However this step needs to be invoked under two conditions: (1) at 4:30 pm, in case no one is in the system, and (2) the more likely case when the last applicant is served after 4:30 pm.

Step 3: Let's first consider the case that the last applicant finishes after 4:30 pm, when no more arrivals are permitted. Applicants exit the system through one of the three sink objects. So at each of the sinks, we need to test whether this entity is the last applicant in the system and that it is after 7.5 hours. We will first create a general process.

- Go to the "*Processes*" tab and click on the "*Create Process*" button, naming the new process **TimeToStop**. Add a `Decide` and `End Run` step as in Figure 8.52.

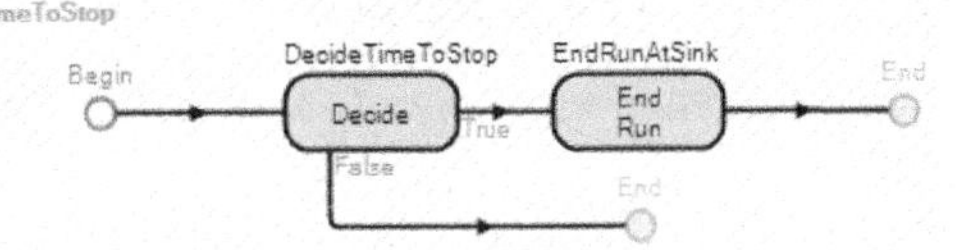

Figure 8.52: Check Last Applicant and End Run

- The `Decide` step is "*ConditionBased*" on the following expression.

```
(EntApplicant.Population.NumberInSystem <= 1) && (Run.TimeNow >= 7.5)
```

This expression states that we will stop the replication if the current applicant being destroyed is the last applicant in the system and the simulation time is greater than or equal to 7.5 hours.

Step 4: Select the "*Destroying Entity*" add-in process trigger property for each the three SINKS and specify the **TimeToStop** process for each.

Step 5: Within the *Run→Run Setup* arbitrarily set the *Run Length* to 12 hours, to be sure the simulation length does not end the run.

Step 6: Save and fast-forward the model to the end of the replication.

Question 47: At what time did the replication stop?

____

Step 7: If you make multiple replications, each replication will stop at a different time. Output statistics are appropriate for collecting statistics on the time the replications stop. Note, output statistics only collect one observation for each replication at the end of the run. Such statistics will allow us to determine on average how long the office stays open. From the *Definitions→Elements* section, insert a new OUTPUT STATISTIC named **OutStatOfficeCloses** with the expression being `Run.TimeNow` and set the *UnitType* to be "*Time*" and *Units* to "*Hours*".

Step 8: To determine on average of how long the office is open, multiple replications of the model need to be executed. From the "*Project Home*" tab, insert a new experiment named **ExpOfficeCloses**.

Step 9: Change the *Required* replications for the first scenario to 100.

Step 10: You generally need to add a response to evaluate each of the experiment scenarios. Under the "*Experiment*" section, click the *Add Response* button to insert a response named **TimeAfterClosing** which has an expression of the `OutStatOfficeCloses.Value - 7.5` which indicates when the simulation ended minus the 7.5 hours which represents the amount of time over the normal day (see Figure 8.53).[119]

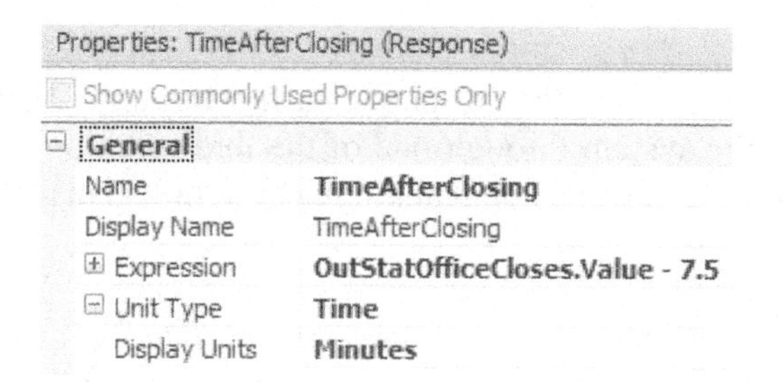
Properties: TimeAfterClosing (Response)

Show Commonly Used Properties Only

General

Name	TimeAfterClosing
Display Name	TimeAfterClosing
Expression	OutStatOfficeCloses.Value - 7.5
Unit Type	Time
Display Units	Minutes

Figure 8.53: Specifying a Response for the Experiment

Step 11: Save the project and execute the experiment and from the Pivot grid answer the following questions.

Question 48: On average how long does the office stay open past the 7.5 hours?

___

Question 49: What is the half width of the time when the office closes (in minutes)?

___

Step 12: Click on the "*Response Results*" tab, to see the (SIMIO Measure of Risk & Error (SMORE) plot for the one response as seen in Figure 8.54. SMORE plots allow you to visual the output response. It concisely displays the minimum, maximum, mean, median, lower and upper percentiles,[120] as well as the confidence intervals of the mean and the lower and upper percentiles. The histogram of values is also displayed because the "Histogram" button is pushed.

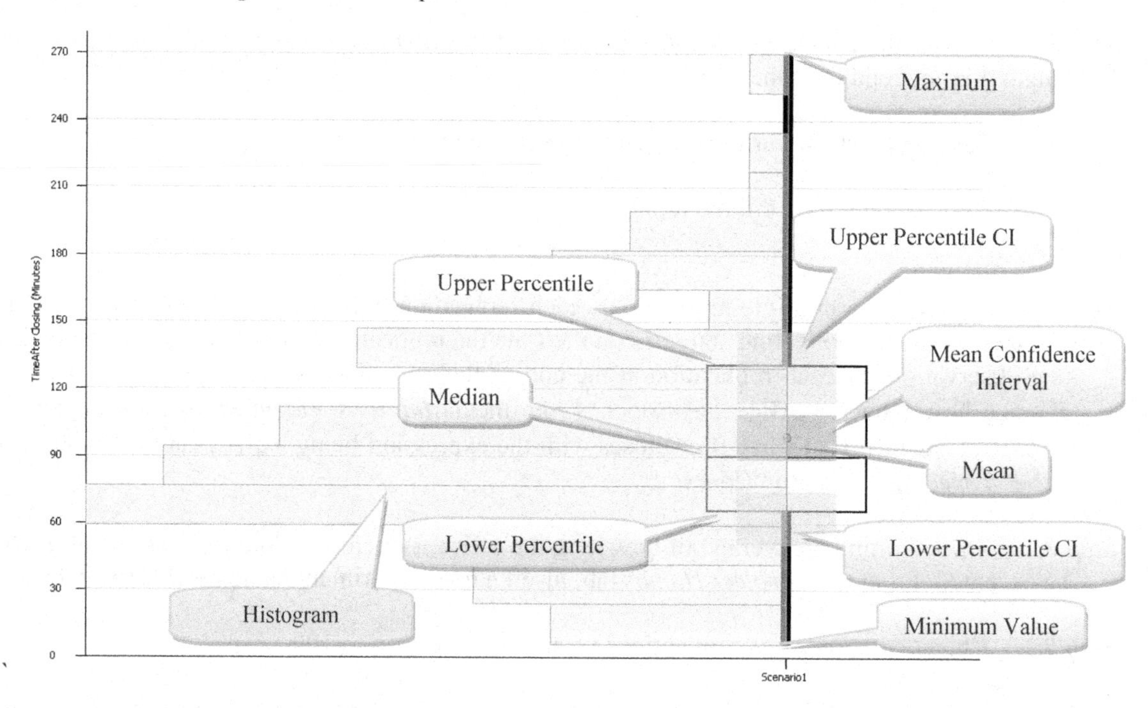

Figure 8.54: SMORE Plot with Histogram for the Additional Time the Office is Open

---

[119] Note we could have made the OUTPUT STATISTIC expression be `Run.TimeNow - 7.5` instead.

[120] By default it is the 25th and 75th percentiles which can be changed on the Experiment properties along with the default 95% confidence interval.

Question 50: Based on the SMORE plot and its "Raw Data", what are the mean and the 95% confidence intervals of the additional time office is open?

____

Question 51: What are the lower and upper percentiles and their values?

____

Question 52: What is the maximum additional time spent in the office and does that make sense?

____

It seems that there are occasions that the maximum additional time spent is 7.5 hours (450 minutes) either via the SMORES plot or the results in the *Pivot Grid*. The simulation does not stop until the 15 hour run length is reached. Note, an observation in a sink must occur before the stopping condition is checked.

Question 53: What happens if the last applicant leaves at 3:45 p.m. but no one arrives before 4:30 p.m.?

____

Step 13: In this situation, the last applicant leaves the system but the current time is still less than the 7.5 hours and therefore the stopping condition is not invoked. The previous stopping method assumes that there would always be applicants in the system at 4:00 p.m. The model needs to check to see if there zero applicants in the system at closing. Return to the Model and insert a new `TIMER` element named **TimerToStop** from the *Definitions→Elements* section and set the *Time Interval* property to 7.5 hours[121].

Step 14: In the *Processes* tab, insert a new process named **StoppingSimulation** which utilizes the **TimertoStop** event as its *Triggering Event Name* property and the *Triggering Event Condition* **as seen in** Figure 8.55. **When the event occurs the process will only run if the triggering event condition is also true.**

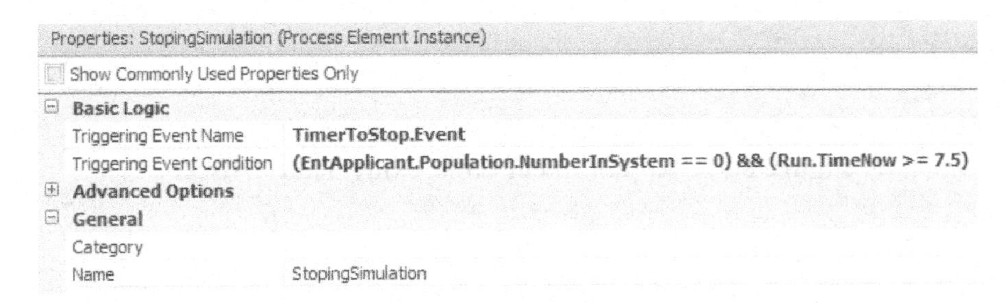

Figure 8.55: Specifying a New Process to be Executed when the Timer fires.

Step 15: Insert the `End Run` step which will stop the simulation run as soon as the step is executed as shown in Figure 8.56.

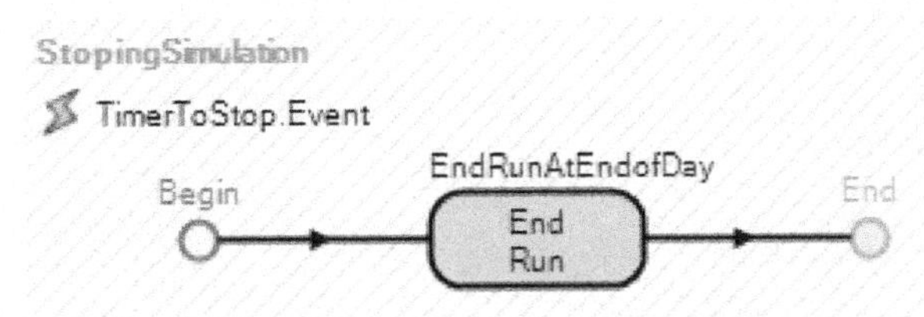

Figure 8.56: Stopping Simulation at Time 7.5 hours Process

Step 16: Save the model and rerun the experiment looking at the "*Response Results*" again.

[121] One timer event will occur at 0.0 since the *Time Offset* is 0.0, but it won't affect the simulation in this case.

Question 54: On average how long does the office stay open past the 7.5 hours?

Question 55: What is the maximum additional time spent in the office and does that make sense?

Question 56: Is it realistic that the office should stay open only 90 minutes past its closing?

Question 57: What can be done to reduce the time the office is open after its closing time?

Part 8.9: Commentary

- In Figure 8.39, the appointment schedule was designated using absolute dates and times. Internally, SIMIO converts the table into a time offset table based on the simulation starting date and time. One can also specify the appointment table directly as time offsets as seen in which represents the exact same table as the one in Figure 8.57. The column type must be a numeric or an expression general property.

Table Applicant | Table Arrivals | Table Arrivals Relative

	Relative Scheduled Time
1	0
2	0.5
3	1.0
4	1.5
5	5.5
6	6.0
7	6.5
8	7.0

Figure 8.57: Appointment table Using Time Offsets

Chapter 9
The Workstation Concept: A Kitting Process

One of the objects in SIMIO not yet described is the WORKSTATION which is rather specialized even though it behaves somewhat like the SERVER object. It was designed to best fit a manufacturing environment, although it is certainly not restricted to those applications. The WORKSTATION can have scheduled capacity but is limited to a capacity of only one. The object can mimic batching by processing an entity as though it were composed of several items, services one or more of the items at a time. There are expanded setup options for the options, even accounting for dependency between setups. Most of these features are found in actual workstations in a production system. Perhaps the most interesting feature of the WORKSTATION it is use of "materials". Materials are those items which are supplied to workstations that are directly required for assembly/production. Typically, a "bill of materials" is used to determine what is needed for any particular operation. Materials are not an object, but an inventory of items which can be consumed and replenished.

Part 9.1: The Kitting Process

A kitting process constructs a "kit" from available components. Orders arrive to a kitting workstation in a Poisson stream with an interarrival time of three minutes as seen in Figure 9.1. An order is for a particular number of identical kits to be produced. The kits are assembled from three components, one SubA subassembly, one SubB subassembly, and two Connectors – this is also referred to as a Bill of Materials (BOM). The assembly of each kit has a processing time described by a Triangular distribution with parameters of (0.1, 0.2, 0.3) minutes. Only one kit can be assembled at the kitting workstation at a given time. Information about the kits produced and components used is needed to analyze the operation, and information on the processing of the orders is also desired.

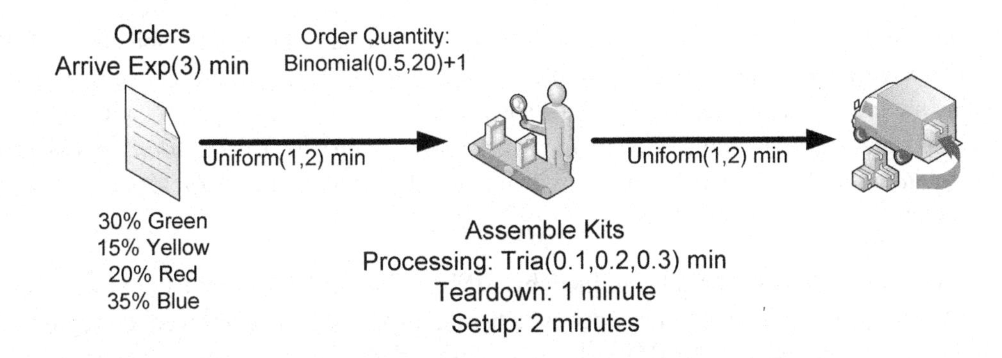

Figure 9.1: The Kitting Operation

There are four types of kits produced from the same BOM, which will be represented by colors: 30% are green, 15% are yellow, 20% are red, and 35% are blue. Following each order that is processed, it takes one minute to tear down for the next kitting order and two minutes to setup for the next order. This assumption will be changed later. Finally, the orders will be assumed to take between one and two minutes to arrive at the kitting workstation as well as to travel to the exit from the kitting workstation.

Step 1: Create a new model that has one SOURCE (**SrcOrdersArrive**), one SINK (**SnkOrdersLeave**), and one WORKSTATION object (**WrkKitOperation**) as seen in Figure 9.2.

- Connect the objects together using TIMEPATHS that take the appropriate `Uniform(1,2)` minutes as the *Travel Time* property.
- The kits' orders arrive with an `Exponential(3)` minute interarrival time.

- The time needed to process each kit at the workstation follows a `Triangular(0.1, 0.2, 0.3)` minutes distribution.
- The workstation's tear downtime is one minute.
- The current setup time is a specific two minutes for the workstation.

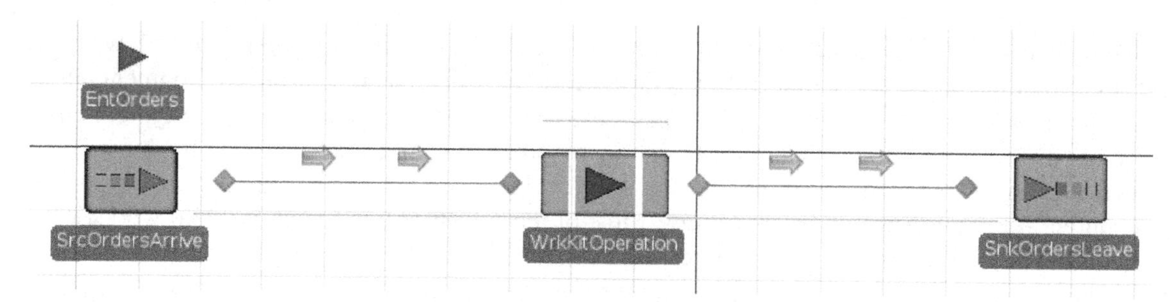

Figure 9.2: Kitting Operation Model

Step 2: Insert a new MODELENTITY named **EntOrders** into the model.

Step 3: Add three additional symbols for the **EntOrders** entity by first selecting the **EntOrders** instance and clicking the *Add Additional Symbol* button. Color the additional symbols "Yellow(1)", "Red(2)", and "Blue(3)"[122]. Note that symbols are numbered starting from zero.

Step 4: All of the SIMIO objects have certain built-in characteristics. Properties for an object (e.g., initial number of in system, maximum arrivals) are established at the time the object is created (usually at the beginning of the simulation) and cannot change during the simulation. State variables are characteristics that can change during the simulation. The SIMIO-defined *Priority* state variable[123] is an example of a characteristic that is often used for ranking. In SIMIO, you have the ability to add your own properties as well as discrete state variables to objects.

Properties and *States* are added through the "*Definitions*" tab of an object. All objects in SIMIO can have their own *Properties* and *States*. A state variable defined in the MODEL will only have one value and can be thought of as having global scope. Defining a state variable on the MODELENTITY will allow each entity to have its own state variable value. Select the MODELENTITY in the [*Navigation*] panel and select the *Definitions→States* section. Each kit order will have a different quantity of kits requested and a state variable is needed to track this information.

- Notice, two variables are already defined by SIMIO. ***Picture*** is defined by SIMIO as a DISCRETE REAL STATE VARIABLE that represents which picture should be displayed to represent the entity. *Animation* is also defined by SIMIO and is a STRING STATE VARIABLE which is used to determine the type of moving animation the entity might do.
- You should add **EStaOrderQuantity** as new DISCRETE INTEGER STATE VARIABLE that will represent the number in the order, as seen in Figure 9.3.

State Variables		
Picture	Real State Variable	Picture
Animation	String State Variable	Animation
EStaOrderQuantity	Integer State Variable	EStaOrderQuantity

Figure 9.3: Add a State Variable to the MODELENTITY Object

[122] Symbol numbers are in parentheses

[123] There is an entity property called *Initial Priority*, which initializes the `Priority` state variable.

Step 5: A table is needed to determine the percentage of time each type of order will occur as well as the symbol picture number. First, select the MODEL in the [*Navigation*] panel to return back to the simulation model. Let's create the **TableSymbol** for the entity pictures as seen in Table 9.1 via the *Data* tab where the **Symbol** column should be an *Integer Standard Property* while the **Percent** could be an *Integer, Real* or *Expression* property (i.e. numeric).

Table 9.1: Order Type Symbol and Percentage

Table Sym...

	Symbol	Percent
1	0	30
2	1	15
3	2	20
4	3	35

Step 6: Next, the order type (i.e., yellow, red, etc.), the order quantity for the arriving order, and the correct picture will be specified using an Add-On process trigger. Select the **SrcOrdersArrive** and double click the *CreatedEntity* Add-On process trigger property which will create the process named be **SrcOrdersArrive_CreatedEntity**. In this process we can make assignments of state variables to the incoming entities as shown in Figure 9.4 after setting the table row.[124]

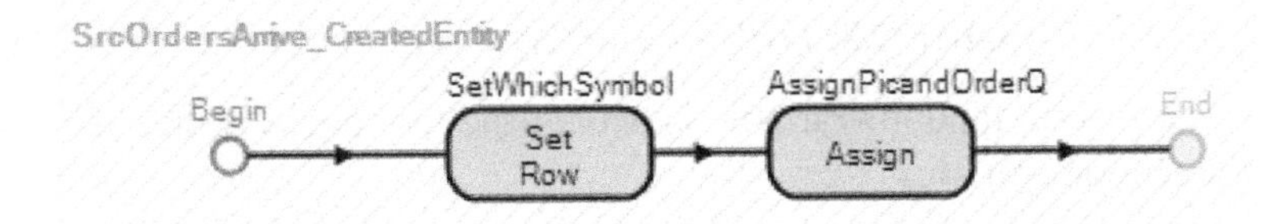

Figure 9.4: Setting the Row of the Data Table and Assigning Picture.

Step 7: Use the `Set Row` process step to randomly assign a row of the **TableSymbol** TABLE to the entity which will be used in the `Assign` step to specify the entity's picture.

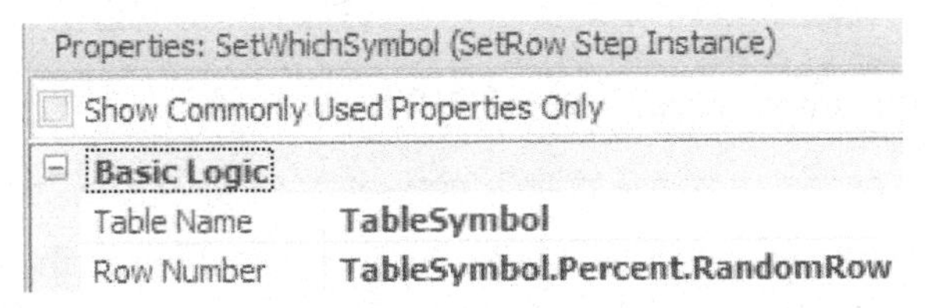

Figure 9.5: Steps for the OrdersArrive_CreatedEntity *Process*

Question 1: Why was the table set in the "*Created*" Add-On Process Trigger versus "*Creating*" as was done in Chapter 5?

____

Step 8: Next assign the correct picture and the order quantity. You can do this in two different assignment steps or in one step using multiple rows by clicking the button on the "0 Rows". You need to add each assignment as seen below in Figure 9.6 through Figure 9.8.

- Use the "*Repeating Property Editor*" to make multiple assignments as shown in Figure 9.6.

[124] Instead of using an add-on process trigger, the *On Created Entity Table Reference Assignment* could have been used to do the same thing along with the *Before Exiting State Assignments*.

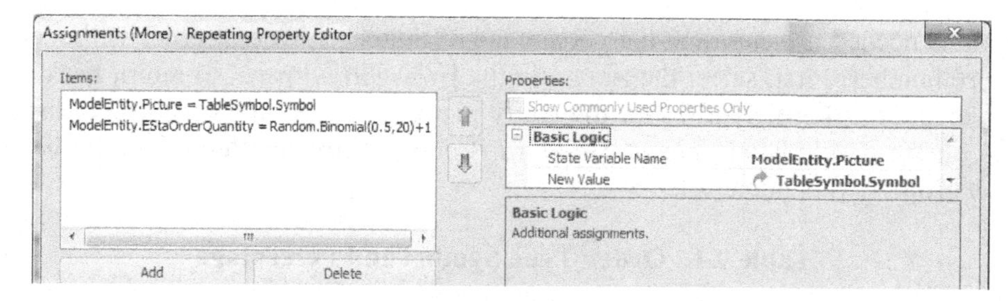

Figure 9.6: Making Multiple Assignments

- Set the picture to represent the symbol associated with the specified order type as in Figure 9.7.[125]

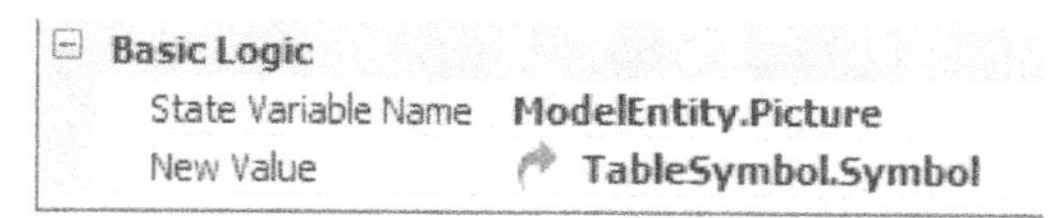

Figure 9.7: Picture Assignment

- Assign the order quantity from a Binomial Distribution as in Figure 9.8.

Basic Logic	
State Variable Name	**ModelEntity.EStaOrderQuantity**
New Value	**Random.Binomial(0.5,20)+1**

Figure 9.8: Order Quantity Assignment

Question 2: Note a `Binomial` random variable is being used to model the number of orders. Why are we adding one to the random variate?

____

Step 9: Save your model with a new name and run the model for eight hours

Question 3: What is the average flow time for the orders?

____

Question 4: How many orders were processed in the 8 hours?

____

Question 5: What are we missing or what assumptions are we making in the current model?

____

Part 9.2: Sequence-Dependent Setup Times

Whenever the kitting workstation must change from one color to another, there is a setup time that depends on the color type that preceded the current order. The changeover time is shown in Figure 9.9.

↓From \To →	Green	Yellow	Red	Blue
Green	0	9	6	3
Yellow	3	0	6	3
Red	3	9	0	3
Blue	9	9	3	0

Figure 9.9: Changeover Setup Times (From/To) in Minutes

____

[125] The **Picture** state variable sets the symbol associated with the object in this case the ModelEntity.

Let's add the sequence-dependent setup times to our model. Under the WORKSTATION properties, you can specify the setup times with the *Specific* constant setup (e.g., two minutes for the kitting operation). But before we can use sequence-dependent setup times, we need to know how the changeover times can be incorporated into the SIMIO model.

Step 1: First, a LIST of identifiers is needed that can be used to create the changeover table (matrix) where in this case, a *List* of *Strings*. Under the "*Definitions*" tab, insert a LIST of Strings named **LstColors**. The contents of the **LstColors** list would correspond to our symbol colors: "Green", "Yellow", "Red", and "Blue" as seen in Figure 9.10. It is important these are put in this order to correspond to the symbols (0, 1, 2, and 3).

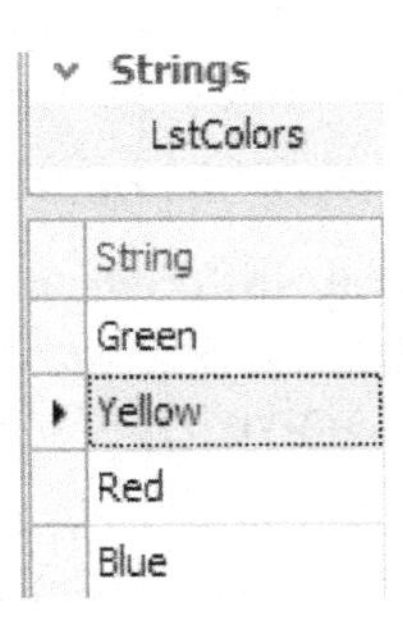

Figure 9.10: List of Colors

Step 2: Now in the "*Data*" tab, add a CHANGEOVERS matrix (this is a "from-to" matrix) named **Changeover** and use the **LstColors** list for the associated list. Change the unit of time to *Minutes* and enter the values into the matrix from Figure 9.9.

Step 3: Next, specify the Setup time logic in the **KitOperation** WORKSTATION as seen Figure 9.11.

- The *Setup Time Type* property: *Sequence Dependent*
- *Operation Attribute*: `TableSymbol.Symbol`
- *Changeover Matrix*: **Changeover**

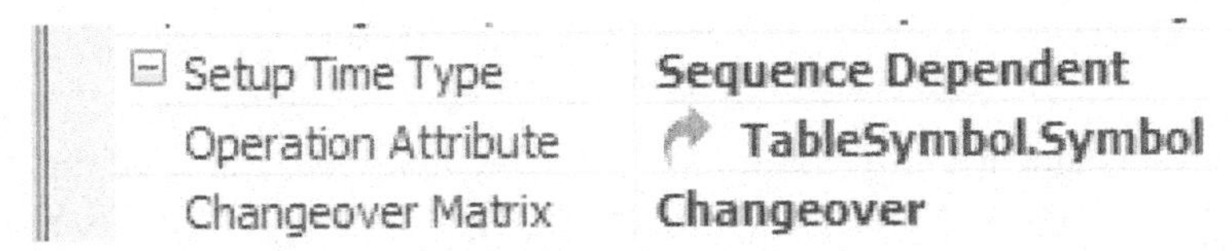

Figure 9.11: Specifying Sequence Dependent Changeovers in the Workstation

Step 4: We can now finish specifying the other elements of the *Process Logic* for the WORKSTATION as seen in Figure 9.12.

- *Operation Quantity* property (i.e., the number of items in each order): `ModelEntity.EStaOrderQuantity`
- *Processing Batch Size*: 1
  (This is the number of items that can be processed at one time. So if the *Operation Quantity* is 10 and the batch size is 2, it will take 5 batches to process the entire order with each batch requiring a different random processing time.)
- *Teardown Time*: 1 minute

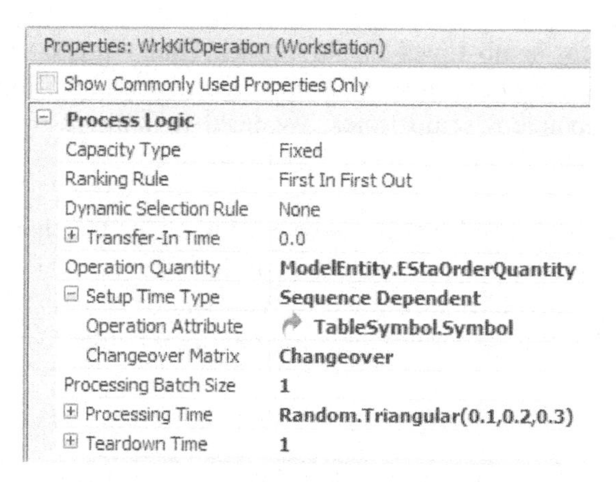

Figure 9.12: Workstation Properties

Step 5: The workstation may be in several different resource states including *starved (i.e., idle)*, *processing*, performing a *setup*, or performing a *teardown*. To see all the nine possible workstation states, click on the workstation and then click the "Active Symbol" to get the display in Figure 9.13.

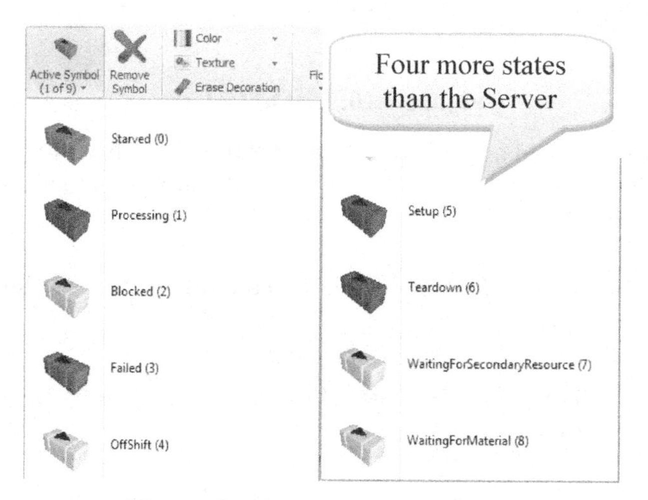

Figure 9.13: Workstation States

Step 6: To see the percentage in these "resource" states evolve over time within the simulation; add a "*Status Pie*" from the *Animation* tab. Specify the *Data Type* property as a *ListState* and the *List State* should be the `WrkKitOperation.ResourceState`.[126]

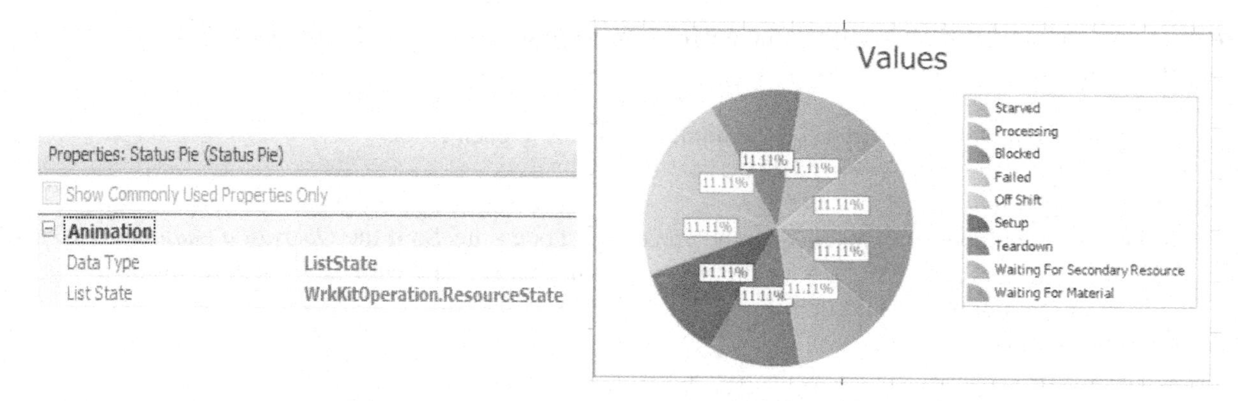

Figure 9.14: Workstation Status Pie Chart

[126] Note, if you select the **WrkKitOperation** object first, the *"Status Pie"* will be associated with the WORKSTATION and you will only need to specify `ResourceState`.

Step 7: Save and run the simulation with animation for an hour. Then run it in fast-forward mode for two, four, and eight hours. After eight hours, answer the following questions.

Question 6: What is average flow time for orders?

____

Question 7: How many orders were processed in the 8 hours?

____

Question 8: What are we still missing in the current model?

____

Part 9.3: Sequence-Dependent Setup Times that are Random

The previous section utilized the CHANGEOVER matrix to handle sequence dependent setup times. However, the matrix is limited to only constant/deterministic times and cannot handle the case when the setup times have variability associated with them (i.e., random distributions). Since SIMIO's CHANGEOVER matrix cannot utilize distributions as the sequence dependent setup times, we will handle this manually using related data tables. Similar to previous examples, a child table will be used to define the setup times for a particular part.

Step 1: Save the current model as a new name to be used again in later section.

Step 2: First we need to declare the relationship (i.e., primary key) for the parent table. From the *Data→Tables* section, insert a new *String Standard Property* named **Color**. Click the *Set Column as Key* button and fill in the field's values as seen in Figure 9.15.[127]

Table Sym... | Table Change Over

		Symbol	Percent	Color
1	⊞	0	15	Green
2	⊞	1	20	Yellow
3	⊞	2	35	Red
4	⊞	3	30	Blue

Figure 9.15: Adding the Color Field as the Primary Key

Step 3: Insert a new DATA TABLE named **TableChangeOvers** as seen in Figure 9.16 which will be the child table.

[127] We could have used the **Symbol** field as the primary key but this will make the setups easier to visualize in the related table.

Table Symbol	Table Change Ov...		
	To Color	From Color	ChangeOver (Minutes)
1	Green	Green	0
2	Green	Yellow	3
3	Green	Red	3
4	Green	Blue	9
5	Yellow	Green	9
6	Yellow	Yellow	0
7	Yellow	Red	9
8	Yellow	Blue	9
9	Red	Green	6
10	Red	Yellow	6
11	Red	Red	0
12	Red	Blue	3
13	Blue	Green	3
14	Blue	Yellow	3
15	Blue	Red	3
16	Blue	Blue	0

Figure 9.16: Creating the Sequence Dependent Setup Times Table

Step 4: From the *Add Column* section, select the *Foreign Key* to insert the **To Color** column. Set the *Table Key* property to **TableSymbol.Color** which will link the parent to the child changeover's table.

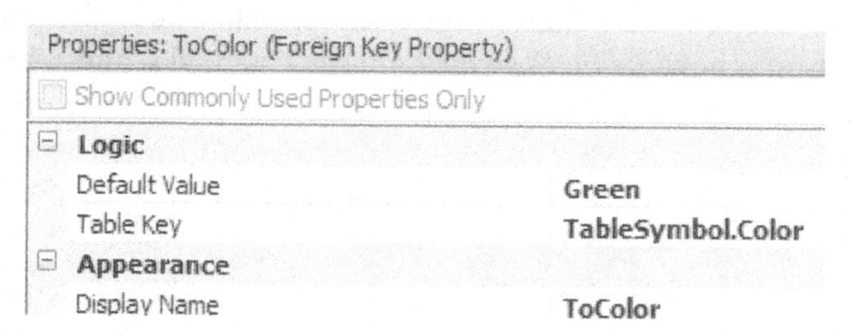

Figure 9.17: Adding the "To Color" Child Foreign Key

Step 5: Insert a *Standard Property* named **From Color** and an *Expression Standard Property* named **ChangeOver**. Make sure to set the *Unit Type* property to "**Time**" and the *Default Units* to "**Minutes**". Since this column is an *Expression* property column, we will be able to specify any SIMIO expression.

Step 6: Insert the values from Figure 9.16 which replicates the previous section's setup times.

Step 7: Recall, when the **EntOrders** are created they are assigned a row to the **TableSymbol** (see Figure 9.4) which was used to determine the picture. Now, they will also be assigned the related portion of the **TableChangeOvers** as seen in Figure 9.18 if the new order was "Red."

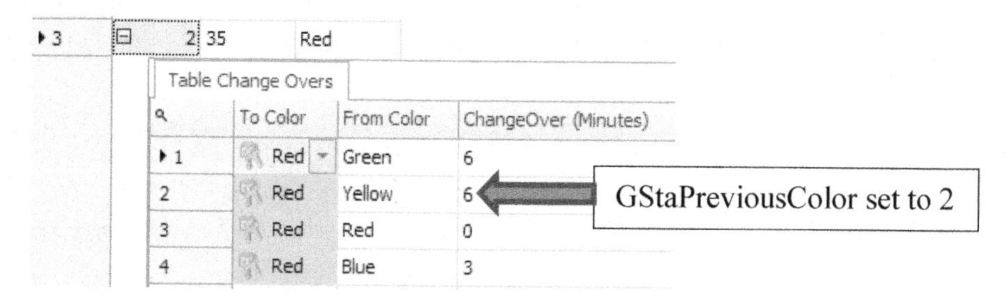
3 | 2 | 35 | Red

Table Change Overs

	To Color	From Color	ChangeOver (Minutes)
1	Red	Green	6
2	Red	Yellow	6
3	Red	Red	0
4	Red	Blue	3

Figure 9.18: Related Records in the ChangeOver Table Associated with Red Orders

Step 8: If the new order "Red" is next to be kitted and the previous order was "Yellow," then the setup time should be six which is associated with the second row of the related table as seen in Figure 9.18. From the *Definitions* tab, insert a new MODEL *Discrete Integer State variable* named **GStaPreviousColor** with a default value of "1" which will indicate the last color that was processed.

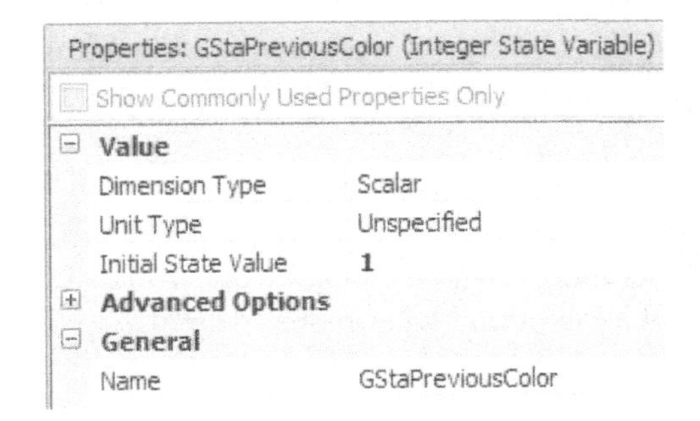

Figure 9.19: Insert a State Variable to Keep Track of the Last Order Processed

Step 9: Insert the *FinishedSetup* Add-on process trigger for the **WrkKitOperation** WORKSTATION which is executed after the setup has completed. Insert an `Assign` step that will set the **GStaPreviousColor** to the current symbol plus one which will be used for the next order. The symbols range from zero to three but table rows start at one.

Figure 9.20: After Setup Completed Set the Previous Color

Step 10: Next, for the WORKSTATION, set the *Setup Time Type* back to "**Specific**" and specify the *Setup Time* to be `TableChangeOvers[GStaPreviousColor].ChangeOver`. The **GStaPreviousColor** indexes into the related table to specify the setup time.

Setup Time Type	Specific
Setup Time	**TableChangeOvers[GStaPreviousColor].ChangeOver**

Figure 9.21: Changing the Setup to Utilize the Sequence-Dependent ChangeOver Data Table

Step 11: Save and run the model comparing the results from the previous section.

Question 9: What is the time in system and the average setup time of the workstation?

____

Step 12: To demonstrate the ability to have variable setup times, change all the setup times that are nine minutes to `Random.Pert(8,9,10)` minutes. Save and run the model comparing the time in system and the average setup time to the previous results.

Question 10: What is the time in system and the average setup time of the workstation now?

____

Part 9.4: Using Materials in the Kitting Operation

The WORKSTATION has more capabilities than a general SERVER. For example, it can be used to keep track of raw materials that have to be consumed, as specified in the "bill of materials" needed to create a part or complete an assembly. If the raw material is not available, then parts (entities) cannot be processed and will wait until the required materials become available.

Step 1: Return to the model from Part 2 of this chapter and save it as a new project.

Step 2: First, we need to declare the existence of raw materials that make up a kit as well as the Bill of Materials that specifies the number of each raw material type needed to make a kit. Five new MATERIAL

"*Elements*" need to be defined via *Definitions→Elements→Workflow* section as seen in Figure 9.22.[128]

- Create **MatSubA**, **MatSubB**, and **MatConnectors** as materials types, all with *Initial Quantities* of zero, which will be changed shortly.
- Next create a **MatBOMKits** as MATERIAL element, but using the *BillofMaterial* Repeating Property Editor as shown in Figure 9.22 to define what materials make up this product (i.e., one **MatSubA** and **MatSubB** and two **MatConnectors**).

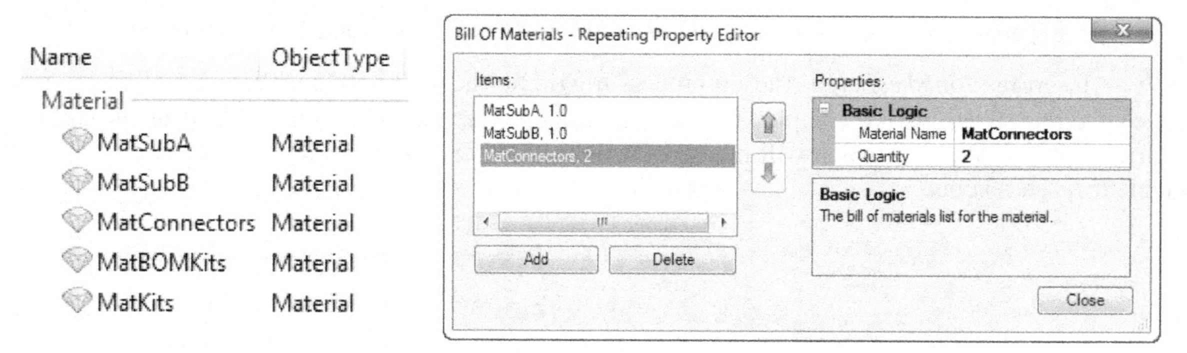

Figure 9.22: Five Materials and the Bill of Materials for the MatBOMKits

- Finally, create **MatKits** as the last "Material" element (with *Initial Quantity* of 0) which will represent the finished product produced from the raw materials in the bill of materials.

Step 3: Since the initial inventory is a fixed characteristic of our model and only needs to be initialized at the start of the simulation, here is an ideal place to use a SIMIO Model "property." From the *Definitions→Properties*" section, add new properties to objects. Create three new "Standard Real Properties" named **InitInvA**, **InitInvB**, and **InitInvC**. Place all of these properties into a new **InvModel** *Category.* The first time, the new category name (**InvModel**) needs to be typed into the window. For the next two properties you can select the category from the dropdown.

Step 4: Return to the material "Elements" and now set the *Initial Quantity* property for each of the three materials (i.e., **MatSubA, MatSubB,** and **MatConnectors**) as a reference property to the previously defined properties. Remember this is done by right-clicking on the drop-down arrow and selecting the referenced property (Initial Quantity | **InitInvC**).

Step 5: As part of the analysis, create some statistics on the consumed materials as well as the produced product. There are three different types of statistics that can be defined. Recall that a "Tally Statistic" is an observation-based statistic (e.g., waiting time at a queue). A "State Statistic" is a time persistent statistic on a state variable (e.g., number in queue). An "Output Statistic" can be used to collect any expression value, once at the end of each replication. Add four OUTPUT STATISTICS elements with the following names and expressions.

- **OutStatSubAConsumed** based on `MatSubA.QuantityConsumed`
- **OutStatSubBConsumed** based on `MatSubB.QuantityConsumed`
- **OutStatConnectorsConsumed** based on `MatConnectors.QuantityConsumed`
- **OutStatKitsProduced** based on `MatKits.QuantityProduced`

---

[128] *Elements* are objects that represent things in a process that change state over time (e.g., TALLYSTATISTIC, STATESTATISTIC, TIMER, etc.)

- For each added "Output" statistic, under the *Results Classification* section set the *Data Source* **to the name of the material (e.g., "MatSubA"), the** *Category* "Total Consumed" or "Total Produced" **and the** *Data Item* "Number" as seen in the example of Figure 9.23 for **MatSubA.**

Results Classification	
Data Source	MatSubA
Category	Total Consumed
Data Item	Number

Figure 9.23: Setting of Results Classification

Step 6: Now return to themodel under the *Facility* tab and for the WORKSTATION **KitOperation** fill in the *Materials & Other Constraints* properties (i.e., *Material Consumption*, *Consumed Material Name*, *Consumed Quantity*, *Material Production*, *Produced Material Name*, and *Produced Quantity*) as seen in Figure 9.24. This will force the WORKSTATION to consume materials based on the bill of materials as it processes the kits.

Materials & Other Constraints	
Material Consumption	Bill Of Materials
Consumed Material Name	MatBOMKits
Consumed Material Quantity	1.0
Material Production	Material
Produced Material Name	MatKits
Produced Material Quantity	ModelEntity.EStaOrderQuantity
Produced Material Transfer Time	0.0

Figure 9.24: Other Material Requirements for a WorkStation

Step 7: To access the new MODEL properties we created, click on the Model "*Properties*" (by right-clicking the "Model" object in the [*Navigation Panel*]).[129] In the **Controls →InvModel** section, set the values of the **InitInvA** to 100, **InitInvB** to 100, and **InitInvC** to 200.

Controls	
InvModel	
InitInvA	100
InitInvB	100
InitInvC	200

Figure 9.25: Setting Model Properties

Step 8: Save and run the new model for eight hours observing the statistics. While you may want to run it for a while looking at the animation, you will need to fast-forward to the end to get the results.

Question 11: After looking at the Kits produced, do you have concerns about the model specifications?

____

Question 12: How many kits were produced in the eight hours?

____

Step 9: Change the initial inventory values to allow kits to be produced throughout the eight hours.

Question 13: How many orders were satisfied over the eight hours?

____

[129] Note you can change the name of the fixed model as well as set other model properties.

Question 14: How many kits were produced now?

Question 15: How many of each raw material **MatSubAs** and **MatSubBs** were needed?

Question 16: How many **Connectors** were needed?

Question 17: What was the cycle time for the orders?

Part 9.5: Raw Material Arrivals during the Simulation

Instead of material being available for the entire eight hour day at the beginning of the simulation, the raw materials will be scheduled to arrive every two hours, beginning at time zero. When the stock arrives, the inventory is returned to a target stock level – implementing an "order-up-to inventory policy". How would this be modeled in SIMIO?

Step 1: First, let's collect some statistics on our "inventory" levels for the three raw materials over a given day: **MatSubA**, **MatSubB**, and **MatConnectors**. Since inventory is a time-based statistic, add three STATE STATISTICS in the *Definitions→Elements* section.

- **StateStatSubAAvailable**: `MatSubA.QuantityAvailable`
- **StateStatSubBAvailable**: `MatSubB.QuantityAvailable`
- **StateStatConnectorsAvailable**: `MatConnectors.QuantityAvailable`
- For each added "State" statistic, under the *Results Classification* section set the *Data* Source to the name of the material (e.g., "MatSubA"), the *Category* "Inventory Level" and the *Data Item* "Number".

Step 2: Save and run the model for eight hours using initial inventory quantities of 1000, 1000, and 2000 for **SubA**, **SubB**, and **Connectors** (specified in the "Model Properties").

Question 18: What was the average **MatSubA available**?

Question 19: What was the average **MatSubB available**?

Question 20: What was the average **MatConnectors available**?

Question 21: How many kits were produced?

Step 3: Now let's implement a replenishment system in which a stocker comes by the kitting station every two hours, beginning at time zero, and fills up the stock for **MatSubA** , **MatSubB,** and **MatConnectors** to particular quantities.

Step 4: Add three order quantity properties for the model under the *Definitions* tab named **OrderQtySubA**, **OrderQtySubB**, and **OrderQtyConn**. Each of these properties should be an *Integer*

Data Format added to the **InvModel** category as shown in Figure 9.26. These properties will be used as the order-up-to quantities that the stocker uses to replenish the raw material inventory.[130]

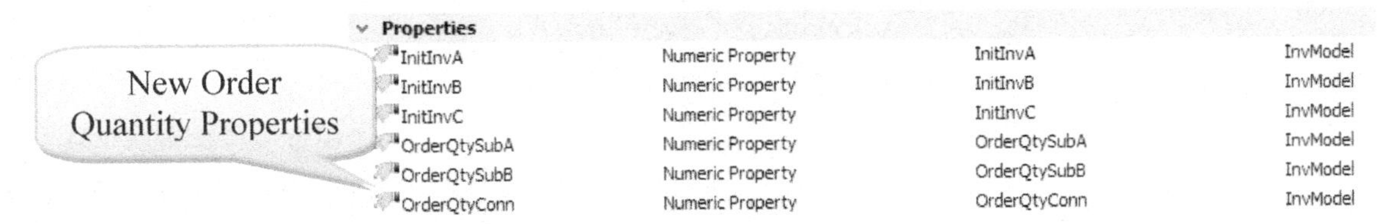

Figure 9.26: Order Quantity Properties

Step 5: Under the *Definitions→Elements* section, add a `TIMER` element which will fire an event based on a time interval (a "timer event" will be an interruption in the simulation that allows a process to execute). See Figure 9.27.

- Start the timer at time 0.0 (i.e., *Time Offset* property should be 0) meaning the first event fires at time zero and it then should "fire" an event every two hours (i.e., *Time Interval* should be "2").
- Name the timer **TimerStockerArrives.**

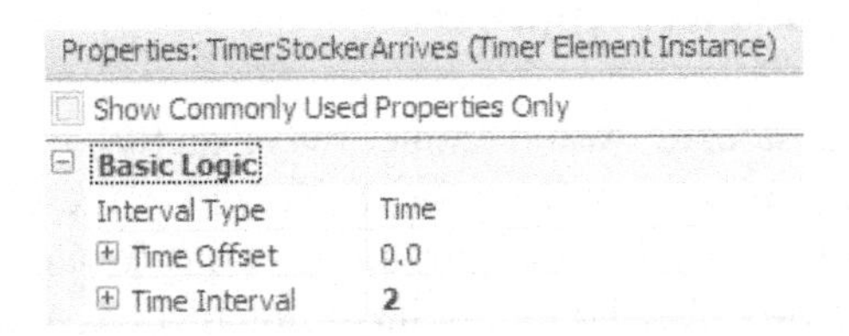

Figure 9.27: Timer for Stocker Arrivals

Step 6: Under the "*Processes*" tab, create a process named **RespondtoTimerEvent** that will respond to the timer event when it expires as seen in Figure 9.28.

- *Triggering Event* property: **TimerStockerArrives.Event**
- Add an `Assign` step that makes the following assignments to the process which always make the quantity available equal to the order quantity (i.e., order-up-to policy).[131]

```
MatSubA.QuantityAvailable = OrderQtySubA
MatSubB.QuantityAvailable = OrderQtySubB
MatConnectors.QuantityAvailable = OrderQtyConn
```

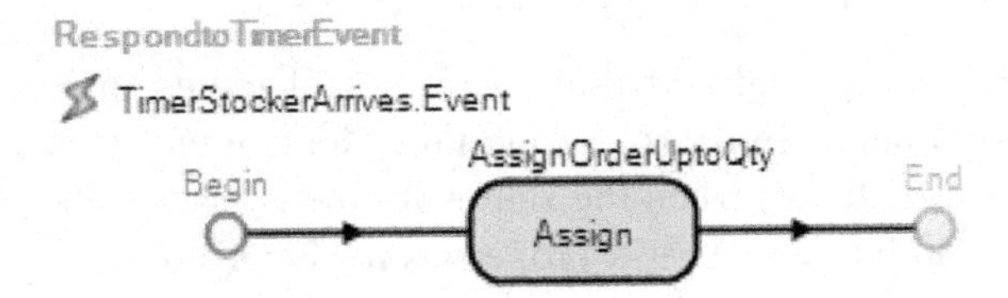

Figure 9.28: Process Used to Respond to the Timer Event

Step 7: Now in the "Model Properties" set the initial values of the raw materials **InitInvA**, **InitInvB**, and **InitInvC** elements to zero, since the first **TimerStockerArrives** event occurs at time 0.0 and will replenish the stock at the beginning of the simulation. Set the order quantities for **OrderQtySubA, OrderQtySubB**, and **OrderQtyConnectors** to 250, 250, and 500 respectively.

---

130 Create one of the properties with the correct category and name. Then copy, paste it and change the name appropriately.

131 SIMIO provides a `Produce` step that essentially does the same thing and a `Consume` step which would perform a subtraction, but the logic using an `Assign` step seems just straight forward. In this example, you would add three `Produce` steps that would produce for example `OrderQtySubA - MatSubA.QuantityAvailable`.

Controls	
InvModel	
InitInvA	0
InitInvB	0
InitInvC	0
OrderQtySubA	250
OrderQtySubB	250
OrderQtyConn	500

Figure 9.29: Setting the Model Properties for Order Quantities and Initial Inventory

Step 8: Save and run the model for eight hours and answer the following questions.

Question 22: What are the average **MatSubA**, **MatSubB,** and **MatConnectors** inventory levels?

____

Question 23: How many kits were produced during the eight hours?

____

Part 9.6: Implementing a Just-In-Time Approach

The previous example assumed a stocker would come by (exactly) every two hours and replenish the inventory up to a certain level (i.e., an order-up-to policy). However, the company is moving toward a more Just-In-Time (JIT) type of operation. They have a supplier that has located a warehouse very near the plant and will respond to orders very quickly. The company has implemented a monitoring system such that when the inventory drops below a certain threshold (i.e., reorder point), an order is placed to the supplier who can supply the products within an hour.

Step 1: Save the current model to a new name. You need to delete the **TimerStockerArrives** TIMER Element and the **RespondToTimerEvent** process or disable the TIMER by setting the *Initially Enabled* property to `False` under the *General* properties of the TIMER.

Step 2: Add a new SINK named **SnkInventory**. Position it underneath the Kit Operation which will be used to model the arrival of new raw materials (**MatSubA**, **MatSubB**, and **MatConnectors**).

Step 3: Add three reorder point properties for the model under the "*Definitions*" tab named **ReorderPTSubA**, **ReorderPTSubB**, and **ReorderPTConn**. Each of these properties should be an "*Integer*" *Data Format* added to the **InvModel** category. These properties will be used as the reorder points for the JIT system to replenish the raw material inventory.

Step 4: Go to the "*Definitions*" tab and add three MONITOR elements underneath the "*Element*" section named **MonitorA**, **MonitorB**, and **MonitorC**. A monitor element can be used to monitor the value of a state variable when it changes value or when the state variable crosses some threshold (either from above or below). Refer to Figure 9.30 for the MONITOR properties for **MonitorA**.

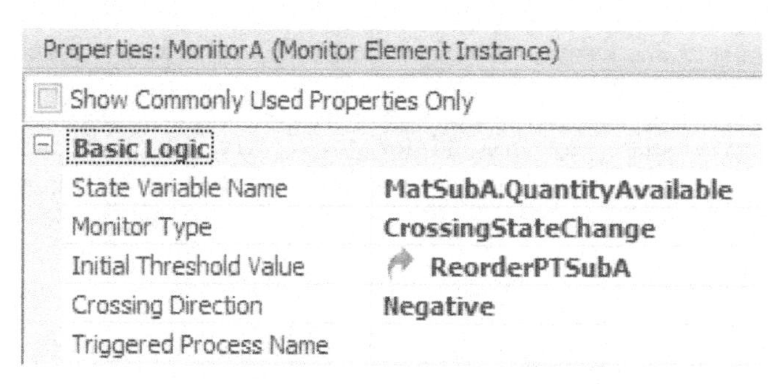

Figure 9.30: Monitor Properties

The following settings for **MonitorA** will monitor the **SubA** inventory level.

- *Monitor Type* should be *CrossingStateChange.*
- *Crossing Direction* should be *Negative* (i.e., 5 to 4 will fire an event if 4 was threshold).
- The *Threshold Value* should set to the new reference property we will call ***ReorderPTSubA.***
- The MONITOR can cause a process to fire just as the TIMER element did but we will not use this feature this time.

Step 5: Repeat the same process for raw material **MatSubB** and **MatConnectors** material using **ReorderPtSubB** and **ReorderPtConn** reference properties respectively.

Step 6: Add three SOURCES (i.e., suppliers) named **SrcSupplierA**, **SrcSupplierB**, and **SrcSupplierC** which will send the various raw materials to the kitting operation as seen in Figure 9.31.

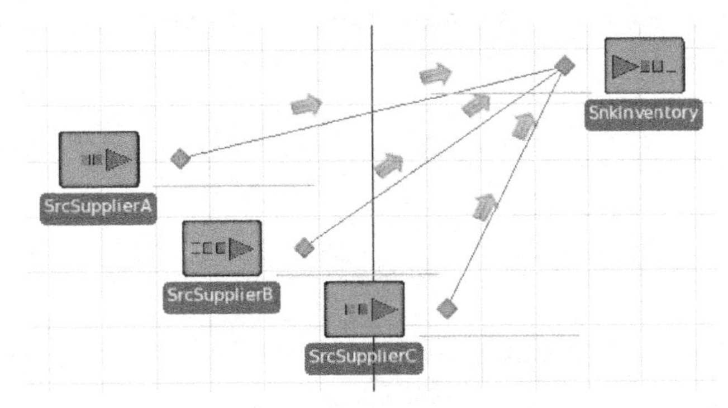

Figure 9.31: Supplier Sources

- Assume that it takes between 50 and 65 minutes for the raw materials to be delivered between the SOURCES and the **SnkInventory** SINK (this is modeled as *Travel Time property of a* TIMEPATH).
- Name each TIMEPATH as **TPA**, **TPB**, and **TPConn** respectively for the three TIMEPATHS.
- Set the *Arrival Mode* property of the suppliers to *On Event* and the *Event Name* to MonitorA.Event, MonitorB.Event and MonitorC.Event respectively for each of the three sources as seen for **SrcSupplierA** in Figure 9.32. Each source will create an entity when its monitor event fires and sends the replenishment entity to the kitting operation.

Properties: SrcSupplieA (Source)	
⊟ **Entity Arrival Logic**	
Entity Type	**EntOrders**
Arrival Mode	**On Event**
Initial Number Entities	0
Triggering Event Name	**MonitorA.Event**
Triggering Event Count	1
Entities Per Arrival	1

Figure 9.32: Utilizing an On Event Arrival Process

Step 7: When the raw material reaches the end of the path, you need to increment the current quantity by the appropriate order quantity using the *ReachedEnd* add-on process trigger for each of the three TIMEPATHS. A *Tokenized Process* will be used instead of three separate processes. From the *Definitions→Token* section, add a custom TOKEN named **TknReplenish**. For this TOKEN, add a "*Material Element Reference*" state variables named **TStaWhichMaterial**, to reference the correct material and "Integer" state variable **TStaQuantity** as seen in Figure 9.33

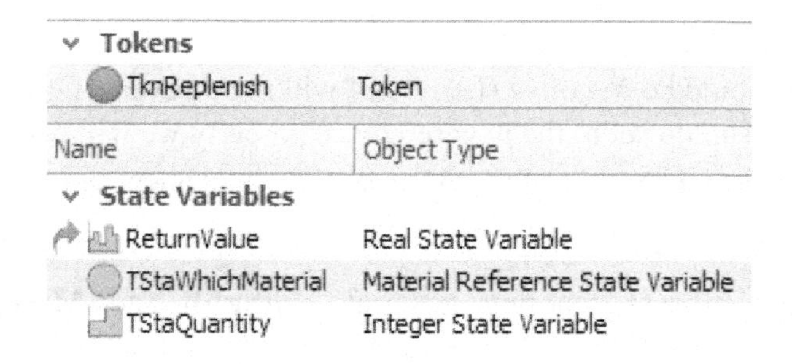

Figure 9.33: Adding the Custom TOKEN to Pass in the Material and Quantity

Step 8: From the *Processes* tab, insert a new process named **Replinish** which uses the **TknReplenish** and specifies two *Input Arguments* properties named **MaterialToProduce** and **QuantityToProduce** as seen in Figure 9.34.

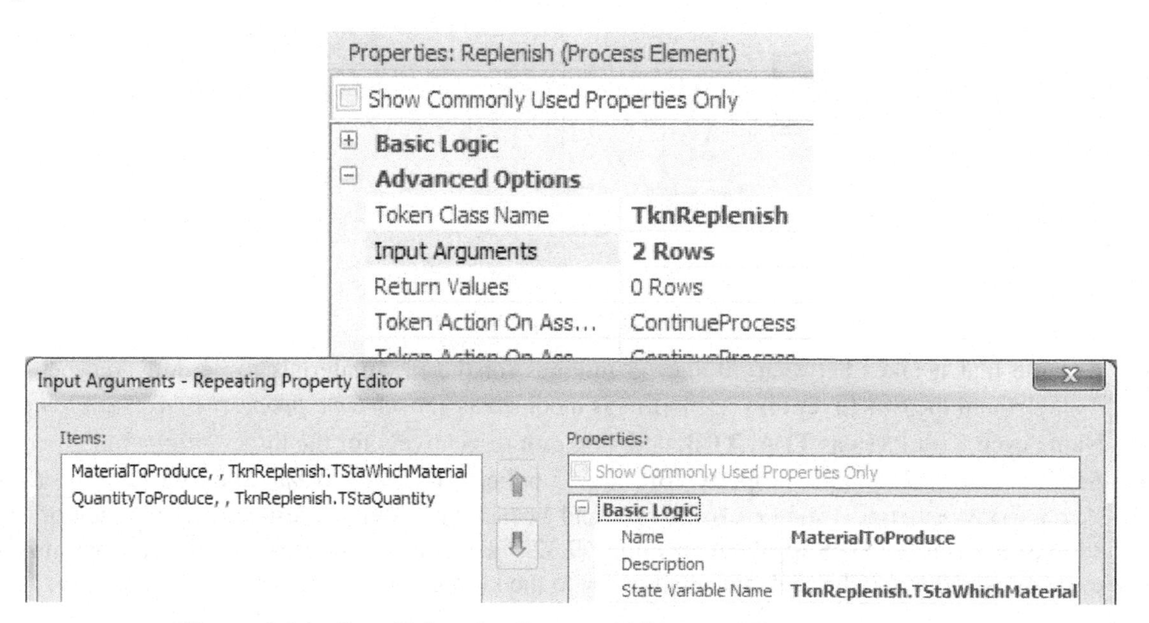

Figure 9.34: Specifying the Custom Token and Input Arguments

Step 9: Insert a `Produce` step (see Figure 9.35) which will be used to add the order quantity to the particular material passed to the process. You will need to physically type the material name property value **TknReplenish.TStaWhichMaterial** directly rather than selecting it from the dropdown list.

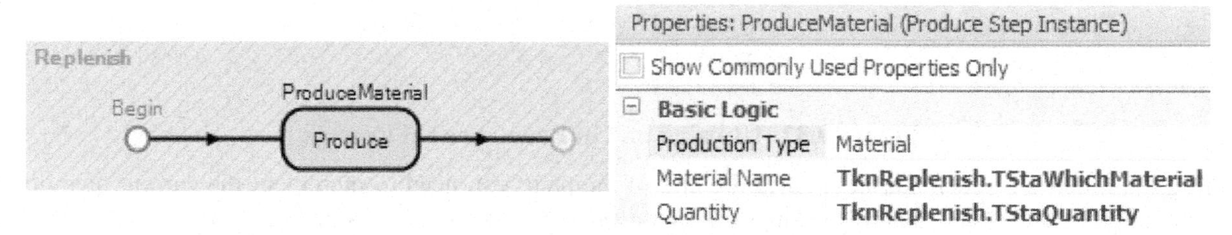

Figure 9.35: Incrementing the Quantity Available

Step 10: For reach of the TIMEPATHS (i.e., **TPA**, **TPB**, and **TPC**), specify the **Replenish** process as the *ReachedEnd* Add-On Process Trigger property using the appropriate material and order quantity. Figure 9.36 shows the example for **TPA** using material **MatSubA** and quantity **OrderQtySubA**.

Add-On Process Triggers
Run Initialized
Run Ending
Entered
Trailing Edge Entered
Reached End — Replenish
Input Arguments
Material To Produce — MatSubA
Quantity To Produce — OrderQtySubA

Figure 9.36: Using the Tokenized Process for Material A Production TPA Path

Step 11: In the Model properties, set the initial values of the raw materials **MatSubA**, **MatSubB**, and **MatConnectors** elements back to 250, 250, and 500 since the replenishment is now being triggered. Set the reorder points and order quantities to the values provided in Table 9.2.

Table 9.2: Initial Inventory, Reorder Point, and Order Quantities

Raw Material	Reorder Point	Order Quantity	Initial Inventory
SubA	125	200	250
SubB	125	200	250
Connector	250	400	500

Step 12: Add six "Status Labels" from the *Animation* tab that will display the available quantities of each of the three raw materials: **MatSubA**, **MatSubB**, and **MatConnectors**, perhaps as shown in Figure 9.37. The first three status labels are just text with the other three labels having an expression as seen in in Figure 9.37 that will display the `QuantityAvailable` for each material.

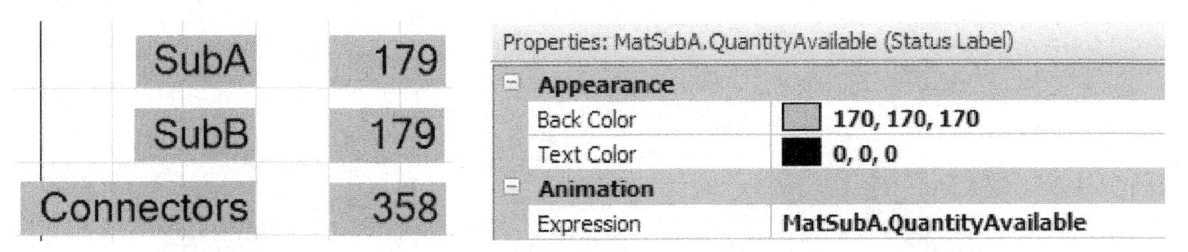

Figure 9.37: Status Labels and the Expression for the Status Label

Step 13: Save and run the model for eight hours and answer the following questions.

Question 24: What are the average **SubAAvailable, SubBAvailable, and Connectors available**?

____

Question 25: How many kits were produced during the eight hours?

____

Part 9.7: Commentary

- This is one of the most interesting chapters in that it illustrates how to model what would be considered a very complex operation composed not only of inventory concerns, but also supply chain issues. It's a powerful lesson and the topic of inventory and supply will be considered in additional chapters along with some of the SIMIO features illustrated here.
- The use of MATERIALS and BILL OF MATERIALS is an extremely powerful modeling concept that has numerous applications.

Chapter 10
Inventories, Supply Chains, and Optimization

Many industries/companies are facing many challenges in their effort to compete in the global marketplace. Demand variations, long lead times, and raw material supply fluctuations can result in excessive finished goods inventory and/or poor customer service levels. Many companies with longer lead times are forced into a make-to-stock policy in order to respond to their customer demands quickly and maintain market share. Too much inventory will create higher costs and may lead to product obsolescence or spoilage while to little inventory may cause stock outs and lower customer service levels. Therefore, designing and optimizing a company's supply chain has become a priority as it is becoming a necessity for the survival of many companies. The main objective of any supply chain is to supply the customer with a product when the customer wants it and at a price that maximizes profit. In order to achieve both objectives, a balance has to be struck between carrying sufficient inventories to meet demand, but not so much as to negatively impact profitability.

Part 10.1: Building a Simple Supply Chain

Figure 10.1 shows a very simple three tier supply chain where customers arrive at a store (which could be a collection of stores) to purchase a product. The stores place orders at the manufacturer DC (Distribution Center) which will supply the product if it is in stock. The DC uses an inventory model to determine reorder policies for the supplier, which is assumed to have a raw material source. It takes between three to six days to truck the product from the supplier to the manufacturer.

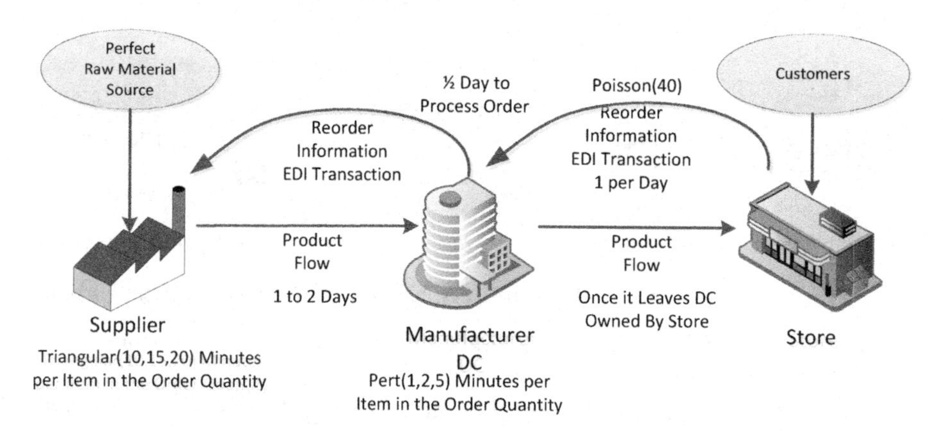

Figure 10.1: Simple Three Tier Supply Chain

The goal is to track the DC's average inventory and service level/fill rate over a year. Therefore, the store customers will not be part of the current model since the manufacturer is interested in determining the reorder point and reorder quantity. The store reorders daily from the DC and the DC utilizes a (s,S) periodic review inventory model as seen in Figure 10.2. In this model, the inventory is reviewed every fixed time interval called the period. An amount s is the reorder point, namely a quantity below which triggers a re-order. The quantity S is determined as the quantity needed after inventory is reordered (an order-up-to amount).

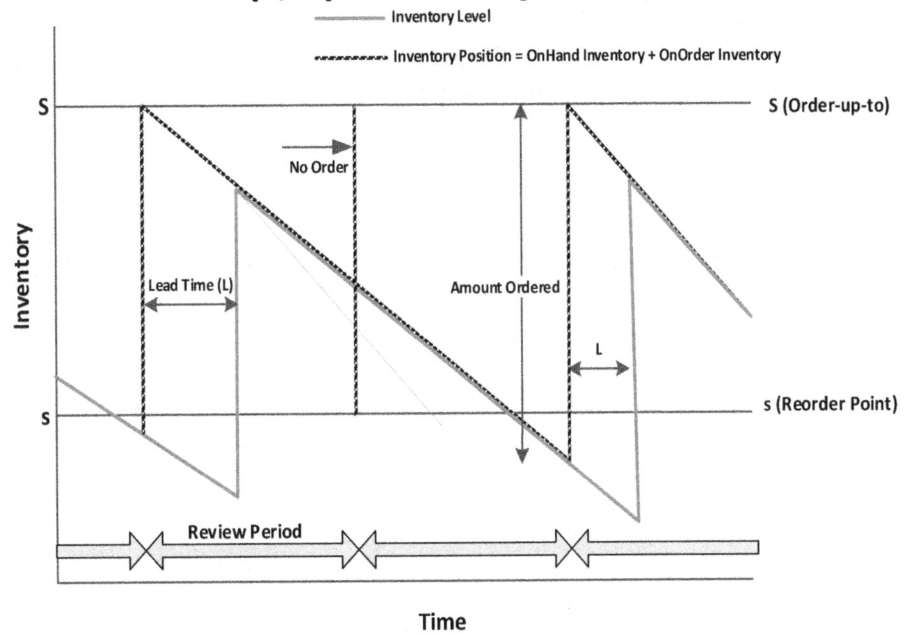

Figure 10.2: (s, S) Inventory Policy

Step 1: Create a new model with two separate systems as seen in Figure 10.3.

- The first system will model the ordering portion of the supply chain (i.e., orders from the store placed at the DC will need to be processed). Insert a SOURCE named **SrcOrders**, a SERVER named **SrvOrderProcessing** and a SINK named **SnkStore**. A new MODELENTITY named **EntOrders** should also be inserted making sure the **SrcOrders** creates these **EntOrders**. Use CONNECTORS to link the objects together, modeling EDI transactions between the stores and the DC.
- The second system will model the flow of products from the back of the chain to the front of the supply chain used to replenish the inventory in the manufacturer DC. Insert two WORKSTATIONS named **WrkSupplier** and **WrkDC** and a SINK named **SnkDCInventory** along with another MODELENTITY named **EntReplenishments**. Connect the **WrkSupplier** and **WrkDC** via a TIME PATH that uniformly takes between one and two days. Use a CONNECTOR from the **WrkDC** to the **SnkDCInventory** since the model is really only concerned with the DC.

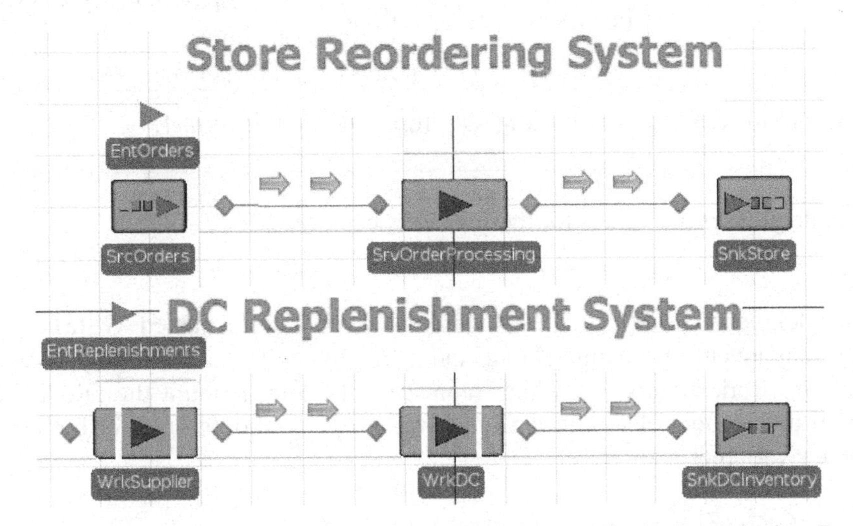

Figure 10.3: SIMIO model of the Three Tier Supply Chain Reordering Process

Step 2: Since we are not going to model every item in the inventory system as a separate entity, add a DISCRETE INTEGER STATE variable named **EStaOrderAmt** to the MODELENTITY[132] to represent the quantity of product for each store order or replenishment reorder.

Step 3: For the Ordering process, set the following information.

- Change the **SrcOrders** to have one order arrive daily.
- The amount of each order that arrives daily from the store is Poisson distributed with a mean of 40. As seen in Figure 10.4, use the *State Assignments→Before Exiting* property of the SOURCE to set the `ModelEntity.EStaOrderAmt` to a `Random.Poisson(40)`.

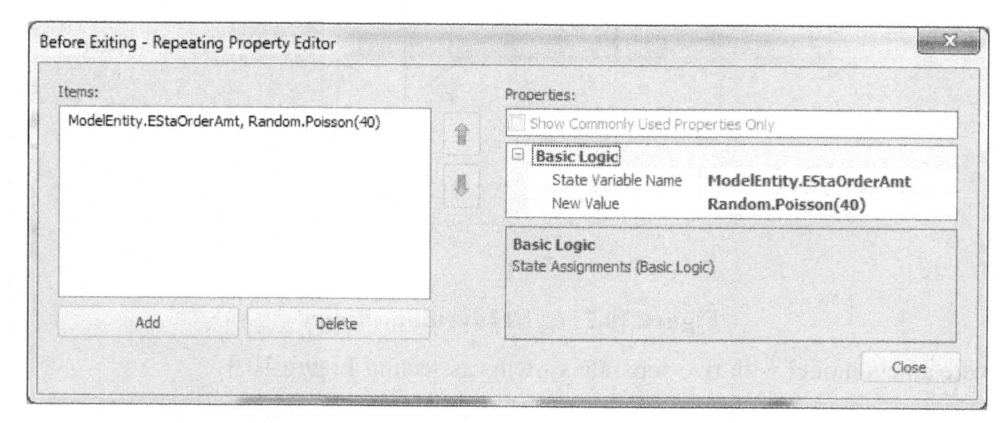

Figure 10.4: Specifying the Order Amount of each Order

Step 4: Next, under the *Definitions→Properties* of the model, insert four *Expression Standard Properties* named **InitialInventory**, **ReorderPoint**, **OrderUptoQty, and ReviewPeriod**. These should all be placed in the "Inventory" Category.[133]

Table 10.1: The Default Values for the Three Properties

Property	Default Value	Category	Unit Type	Description
InitialInventory	700	Inventory	Unspecified	The initial inventory at the start of the simulation.
ReorderPoint	300	Inventory	Unspecified	Determines the point to reorder.
OrderUptoQty	700	Inventory	Unspecified	The maximum inventory to order up to quantity.
ReviewPeriod	5	Inventory	Time (Days)	The periodic review period

Question 1: What are the advantages of setting up properties for this model?

____

Question 2: How do you set the values for these properties of the Model?

____

Step 5: In the Model, insert a DISCRETE INTEGER STATE variable named **GStaInventory** which will represent the current inventory amount ("OnHand") of the product at the DC and another one named **GStaOnOrder** (i.e., work in process) which represents the total amount the DC has currently ordered ("OnOrder") from the supplier. The *Initial State Value* property should be 0 for both the **GStaInventory** and **GStaOnOrder** variables respectively.

[132] Select the MODELENTITY in the [*Navigation*] panel and go the *Definitions→States* section to add the state variable.

[133] The first time, you will need to type the category name "Inventory" and then it can be selected from the dropdown box.

Step 6: Next, we need to insert several ELEMENTS into the model as seen in Figure 10.5.

- Insert a TALLY STATISTIC named **TallyStatSL** to track the service level performance.
- Insert a STATESTATISTIC named **StateStatInv** to track the inventory level performance. Specify the *State Variable Name* property to be the **GStaInventory** variable.
- Insert a TIMER named **TimerReview** to model the review period. The DC currently reviews their inventory level every five days and should utilize the **ReviewPeriod** property (see Figure 10.5).

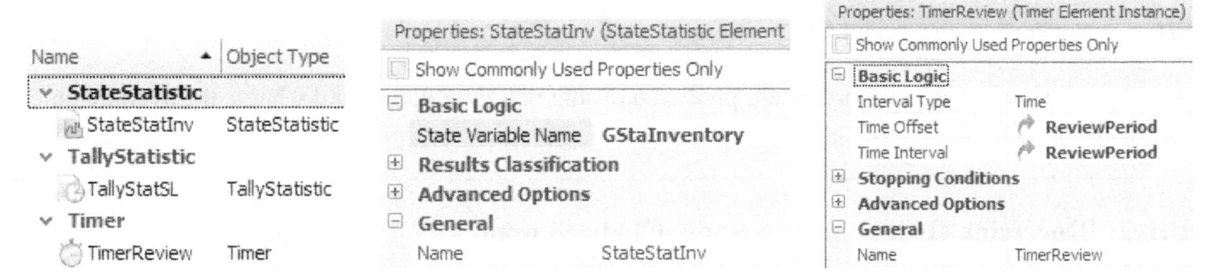

Figure 10.5: Setting up the Statistics and Periodic Review Timer

Step 7: Change the *Processing Time* property of **SrvOrderProcessing** such that orders take a ½ a day to be processed before an attempt is made to fill the order with the available inventory.

Step 8: The supplier will take between 10 and 20 minutes with the most likely time of 15 minutes to process each product. Therefore, the total processing time for one order would be the sum of a number of triangular distributions (i.e., an order amount).[134] The workstation handles the processing of the batch correctly by sampling from the distribution the appropriate number of times. For the **WrkSupplier**, specify the *Operation Quantity* property to be `ModelEntity.EStaOrderAmt`, the *Processing Time* to be the individual process time (in minutes), and the *Processing Batch Size* property to "`1`" as seen in Figure 10.6.[135] Since the batch size is one, it will produce a set of sequential batches.

Operation Quantity	**ModelEntity.EStaOrderAmt**
⊟ Setup Time Type	Specific
⊞ Setup Time	0.0
Processing Batch Size	**1**
⊞ Processing Time	**Random.Triangular(10,15,20)**
⊞ Teardown Time	0.0

Figure 10.6: Specifying the Operation Quantity and Batch Size for Processing of an Order

Step 9: Once the product arrives at the **DC**, the entire lot must be individually packaged before it is available to satisfy demands from the store. Therefore set the **WrkDC's** *Processing Time* property to the following expression `Random.Pert(1,2,5) in minutes`, the *Processing Batch Size* property to "`1`" and the *Operation Quantity* property to be `ModelEntity.EStaOrderAmt` just like the **WrkSupplier**.

---

[134] One maybe inclined to use `ModelEntity.OrderAmt*Random.Triangular(10,20,30)` for the *Processing Time* property. However, this will not correctly model the summed distribution as seen in the example in Appendix A.

[135] If one does not specify the *Processing Batch Size*, it is assumed to be same as the *Operation Quantity* therefore processing the entire quantity as only one batch rather a set of sequential batches.

Operation Quantity	**ModelEntity.EStaOrderAmt**
⊟ Setup Time Type	Specific
⊞ Setup Time	0.0
Processing Batch Size	**1**
⊞ Processing Time	**Random.Pert(1,2,5)**
⊞ Teardown Time	0.0

Figure 10.7: Setting up the DC WORKSTATION Properties to Process the Batch

Step 10: Save and run the model for 26 weeks.

Question 3: How many store orders are produced during that time period and what is the average time in the system?

____

Part 10.2: Processing Orders in the Supply Chain System

Recall, the initial inventory level property(i.e., ***InitialInventory***) was created to allow for changing the initial condition at the model level.

Step 1: Since state variable default values cannot be reference properties, the state variable **GStaInventory** needs to be initialized at the beginning of the simulation via the *OnRunInitialized* process of the model. Insert this process by selecting it via the *Processes→Select Process→OnRunInitialized* drop down menu. Insert an `Assign` step that assigns the **GStaInventory** state variable the ***InitialInventory*** as seen in Figure 10.8.

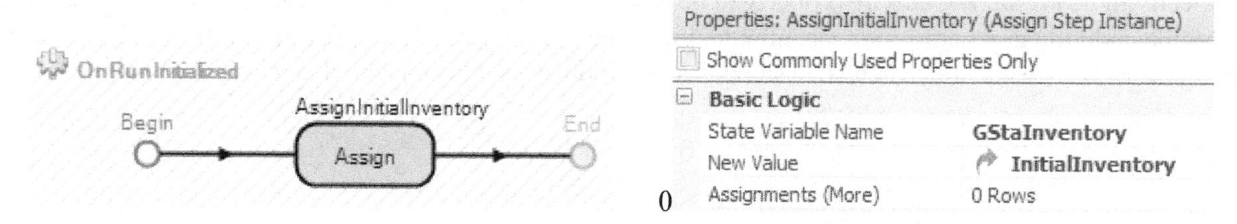

Figure 10.8: Assigning the Initial Inventory Level

Step 2: When **EntOrders** arrive at the DC and have been processed, they need to be filled based on the current inventory position.[136] Therefore, insert a new "*After Processing*" add-on process trigger for the **SrvOrderProcessing** SERVER to handle the behavior (i.e., update the current inventory position and the current service level).

- Add a `Tally` step which will update the TALLY STATISTIC **TallyStatSL** with the value equal to `Math.Min(1,GStaInventory/ModelEntity.EStaOrderAmt)`.[137]
- Next, insert an `Assign` step that will update the current **GStaInventory** level by subtracting the **EStaOrderAmt** of the particular order from it. However, if the **GStaInventory** level is less than the order amount it would be negative and since we will not allow backorders set the level to the following expression as seen in Figure 10.10.
    - `Math.Max(0,GStaInventory-ModelEntity.EStaOrderAmt)`

____

[136] Inventory position is defined as on-hand inventory plus on-order inventory (to account for orders that haven't arrived).

[137] Some may argue this service level calculation is really a fill rate, if the inventory is greater than the order amount then 100% of the demand appears to be satisfied.

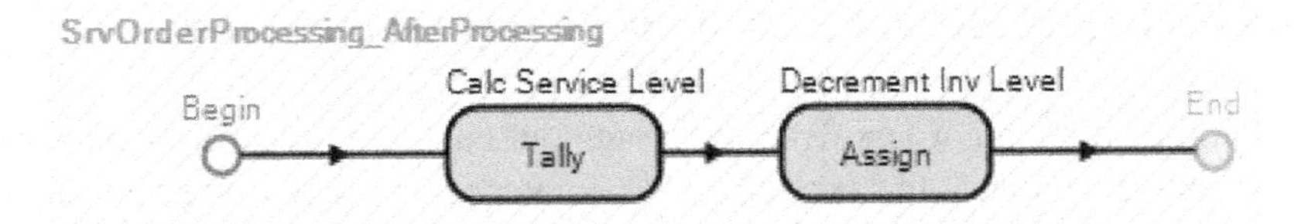

Figure 10.9: Process to Fill Demands from the Stores

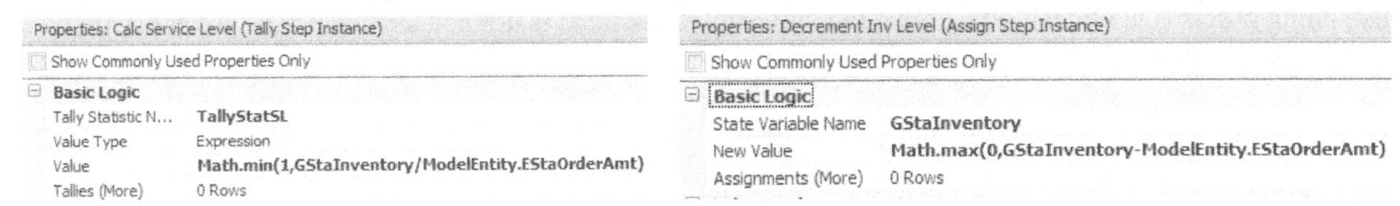

Figure 10.10: Properties of the Steps

Step 3: To track the information while the simulation is running insert eight status labels from the "*Animation*" tab, as seen in Figure 10.11, that keeps track of the current level, the average inventory and service level. The first column of labels is the description while the second column uses the expressions `GStaInventory`, `GStaOnOrder`, `StateStatInv.Average`, and `TallyStatSL.Average`.

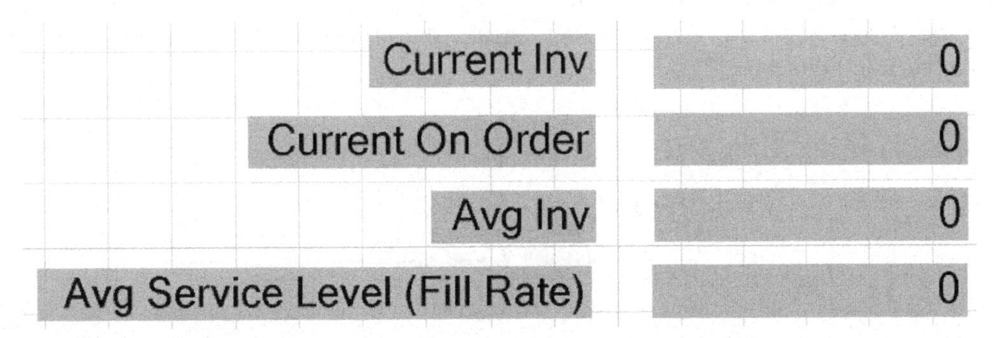

Figure 10.11: Status Labels to Show Current Inventory Level and Average Inventory and Fill Rate

Step 4: From the "*Animation*" tab, insert a new STATUS PLOT that will track the current onhand inventory level, the replenishment on order, and the inventory position which represents the onhand plus the on order. Using the *Additional Expression* property, add three items to the plot as seen in **Figure 10.12** and make the Time Range equal to "1" week.

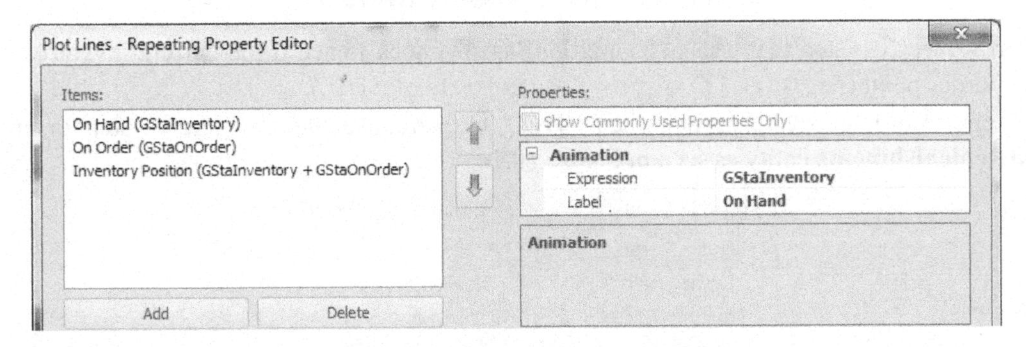

Figure 10.12: Setting up the Status Plot to Track Inventory and On Order Amounts

Step 5: Save and run the model (you may need to adjust the speed factor to 5000 or fast forward).

Question 4: What is the current inventory level?

____

Question 5: What is the average inventory level and service level?

____

Part 10.3: Creating the Replenishment Part of the Supply Chain System

The previous section handled demand orders from the stores but once the inventory was depleted, there was no replenishment of the inventory. Hence, the service level was very poor. The timer was setup to fire every five days, at which time if the current inventory is below the reorder point, a replenishment order should be sent to the supplier based on the order-up-to quantity, the current inventory level and the current number of outstanding orders (On Order). Recall that the "inventory position" is defined as the current inventory on hand plus the inventory represented in outstanding orders.

Step 1: Create some statistics on the replenishment procedure:

- Insert a `STATE STATISTIC`, called **StateStatOnOrder,** to keep track of the amount of inventory on order over time. This statistic should reference the **GStaOnOrder** state variable.
- Insert a `TALLY STATISTIC`, called **TallyStatAmtOrdered**, to record the amount that is ordered from the supplier.

Step 2: Setup the periodic review by creating a process to respond to the timer **TimerReview**. From the "*Processes*" tab, create a new process named **InventoryReview.** Set the *Triggering Event* property to respond to the `TimerReview.Event` which causes the process to be executed every time the timer event fires as seen in Figure 10.13.

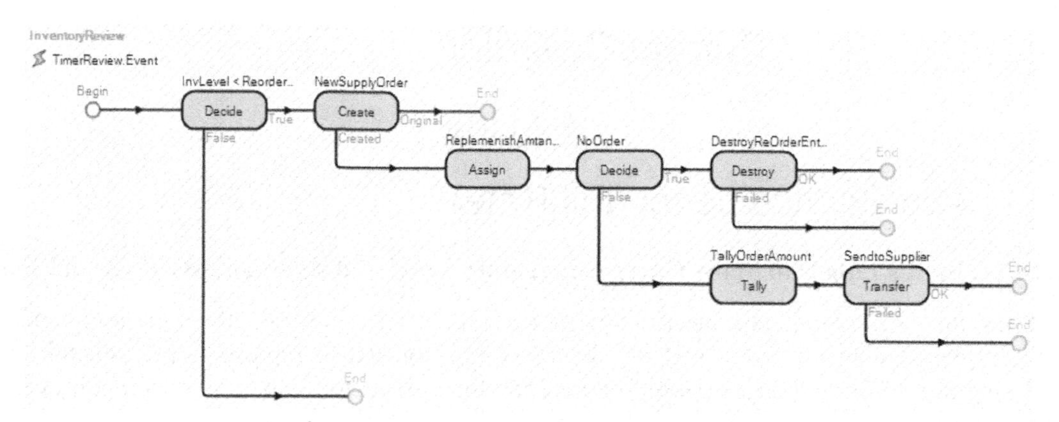

Figure 10.13: Process to Reorder

- Insert a *`Decide`* step that uses a "ConditionBased" check to see if the current inventory is less than the reorder point (i.e., `GStaInventory < ReorderPoint`).
- If a new reorder appears needed (i.e., "True" branch), use the *`Create`*[138] step to create a new **EntReplenishments** entity as seen in Figure 10.14.

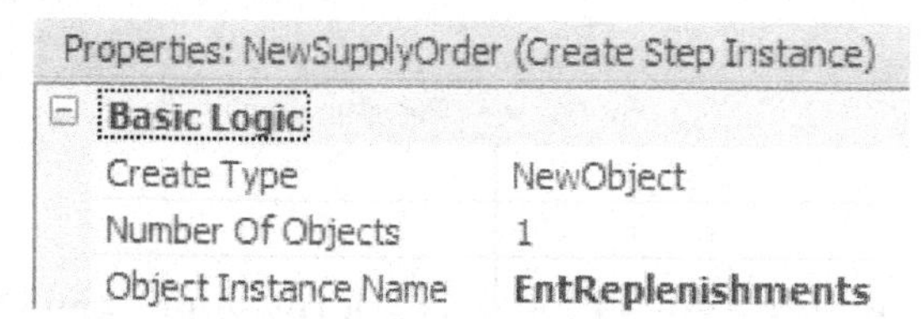

Figure 10.14: Creating a New ReOrder to Send the Supplier

- For the newly created **EntReplenishments**, use an *`Assign`* step to assign the order amount and then increase the **StaOnOrder** value based on the order amount.

---

[138] The *`Create`* step can create new objects or copies of the associated or parent objects. Each of the created objects will be executed by their own token. Note that created tokens are updated in SIMIO before the original.

Table 10.2: Determining the Order Amount and Updating the WIP

State Variable	New Value
ModelEntity.EStaOrderAmt	`Math.Max(0,OrderUpToQty -GStaInventory +GStaOnOrder)`
GStaOnOrder	`GStaOnOrder + ModelEntity.EStaOrderAmt`

- In the second `Decide`, determine if the **EStaOrderAmt** is zero. Namely, no replenishment order is needed (an assumption here is that there are never any zero quantity replenishment orders)
- If no replenishment order is needed, then the replenishment order entity is destroyed using the `Destroy` step.
- Otherwise, the amount of the replenishment order **(EStaOrderAmt)** is tallied via the **TallyStatAmtOrdered.**
- Once the **EntReplenishment** has been tallied, it needs to be sent to the supplier by using a `Transfer` step to move it from "FreeSpace" to the **Input** node of the **WrkSupplier.**[139]

Properties: SendtoSupplier (Transfer Step Instance)

Basic Logic	
From	FreeSpace
To	Node
Node Name	**Input@WrkSupplier**

Figure 10.15: Sending the New Reorder to the Supplier

Step 3: Now, when the products arrive back at the DC and have been packaged, the **GStaInventory** variable needs to be increased by the order amount while the **GStaOnOrder** variable is reduced by the same value. Insert the "*FinishedGoodOperation*" add-on process trigger for the **WrkDC** as seen in Figure 10.16. Insert an `Assign` step that updates the values in Table 10.3.

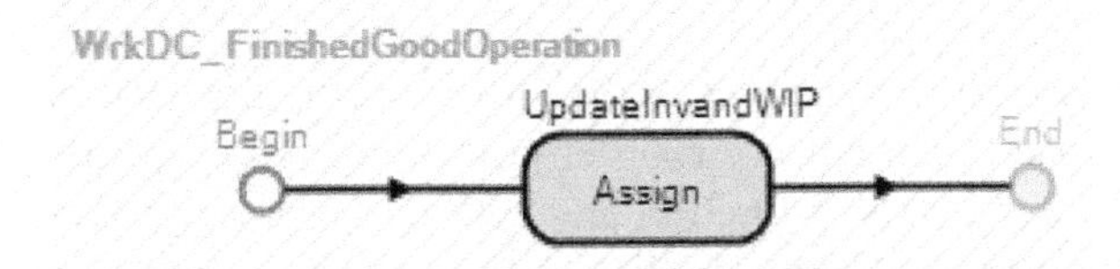

Figure 10.16: Updating the Inventory and WIP Values

Table 10.3: Updating Inventory and DC once Product Arrives

State Variable	New Value
`GStaInventory`	`GStaInventory + ModelEntity.EStaOrderAmt`
`GStaOnOrder`	`GStaOnOrder - ModelEntity.EStaOrderAmt`

Step 4: Save and run the model (you may need to fast forward the simulation) for 52 weeks.

Question 6: What is the average inventory level, the inventory on order and the service level?

____

Question 7: What is the average amount ordered?

____

Question 8: Is this an adequate system?

____

[139] When entities are created, they are placed in "FreeSpace" and, typically, will need to be "transferred" from there to one of the model nodes or stations.

Step 5: You may notice that the status labels display as many significant digits that are available for both the average inventory and the service level/fill rate. SIMIO exposes many of the .Net string manipulation functions. For example, `String.Format` function can be used to format the output.[140] Replace the two status labels for the average and service level with the ones in Table 10.4.

Table 10.4: Formatting the Output of the Status Labels

Status Label	Expression
Average Inventory	`string.format("{0:0.##}",StateStatInv.Average)`
Average OnOrder	`string.format("{0:0.##}",StateStatOnOrder.Average)`
Service Level Status	`string.format("{0:0.00%}",TallyStatSL.Average )`

Step 6: Save and run the model (you may need to adjust the speed factor to 100 or fast forward the simulation).

Question 9: Was the average inventory level, on order, and service level now formatted correctly?

____

Part 10.4: Using an Experiment to Determine the Best Values

Simulation is often used to improve a system. Can we find a better order-up-to quantity and re-order point that improves the overall performance of the system? Can SIMIO provide help with this kind of question? Or can SIMIO help in searching for an improved system? The answer is yes to all these questions.

Step 1: From the *Project Home→Create* section, insert a new "*Experiment*" named **FirstExperiment** which will create a new experiment window and automatically create the first scenario which is the base model as seen in Figure 10.17. SIMIO automatically added the four properties as control variables.[141]

Step 2: Insert eight more scenarios by clicking the little box in the last row. This will copy all of the parameters from the base case. Then change reorder points and order up to quantities to do a full factorial experiment of the three reorder points (200, 300, and 400) and order up to quantity (600, 700, 800) as seen in Figure 10.17. Change the names of the scenarios so they can be interpreted.

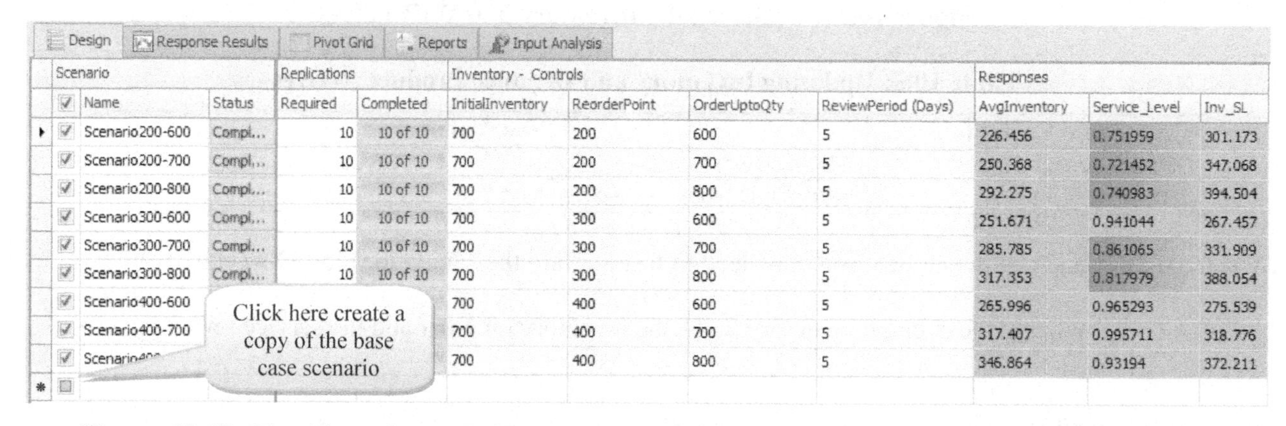

Name	Status	Required	Completed	InitialInventory	ReorderPoint	OrderUptoQty	ReviewPeriod (Days)	AvgInventory	Service_Level	Inv_SL
Scenario200-600	Compl...	10	10 of 10	700	200	600	5	226.456	0.751959	301.173
Scenario200-700	Compl...	10	10 of 10	700	200	700	5	250.368	0.721452	347.068
Scenario200-800	Compl...	10	10 of 10	700	200	800	5	292.275	0.740983	394.504
Scenario300-600	Compl...	10	10 of 10	700	300	600	5	251.671	0.941044	267.457
Scenario300-700	Compl...	10	10 of 10	700	300	700	5	285.785	0.861065	331.909
Scenario300-800	Compl...	10	10 of 10	700	300	800	5	317.353	0.817979	388.054
Scenario400-600				700	400	600	5	265.996	0.965293	275.539
Scenario400-700				700	400	700	5	317.407	0.995711	318.776
Scenario4				700	400	800	5	346.864	0.93194	372.211

Figure 10.17: First Experiment Trying Different Reorder Points and Order Up to Quantities

[140] The syntax for the format function is `String.Format(string, arg1, arg2, ...)` where the format "string" can contain place holders that are enclosed by {} which start with the 0th argument (i.e., {0}). For example, `String.Format("SL = {0} and the Avg = {1}", 0.45, 89.4)` has two arguments {0} and {1} and replaces those with the numbers 0.45 and 89.4 respectively to produce a 'SL=0.45 and the Avg = 89.4'. Along with the arguments, one can specify formatting information especially as applied to numbers by using a colon on the argument then specifying the number format. For example, {0:#.##} will display only two significant digits while {0:#.00%} will display a percentage and force there to always be two significant digits.

[141] Other control variables can be defined.

Step 3: Using the *Design→Experiment→Add Response* button, insert three responses which will be used (see Figure 10.17) to select the best scenario with the following parameters specified in Table 10.5. The first two responses seem very logical. In this model these responses are typically conflicting (i.e., as the inventory level is minimized the service level goes down). Therefore, the third one converts the multi-objective problem into a single response. As the service level goes up and the inventory goes down, this new response goes down. The lower and upper bounds on the responses will be used to flag responses outside a stipulated range.[142] In this case we would like to have a service level greater than 90%.

Table 10.5: Experimental Responses

Name	Expression	Objective	Lower Bound
AvgInventory	`StateStatInv.Average`	Minimize	
Service_Level	`TallyStatSl.Average`	Maximize	0.90
Inv_SL	`StateStatInv.Average/TallyStatSL.Average`	Minimize	

Step 4: Figure 10.18 shows the basic experiment parameters where you can define the warm up period, the default number of replications[143] that all scenarios will run. For the SMORE plots and optimization, the confidence interval and upper and lower percentiles displayed. Also, you need to define the primary response that will be used by the ranking and selection and Optquest™ add-ins (see next sections).

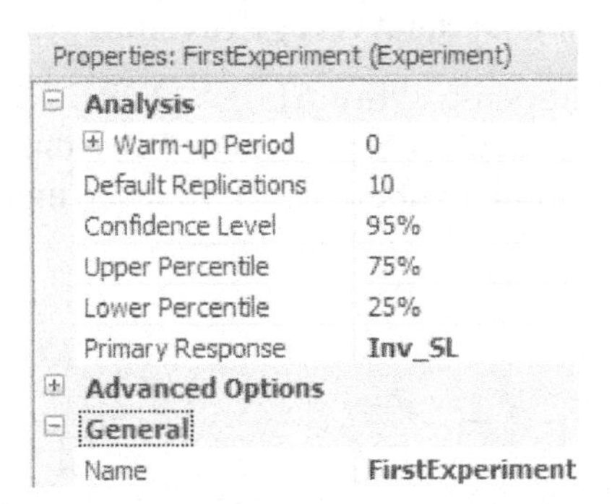

Figure 10.18: Setting up the Experiment Parameters

Step 5: Save the model and run the experiment.

Question 10: Based on the responses, which reorder point and order up to quantity seems to be the best?

____

Question 11: How did you make this determination?

____

Part 10.5: Using SMORE Plots to Determine the Best Values

In the previous section, we just used the average of the ten replications to assist you in choosing the best selection. The difficulty is that the variability of the system is not accounted for when choosing the best scenario. "SIMIO Measure of Risk & Error" (SMORE) plots offer the ability to see the variability, along with confidence intervals. They show the major percentiles (median or 50th percentile, lower chosen percentile and upper chosen percentile) and mean, and confidence intervals for the mean and the percentiles.

Step 1: Select the "Response Results" tab from the previously run model to see the SMORE plots. The primary response (Inv_SL) will be displayed by default but the other response can be selected from the dropdown as seen in Figure 10.19. Quickly we can see that from the "AvgInventory" response

[142] The values in the response that don't fall within the bounds are shown in a red gradient color.

[143] The default is ten replications, although more replications will increase your confidence in the results.

Scenario200-600 seems to be the best and its variation is clear of the other plots. Scenio300-600 appears to be the second best; however its variation overlaps with Scenario200-700. However when looking at the service level with the limits turned on, only Scenario300-600, Scenario400-600, Scenario400-700 and Scenario400-800 meet the 90% service level limit with the Scenario400-700 seeming statistically the best. Considering the two plots together, Scenerio300-600 appears the better choice.

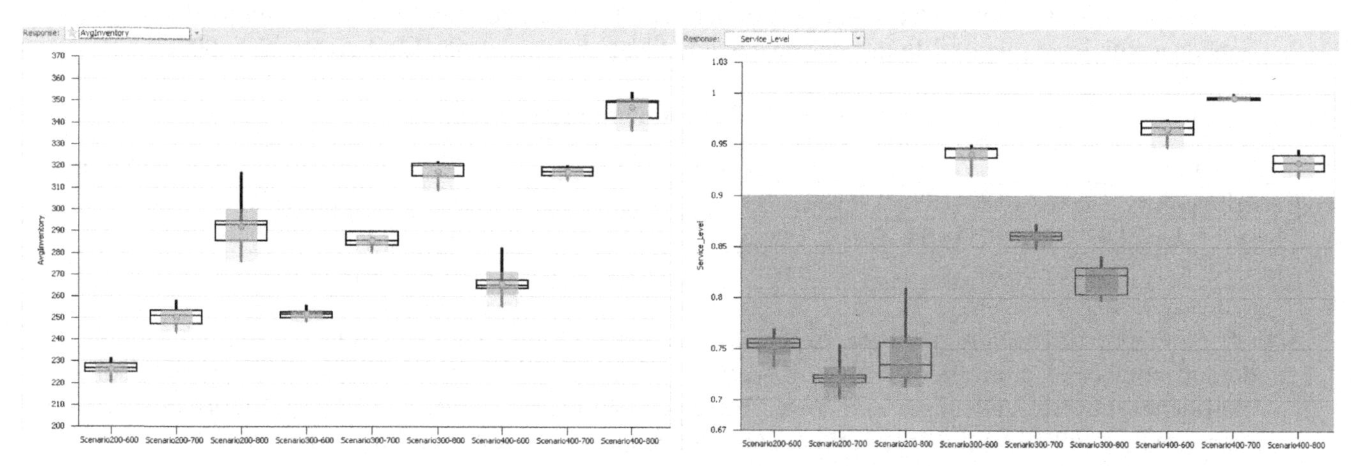

Figure 10.19: SMORE Plots of the Average Inventory and Average Service Level

Step 2: Looking at the Primary Response ("Inv_SL") SMORE Plot in Figure 10.20, Scenario300-600 appears to be statistical the best, based on the combined response. Its SMORE plot does, however, overlap with Scenario400-600 **minimally** but not the confidence intervals.[144]

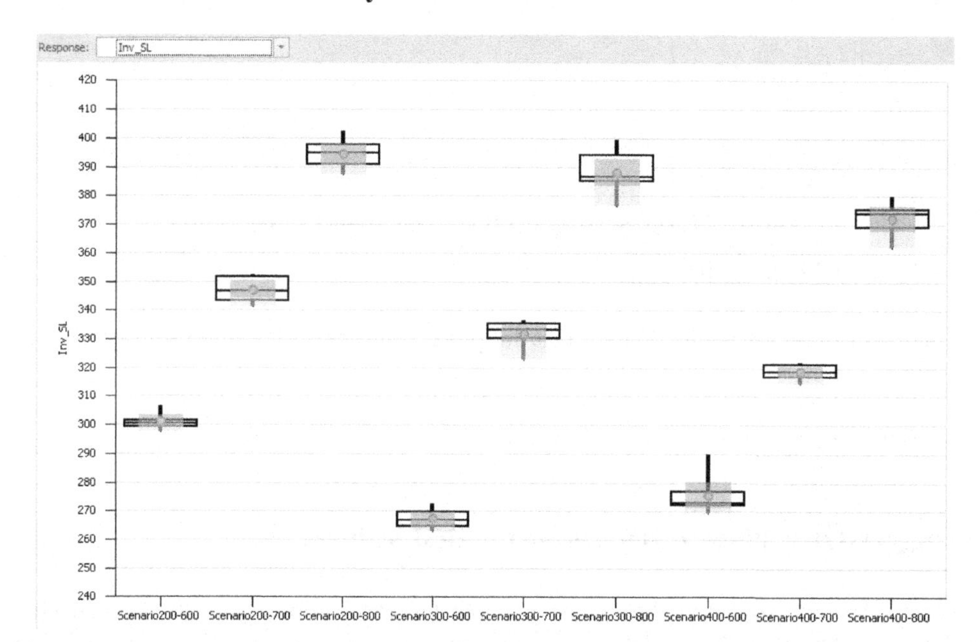

Figure 10.20: SMORE Plots of the Combined Inventory/Service Level Response

Step 3: By clicking the "*Subset Selection*" option in the "*Analysis*" section of the ribbon in the "*Design*" tab, you can see the choices made by a series of algorithms within SIMIO. Scenarios for each response are divided in to the "best possible group" and the "rejects group", as shown in Figure 10.21.

[144] Refer back to Chapter 8 for more explanation of SMORE plots and their meanings.

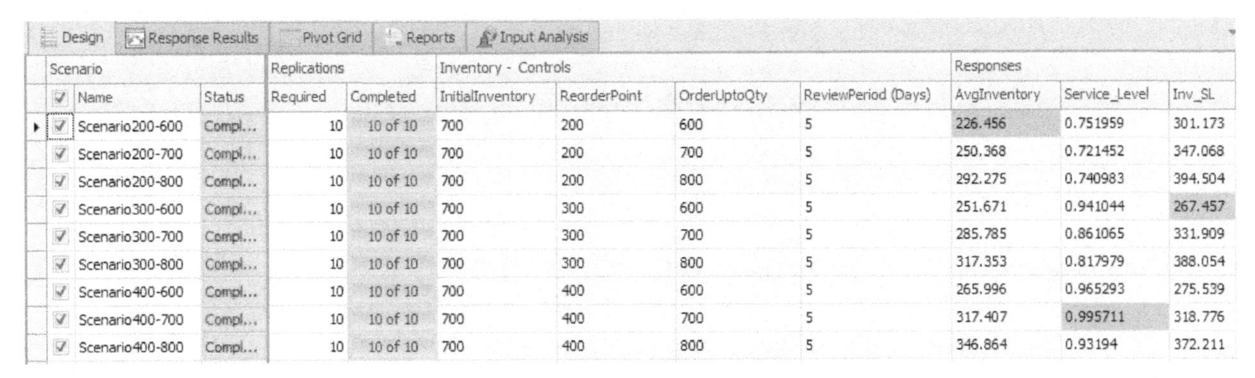

Scenario			Replications		Inventory - Controls				Responses		
✓	Name	Status	Required	Completed	InitialInventory	ReorderPoint	OrderUptoQty	ReviewPeriod (Days)	AvgInventory	Service_Level	Inv_SL
✓	Scenario200-600	Compl...	10	10 of 10	700	200	600	5	226.456	0.751959	301.173
✓	Scenario200-700	Compl...	10	10 of 10	700	200	700	5	250.368	0.721452	347.068
✓	Scenario200-800	Compl...	10	10 of 10	700	200	800	5	292.275	0.740983	394.504
✓	Scenario300-600	Compl...	10	10 of 10	700	300	600	5	251.671	0.941044	267.457
✓	Scenario300-700	Compl...	10	10 of 10	700	300	700	5	285.785	0.861065	331.909
✓	Scenario300-800	Compl...	10	10 of 10	700	300	800	5	317.353	0.817979	388.054
✓	Scenario400-600	Compl...	10	10 of 10	700	400	600	5	265.996	0.965293	275.539
✓	Scenario400-700	Compl...	10	10 of 10	700	400	700	5	317.407	0.995711	318.776
✓	Scenario400-800	Compl...	10	10 of 10	700	400	800	5	346.864	0.93194	372.211

Figure 10.21: Using Subset Selection

While the best possible group may consist of scenarios that cannot be proven statistically different from each other, it can be shown that the best possible group scenarios will be statistically better than the rejects group (shown in muted color).

Question 12: Which scenario(s) is(are) statistically selected as the best for the three responses?

Part 10.6: Using Ranking and Selection to Determine the Best Scenario

The analysis is quite simple (focused on only two variables within the model) and ten replications seemed to be enough to choose the best. However, using the SMORE plots still requires a judgment or visual identification. Ranking and selection methods allow you to determine statistically which scenario is the best. One of the advantages of SIMIO is the built in state-of-the-art ranking and selection method based on research by Kim and Nelson (called "KN").[145]

Step 1: Save the current model and copy the **FirstExperiment** by right clicking it in the [*Navigation*] panel and choosing the *Duplicate Experiment* menu item. Name the new experiment **KNExperiment**.

Step 2: Next, choose the "*Select Best Scenario using KN*" from the *Design→Add-Ins* section seen in Figure 10.22.

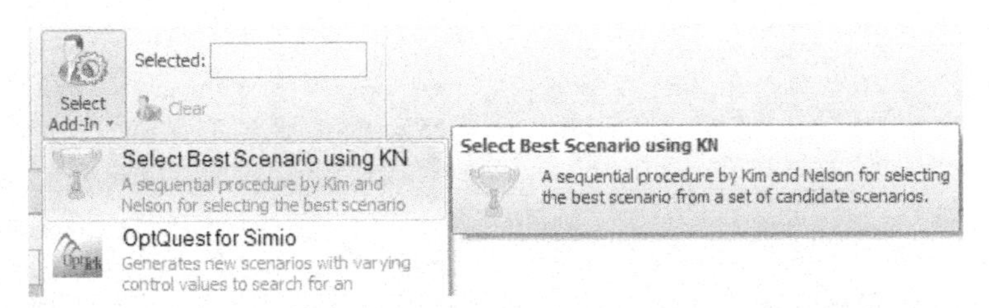

Figure 10.22: Selecting the KN Algorithm Add-in

Step 3: This will add the KN algorithm to the experiment (see Figure 10.23) as well as give you additional properties.[146] Set the *Indifference Zone* property to five, which stipulates that one solution, can be determined better than another only if the two solutions differ by more than five. Since the ranking and selection scheme will change the number of replications during its execution, the *Replication Limit* property specifies the maximum number that can be run.

---

[145] S. Kim and B. L. Nelson, "A Fully Sequential Procedure for Indifference-Zone Selection in Simulation," *ACM Transactions on Modeling and Computer Simulation* 11 (2001), 251-273.

[146] To access the EXPERIMENT properties, right click the experiment in the [*Navigation*] panel.

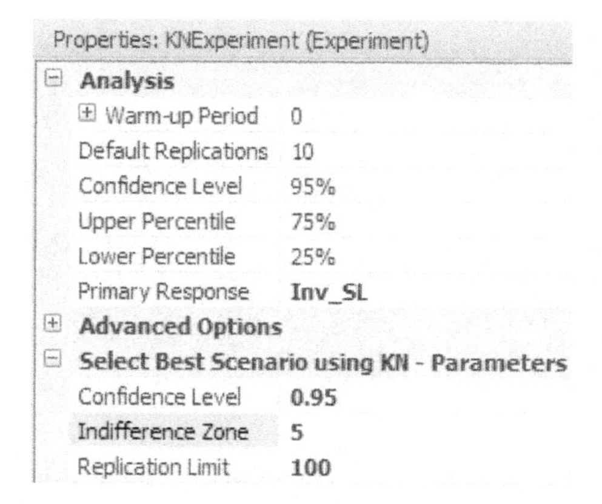

Figure 10.23: Setting up the KN Parameters

Step 4: Set the number of "Required" replications (under the Replications category) each to one and make sure that each scenario is "checked" (in left column) and "Reset" the experiment.

Step 5: Run the experiment and notice that the number of replications required for each scenario increases to ten. The KN algorithm requires at least ten replications in order to reliably detect a difference among scenarios.

Question 13: Based on the differences among the scenarios **Inv_SL**, do you think a large number of replications are needed to determine differences greater than five?

____

Question 14: What does running more replications do for the analysis?

____

Step 6: The KN algorithm has selected Scenario300-600 as the best (it is the only one checked), as seen in Figure 10.24. In more complex models with scenarios that are more similar, more replications than ten will be needed.

Scenario			Replications		Inventory - Controls				Responses		
☐	Name	Status	Required	Completed	InitialInventory	ReorderPoint	OrderUptoQty	ReviewPeriod (Days)	AvgInventory	Service_Level	Inv_SL
☐	Scenario200-600	Compl...	10	10 of 10	700	200	600	5	226.456	0.751959	301.173
☐	Scenario200-700	Compl...	10	10 of 10	700	200	700	5	250.368	0.721452	347.068
☐	Scenario200-800	Compl...	10	10 of 10	700	200	800	5	292.275	0.740983	394.504
☑	Scenario300-600	Compl...	10	10 of 10	700	300	600	5	251.671	0.941044	267.457
☐	Scenario300-700	Compl...	10	10 of 10	700	300	700	5	285.785	0.861065	331.909
☐	Scenario300-800	Compl...	10	10 of 10	700	300	800	5	317.353	0.817979	388.054
☐	Scenario400-600	Compl...	10	10 of 10	700	400	600	5	265.996	0.965293	275.539
☐	Scenario400-700	Compl...	10	10 of 10	700	400	700	5	317.407	0.995711	318.776
☐	Scenario400-800	Compl...	10	10 of 10	700	400	800	5	346.864	0.93194	372.211

Figure 10.24: Results of running the KN

Step 7: Verify that Scenario300-600 is selected and examine its SMORE plot in the "Response Chart".

Question 15: What is the 95% confidence interval on the mean?

____

Step 8: Reset the experiment and make sure to select all the scenarios again. Change the indifference zone to one, and rerun the experiment to see the impact.

Question 16: Did KN need more than ten replications to distinguish the best scenario and which scenario did it select this time.

____

Part 10.7: Using OptQuest™ to Optimize the Parameters

The KN algorithm used in the previous section determined the optimal scenario from a fixed set of scenarios defined by the user. SIMIO has another add-in optimizer OptQuest™ that can perform traditional optimization by trying other values for the decision variables not specified. It will also allow constraints on the decision variables and output responses.[147]

Step 1: Save the current model and duplicate the **FirstExperiment** and then rename the new experiment **OptQuestExperiment**.

Step 2: Use the "*Clear*" button to clear out the Select Best Scenario add-in and then select the "OptQuest for SIMIO" add-in if you duplicated the **KNExperiment** instead.

Step 3: Since the optimizer will try different values for our control (decision) variables, we need to define their valid ranges. Select appropriate ranges and increment values to speed up the convergence and potentially the solution quality. Select each of the control variables and set their values according to Figure 10.25. As you can see, the **InitialInventory** is not included in the optimization. An *increment* value of 25 has been chosen for the other two variables to allow the algorithm to only select values in increments of 25.[148]

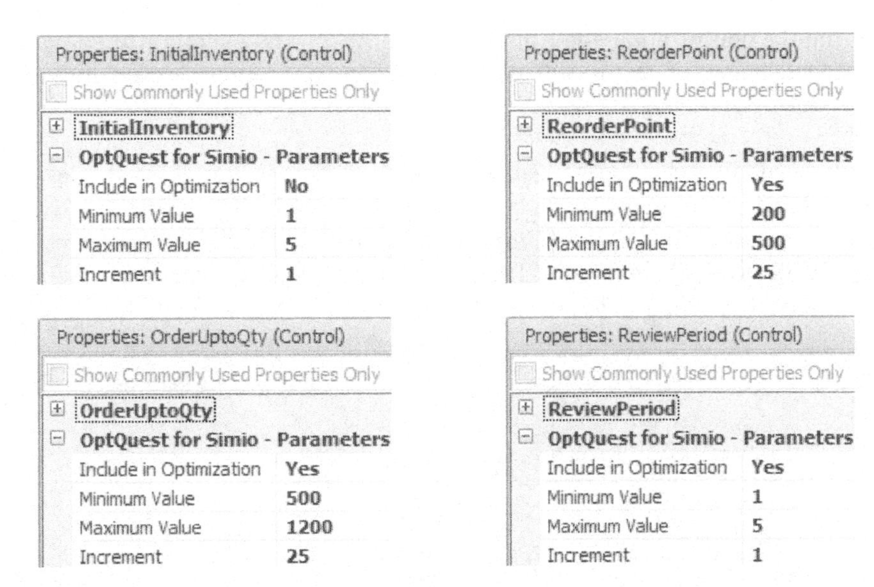

Figure 10.25: Setting the Lower and Upper Bounds on the Decision Variables

Step 4: Next, choose the "OptQuest™ for SIMIO" from the *Design→Add-Ins* section to insert the optimization algorithm. It will add a setup section in the EXPERIMENT properties as seen in Figure 10.26. The algorithm does use a confidence level to distinguish among solutions. The relative error percentage of the confidence level is expressed as a percent of the mean. Notice, that a range of replications is specified and a maximum number of scenarios is specified. These will limit the time OptQuest will spend computationally in the simulations and search for a better solution.

[147] OptQuest is an additional add-in that has to be purchased.

[148] Finer resolutions (i.e., smaller values) may lead to better solutions. However, this can affect the solution quality of the optimization algorithm. Having a course resolution quickly allows the algorithm to find a good region. A second optimization with a finer resolution could be run around the point found by the first optimization problem.

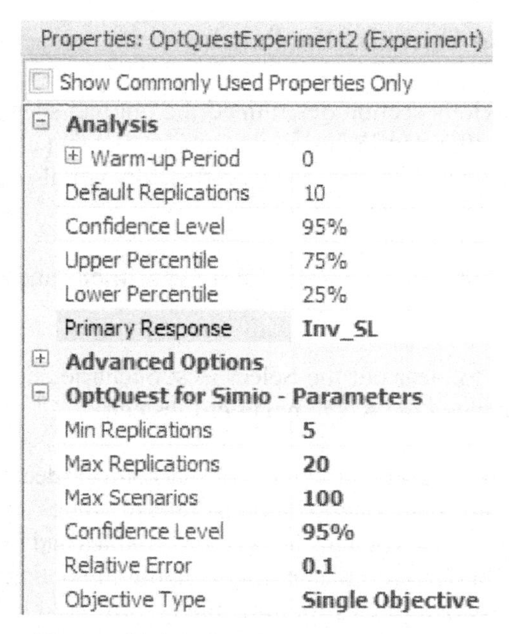

Figure 10.26: OptQuest Parameters

Step 5: Save and run the optimization algorithm. Notice how it tries several different variable values of the controls. It will create a maximum of 100 scenarios or decision points as specified the *Max Scenarios* property.

Step 6: Notice only scenarios which are feasible with respect to constraints on the responses (i.e., service level greater than 90%) are checked. Click on the little funnel ( ) on the upper right corner of the **Service_Level** response and choose the [*Custom*] filter option. Setup the filter to only allow feasible scenarios to be visible as seen in Figure 10.27. Then sort the scenarios by the **Inv_SL** column in ascending order to see those with minimum **Inv_SL** ratios by right clicking on the column heading.

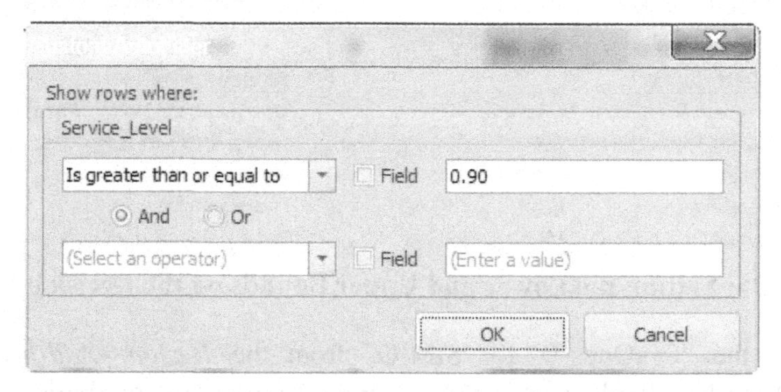

Figure 10.27: Filtering only Solutions which meet the Threshold of 90% Service Level

Question 17: What appears to be the best reorder point and order up to quantity values considering not only the **Inv_SL** ratio but the service level?

____

Step 7: Since the OptQuest™ only used five replications; it is possible that the variability may be a problem. Keep the top ten (or some other number) scenarios and delete the rest. Since the optimization parameters (i.e., bounds on the control variables will be lost), first duplicate the experiment and name it **OptQuestExperiment2** by right clicking on the name in the [*Navigation*] panel.

Step 8: Select the **OptQuestExperiment** from the [*Navigation*] panel and clear the *OptQuest* add-in and add in the *Select Best Scenario* and make sure the *Primary Response* is **Inv_SL**. Use an *Indifference Zone* property of "1" and then rerun the KN algorithm.

Question 18: What did you get for the best Reorder Point and Order Up to Quantity?

Question 19: What alternative do you choose if you want the service level to be at least 90%?

Step 9: To see that one is better than the others, examine the *Response Chart* for the one scenario KN selected, as seen Figure 10.28. Note you must "check" the alternative scenarios to see them in the *Response Chart*. In our example, Scenario 052 (i.e., 300 reorder point and 550 (order-up-to quantity) was deemed to be the best with one as the indifference zone.

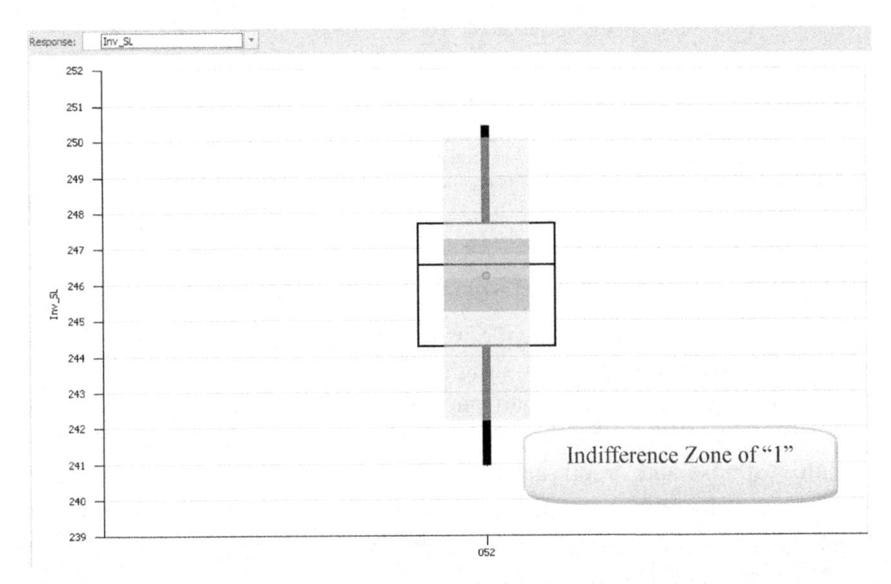

Figure 10.28: Viewing the Best Scenaro

Part 10.8: Multi-objective and Additional Constraints using OptQuest™

While the KN selection algorithm can statistically determine which scenario is the best, it cannot consider constraints on the responses. Constraints on the control variables should be handled since the scenarios selected should be feasible. Unfortunately, many problems often have multiple conflicting objectives and it is rare that a single point optimizes all the objectives. Figure 10.29 depicts the Efficiency Frontier of a problem which is optimizing both service level and inventory level at the same time. The frontier represents all non-dominated points that a decision maker could choose. From the figure, the inventory level could be reduced in half if they would be willing to go from 98% service level to around 90% service level.

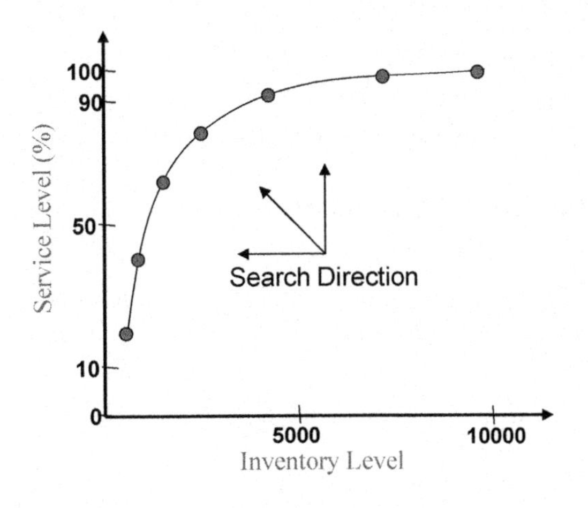

Figure 10.29: Efficiency Frontier

Finding the frontier is very difficult. To help alleviate the problem, many people try to convert the multi-objective problem into a single objective ($g(x)$) problem. The most common way and easiest way is to aggregate the objectives into a single objective by summing up the weighted objectives (i.e., $g(x) = \sum_{i=1}^{k} w_i f_i(x)$). The decision maker can determine the importance of each objective by choosing the weights accordingly. SIMIO has the ability to automatically create the combined function by specifying the appropriate weights of the responses and the choosing the **Multi-Objective Weighted** value as the *Objective Type* property of the EXPERIMENT as seen in Figure 10.30.

Properties: AvgInventory (Response)	
OptQuest for Simio - Parameters	
Weight	1
General	
Name	**AvgInventory**
Display Name	AvgInventory
Expression	**StateStatInv.Average**
Unit Type	Unspecified
Objective	**Minimize**
Lower Bound	
Upper Bound	

Properties: Service_Level (Response)	
OptQuest for Simio - Parameters	
Weight	2
General	
Name	**Service_Level**
Display Name	Service_Level
Expression	**TallyStatSL.Average**
Unit Type	Unspecified
Objective	**Maximize**
Lower Bound	
Upper Bound	

OptQuest for Simio - Parameters	
Min Replications	5
Max Replications	**20**
Max Scenarios	**150**
Confidence Level	**95%**
Relative Error	**0.1**
Objective Type	**Multi-Objective Weighted**

Figure 10.30: Using the Multi-Objective Aggregate Optimization Method

In this example, the decision maker is putting twice as much importance on service level than inventory. The difficulty with this multi-objective approach is scaling the weights to find solutions. In this example, inventory is in the hundreds while service level is less than one. In many cases, the optimization algorithm will only optimize the dominate objective. While this may be a convenient method, one can accomplish the same thing by creating an aggregate response and specifying the expression to be `StateStatInv.Average + 2 * TallyStatSL.Average`. In order to avoid the scaling problem, we created an aggregate response that was minimized by taking the inventory level divided by the service level to try to force the algorithm to minimize the inventory while maximizing the service level.

The other way is to optimize one primary objective (e.g., average inventory) while setting thresholds (i.e., lower and/or upper bounds) on the other objectives (e.g., lower bound of 0.90 on service level) which would create one point on the frontier which we did in the previous section. There are two categories of constraints in the OptQuest for SIMIO: constraints on the controls (inputs) and constraints on the responses (outputs).

Suppose we consider a slightly different optimization problem, namely minimizing the average inventory level while keeping the service level above 95%. We can write this more formally as the following equations. Of the three constraints, the last two are simply bounds on the controls. The first constraint restricts a response of the simulation (i.e., Service Level).

Minimize: AvgInventory

Subject to: Service Level >= .95

200 <= ReorderPoint <= 500

500 <= Order UpToQty <= 1200

Step 1: Select the **OptQuestExperiment2** and change the primary objective to be the **AvgInventory** response under the experiment properties.[149]

[149] Right click the experiment name in the [*Navigation*] window to access the properties of the experiment.

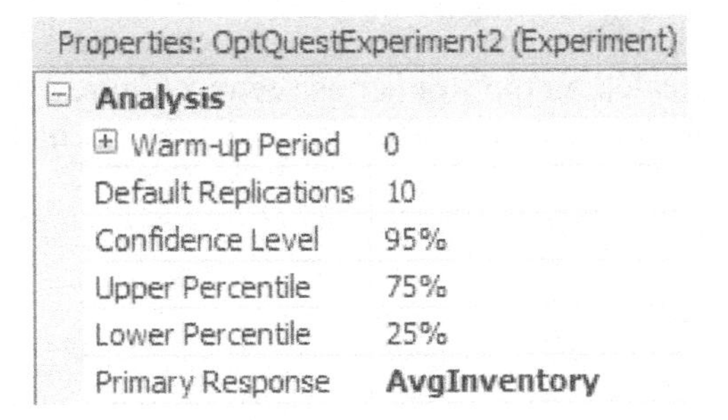

Properties: OptQuestExperiment2 (Experiment)	
Analysis	
Warm-up Period	0
Default Replications	10
Confidence Level	95%
Upper Percentile	75%
Lower Percentile	25%
Primary Response	**AvgInventory**

Figure 10.31: Change the Primary Response of the Experiment Properties

Step 2: Save and run the optimization algorithm and compare the results to the previous experiment.

Step 3: Sort the scenarios in the average inventory column in ascending order to see those with minimum average inventory as well as filter the service levels to only display feasible solutions.

Question 20: What appears to be the best reorder point and order up to quantity values considering not only the average inventory but the service level?

____

Step 4: All of the previous methods performed the decision before the search. OptQuest can discover the efficiency frontier as well. Duplicate the **OptQuestExperiment2** and name it **OptQuestExperiment3**.

Step 5: Remove the **Inv_SL** response by selecting it and clicking the *Remove Response* button. SIMIO will generate the frontier based on all responses present.[150]

Step 6: Change the *Objective Type* property under the **OptQuestExperiment3** properties to *"Pattern Frontier" and the Max Scenarios to "200"* as seen in Figure 10.32.

OptQuest for Simio - Parameters	
Min Replications	**5**
Max Replications	**20**
Max Scenarios	**200**
Confidence Level	**95%**
Relative Error	**0.1**
Objective Type	**Pattern Frontier**

Figure 10.32: Changing the Objective Type to Pattern Frontier

Step 7: Select the **Service_Level** response and change the lower bound to 0.5 so we will generate points above 50% service level.

Properties: Service_Level (Response)	
OptQuest for Simio - Parameters	
General	
Name	**Service_Level**
Display Name	Service_Level
Expression	**TallyStatSL.Average**
Unit Type	Unspecified
Objective	**Maximize**
Lower Bound	**0.5**

Figure 10.33: Change the Lower Bound on the Service Level

[150] The frontier for two responses is a line and difficult generate while three would need a plane needing many scenarios.

Step 8: Save and run the model. After the model runs, sort the checked column in descending order to show all the points that were created on the frontier as seen in Figure 10.34 while Figure 10.35 shows the frontier graphed by copying the data to Excel. The data in Excel has had its duplicates removed and been sorted in descending order.

Scenario		Replications		Inventory - Controls				Responses	
Name	Status	Required	Completed	InitialInventory	ReorderPoint	OrderUptoQty	ReviewPeriod (Days)	AvgInventory	Service_Level
200	Compl...	20	20 of 20	700	275	550	5	222.299	0.903114
199	Compl...	20	20 of 20	700	250	550	5	221.618	0.895011
197	Compl...	20	20 of 20	700	255	500	5	194.562	0.83744
151	Compl...	20	20 of 20	700	330	550	5	225.123	0.912277
145	Compl...	20	20 of 20	700	370	575	5	242.496	0.941197
140	Compl...	20	20 of 20	700	340	550	5	226.443	0.916291
139	Compl...	20	20 of 20	700	365	550	5	247.802	0.945026
129	Compl...	20	20 of 20	700	345	550	5	226.822	0.91666
125	Compl...	20	20 of 20	700	485	500	5	249.534	1
118	Compl...	20	20 of 20	700	345	575	5	238.42	0.933309
105	Compl...	20	20 of 20	700	315	575	5	238.152	0.92974

Figure 10.34: Part of the Points Generated on the Pattern Frontier by SIMIO

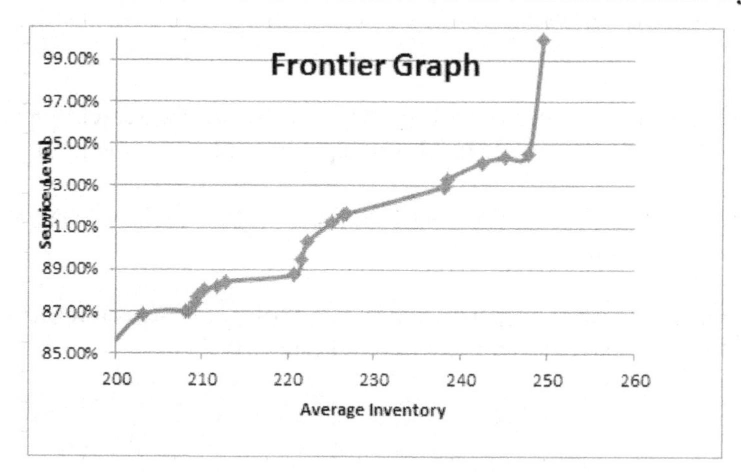

Figure 10.35: Pattern Frontier Graphed in Excel

Step 9: Note that more complicated constraints can be composed on the inputs (controls). In fact, these constraints are more common. A nonsensical example of a constraint on an input in this model could be (`ReorderPoint + 3*OrderUpToQty >=1000`) which is inserted by selecting the *Add Constraint* button and entering the equation as seen in Figure 10.36.

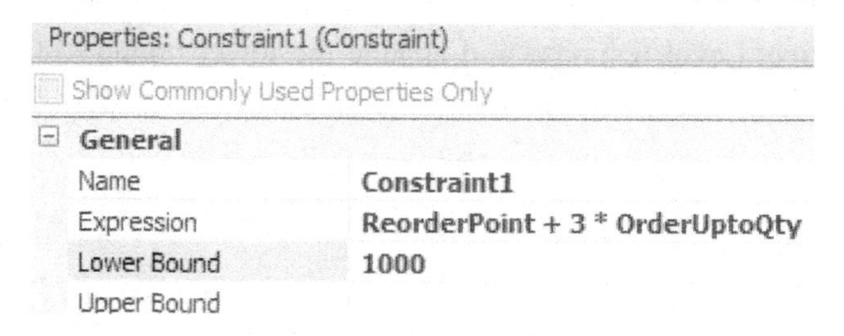

Figure 10.36: Example of Constraint on Decision Variables

Part 10.9: Commentary

- The use of the subset selection facilitates finding a good collection of solutions. For a large number of control variables and scenarios, the KN ranking and selection scheme is quite good in determining statistically the best scenario.
- OptQuest is a very good optimizer given enough time and replications. However the SMORES, subset selection, and KN add greatly to determining what "solution" you think is best.
- With the Cost Financials added to SIMIO, one could have optimized cost as well.

Chapter 11
Simulation Output Analysis

You have carefully created a model and paid close attention to the present operational system. Your model produces appropriate performance measures. You may have done some considerable re-modeling as you learned more about the system being modeled and its important aspects. You have been concerned about modeling errors and checked your model often and *verified* it is working the way you want. Furthermore, you have attended to the question of model *validity* as you have been checking the simulated performance measures against actual results. Through all your efforts, you have been thinking about the ways to improve the system performance and have an accumulated set of alternative scenarios you want to try. You are now satisfied that you can begin to examine these alternative scenarios and begin the process of developing recommendations.

The simulation model produces outputs but we know they are sets of random variables. But how do we best understand these results. We can generate averages of observed outputs, but these averages exhibit variation. Can we simply make multiple replications of our model, create confidence intervals, and draw conclusions?

Part 11.1: What can go wrong?

There are two overriding statistical issues that should concern us. The first is bias in our estimates and biased estimates are potentially wrong estimates. The second is the "independence and identicalness" assumptions which plague us in calculating measures of variability and ultimately comparing systems.

Bias

Perhaps the biggest concern is the potential "*bias*" in our computed averages. Bias means that we computed a value that is systematically different than the performance measure of interest. In other words, we may be computing a value that simply isn't the value of interest. The bias can occur in our method of calculation, but more likely it is in the values being used. For instance, consider the computation of the average waiting time in the queue of the ice cream store. Now, we know that the simulation starts "empty and idle", so that the early waiting times are going to be considerable smaller than those during the day. In fact, expected waiting time changes throughout the day. Since the arrival process uses a time-varying arrival, the waiting times grow and shrink with strong influence from the arrival rate. Other statistics like resource utilization will vary as well.

Bias may or may not be a problem. If we need to know the waiting time, say during the busiest time in the ice cream store, computing the waiting time during the entire day will clearly underestimate that waiting time. In such a case we need to collect our own statistics during the periods of interest. On the other hand, an overall average waiting time computed over the whole day, can be useful. While it may underestimate our intuition about the congestion, it is a systematic and consistent computation that can be useful in comparing alternative scenarios. Because the waiting time will be computed in the same manner in all scenarios, the observed improvement or deterioration will be "relatively" correct. In other words, our absolute estimate of waiting times may be an underestimate, but that same degree of underestimation will be present in all scenarios explored.

This idea that relative differences are the important perspective that allows us use many simulation models that underestimate many performance measures simply because all the "small" factors that impact the performance measure is being ignored. For instance, our simulation model may only account for 50% of a resource's duties when we know their workload is closer to 80%. But because all our scenarios ignore those same small factors (e.g., meetings, breaks, interruptions), if we find a scenario that enhances the resource to 60% utilization, then we believe that the base utilization of the resource will be increased by 10% by employing this scenario. So the recommendation here is to use a relative difference perspective rather than an absolute perspective when possible.

Independent and Identical

In order to measure variability associated with any simulation output, one needs data to be independent and identically distributed (IID). The reason is that the computation of the confidence interval for an expected value is based on the central limit theorem, which allows us to compute the standard deviation of the mean from the standard deviation of the observation. And unfortunately, variability is present in almost all our simulation output, so we had to "construct" IID observations. We recognized early in our study of simulation, that the observations within a single replication do not satisfy the IID requirement. So we rejected the observed waiting times within a simulation replication as unsuitable for estimating the variance of an expected value. Instead we compute the replication average as an observation and that works for most of the cases of interest.

There are a couple of situations where getting the IID observations maybe a cause for concern. One case occurs when a simulation replication takes a very long time. It becomes very costly in terms of computer time)to do the appropriate replications needed. A second case is when the simulation starts from, for example, empty and idle and there is a lengthy period when the performance measures are poorly estimated due to the startup of the simulation. In both cases, using replications (i.e., one observation per replication) becomes problematic since we have limited opportunity to do many replications.

Part 11.2: Types of Simulation Analyses

There are two general types of simulation analyses: *terminating* and *steady-state*. A terminating simulation typically models a terminating system that has a natural opening and closing. The ice cream store is a good example because it opens empty and idle in the morning and then closes in the afternoon, sometimes after the last customer is served. Terminating simulations are characterized by their opening and closing. Thus any performance measures computed are implicitly dependent on how the simulation is started and stopped. One replication is generally the time between the opening and closing. A terminating simulation is sometimes called a "transient" simulation because its state is constantly changing throughout the replication. Although bias in the estimates of the averages may occur because of the changing state, the estimation of variance is facilitated by the use of replications.

Question 1: What is an example of a simulation that should be examined as a terminating simulation?

____

A steady-state simulation has no "natural" opening or closing. In a steady-state simulation, we want our performance measures to be independent of how we started or stopped the simulation. In other words, we want a kind of long-term average for the expectations. However, while there is considerable appeal for performance measures that independent of starting and stopping, just how to obtain such a simulation is not clear. There are two fundamental problems. The first is how to eliminate the effects of startup or warmup. The influence of the way the simulation starts can extend far into the simulation. The second problem is how long a replication should last, since there is no natural stopping state. Neither of these problems can be easily answered, so the tendency is to resort to "rules of thumb" that give some general guidance, but for which there is limited theory.

Question 2: What is an example of a simulation that should be examined as a steady-state simulation?

____

Part 11.3: Output Analysis

To provide a context for an output analysis, re-consider the work cell problem of Chapter 5, section 6. Recall that this problem is a manufacturing cell consisting of three workstations (i.e., A, B, C) through which four part types are processed. The parts have their own sequences and processing times depended on the part type and the station visited. We defined two statistics of interest: a "state statistic" on the number of Part type 3's in the system during the simulation and a "tally statistic" on the time Part type 3's are in the system.

Very little analysis of the problem was undertaken. We ran the simulation (interactively) for 40 hours and examined the output. The average number of Part type 3's in the system was 0.40 parts and the average time for Part type 3's was in the system was 0.26 hours. At that time, our focus was on the construction of the model not necessarily the interpretation of the output which will explore now.

Since we have more experience with simulation output, we may be considering an SIMIO experiment and making multiple replications. By making multiple replications, we obtain an estimate of the variability associated with the observed averages and could compute a confidence interval. But before doing that analysis, let's think about our simulation intent. How do we want to use the output from the simulation? Do we want to say something about the "short-term" or the "long-term" outcomes? We probably chose a 40 hour simulation length because it represents a "weeks' worth" of work. However, we also used the default "empty and idle" state for the start of the week. Is that representative of a real week? If the answer is yes, then we can proceed to making multiple replications and the creating confidence intervals for our output measures.

Question 3: In your judgment, how likely is it that you are interested in one week statistics being empty and idle at the beginning of the week?

____

By doing replications we should obtain approximate IID observations, but a more fundamental concern is with the bias. We have two sources of bias. The first is the "initialization" or "warmup". Starting empty and idle certainly affects our statistics. The second is whether the 40 hours is sufficient time to observe the system. To limit our let's focus primarily on the two user-defined statistics: the number of Part type 3's in the system during the simulation and the time Part type 3's are in the system.

Initialization Bias

When we describe the initialization bias, we often use the term "warmup" implying that the simulation output goes through a phase change like a car that needs to warm-up in cold weather before turning on the heater. We think of a "transient phase" followed by a transition to the "steady-state phase." And usually, it's the steady-state phase that interests us since it represents the "long-term" perspective.

Three "obvious" ways to deal with the initialization bias are: (1) ignore it, trusting that you have enough "good" statistical contributions from the steady-state phase that the "poor" statistics at the beginning (during the transient period) are completely diluted, (2) "load" your system with entities at the beginning of the simulation, so that the transition to steady-state behavior occurs early, and (3) after a while (i.e., warm up period), clear (truncate) the statistics in an attempt to clean out the transient statistics so they can no longer influence the statistics collection.

The problem with ignoring the presence of the transient statistics is that their influence on the final statistics is really unknown. The problem with loading the system with entities is that we don't know where the loading should take place and furthermore, we don't know that our loading really hastens the transition to steady-state. Finally the problem with deleting statistics is that we delete "good" as well as "bad" statistics and simulation observations are important usually somewhat hard to obtain. So let's consider a more thorough examination of initialization bias in the context of the work cell problem.

Step 1: Open up the simulation from Chapter 5, section 6. For purposes of comparison, run the model interactively for 40 hours.

Question 4: What is the average number of parts of type three in the system during the 40 hours?

____

Question 5: What are the average time parts of type three are in the system during the 40 hours?

____

Step 2: Add a SIMIO Experiment, specifying 10 replications, each replication being 40 hours. We will use these results also for comparison.

Question 6: What is the 95% confidence interval on the average number of parts of type three in the system during the 40 hours?

____

Question 7: What is the 95% confidence interval on the average time parts of type three are in the system during the 40 hours?

____

Question 8: Do the averages from a single replication fall with the confidence intervals?

____

Question 9: Why wouldn't you be concerned if they didn't?

____

Step 3: Next, let's add some plots of the two user-statistics so we can visualize the change in these over time due to the initialization bias. Keep in mind this simulation starts empty and idle. Add two Animation Plots from the Animation tab: one for the number of parts of type three in the system and the second for the time parts of type three are in the system. Figure 11.1 shows the specifications for the average number of part type three in the system. Figure 11.2 shows the specifications for the average time part type 3 are in the system

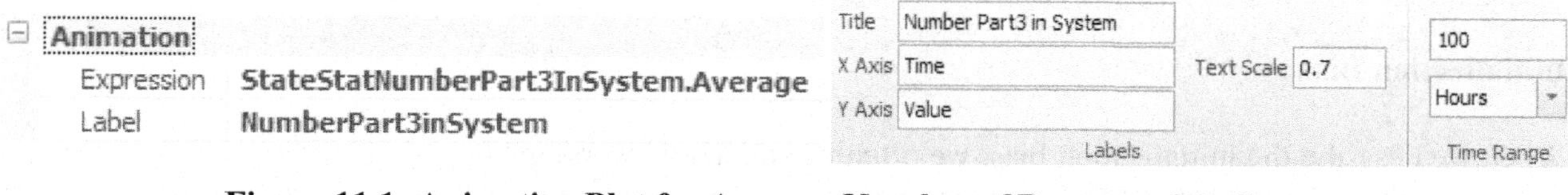

Figure 11.1: Animation Plot for Average Number of Part type 3 in System

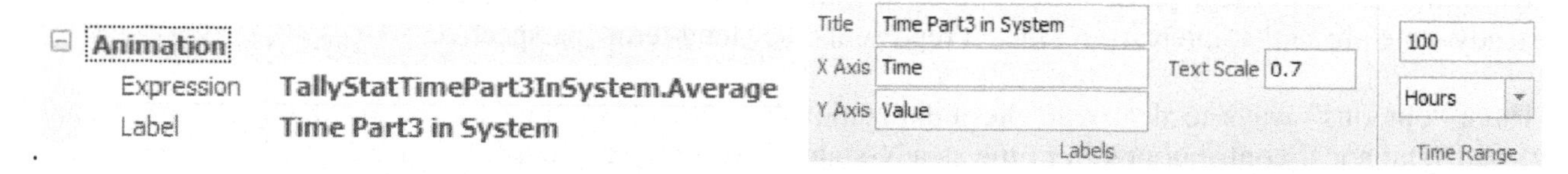

Figure 11.2: Animation Plot for Average Time Part type 3 in System

Step 4: Be sure that the *Time Range* on the plots is 100 hours, so we have enough time to see the changes.

Step 5: Change the run length of the simulation to 100 hours and run the simulation interactively (set the speed factor to 1000) and observe the plots.

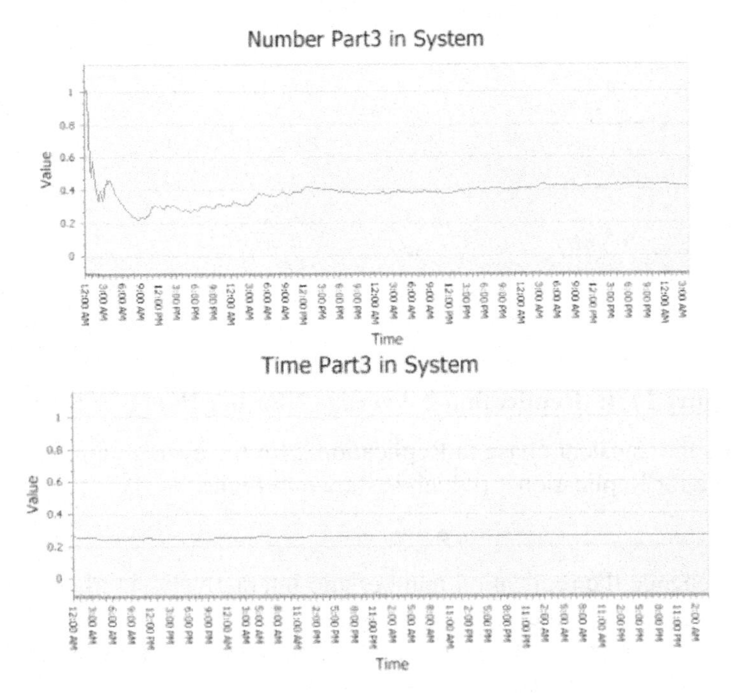

Figure 11.3: Average Number and Time in System

Step 6: Notice how the number in system plot changes rather markedly during the first 50 hours, while the time in system plot doesn't change much.

Question 10: During the initial 50 hours, how long does the transient period seem to last (recall this is the time during which the behavior is erratic)?

____

Question 11: After the transient period, there is a period of transition as the state of the statistics moves from being highly unstable to becoming stable. How long do you think this transition period lasts?

____

Question 12: How are the animation plots changing during the last 50 hours?

____

When these plots stabilize, we say the statistics are in steady-state. Notice that we talk about the statistics, not the system because it's the statistics that interest us, even though it's the system that is changing.

Question 13: Which statistic enters steady-state first? Number in system or time in system?

____

Question 14: Since we can specify only one "warmup" period, what do you suggest?

____

Step 7: We have seen only one replication during the 100 hour simulation. Under the *Advanced Options* within the **Run Setup** section of the ***Run*** tab, click on *Randomness* and select the *Replication Number* option. Set the replication number to "2" and re-run the 100 hour simulation. Now the time in system continues to be relative stable, while the number in system shows a different transient behavior. Figure 11.4 shows the average number of part type three's in the system during the first 100 hours.

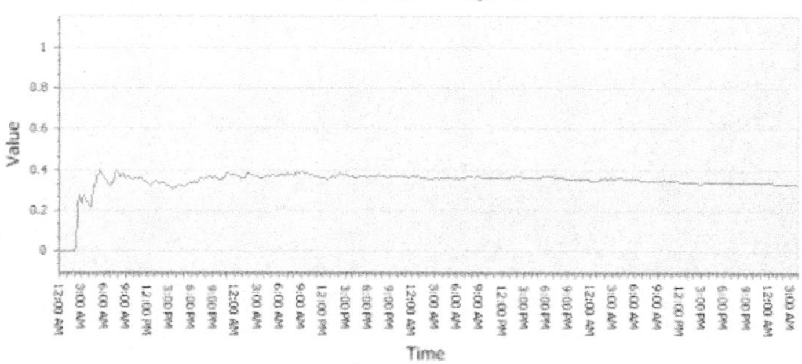

Figure 11.4: Replication 2 Average Number Part 3 in System

Question 15: How does the transient phase in Replication 2 for the average number of part 3 in the system differ from the one for Replication 1 (which is shown in Figure 11.3)?

___

Step 9: Now continue to change the replication number and interactively simulate, observing the behavior of the average number of part 3's in the system during the first 50 hours. Figure 11.5 shows the results for replications 3, 4, and 5. We focus on number in system, but be sure to look at the time in system also.

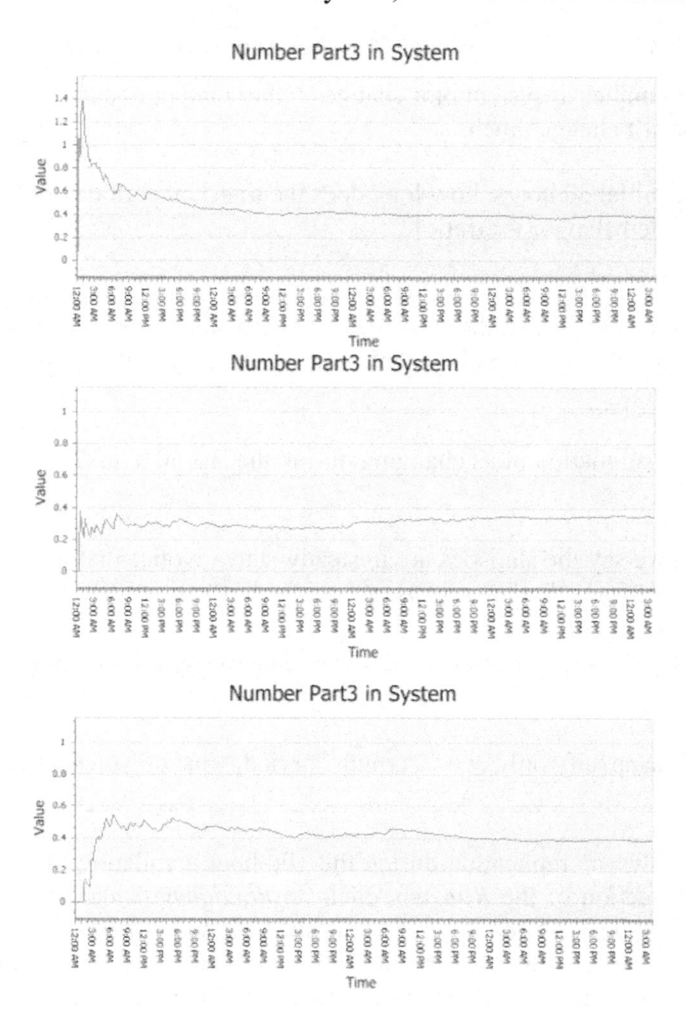

Figure 11.5: Replications 3, 4, and 5 Average Number of Part Type 3's in System

Step 10: With these five replications, we can see a range of behaviors at the beginning of the simulation. In all cases the statistics appear converge to a common steady-state value, but there was a consistent transient period during which there was no consistent behavior

Step 11: Now simulate Replication 5 for 40 hours interactively and compare to *Step 1*.

Question 16: What is the average number of parts of type three in the system during the 40 hours?

____

Question 17: What are the average time parts of type three are in the system during the 40 hours?

____

Question 18: In comparing 40 hours with 100 hours simulations, do you think the differences are important?

____

Step 12: While the differences are probably not important in this case, you can see that they may be in other cases. In any case we our simulation would benefit by the elimination of the initialization bias. Now the question is how large should the warm-up be? Clearly we want to eliminate the transient period, which after observing five different replications, lasts for about 20 hours. The transition period still shows some effect, but we don't want to throw away observations unless we think they are really bad. So let's make the warmup period 15 hours.

Replication Length

Now, as a rule of thumb, we do not want the warmup period to exceed 10-20% of the length of a replication. In fact, if time permits, 5% would be a target.

Step 13: So let's run the simulation experiment with a run length of 150 hours with a warmup of 15 hours with a total of 10 replications.

Question 19: What is the 95% confidence interval on the average number of parts of type three in the system?

____

Question 20: What is the 95% confidence interval on the average time parts of type three are in the system?

____

Question 21: How do these confidence intervals compare to the ones from *Step 2* previously?

____

In summary, it's important to remove the initialization bias by using the warmup period. However you should look at several replications of the warmup period in order to understand the range of behavior. Also you need to examine all the statistics of interest, so your warmup covers the largest transient behavior. With the warmup period established, be sure that the warmup period is less than 20% of the replication length. It is generally a good rule to multiply the warmup period by 10 to obtain the replication length. Use a multiplier by 20, if your model runs quickly enough.

Part 11.4: Automatic Batching of Output

To obtain our previous confidence interval, we simulated (150 hours * 12 replications) for a total of 1800 hours. Of that total we used 450 hours for the warmup. Although this small problem runs through that simulation time rather quickly, a larger problem may take hours of computer time to provide such a confidence interval. So we may be interested in obtaining a confidence interval in less time. We "wasted" lots of observations during warmup and we created fairly lengthy simulation replications.

An alternative to using replications is to employ a method known as "***batch means***". Using batch means, we attempt to create IID observations within one long replication. Since only one run of the simulation is used, we only lose the warmup observations once and we are able to invest our simulation effort in one long run, rather than several. Hopefully, all this means that we save simulation time. The batch means method creates "batches" of observations for which the average for the batches satisfy our IID requirement for requirements. For example, the time-persistent statistics (state statistics), like number in the system may have a mean computed every 25 hours if the means from every 25 hours appear to be IID. For observations-based statistics (tally statistics), we might compute an average waiting time from each 50 customers, if the means from each 50 customers appear to be IID. The 25 hours for state statistics and the 50 customers for tally statistics are called the "***batch size***". There will be two batch sizes: one for state statistics and one for tally statistics. The number of batches created during the simulation corresponds to the number of observations when computing a confidence interval.

SIMIO will automatically employ a version of batch means to compute a confidence interval when: (1) user statistics are defined, and (2) an SIMIO experiment with one replication is employed. The only SIMIO-defined statistic that will be automatically batched are the sink node statistics.

Step 1: Use the simulation model from the previous section. Change the number of required replications to one. Leave the warm-up at 15 hours and leave the replication length at 150 hours.

Step 2: When you execute the simulation, you will see a new *row*, called "HalfWidth" in the two user-defined statistics for the average number of part type 3 in the system and the average time part type 3 are in the system. You will also see that new row in the **SnkPartsLeave** statistics on flowtime. The output for the two user-defined statistics are shown in Figure 11.6.

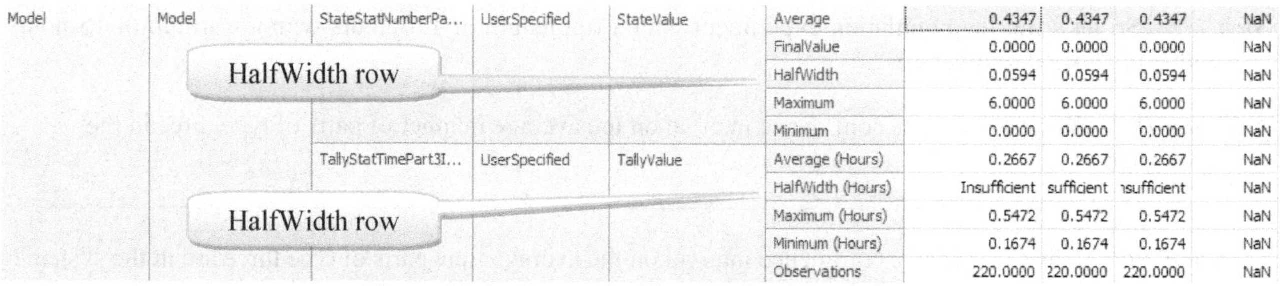

Model	Model	StateStatNumberPa...	UserSpecified	StateValue	Average	0.4347	0.4347	0.4347	NaN
					FinalValue	0.0000	0.0000	0.0000	NaN
	HalfWidth row				HalfWidth	0.0594	0.0594	0.0594	NaN
					Maximum	6.0000	6.0000	6.0000	NaN
					Minimum	0.0000	0.0000	0.0000	NaN
		TallyStatTimePart3I...	UserSpecified	TallyValue	Average (Hours)	0.2667	0.2667	0.2667	NaN
					HalfWidth (Hours)	Insufficient	sufficient	ısufficient	NaN
	HalfWidth row				Maximum (Hours)	0.5472	0.5472	0.5472	NaN
					Minimum (Hours)	0.1674	0.1674	0.1674	NaN
					Observations	220.0000	220.0000	220.0000	NaN

Figure 11.6: Output for User-Defined Statistics

Step 3: In the case of the time in system we see an "Insufficient" meaning we have not run the simulation long enough for SIMIO to obtain enough batches to compute a confidence interval. It is also possible to get a "Correlated" message meaning that the obtained batches do not satisfy the IID requirement. Figure 11.7 is an example of obtaining this message.

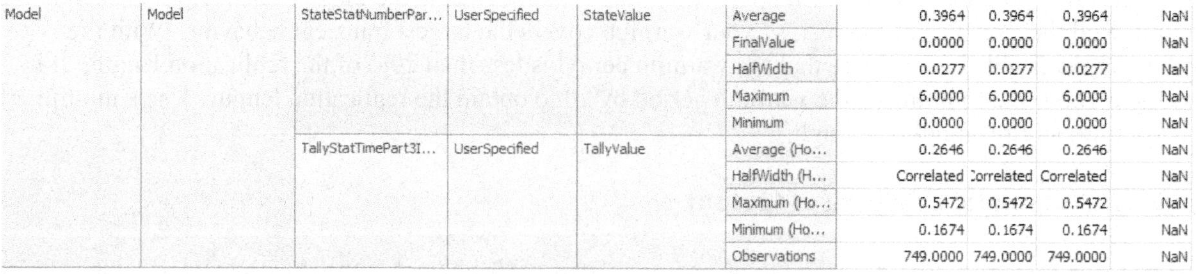

Model	Model	StateStatNumberPar...	UserSpecified	StateValue	Average	0.3964	0.3964	0.3964	NaN
					FinalValue	0.0000	0.0000	0.0000	NaN
					HalfWidth	0.0277	0.0277	0.0277	NaN
					Maximum	6.0000	6.0000	6.0000	NaN
					Minimum	0.0000	0.0000	0.0000	NaN
		TallyStatTimePart3I...	UserSpecified	TallyValue	Average (Ho...	0.2646	0.2646	0.2646	NaN
					HalfWidth (H...	Correlated	:orrelated	Correlated	NaN
					Maximum (Ho...	0.5472	0.5472	0.5472	NaN
					Minimum (Ho...	0.1674	0.1674	0.1674	NaN
					Observations	749.0000	749.0000	749.0000	NaN

Figure 11.7: Example of Correlated Batch Means

Question 22: Why do the columns for "Average", "Minimum", and "Maximum" all have the same value?

____

Question 23: Why does the "HalfWidth" show "NaN" (Not a Number)? Remember we are using only one replication.

____

Step 4: Now change the replication length to 300 hours which is twice the length of the replication used earlier.

Question 24: Do you obtain any "Insufficient" or "Correlated" messages on the statistics?

____

Question 25: What is the 95% confidence interval on the average number of parts of type 3 in the system?

____

Question 26: What is the 95% confidence interval on the average time parts of type 3 are in the system?

____

Question 27: How do these confidence intervals compare to the ones from ***Section 3*** previously?

____

The automatic batching of statistics by SIMIO was able to produce a "legitimate" confidence interval with only 300 hours of simulation, as opposed to the 1800 hours previously. In using the automatic batching, we still needed the warmup analysis.

Part 11.5: Algorithms used in SIMIO Batch Means Method

SIMIO automatically attempts to compute a confidence interval via a batch means method for only the user-defined tally and state statistics (the only exception is that the batch means method is also applied to sink statistics) for a SIMIO Experiment having only one replication. It displays the results in the output in the ***row*** called "HalfWidth", which is one half the confidence interval. It can fail to make the computation if it believes there is insufficient data or if the batched data remains correlated. SIMIO does not include data obtained during the *Warm-up Period* specified in the Experiment.

SIMIO's batching method is done "on the fly," meaning the batches are formed during the simulation. Thus batches can be enlarged, but not re-batched.

Minimum Sufficient Data

- For a Tally statistic, SIMIO requires a minimum of 320 observations
- For a State statistic, there must be at least 320 changes in the variable during at least 5 units of simulated time.
- For statistics that do not satisfy this minimum requirement, the half width is labeled "Insufficient".

Forming Batches

- Form 20 batch means
  - A Tally batch is the mean of 16 consecutive observations
  - A State batch will be the time average over 0.25 time units
- Continue forming batches until there are 40 batches
  - At 40 batches, the batch size is doubled (each batch is "twice as big"), and the number of batches is reduced to 20
- Continue this procedure as long as the simulation runs. Notice that the number of batches will be between 20 and 39

Test for Correlation

- SIMIO uses the VonNeuman's Test[151] on the final set of batches to determine if the assumption of independence can be justified.
  - If the test fails, the half width is labeled "Correlated".
  - If the test passes, the half width is computed and displayed.

Cautions

The SIMIO batch means method is only for steady-state analysis – it should be ignored completely for terminating systems. There is no automatic determination of warmup period. All rules for collecting batches are somewhat arbitrary, so some care needs to be exercised in using it.

Part 11.6: Input Analysis

It may seem strange to include "input analysis" in a chapter devoted to "output analysis" but it must be remembered that the analysis of the input is based on the behavior of the output. SIMIO provides two input data analysis: (1) **Response Sensitivity Analysis** and (2) **Sample Size Error Analysis**. Both analyses require that at least one of your inputs to be an *Input Parameter* and that you have defined at least one *Experiment Response*. Although we briefly introduced response sensitivity analysis previously, we will describe it as well as the sample size error analysis within the same context, namely the workcell problem used throughout this chapter.

Recall that "input parameters" are defined from the "*Data*" window tab in three possible ways: (1) through a DISTRIBUTION you hypothesized, (2) with a TABLE of observations that you have collected, and (3) using a SIMIO EXPRESSION. Any defined input parameter can be used anywhere in your model and various objects can reference the same input parameter. Using distributions have the advantage that SIMIO will provide you with a histogram based on 10,000 samples to show you what that distribution looks like.

Step 1: Save the current model and delete the animation plots as they will not be needed.

Step 2: Rename the model state variable to **GStaNumberInSystem**. Modify the two user-defined statistics so we have some possible responses for experimentation. Call the state statistic **StateStatNumberInSystem** and the tally statistic **TallyStatTimeInSystem**. Be sure that the state statistic references the **GStaNumberInSystem state variable and that the tally statistic has Time as the** ***Unit Type*****.**

Step 3: Fix the *State Assignment* at the **SrcParts** and at the **SnkPartsLeave** so that **GStaNumberInSystem** is always incremented (add one) at the source and decremented (subtract one) at the sink. Modify the *Tally Statistics* collection at the **Input@SnkPartsLeave** so that it corresponds to Figure 11.8.

⊟ Basic Logic	
Tally If	Entity Entering
Tally Statistic Name	**TallyStatTimeInSystem**
Value Type	Expression
⊞ Value	**ModelEntity.TimeInSystem**

Figure 11.8: Tally All Entities Time in System

Step 5: Next, we need to create the input parameters, which will correspond to, in this case, all the distributions used in the model. These are created by specifying a "*Distribution*" with the *Input*

[151] This is a statistical test of independence.

Parameters section from the "*Data*" Tab. Make sure to change the *Unit Type* property to "Time" and specify minutes. Figure 11.9 shows the inputs and these are the same the **SequencePart** table.

Name	Object Type	Summary
Input Parameters		
InpDistPart1StationA	Distribution Input Parameter	Pert(2 Minutes,5 Minutes.8 Minutes)
InpDistPart1StationC	Distribution Input Parameter	Pert(2 Minutes,6 Minutes.11 Minutes)
InpDistPart2StationA	Distribution Input Parameter	Pert(1 Minutes,3 Minutes.4 Minutes)
InpDistPart2StationB	Distribution Input Parameter	Uniform(5 Minutes,11 Minutes)
InpDistPart2StationC	Distribution Input Parameter	Uniform(2 Minutes,11 Minutes)
InpDistPart3StationA	Distribution Input Parameter	Triangular(2 Minutes,5 Minutes,8 Minutes)
InpDistPart3StationB	Distribution Input Parameter	Triangular(5 Minutes,9 Minutes,11 Minutes)
InpDistPart4StationB	Distribution Input Parameter	Pert(5 Minutes,9 Minutes.11 Minutes)
InpDistPart4StationC	Distribution Input Parameter	Triangular(2 Minutes,6 Minutes,11 Minutes)
InpDistPartsArrive	Distribution Input Parameter	Exponential(10 Minutes)

Figure 11.9: Definition of Input Parameters

Step 6: Select the **SrcParts** and use the **InpDistPartsArrive** as the *Interarrival Time*.

Step 7: Also substitute the appropriate input distribution into the **SequencePart** *Data Table* for each of the **ProcessingTimes** values as seen in the partial view of the table in Figure 11.10.

Table Parts | Sequence Part

	Sequence	ProcessingTimes (Minutes)	ID
1	Input@SrvStationA	InpDistPart1StationA	0
2	Input@SrvStationC	InpDistPart1StationC	0
3	Input@SnkPartsLeave	0	0
4	Input@SrvStationA	InpDistPart2StationA	1
5	Input@SrvStationB	InpDistPart2StationB	1
6	Input@SrvStationC	InpDistPart2StationC	1

:

Figure 11.10: Partial View of the Sequence Part Table using the New Input Distributions

Step 8: Next select the **Experiment** and add two experiment responses as shown in Figure 11.11.

General	
Name	**NumberInSystem**
Display Name	NumberInSystem
Expression	**StateStatNumberInSystem.Average**
Unit Type	Unspecified
Objective	None

General	
Name	**TimeInSystem**
Display Name	TimeInSystem
Expression	**TallyStatTimeInSystem.Average**
Unit Type	**Time**
Display Units	**Minutes**
Objective	None

Figure 11.11: Experiment Responses for Number and Time in System

Step 9: Run the experiment with a *Warm-up Period* of 15 hours and a run length of 150 hours with 20 replications which will be used as a base reference for the results.

Question 28: What did you get for the 95% confidence interval on the average number in the system?

____

Question 29: What did you get for the 95% confidence interval for the average time in the system?

____

Response Sensitivity Analysis

The response sensitivity analysis measures how each response is affected by changes in the input parameters. It will be computed for each scenario in the experiment. A linear regression relates each input parameter to each response, so it is necessary that the number of replications be greater than the number of input parameters.

Step 10: Our prior model has ten input parameters and we are using 20 replications. Select the *Input Analysis* tab and then the *Response Sensitivity* icon and run the model (it may be necessary to Reset it first). The results of the sensitivity analyses are shown in a Tornado Chart, a Bar Chart, and a Pie Chart. The raw data for each of the regressions are also available.

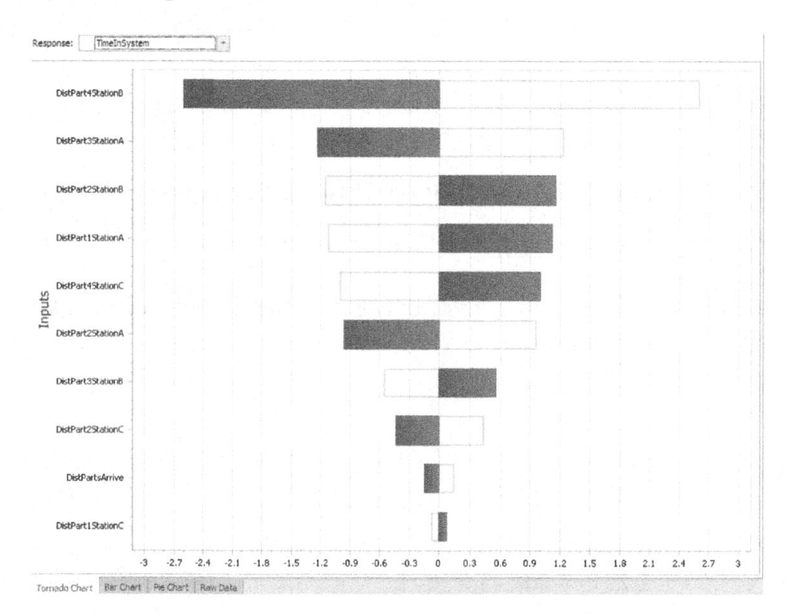

Figure 11.12: Tornado Chart for Time in System

Step 11: Refer to the Tornado Chart in Figure 11.12 and pass your cursor over the bars. If an input parameter has a negative coefficient, then when it is increased, it reduces the response. If the input parameter has a positive coefficient, then when it is increased, it increases the response. The magnitude of the coefficient determines it's important relative to the other input parameters.

Question 30: What input parameter will reduce the time in system the most when its distribution is increased?

____

Question 31: What input parameter will increase the time in system the most when its distribution is increased?

____

Question 32: What input parameter reduces the number in system the most when its distribution is increased?

____

Question 33: What input parameter increases the number in system the most when its distribution is increased?

____

Question 34: What surprises you about these results?

____

Question 35: What is the primary value of the pie chart relative to the sensitivity analysis?

____

Sample Size Error Analysis

The sample size error analysis attempts to assess the uncertainty in the responses relative to the uncertainty in the experiment and the uncertainty in the input parameter data. To perform this analysis, it is assumed that the input parameters were estimated from collected data samples. In other words you actually have a record of observations that you used for a given input parameter estimation (or distribution fitting).

Step 12: Pretend now you collected data for the input parameter estimation. From the Mode and the *Input Parameters* on the "*Data*" tab. Select a defined input parameter. Change the *Number of Data Samples* property according to Table 11.1 and be sure that the *Include Sample Size Error Analysis* property is **True**.

Table 11.1: Data Samples for Each Input Parameter

Input Parameter	Data Samples
DistPartsArrive	125
DistPart1StationA	85
DistPart1StationC	85
DistPart2StationA	100
DistPart2StationB	100
DistPart2StationC	100
DistPart3StationA	60
DistPart3StationB	60
DistPart4StationB	45
DistPart4StationC	45

Step 13: Go to the *Input Analysis* tab within the EXPERIMENT. Select the *Sample Size Error* icon and run the analysis (after doing Reset). This may take a little bit of time to complete.

Step 14: Next run the experiment from the design view to obtain a SMORE plot.

Step 15: The two important tabs are: (1) Contribution to Uncertainty and (2) Benefit of Additional Samples. Look first at the contribution to uncertainty, shown in Figure 11.13.

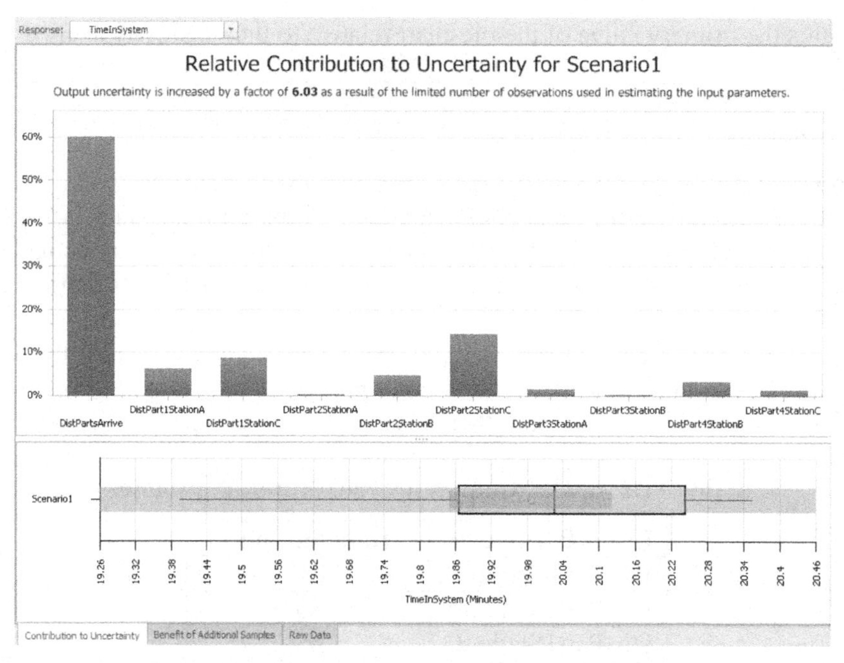

Figure 11.13: Uncertainty in Time In System

Question 36: Which factor has the biggest impact on time in the system?

Question 37: The SMORE plot illustrates the confidence interval for the mean (in tan), with the expanded half width due to input uncertainty in blue (and stated in text). How much does the uncertainty in the input parameters "expand" the output uncertainty?

Question 38: Which uncertainty then, input or output, dominates the overall uncertainty in this scenario?

Step 16: Next look at the tab: Benefit of Additional Samples as shown in Figure 11.1.

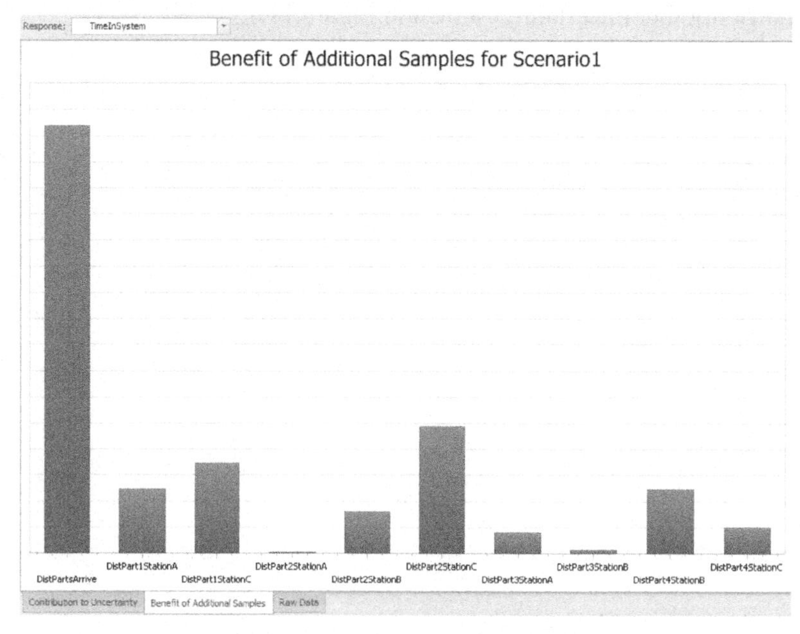

Figure 11.14: Benefit of Additional Samples

Question 39: Which input parameter would benefit the most from collecting additional observations for time in the system?

____

Step 17: Go back to the Input Parameters and change the *Number of Data Samples* property for **InpDistPartsArrive** to 250. Next repeat the Sample Size Error analysis again.

Question 40: How does the Contribution to Uncertainty change?

____

Question 41: What happens to the Benefit of Additional Samples?

____

Question 42: Would you now put more time into observing **InpDistPartsArrive or would you allocate observations to other input parameters**

____

Part 11.7: Commentary

- Sometimes the modeling effort consumes so much of a simulation project that the analysis of the model is almost overlooked. In particular, it is easy to forget that simulation output can be difficult to interpret and requires time and care. The issue of bias is often ignored, but can be a critical mistake.

- Replications permit the computation of confidence intervals and confidence intervals are our best way to measure of mean and variance of the expected value.

- Our experience is that most simulations can be run quickly, which means that elimination of the warmup and the use of replications becomes feasible even for steady-state simulations. However there are some steady-state simulations that take excessive computer time and so the idea of using batch means becomes attractive.

- The availability input analysis features in SIMIO should encourage you to consider the importance of the simulation inputs relative to your performance measures. You can judge the value of the various inputs and consider the amount of data collection devoted to their specification.

Chapter 12
Materials Handling

In the STANDARD LIBRARY, you have already noticed that there is a VEHICLE object and a CONVEYOR path. Perhaps the term "vehicle" already conjures up the idea of moving materials or people from place to place and your intuition would be correct. The VEHICLE object can pick up and delivery entities which can be used to act as a fork lift truck in a manufacturing operation, a human stock handler in a warehouse, a taxi cab for picking up and delivering people as well as a bus, a train, or other type of people mover to move entities either individually or in a group. The vehicle is an individual object whose routing is based on either "demand" or on a "fixed path". A conveyor in SIMIO, on the other hand, provides a continuous platform for conveying items or people. It can represent a wide range of industrial devises like gravity conveyors, belt conveyors, powered overhead conveyor, and even a power and free conveyor. Often materials handling in industry will employ a variety of materials handling devises. A distinction among conveyors used by SIMIO is whether the conveyor "accumulates" or doesn't ("non-accumulating). An accumulating conveyor allows entities to queue up at its end. A non-accumulating conveyor will stop when an entity gets to the end and won't start up until that entity is removed.

A key to modeling materials handling problems in SIMIO is the use of the TRANSFERNODE and to somewhat a lesser extent, the external, OUTPUTBUFFER which are attached to many of the objects like the SOURCE and the SERVER. The TRANSFERNODE (unlike the BASICNODE) provides the "*Transport Logic*" which means you can request a vehicle to move the entity to its next designation. If that transport device is not available, then the entity must wait. The TRANSFERNODE can be used alone and entities can wait there for pickup at that node but they will not queue up in the usual fashion. For other objects, the OUTPUTBUFFER provides a place for the entities to wait for transportation.

Part 12.1: Vehicles: Cart Transfer in Manufacturing Cell

We will reconsider the original problem of Chapter 5, whose layout and information are shown in Figure 12.1 Recall there are four part types, each with its own source and each part type has its own routing through the manufacturing cell. However, we now learn that instead of leaving the system, parts actually have to be put on a cart and wheeled to the warehouse building 300 yards away (see Figure 5.9). The cart can carry a maximum of three parts and travels at four feet per second. After the cart has finished dropping off parts, it will return to the pickup station for the next set of parts or wait there until a part is ready to be taken. The cart does not have to be filled for it to be taken to the warehouse (i.e., will carry one part if it is available).

Step 1: Open up the last model from Chapter 5 and save it as chap12.1.

Step 2: To facilitate a cart taking the parts to the warehouse, delete the CONNECTOR between the SINK object and the prior BASICNODE. Next, insert a TRANSFERNODE in between the **SnkPartsLeave** named **TNodeCart** as seen in Figure 12.1.

Step 3: Select the **Input@SnkPartsLeave** and remove the *Tally Statistics* as VEHICLES will be entering instead. To calculate the same statistics, we could insert a `Tally` step in the *Entered* add-on process trigger.

Step 4: Connect the BASICNODE to the TRANSFERNODE with a PATH with a logical length of 10 yards. Finally, connect the **TNodeCart** TRANSFERNODE to the SINK with a PATH making sure it's *Type* is set to "*Bidirectional*" so our cart can move in both directions. Set the path's logical length to 300 yards.

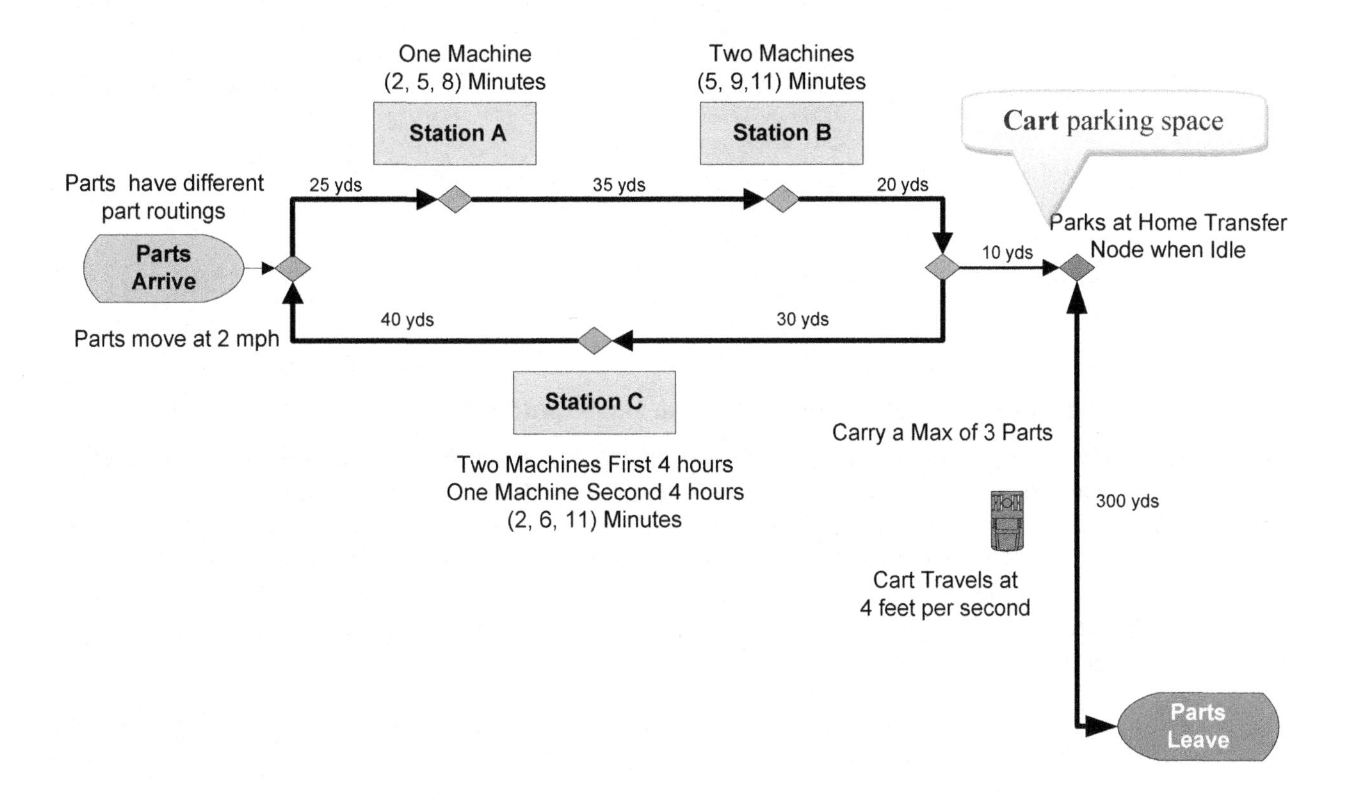

Figure 12.1: Adding a Vehicle to Carry Part to the Exit

Step 5: Drag and drop a VEHICLE object from the [*Standard Library*] to the modeling canvas in the facility window named **VehCart** similar to Figure 12.2. Make sure to increase the size of the RIDESTATION queue which animates the parts being transported so it can hold three parts.

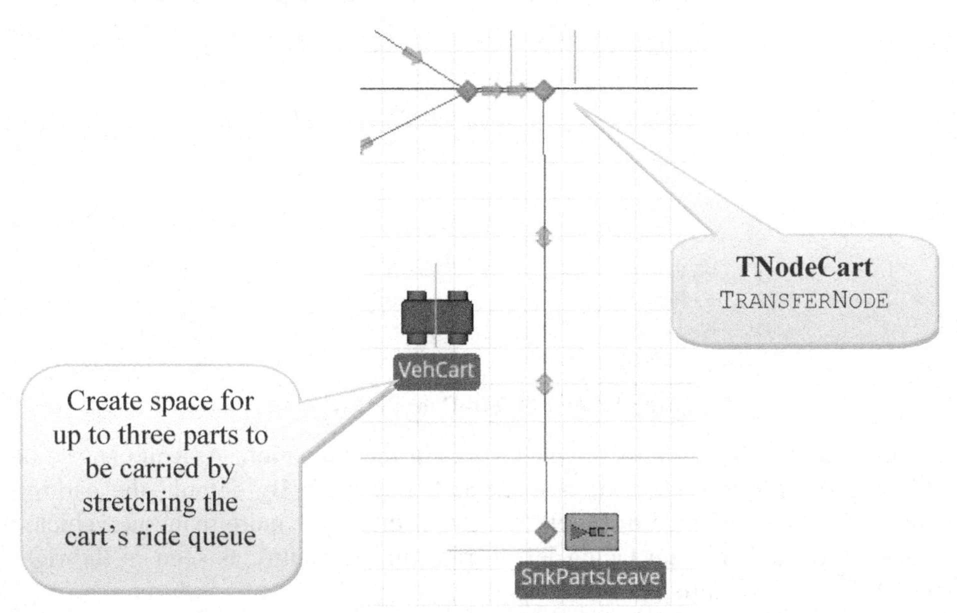

Figure 12.2: Adding a Vehicle

Step 6: Change the **TNodeCart** TRANSFERNODE's *Ride On Transporter* property to "True" and specify the "VehCart" as the *Vehicle Name* which was just added (see Figure 12.3). Every time an entity passes through this node, it will now generate a ride request to be sent to the vehicle. The entity cannot continue until the vehicle is there to pick it up.

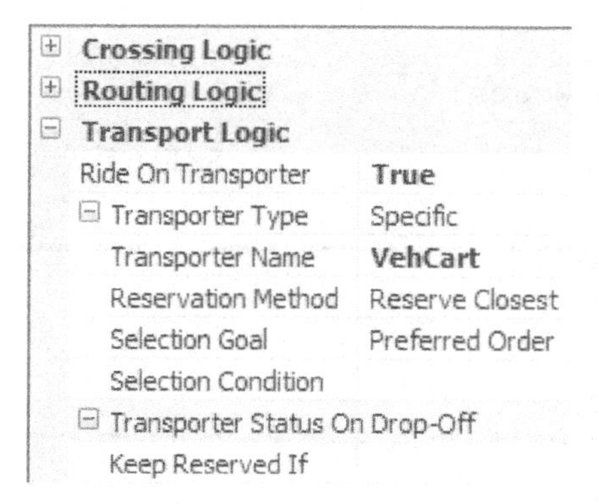

Figure 12.3: Setting up the Node to Have Entities Utilize a Vehicle

Step 7: Use Figure 12.4 as a guide to set the **VehCart**'s *Initial Desired Speed* to four feet per second and *Ride Capacity* to three. Also, make sure to set the *Initial Node(Home)* to the **TNodeCart** node which was just added, as this will be our vehicle's home node. Leave the *Routing Type* specified as "*On Demand*", meaning it will respond to requests for transportation from the stations as needed. Now change the **VehCart**'s *Idle Action* property to "*Park At Home*" so that the vehicle will always go to the home node if there are no entities with ride requests.[152] Otherwise the VEHICLE would remain at the SINK until a ride request is generated.

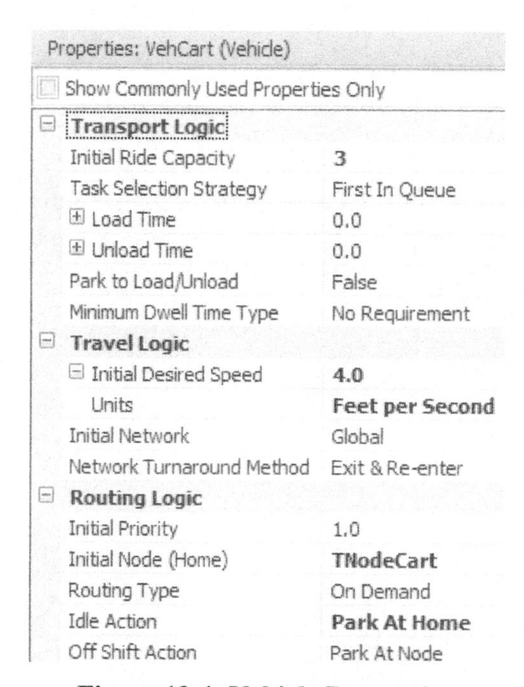

Figure 12.4: Vehicle Properties

Step 8: Run the simulation now to see how the cart works in the animation. As you can see the **VehCart** originally is parked at the **TNodeCart** node as seen in Figure 12.5. By default, the parking animation queue is above the node[153] and oriented from left to right. If there was more than one vehicle they would be stacked on top of one another again facing left to right in the facility as seen in the right picture of Figure 12.5 where there are three carts in the system.

[152] The *Network Turnaround Method* can be changed for example to "Reverse" so the cart would go forward to the exit and then look as if it backs up to the **TNodeCart** node. By default it will turn around and move forward back to the home **TNodeCart** node.

[153] Note the familiar green line that represents the animated queue line is not visible in this case.

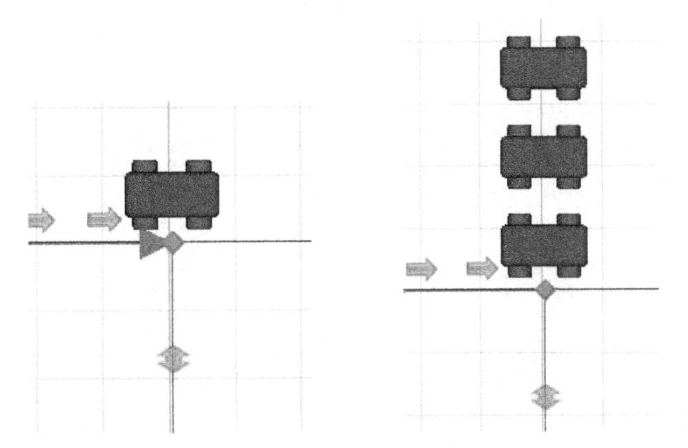

Figure 12.5: Vehicle is parked at the Home Node using Default Parking Station

Question 1: Does the Cart pick up more than one part?

Question 2: What do you notice about how a single part or two parts ride on the vehicle?

Step 9: To fix the issue of a single part riding out to one side, select the RIDE STATION queue on the VEHICLE and change the *Alignment* from "None" to "Point" under the *Appearance* tab. Add an additional vertex to the queue to accommodate three parts and move the first orientation point which is on the bottom to the middle as seen in Figure 12.6.

Figure 12.6: Changing the Orientation of the Riding Queue

Step 10: Run the simulation now to see how the cart works in the animation.

Step 11: By default, a parking location for a node is automatically added above the node as seen in Figure 12.7. However, you should generally not utilize the default one since the location and/or shape of the parting station cannot be adjusted as well as the orientation of the VEHICLES. Select the **TNodeCart** node by clicking it and then toggle the *Parking Queue* option in the *Appearance* tab to be clear. Next, a parking location is needed for the vehicle or the vehicle will vanish when parked.[154] Click on the **TNodeCart** node and select "*Draw Queue*" from the *Appearance* Tab. Click on the `ParkingStation.Contents.` A cross cursor will now appear in the facility window. Left click where you want the front of the queue to be and then right click where you want it to end (before you right click to end the queue you could also continue left clicking to add more vertices). Place it above the node so the Cart will be facing downward as seen in Figure 12.7. Click on the queue and in the properties window make sure the *Queue State is* `ParkingStation.Contents.` This queue will animate the parking station of the **TNodeCart** TRANSFERNODE for vehicles parked at the node. It can be oriented and placed at any location. Figure 12.7 shows the same scenario as Figure 12.5 but this time we are able to orient the vehicles to pointing in the same direction as the path.

---

[154] The parking queue is again just for the animation since the vehicle is physically at the node even though it cannot be seen.

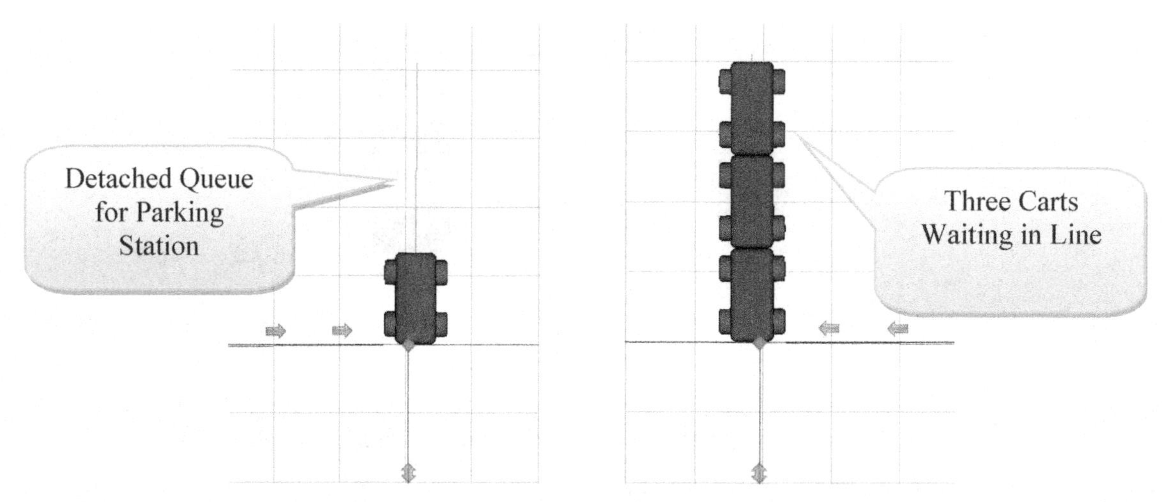

Figure 12.7: Vehicle is parked at the Home Node using the New Station

Step 12: Run the simulation now to see how the cart works in the animation.

The model does not show how many parts are waiting for the cart at the transfer node so it may be helpful to see how many are there at any given time in the simulation. To accomplish this we can add an animated "Detached Queue" for the **TNodeCart**.

Step 13: Click on the **TNodeCart** node and select "*Draw Queue*" from the "*Appearance*" Tab. Click on the "`RidePickupQueue`".[155] A cursor will now appear in the facility window. Left click where you want the front of the queue to be and then right click where you want it to end (before you right click to end the queue you could also continue left clicking to add more vertices). As seen in Figure 12.8, arrange this detached queue in an "L" shape. Otherwise an entity is shown trying to enter the transfer node will be visible[156] and by overlaying the "L" it shows all the entities waiting for transportation.[157]

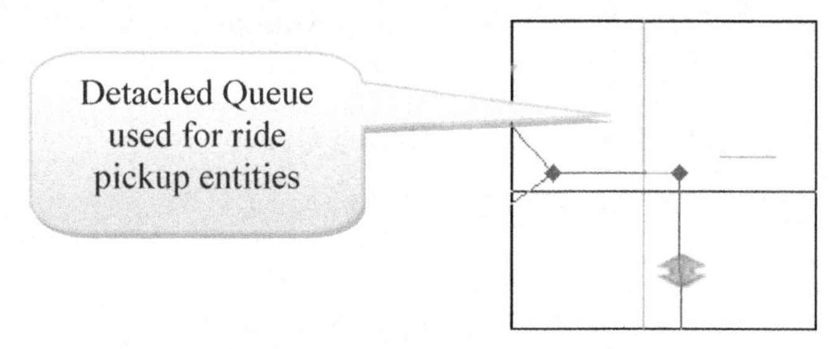

Figure 12.8: Ride Pickup Queue

Step 14: Run the simulation to see how the DETACHED QUEUE works.

Question 3: What is the utilization of the **VehCart** during a 40 hour replication?

____

[155] If added just generic animated queue now, change the *Queue State* property to **TNodeCart.RidePickupQueue** where **TNodeCart** is the name of the transfer node where the parts are being picked up. Your transfer node's name may be different.

[156] You can actually see the entity attempting to enter the transfer node if you watch that node carefully.

[157] These queues are for animation purposes only. The parts physically reside at the entrance of the node and one will always be visible as the others will stack up underneath. In later chapters, the notion of a STATION to hold entities will be discussed.

Part 12.2: Cart Transfer among Stations

Now, suppose that the transportation of parts among the stations, including entry and exit are handled by two similar vehicles. These vehicles are different than the **VehCart** which carries the parts to the warehouse SINK. These carts will wait at the beginning of the circular network until they are needed.

Step 1: Convert the one BASICNODE that connects the four SOURCES into a TRANSFERNODE by right clicking and selecting the *Convert to Type* menu item. This will be necessary in order for parts entering the network from the sources the ability to request transportation on the inside carts. Next, name this new TRANSFERNODE **TNodeStart**.

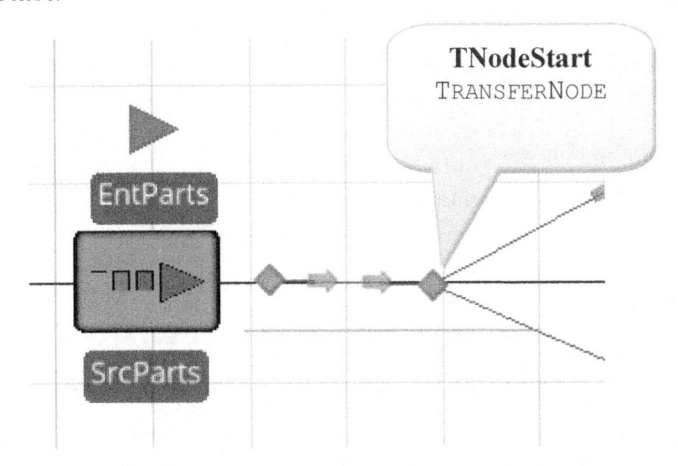

Figure 12.9: Changing the Start Node to a TRANSFERNODE

Step 2: Now add the inside carts by inserting another VEHICLE from the [*Standard Library*] into the model naming it **VehInsideCart**. Shrink the vehicle leaving the ride station queue in the same position.

Step 3: Make the **VehInsideCart** have an initial *population* of two[158] which will allow you to see the simultaneous behavior of multiple vehicles on the same path. The new vehicle will have the characteristics, as shown in Figure 12.10 (i.e., the *Initial Node* should be **TNodeStart** and the *Idle Action* property should be set to "Park at Home"). Select the "Transporting State" from the "*Additional Symbols*" and color it orange.

Notice that the VEHICLE object is similar to the ENTITY object in that multiple instances, which we call "run-time" instances can be created from the "design-time" instance of the **VehInsideCart** .[159] "Design-time" occurs while you are putting instances of the objects onto the modeling. All the objects in the FACILITY window are initialized just before the simulation begins to execute. "Run-time" occurs when the simulation executes.

[158] The fact that the property description says *Initial number in System* should indicate that vehicles can be added and removed from the simulation during the simulation execution (something that is left to a later chapter).

[159] We will see that the WORKER object can also have run-time instances.

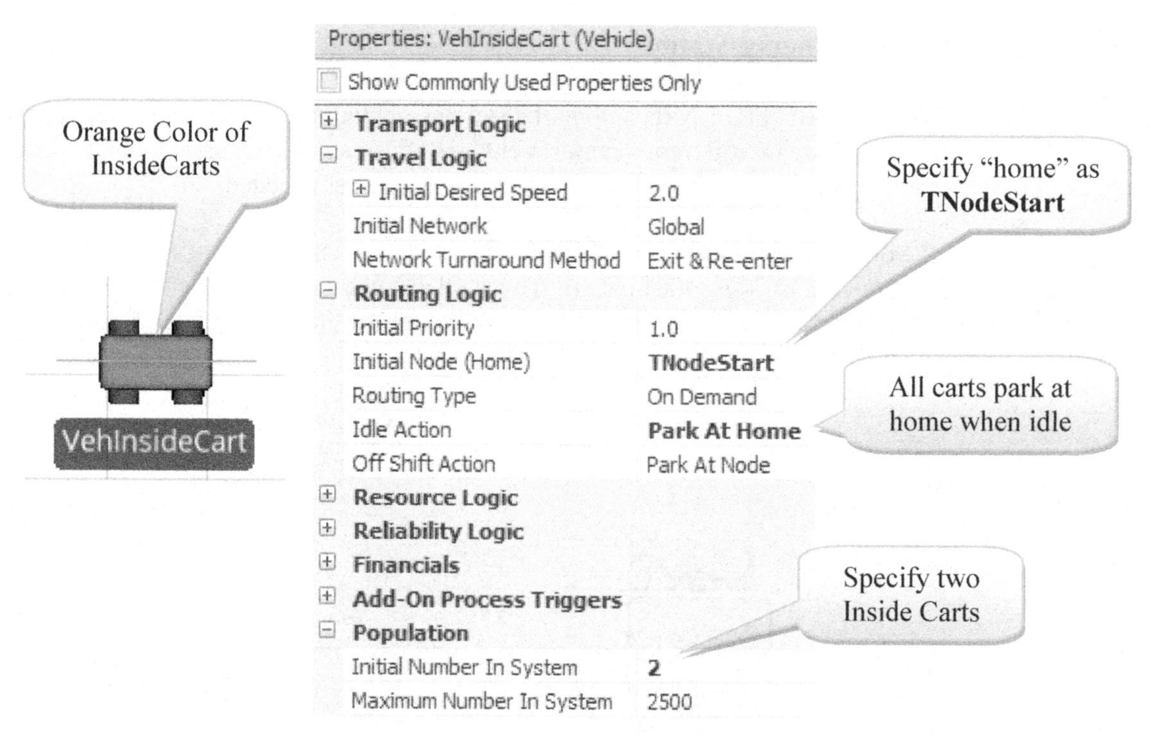

Figure 12.10: Properties of the InsideCart

Step 4: Now we need to make sure that the inside carts can drop off parts and then return back to the circular path. Since they obviously cannot go through the SERVERS, the inside carts would move to the input node of a SERVER, drop off the part and then become deadlocked (i.e., stuck).[160] The easiest solution is to allow the carts to maneuver around the stations (i.e., add a CONNECTOR from the input to the output nodes of each station as shown in Figure 12.11 making sure the direction of the link is correct.[161]

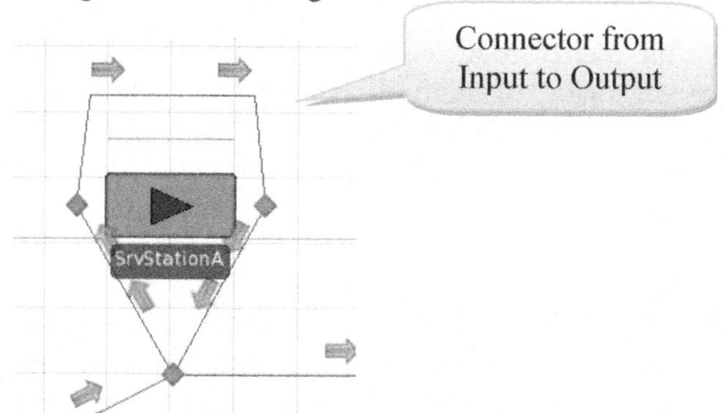

Figure 12.11: Adding a Connector at a Station

Step 5: Next, select the **TNodeStart** and the three output nodes (i.e., TRANSFERNODES) of the station servers and specify the entities leaving these nodes to ride on an **VehInsideCart** as seen in Figure 12.12.

[160] By default, WORKERS and VEHICLES will not enter fixed objects like SERVERS, SINKS, COMBINERS, etc. One solution is to create a connector leaving the input node of the server to the transfer node as well as another one from the transfer node to the output node to allow pickup.

[161] Recall, CONNECTORS take up zero time as well as have zero length in the model. Basically, the inputs and outputs of the server are really the transfer nodes of the circular path.

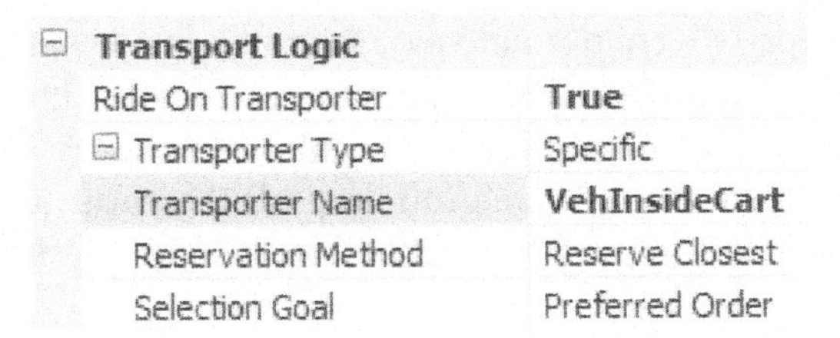

Transport Logic	
Ride On Transporter	**True**
Transporter Type	Specific
Transporter Name	**VehInsideCart**
Reservation Method	Reserve Closest
Selection Goal	Preferred Order

Figure 12.12: Use the Inside Cart

Step 6: Finally change the path between the exit node on the inside loop and the **TNodeCart** node to two one-way paths (i.e., add an additional path from the **TNodeCart** back to the basic node making sure to make the path ten yards long). Also change the path between the **TNodeCart** and the **SnkPartsLeave** to two one-way paths as well. Now your overall model should look like the one in Figure 12.13.[162]

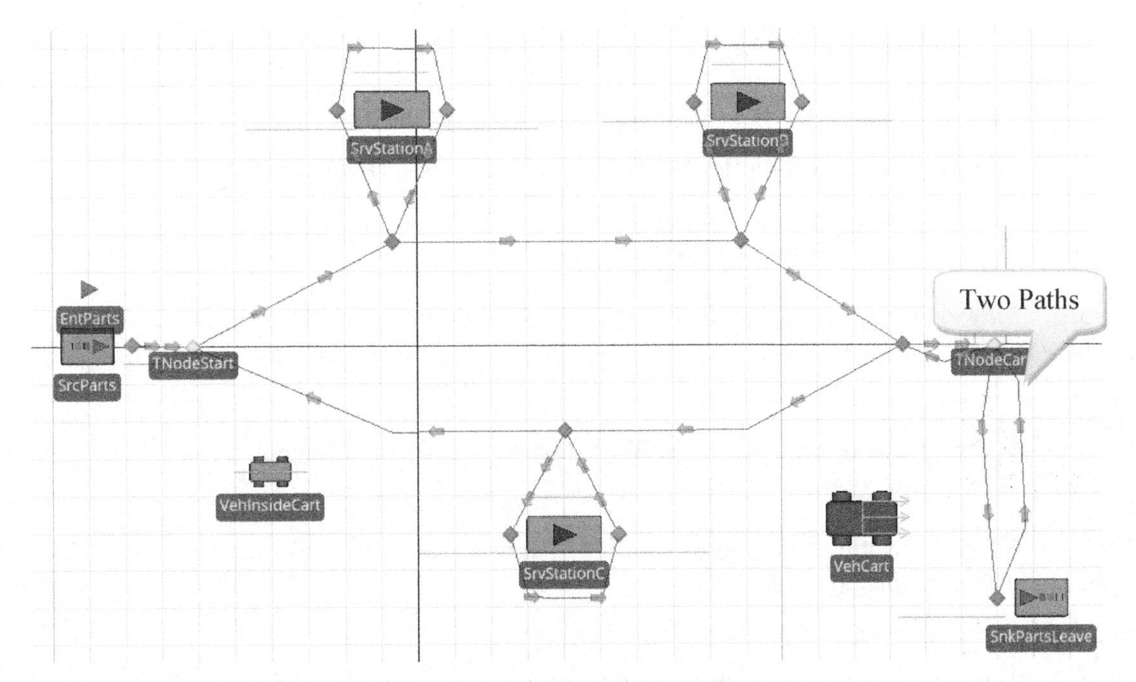

Figure 12.13: Model with Vehicles

Step 7: Also, change the **VehCart** *Population* property *Initial Number in the System* to "2" so two carts will be available for the outside transportation. The **TNodeCart** node should also be specified as shown in Figure 12.14.

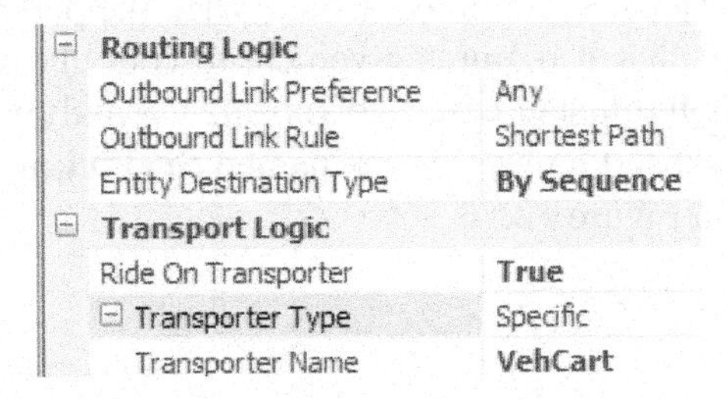

Routing Logic	
Outbound Link Preference	Any
Outbound Link Rule	Shortest Path
Entity Destination Type	**By Sequence**
Transport Logic	
Ride On Transporter	**True**
Transporter Type	Specific
Transporter Name	**VehCart**

Figure 12.14: Specifying TNodeCart

Step 8: Save and run the model observing what happens.

Question 4: What did you notice about the utilization of the outside carts?

____

[162] One can use bidirectional paths but this can lead to bottlenecks and deadlock as vehicles, workers, and entities can only be going in one direction and must wait for all items to clear path before the direction can be reversed. It is always better to use to separate paths.

Step 9: The difficulty was the parts were not unloaded from the inside cart to be transported by one of the two outside carts to the SINK. The issue is once the part has finished its last processing on a station, the next sequence is the exit. Therefore, the cart takes the part to its next destination (i.e., **Input@SnkPartsLeave**). To force the inside cart to drop off the parts at the **TNodeCart** node, this node should be inserted into the four part sequences right before the **Input@SnkPartsLeave** row for each type as seen in Figure 12.15.[163]

Table Parts | Sequence Part

	Sequence	ProcessingTimes (Minutes)	ID
1	Input@SrvStationA	Random.Pert(2,5,8)	1
2	Input@SrvStationC	Random.Pert(2,6,11)	1
3	TNodeCart	0.0	1
4	Input@SnkPartsLeave	0.0	1
5	Input@SrvStationA	Random.Pert(1,3,4)	2

Figure 12.15: Adding the Transfer Cart to Force a Drop off of the Parts

Step 10: Once all four sequences have been modified, save and execute the model. Observe the interacting behavior of the carts as well as notice how the carts park.

Question 5: How do parts get transported from the inside workstations to the outside cart for transfer to the sink where the parts leave?

___

Question 6: Is the queue ride parking for the **TransferCart** node needed anymore?

___

Question 7: What is the utilization of both (outside) **Carts** as well as the utilization of both **InsideCarts** at the end of the simulation (24 hours)?

___

Question 8: How are the individual vehicles within a cart type distinguished in the output from each other?

___

___

Step 11: There are a few anomalies that occur within the model. Notice how the inside carts will wait while they drop off a part when an outside cart is not available. When the part starts the drop off process, it visits the node which informs the part that it requires a vehicle to continue and therefore never completes the drop off process which releases the inside cart. To fix this anomaly along with other animation issues, delete the **TNodeCart** node and insert a new SERVER named **SrvPickup** with an infinite capacity and zero for the processing time as seen in Figure 12.16.

___

[163] The easiest way to insert the node into the sequence is to select the **Input@SnkPartsLeave** row and then click the *Insert Row* button in the *Data* section of the *Table* tab.

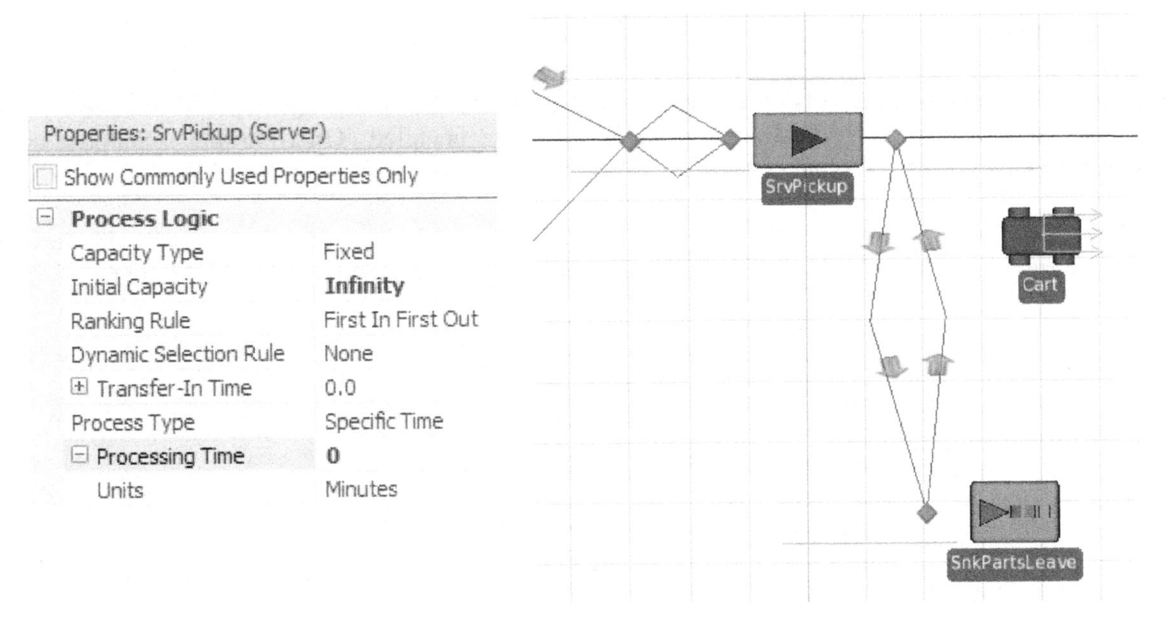

Figure 12.16: Adding a new Server as Pickup Point

Step 12: Reconnect the two 10 yard paths from the path network to the input of the **SrvPickup** and the two 300 yard paths from the output node to the SINK.

Step 13: For the sequence table, change the TransferCart node to the **Input@SrvPickup** as well as make the home node for the **Cart the Output@SrvPickup node** as seen in Figure 12.17.

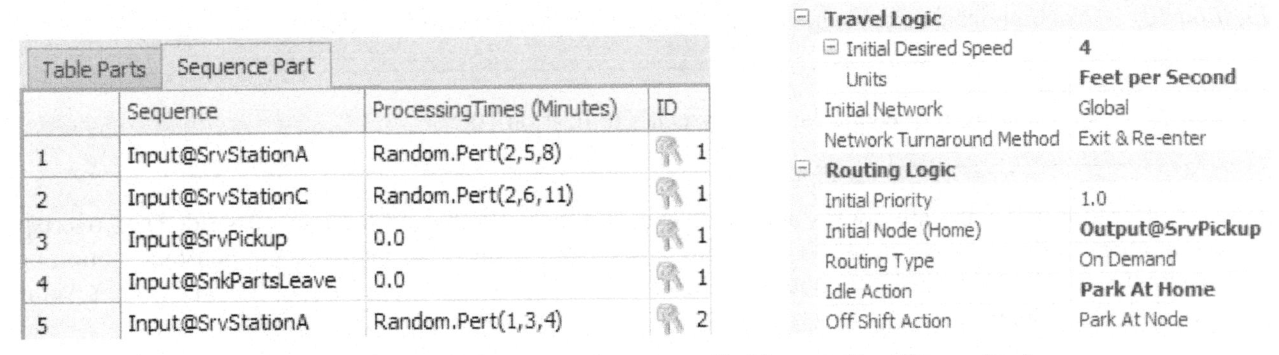

	Sequence	ProcessingTimes (Minutes)	ID
1	Input@SrvStationA	Random.Pert(2,5,8)	1
2	Input@SrvStationC	Random.Pert(2,6,11)	1
3	Input@SrvPickup	0.0	1
4	Input@SnkPartsLeave	0.0	1
5	Input@SrvStationA	Random.Pert(1,3,4)	2

Figure 12.17: Changing the Sequence Table and Cart Home Node

Step 14: Finally, request for transportation on the **Output@SrvPickup** transfer node using the same information from Figure 12.14.

Step 15: Save and run the model observing what happens with the two types of carts.

Question 9: Now what is the utilization of both (outside) **Carts** as well as the utilization of both **InsideCarts** at the end of the simulation (24 hours)?

____

Part 12.3: Other Vehicle Travel Behaviors (Fixed Route and Free Space Travel)

Now let's experiment with different vehicle specifications to create different cart behaviors. We will now explore both a fixed route travel (i.e., bus) as well as travel through free space.

Fixed Route Travel

Currently the **VehInsideCart** travels to wherever transport is needed among the stations including entry and pickup. That *Routing Type, specified on the vehicle definition,* is called "**On Demand**." Suppose now we want the cart to travel in a fixed sequence around all the service points in the model.

Step 1: From the Data tab, insert a SEQUENCE TABLE for the vehicle *Route Sequence named* **SeqInsideCart** as shown in Figure 12.18. Note the sequence includes both input and output nodes of the stations.

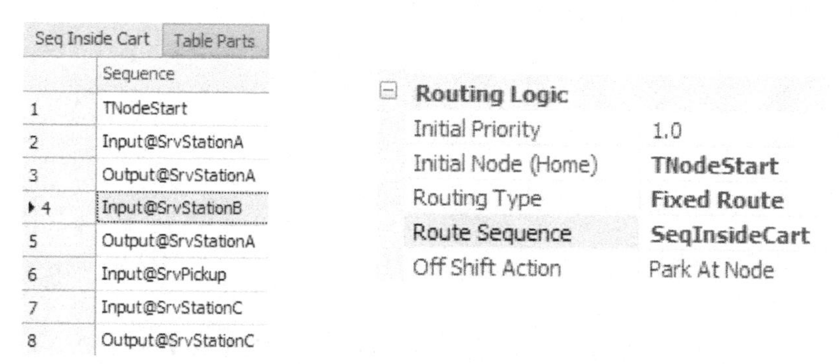

Figure 12.18: New Route Sequence for InsideCart

Step 2: Select the **VehInsideCart** and set the *Routing Type* property to "**Fixed Route"** and the *Route Sequence* to the **SeqInsideCart** sequence table as seen in the right picture of Figure 12.18.

Step 3: Run the model and observe both the behavior of the cart(s) and the statistics on the cart performance.

Question 10: Are both inside carts behaving as expected? What is their utilization?

____

Question 11: Can you see the second cart? Why is it's utilization 100%?

____

Step 4: Because both carts are always active, their utilization is 100% and the two carts ride (i.e., overlap one another) in the same location. So if we are to see carts we need to interrupt the constant motion or have the two carts start at different times (i.e., have a Source create them at different times). There several ways to achieve the result.

Step 5: Change the *Load Time* of the **VehInsideCart** to an `Exponential(0.5)` minutes. Save and run the model observing the cart behavior.

Question 12: What happens to both inside carts?

____

Free Space Travel

Rather than travel on a specific path between stations, vehicles, workers and entities can travel in "free space." To do this, SIMIO needs the "location" of the various nodes. In other words, it's going to assume that your model is "to scale" as it might be if the animation layout is based on computer-aided drawing (CAD) software. It's important to recognize that the time it takes parts to be moved by vehicles depends on the distance the vehicle travels and its travel speed[164]. To illustrate the free space travel, let's treat our animation as though it came from a CAD drawing.

---

[164] When vehicles move entities, it's the speed of the vehicle that determines how fast it traverses a distance and not the entity speed.

Step 6: Save the current model as Chap12.3Step6 and eliminate all the paths and nodes between the stations and the **SrvPickup** so that travel is unrestricted, as seen in Figure 12.19.

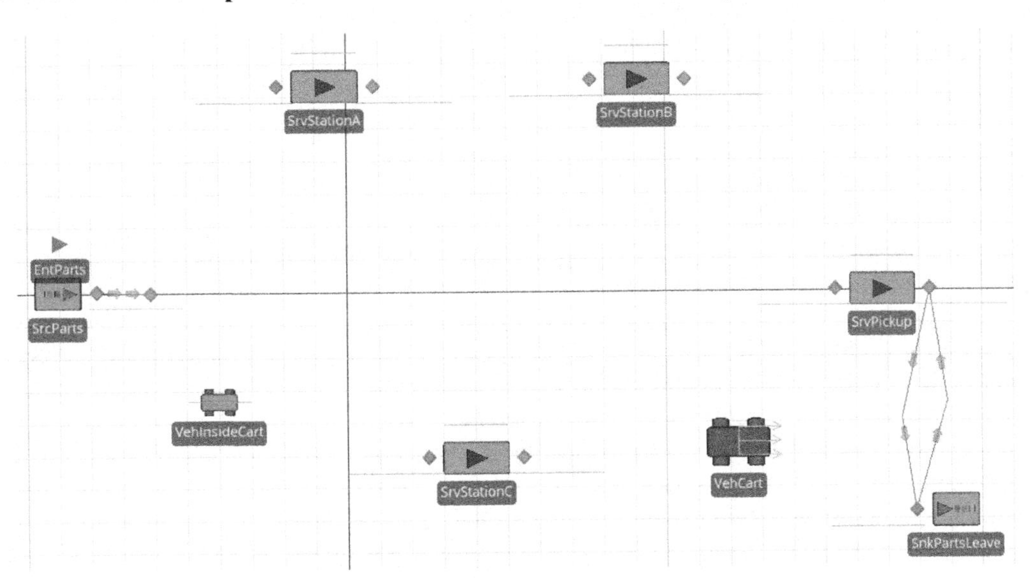

Figure 12.19: Layout to Scale: No Paths

Step 7: Modify the **VehInsideCart** "*Travel Logic*" and "*Routing Logic*" as Figure 12.20. In this case we are showing there is no travel network – meaning the vehicle must use "*Free Space*". We also are routing **On Demand** and the *Idle Action* is to **Remain in Place**.

Properties: VehInsideCart (Vehicle)

☐ Show Commonly Used Properties Only

⊞ **Transport Logic**	
⊟ **Travel Logic**	
⊞ Initial Desired Speed	2.0
Initial Network	**No Network (Free Space)**
Network Turnaround Method	Exit & Re-enter
⊟ **Routing Logic**	
Initial Priority	1.0
Initial Node (Home)	**TNodeStart**
Routing Type	On Demand
Idle Action	**Remain In Place**
Off Shift Action	Park At Node

Figure 12.20: Travel Logic and Routing Logic

Step 8: Run the model and observe the travel of the inside carts.

Question 13: When would this type of travel be most appropriate and what do you observe about the parking?

____

Part 12.4: Conveyors: A Transfer Line

SIMIO provides for two kinds of conveyor modeling concepts: *accumulating* and *non-accumulating*. An accumulating conveyor allows parts to queue at its end, but the conveyor keeps moving. This approach might model a belt conveyor whose parts rest on "slip sheets" that allows the belt to continue to move. A non-accumulating conveyor must stop until the on-off operation completes. This concept might apply to an overhead conveyor with parts on carriers. SIMIO conveyors do not have on/off stations, so that a transfer line will have to be modeled as a series of individual conveyors. Since the transfer line stops, the non-accumulating concept applies. However the individual conveyors will need to be synchronized.

Suppose now that the transfer among stations in the manufacturing cell is handled by a transfer line which is a continuous conveyor, possibly in a loop, that has on/off stations. When a part is at an on/off station, the entire line pauses until the part has been placed on or taken off the conveyor. It takes between 0.5 and 1.5 minutes to load/unload the parts from the nodes.

Step 1: Reopen the last model from Chapter 5 and save it as a new name. Change the arrival rate of the source to be Exponential(15) now.

Step 2: Change all the paths on the interior of the model to conveyors by selecting the path and accessing the *Convert To Type* submenu via a right mouse click.[165] To embellish the model, add path decorators for conveyors as seen in Figure 12.21 from the *Path Decorators* tab.

Step 3: Also modify the symbols for the parts so they are one meter wide by one meter in length by one-half meter in height. Select the **EntParts** MODELENTITY and for each *Active Symbol*, change the size under the *General→Physical Characteristics→Size* properties. After selecting a different active symbol, you will need to deselect it and the reselect it to see the size for that symbol. We will need to be a little more specific about these sizes as we size the conveyors. We created these symbols from the "*Project Home*" tab in the "*Create*" section, clicking on "New Symbol" and selecting "*Create a New Symbol*". When creating the symbol we paid close attention to its size (the grid is in meters).

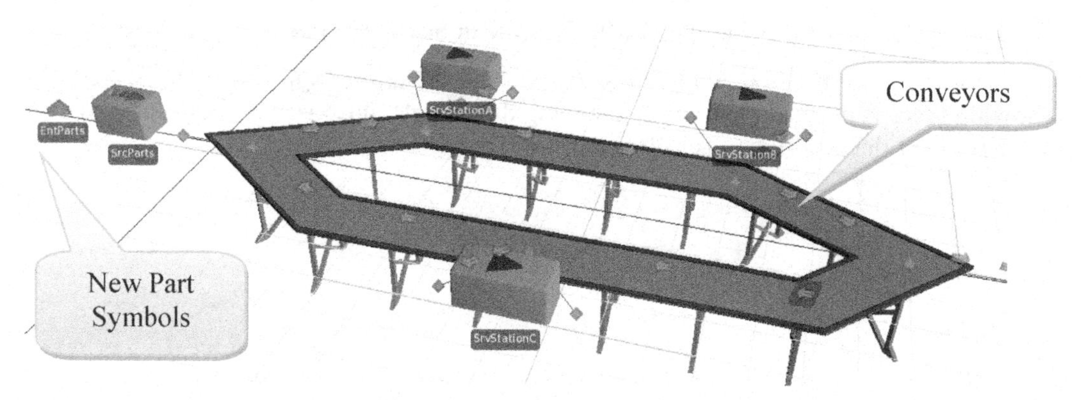

Figure 12.21: 3D View with Conveyors

Step 4: Next, properties of the conveyors need to be specified. Here it is convenient to use the "Spreadsheet View" by right-clicking on one of the conveyors and selecting *Open Properties Spreadsheet View for all objects of this type*. The lengths from the previous paths (recall they were 25, 35, 20, 30, and 40 yards) do not need to be modified. All five conveyors should have a conveyor speed of 0.5 meters/second for each conveyor. Set the *Accumulating* property to ***False*** for each conveyor (i.e., the selection boxes should be unchecked).

Step 5: In the *Entity Alignment* property, choose the *Cell Location* option. Choosing this option causes the conveyor to be conceptually divided into cells. Parts going onto the conveyor will need to be placed into a cell. Therefore choose the number of cells to be equal to number associated with the conveyor length. Otherwise, if you select *Any Location* for the *Entity Alignment*, a part can go onto the conveyor at any place, even interfering with another.

---

[165] Note you can use the *Ctrl* key to select all five paths and convert them all at once.

Instance Name	Initial Traveler Capacity	Entry Ranking Rule	Entry Ranking Expression	Initial Desired Speed	Units	Entity Alignment	Cell Spacing Type	Number Of Cells	Cell Size	Units	Auto Align Cells	Drawn To Scale	Logical Length	Units
Conveyor1	Infinity	First In First Out	Entity.Priority	2.0	Meters per Second	**Cell Location**	Fixed Cell Size	1	**1**	Meters	With First Entity		**25**	Yards
Conveyor2	Infinity	First In First Out	Entity.Priority	2.0	Meters per Second	**Cell Location**	Fixed Cell Size	1	**1**	Meters	With First Entity		**35**	Yards
Conveyor3	Infinity	First In First Out	Entity.Priority	2.0	Meters per Second	**Cell Location**	Fixed Cell Size	1	**1**	Meters	With First Entity		**20**	Yards
Conveyor4	Infinity	First In First Out	Entity.Priority	2.0	Meters per Second	**Cell Location**	Fixed Cell Size	1	**1**	Meters	With First Entity		**30**	Yards
Conveyor5	Infinity	First In First Out	Entity.Priority	2.0	Meters per Second	**Cell Location**	Fixed Cell Size	1	**1**	Meters	With First Entity		**40**	Yards

Figure 12.22: Specifying Properties for all Five Conveyors of the TransferLine (Spreadsheet View)

Step 6: Since the transfer line consists of multiple conveyors rather than a continuous one, the conveyors have to be synchronized so that when parts enter an on/off location on the transfer line, the entire line which is a set of conveyors stops to await the on/off operation.[166] To do this, first define a new `DISCRETE STATE VARIABLE` of type integer (from the *Definitions* tab) for the **Model** named **GStaNumberNodesWorking**. This variable will represent the number of on/off locations that are currently either putting parts onto or taking parts off the transfer line. When this state variable is zero, the transfer line is running and when this variable is greater than one, the transfer line should be stopped.

Step 7: Let's now insert a new "process." From the "*Processes*" tab, select "*Create Process*" and name the new process **On_Off_Entered**. The process will consist of the steps shown in Figure 12.23.

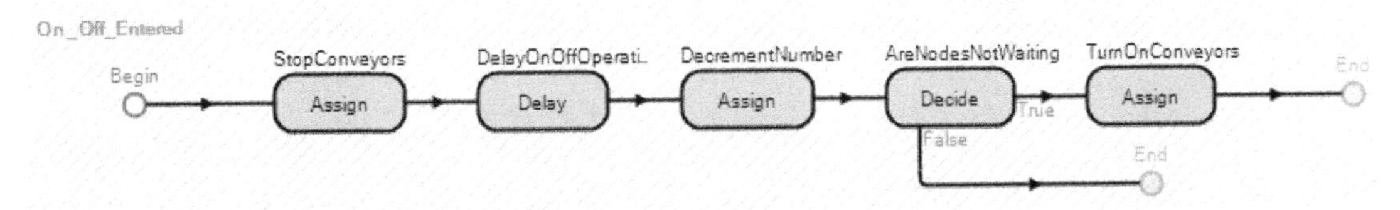

Figure 12.23: Process when Entering/Leaving the Transfer Line

- When a part reaches an on/off operation, the `Assign` step needs to increment the **GStaNumberNodesWorking** by one and all conveyors need to be stopped by setting their desired speed to zero (see Figure 12.24).[167]

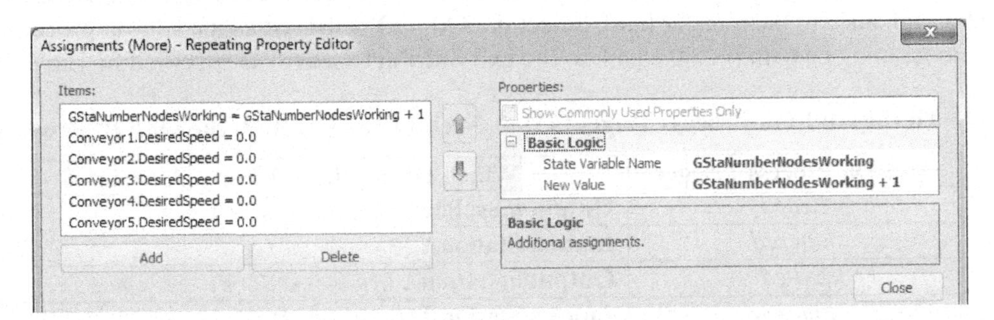

Figure 12.24: Incrementing the Number Nodes Working and Shutting Down the Conveyors.

- The `Delay` step will be used to model the on/off time, which is assumed to be uniformly distributed with a minimum of 0.5 and a maximum of 1.5 minutes.
- The second `Assign` step will now decrement the **GStaNumberNodesWorking** by one.
- The `Decide` step is based on the condition that `GStaNumberNodesWorking==0`, meaning no other on/off stations are busy loading/unloading the transfer line.

[166] The synchronization is not exact since the occupied cell of the upstream conveyor is not known to the downstream conveyor and parts from a cell may conflict with parts arriving from the upstream conveyor (i.e., **SrcParts**).

[167] You will need to select the *Assignments (More)* button in the `Assign` step.

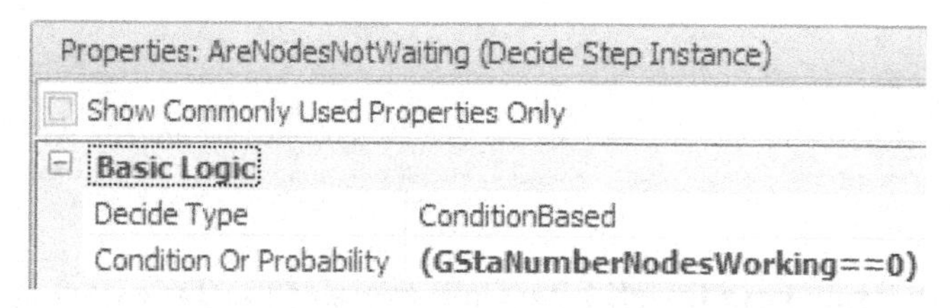

Figure 12.25: Decide Step Option

- The last `Assign` step will turn the conveyors back on. SIMIO stores all the times internally in meters per hour. If the state variable has units these can be specified in the assignment. Make sure you change the units to meters per second or an assignment of 0.5 would be treated as 0.5 meters per hour.[168] You can convert the original specification of 0.5 meters per second to 1800 meters per hour to be consistent with how SIMIO will interpret the assignment and avoid having a conversion each time. Figure 12.26 shows the conveyors being turned back on.

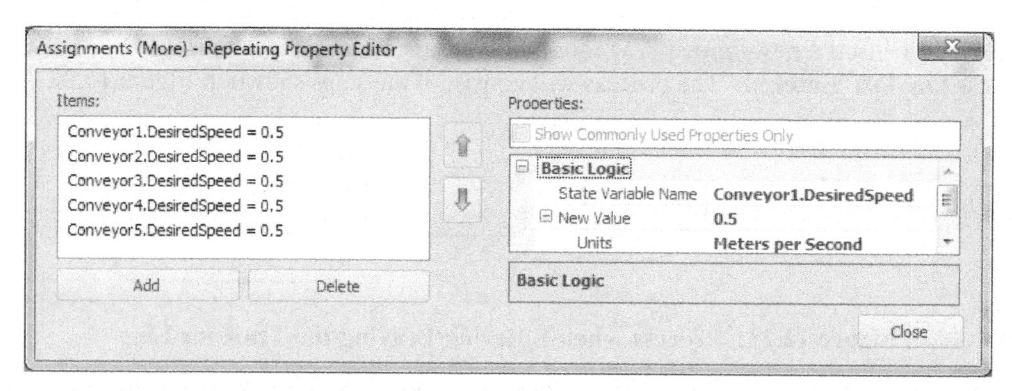

Figure 12.26: Turning the Conveyors Back On

Step 8: Finally, the **On_Off_Entered** generic process needs to be invoked appropriately (i.e., every time a part is loaded or unloaded to the transfer line). Select the **On_Off_Entered** as the add-on process trigger for all the nodes and one path (i.e., the path to the exit **SnkPartsLeave**) as specified in Table 12.1.

Table 12.1: All the Add-On Process Triggers that Need to Go Through the On/Off Process Logic

Add-On Process Trigger	Which Object the Trigger is Associated
Exited	**Output@SrcParts** (TRANSFERNODE)
Entered	**Input@StationA** (BASICNODE)
Exited	**Output@StationA** (TRANSFERNODE)
Entered	**Input@StationB** (BASICNODE)
Exited	**Output@StationB** (TRANSFERNODE)
Entered	**Input@ SnkPartsLeave** (BASICNODE)
Entered	**Input@StationC** (BASICNODE)
Exited	**Output@StationC** (TRANSFERNODE)

Question 14: Why did we increment and decrement **GStaNumberNodesWorking** by one rather than just setting it to one or zero which would eliminate the need for the `Decide` step?

____

Question 15: Why not simply use the **On_Off_Entered** process at the BASICNODE of the entry/exit?

____

[168] The presumption of a standard internal time probably represents a choice of efficiency over ease of use, since determining the original time units would represent some additional computational effort.

Step 9: Run the simulation for 40 hours at a slow speed to be sure each of the parts follows their sequence appropriately and answer the following questions.

Question 16: How long are all part types in the system (enter to leave)?

____

Question 17: What is the utilization of each of the SERVERS?

____

Question 18: What is the average number on the **Conveyor2** link?

____

Question 19: What is the average time on the link for **Conveyor3** link?

____

Part 12.5: Machine Failures in the Cell

Failures are a part of most mechanical and other systems. SIMIO provides methods to model these failures within the "*Reliability Logic*" section of the SERVER (and other objects). In the case of the transfer line, we also need to shut down the line while a machine is in the state of being repaired.[169] Let's assume that **SrvStationC** is the unreliable machine. In particular, we will assume that the MTBF (Mean Time Between Failures) is exponentially distributed with a mean of ten hours, while the MTTR (Mean Time To Repair) is modeled with a Triangular distribution (0.5, 1, 2) hours.

Step 1: In the "*Reliability Logic*" section for the Station C SERVER, specify the *Failure Type* to be "*Calendar Time Based", the Uptime Behavior* be `Random.Exponential(10)` with units of hours, and the *Time To Repair* should be `Random.Triangular(0.5,1,2)` with units of hours as seen in Figure 12.27. Remember that the server is failing and that means that all the capacity associated with this server fails (the capacity for this server is scheduled, but it changes between 1 and 2).

⊟ **Reliability Logic**	
Failure Type	**Calendar Time Based**
⊟ Uptime Between Failures	**Random.Exponential(10)**
Units	Hours
⊟ Time To Repair	**Random.Triangular(0.5,1.0,2)**
Units	Hours

Figure 12.27: Specifying a Failure and Repair

Step 2: When the **SrvStationC** fails, the entire transfer line should be shut down and then started back up once the station has been repaired. Select **SrvStationC** and create new Add-On Process Triggers, "*Repairing*" and "*Repaired*" to produce new processes **SrvStationC_Failed** and **SrvStationC_Repaired**. The "*Failed*" trigger will be invoked at start of the failure while the "*Repaired*" trigger will be invoked after the SERVER has been repaired. In **SrvStationC_Failed,** use an *`Assign`* step to turn off all of the conveyors as by setting their `DesiredSpeed` to zero. In the **SrvStationC_Repaired**, use an *`Assign`* step to turn back on the conveyors if there is currently no nodes loading/unloading (i.e., setting the `DesiredSpeed = 0.5` meters per second for all conveyors) as seen in Figure 12.28.[170]

---

[169] The tremendous cost of downtime due to everything on the line becoming idle is one of the reasons why transfer lines and similar highly automated processes are becoming relatively rare.

[170] Copy the *`Decide`* and *`Assign`* steps from the **On_Off_Entered** process and remove the increment of the **GStaNumberNodesWorking**.

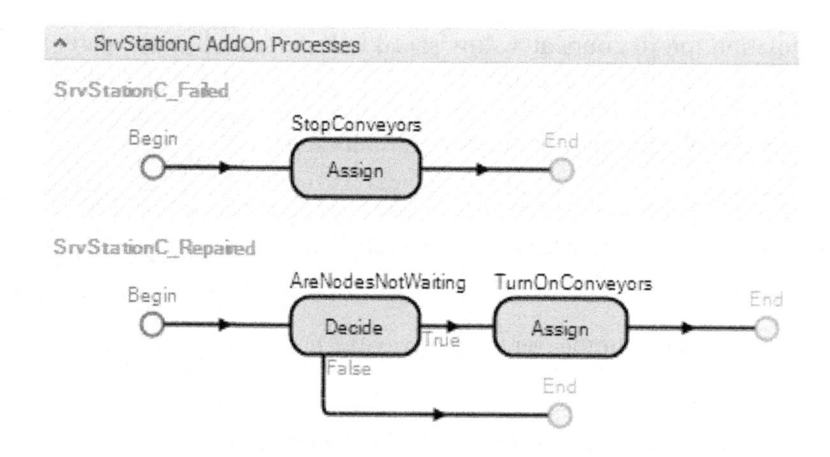

Figure 12.28: Failed and Repairing Process Triggers

Step 3: Also, the **On_Off_Entered** process should not start up the transfer line if the **SrvStationC** has failed and is currently being repaired. To do this change the `Decide` step in the **On_Off_Entered** process so the conditional expression becomes the following.[171]

```
(GStaNumberNodesWorking==0) && (SrvStationC.Failure.Active==False)
```
[172]

Step 4: Note that the **SrvStationC** will change to the color red when it is down, indicating that the server (and all its capacity) is in the "Failed" state. Run the model and make sure the transfer line stops when the station is also under repair.

Question 20: Is there anything about the final animation that troubles to you?

Part 12.6: Sorting Conveyors

Now we find out that the output from the conveyor system is another system of conveyors which sorts the parts into destinations. The parts that exit the manufacturing cell are routed to an accumulating conveyor at a labeling machine. From the labeling machine, the parts are conveyed to one of four sinks corresponding to the specific part type as seen in Figure 12.29. The four conveyors directly in front of the sinks are accumulating conveyors. These conveyors keep moving and parts are allowed the queue, although in this case the sinks pose no barrier to exit. The rest of the conveyors are non-accumulating.

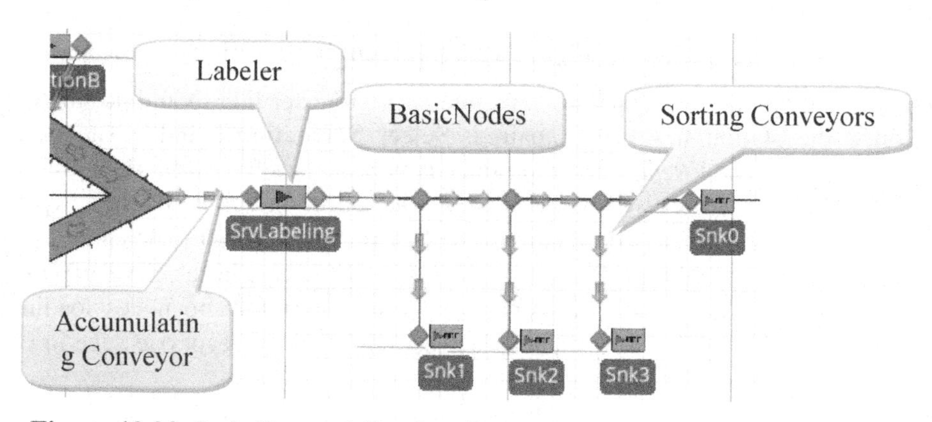

Figure 12.29: Labeling and Sorting Conveyors

---

[171] `Object.Failure.Active` function will return true if the current object has failed.

[172] Notice that SIMIO utilizes C# logical operators (i.e., == for equality, && for "And" statement, and || for the "Or" statement).

Step 1: Remove the original **SnkPartsLeave** and replace it with a **SrvLabeling**. The labeling processing time follows a Triangular(1,2,3) minutes and has a capacity of "1." Since the labeler will be supplied by an accumulating conveyor from the inside set of conveyors (i.e., the BasicNode), set the *Input Buffer* capacity property to zero because it is not needed and there is no delay in off-loading to this conveyor.

Step 2: Add three BasicNodes to follow the **SrvLabeling** and name them **Basic1**, **Basic2**, and **Basic3**, in that order. These are conveyor diverters that will send the sorted parts to their destinations. Add three sinks, **Snk1**, **Snk2**, **Snk3**, and **Snk0** to represent the destinations for parts of type 1, 2, 3, and 0 respectively. Connect up the objects with conveyors according to Table 12.1.

Table 12.2: Conveyor Properties

From/To	Conveyor Type	Conveyor Length	Conveyor Speed
BasicNode4 to SrvLabeling	Accumulating	10 meters	1.0 meters per second
SrvLabeling to Basic1	Non-Accumulating	10 meters	0.5 meters per second
Basic1 to Snk1	Accumulating	10 meters	0.1 meters per second
Basic1 to Basic2	Non-Accumulating	10 meters	0.5 meters per second
Basic2 to Snk2	Accumulating	10 meters	0.1 meters per second
Basic2 to Basic3	Non-Accumulating	10 meters	0.5 meters per second
Basic3 to Snk3	Accumulating	10 meters	0.1 meters per second
Basic3 to Snk0	Accumulating	10 meters	0.1 meters per second

Step 3: For the part's. sequence table, replace the **Input@SnkPartsLeave** entries with **Input@SrvLabeling**. Next add the sink node destinations associated with the part types. Figure 12.30 shows part of the table and you can insert the last placement at the end to maintain the sequence.

10	Input@SrvLabeling	0	2
11	Input@SrvStationB	Random.Pert(5,9,11)	3
12	Input@SrvStationC	Random.Triangular(6,9,11)	3
13	Input@SrvLabeling	0.0	3
14	Input@Snk0	0.0	0
15	Input@Snk1	0.0	1
16	Input@Snk2	0.0	2
17	Input@Snk3	0.0	3

Figure 12.30: Changes to the Sequence Table

Next change the **Output@SrvLabeling** node ***Entity Destination*** to ***By Sequence***. There is no need to add intermediate nodes (the BasicNodes) since SIMIO can find the designations from the **SrvLabeling**.

Step 4: Select all the conveyors and add the conveyor path decorators as seen in Figure 12.31.

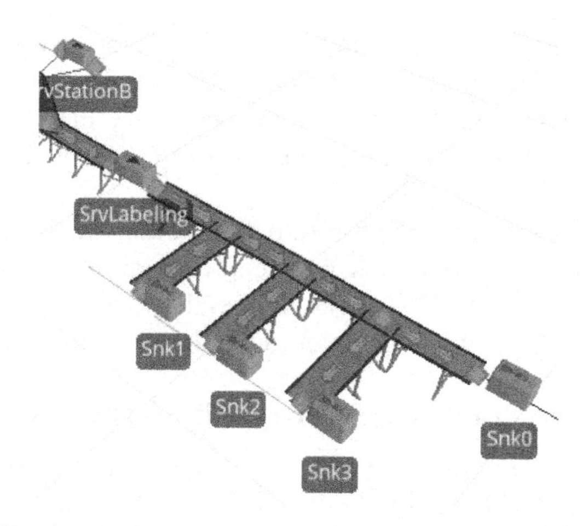

Figure 12.31: Conveyor with Path Decorators

Step 5: Save and run the simulation for a while observing the animation which may need slowing down.

Question 21: Are the parts being conveyed as expected and do you see accumulation at the labeler?

___

Step 6: Run the simulation for 40 hours.

Question 22: What is the average number of parts that accumulate on the conveyor in front of the labeler?

___

Question 23: What is the Utilization of the labeler?

___

Question 24: Why does the OUTPUTBUFFER of the labeler have content?

___

Question 25: What is the average number on the conveyor between **Basic2** and **Basic3**?

___

Question 26: Do you have any concerns about the operation of the sorting conveyor system?

___

Part 12.7: Commentary

- Material handling can be a substantial cost of production and thus is of great interest to many companies. Vehicles and conveyors provide considerable flexibility in modeling material handling.
- The modeling of cranes has not been included in this chapter. These are a complex category of materials handling. However, SIMIO has developed a special library for cranes that can represent a wide variety of overhead and floor cranes. In particular, there is a 3D animation of the cranes to reflect the fact that cranes operate in three dimensions. This package is highly recommended.

Chapter 13
Management of Resources: Veterinary Clinic

The RESOURCE object provides a flexible means to model individual resource needs. Unlike the SERVER, the RESOURCE does not provide a processing capability with its input and output buffers at a specific location. Further because the RESOURCE object is not location restricted, as the units of its capacity can be used at several places within a model. Because the resource is flexible, its capacity can be allocated (seized and release) in a complex fashion. The management of a firm's resources is one of main challenges of any organization. A simulation of that management can greatly assist in determine how those resources can be best used.

To illustrate part of the resource management capability, a Veterinary Clinic is modeled. The veterinary clinic provides care for many different kinds of animals. The office wants to provide efficient care through the employment of efficient resources. For simplicity we model primarily the activities of the veterinary staff. The staff consists of four veterinarians named (first names) Billy, Jim, Maisy, and Ellie. This clinic service is complicated by several factors. There are times when multiple types of resources, such as a Veterinary staff member and an X-ray machine, are required before the patient serviced. There are certain preferences for resources such as the fact the Jim is the only veterinarian who can service iguanas. The SERVER object cannot handle these cases easily. For instance, you may need a nurse and doctor to perform a procedure and then need the nurse for some after-procedure service.

Part 13.1: Utilizing the Fixed Resource Object

The RESOURCE object is a capacitated "Fixed object" (i.e., it does not visibly move in the network like the TRANSFERNODE or SOURCE objects), and which can be seized and released. It does provide a generic object that can be used to constrain entity flow. RESOURCES have reliability capabilities as well as the ability to statically and dynamically rank and select the entities for processes (i.e., process all critical ill patients first) as well it can choose to remain idle under some conditions. (i.e., refuse to be seized).

Step 1: Create a new fixed model ("Project Home" tab). Note, you can have multiple models in the same project, but only one will run at a given time unless you use them "inside" of each other.

Step 2: Add a new SOURCE named **SrcPatients**, SERVER named **SrvVetClinic**, and a SINK named **SnkExit**. Add a MODELENTITY named **EntPatients**. Connect the objects with paths that are logically 30 meters long.[173]

- Patients arrive Exponentially with an interarrival time of 6.5 minutes.
- Each patient takes between 10 and 30 minutes to be seen by the staff with most patients taking 15 minutes to be served (i.e., use a Pert distribution to model the processing time).

Step 3: From the Standard Library section, add a new fixed RESOURCE[174] named **ResBilly.**

- Since RESOURCES have reliability logic set up emergency phone calls for the doctors. Specify the *Uptime Between Failures* to follow an exponential distribution with a mean of six hours and *Time to Repair* expression equal to a `Random.Pert(40,55,60)` minutes as shown in Figure 13.1.

[173] Recall, if you select the output node of the SOURCE while holding down the *Shift* and *Ctrl* keys, SIMIO will allow you draw LINKS without first selecting a LINK type from the Standard Library. This key sequence can be more efficient. Once the two PATHS have been drawn, select them both and change their *Drawn to Scale* properties to "False" and the *Logical Length* properties to 30 meters.

[174] The default symbol looks like a gray eye with a blue iris.

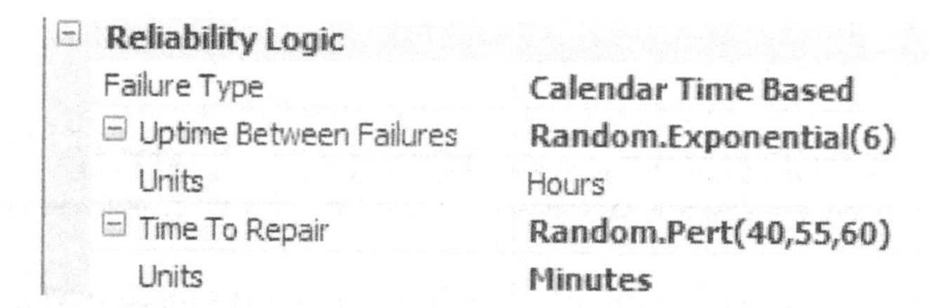

Figure 13.1: Reliability Logic to Handle Emergency Call from the Veterinarian

- Like the servers, the resource symbol changes as its state changes. If you click on **ResBilly** and examine the "active symbol", there are seven resource states. The first five states are the same as those for servers, but there are two additional states (i.e., "Failed Busy" and "Off Shift Busy"). The symbols change colors, while the iris remains blue (i.e., the color is red when a resource fails). Similar to the server states (see Chapter 8), the resource states have a name, color, and state number.
- Like the servers, the resource is idle (state 0) when all the capacity of the resource is idle, is busy when any of the capacity is being used, blocked whenever the resource is presented from disengaging from an entity, failed whenever the reliability causes a failure of the resource, and offshift when the capacity of the resource is zero.

Step 4: Copy the **ResBilly** resource instance three times naming them **ResEllie**, **ResJim**, and **ResMaisy** respectively to represent the other three staff members as seen in Figure 13.2.

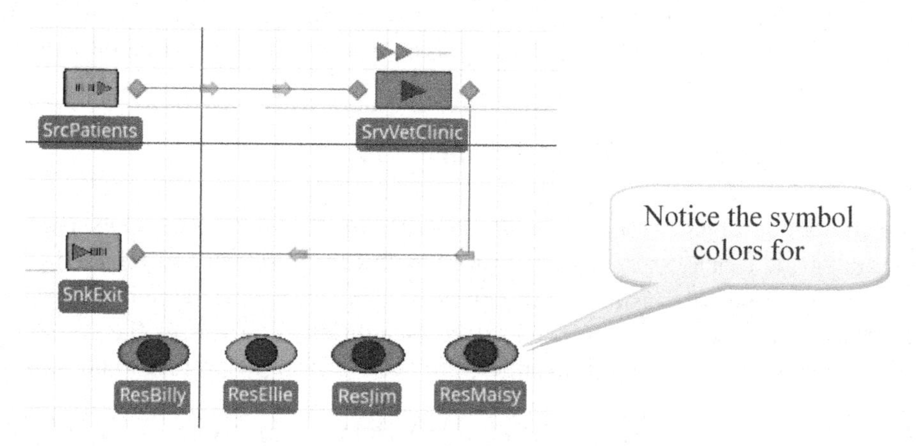

Figure 13.2: Model of the Veterinary Clinic Using Four Resources

Step 5: Change the capacity of the SERVER **SrvVetClinic** to be *Infinity* so the SERVER's capacity does not constrain the system. We want the new resource objects to be the constraining requirement.

Step 6: At this point, we can seize an individual RESOURCE but our model will require selecting among any available staff member rather than a specific resource. Go to the "*Definitions*" tab of the model and create a new LIST of OBJECTS named **ListResources** and add all four resources to the list as seen in Figure 13.3. This list will be used by patients to select an available veterinary staff.

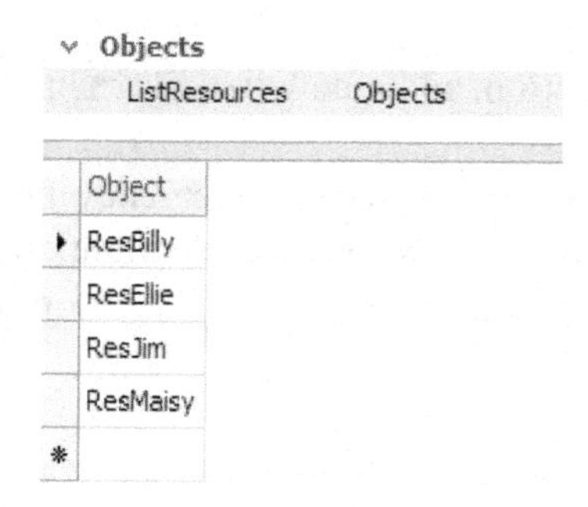

Figure 13.3: Creating a List of Resources to be Seized

Step 7: The basic model has been set up with the fixed RESOURCES; however, the patients need to request (seize) an available staff member to be serviced and then, once they have completed service, need to release the staff member to work on the next patient. Patients will request a staff member when they arrive to the Vet Clinic. Two *Add-On Process Triggers* for the SERVER will be used.

Step 8: The patient entity needs to seize one of the available resources before entering processing (i.e., the "*Before Processing*"[175] trigger). To invoke the "*Before Processing*" add-on process, expand the "*Add-On Process Triggers*" in the properties window and double click on the label to invoke/create a process.[176]

Step 9: Since the patient wants to seize the resource; insert a `Seize` step by selecting from the list of possible steps and dragging it between the *`Begin`* and *`End`* nodes which represent where the process starts and ends as seen in Figure 13.4.

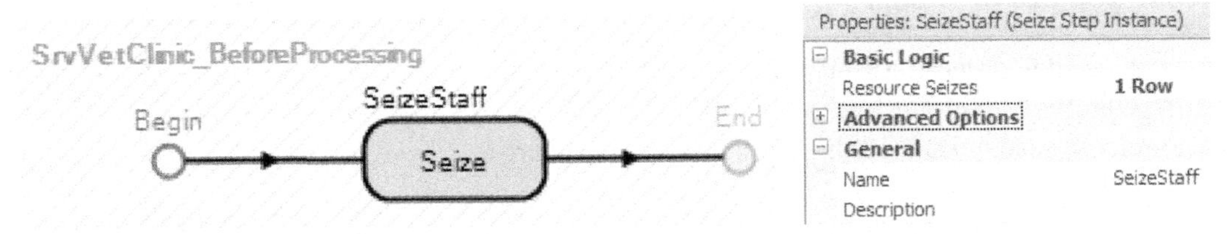

Figure 13.4: Using a Seize Step to Request a Veterinary Staff Member

- Under the "*General*" properties of the process step as seen in the right side of Figure 13.4, set the *Name* to **SeizeStaff**. The resources to try to seize will be specified under the "Basic Logic". Click on the [...] button in the property window in the *Seizes* property row which currently indicates that one type of resource has been specified to be seized.

- The *Seize – Repeating Property Editor* (see Figure 13.5) will pop up and allow you to specify as many resources that need to be seized before the entity can be processed. In this example, we will specify one resource type and that the *Object Type* will be "*FromList*"[177] can be selected from any in a list of resources (i.e., **ListResources**). We do not want to select a resource that is currently failed so the *Selection Condition* property should be `!Candidate.Resource.Failure.Active`.

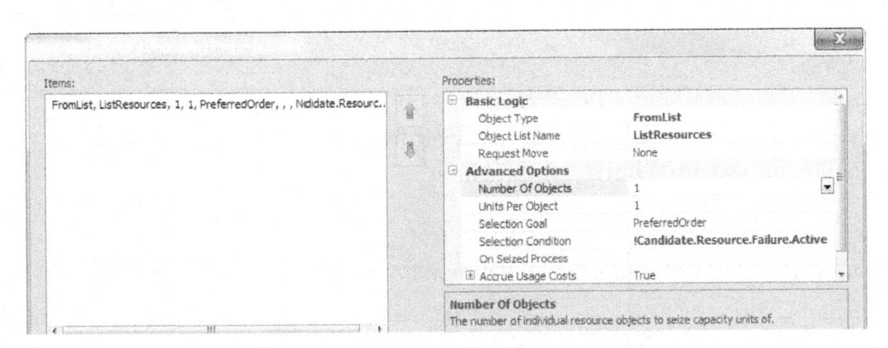

Figure 13.5: Seize Property Editor

---

[175] The *Before Processing* add-on process trigger is invoked after the entity has seized capacity of the server but before the entity is physically transferred to the processing station while the *Processing* add-on trigger is invoked after the entity has been transferred and directly before the delay for processing. The *After Processing* add-on process trigger is invoked directly after the processing has finished.

[176] Note, we can utilize the secondary resources to seize and release additional resources (i.e., *Resource for Processing*) using the same list. However, utilizing processes gives you more flexibility and control.

[177] The *Object Type* for the`seize` can be "Specific", "From List", or "Parent Object". The "Specific" option allows you to select an individual RESOURCE (e.g., **Jim**) or any object that has capacity (e.g., Servers, Paths) or any of the Stations inside objects which contain the capacity. The "Parent Object" will allow you to seize capacity of the object that is running the process which for the SERVER in this process is the **Processing** STATION. The list option will specify any list that contains objects that can be seized.

Step 10: The patients are now able to `Seize` the clinic staff but once the patient has finished processing, the entity (patients) will need to release the staff members (resource). There are two potential triggers (i.e., "*After Processing*" or "*Exited*"). Insert a new "*After Processing*" add-on process, by double clicking on the label to invoke/create a process which will be executed once the process has completed before exiting.

- Insert a `Release` step into the process between the `Begin` and `End` nodes which will allow the patient to release the resource seized, as seen in Figure 13.6.

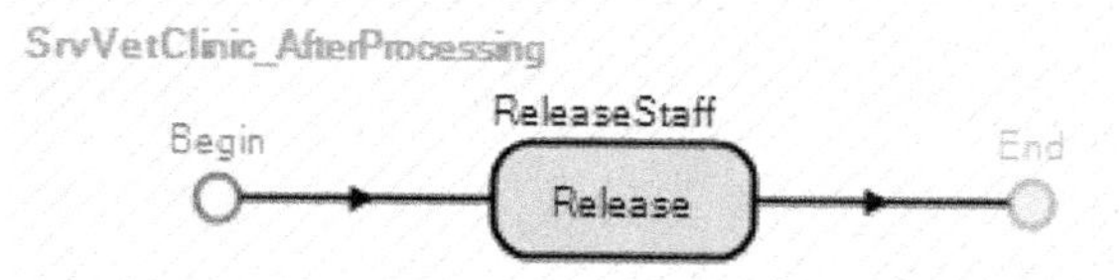

Figure 13.6: Using a Release Step to Release a Veterinary Staff Member

- Repeat a similar process to the `Seize` step of naming the step as **ReleaseStaff** and specifying the *Releases* property. Similar to the Seize Editor, the *Releases* → *Repeating Property Editor* will allow one to release resources according to the release properties.

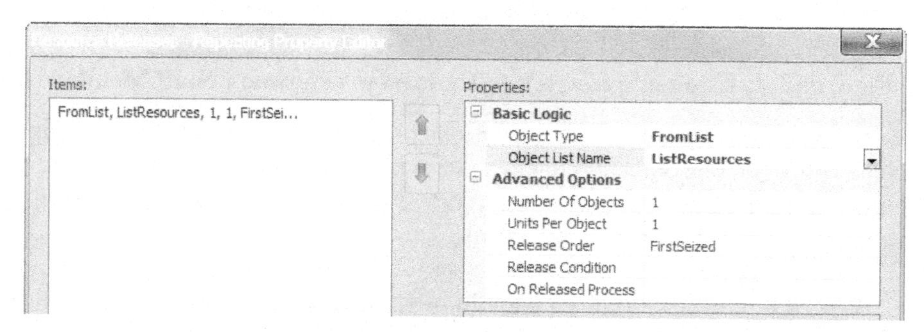

Figure 13.7: Release Property Editor for the Release Step

Step 11: Save and run the model for 40 hours and observe what happens.

Question 1: What is the overall average utilization for each of the RESOURCES and why does Billy have the highest utilization?

____

Question 2: What is the total time a patient has to wait for a staff person as and what is the average number in queue for the input buffer of the clinic?

____

Question 3: What is the total cycle time for the patients?

____

Part 13.2: Different Resource Needs based on Different Patient Types

The previous model assumed all patients can be seen by all clinic staff members. However, suppose Billy and Maisy have expertise with birds while Jim has expertise with iguanas, but if he is busy Ellie can provide the service. Horses and cows are troublesome animals owing to size and require any two veterinarians. All other animals (cats and dogs) can be seen by any one of the veterinarians.

Step 1: Create a new data table named **TableAnimals** that has a "String" column(property) for the animal name, an "Integer" column for animal type, and a "Real" column for the percentage of time this type of animal typically arrives to the clinic. The entry values are shown in Table 13.1. For example, 25 percent of the time large animals representing cows and horses will arrive at the clinic.

Table 13.1: Table with Animal Information and Types

Table Anim...

	Animal	Type	Percent Type
1	CatsDogs	0	47
2	CowsHorses	1	25
3	Birds	2	20
4	Iguanas	3	8

Step 2: Create two new OBJECT lists named **ListBird** and **ListIguana** in a similar manner as **ListResources** in Figure 13.3 that represent the staff that can service these specialty animals as seen in Figure 13.8.

Object
ResBilly
ResMaisy

Properties: ListBird (ObjectList Instance)

General

Name	ListBird
Description	

Object
ResJim
ResEllie

Properties: ListIguana (ObjectList Instance)

General

Name	ListIguana
Description	

Figure 13.8: Object Lists for the ListBird and ListIguana

Step 3: Add three additional symbols to the patient entity to distinguish between the various animal types. One can use four different colors but you can also use interesting pictures of these creatures by downloading symbols from Trimble 3D Warehouse as seen in Figure 13.9 which shows the four symbols horizontal (instead of vertical). Note the order of the symbols should correspond to the type column in the Table 13.1.

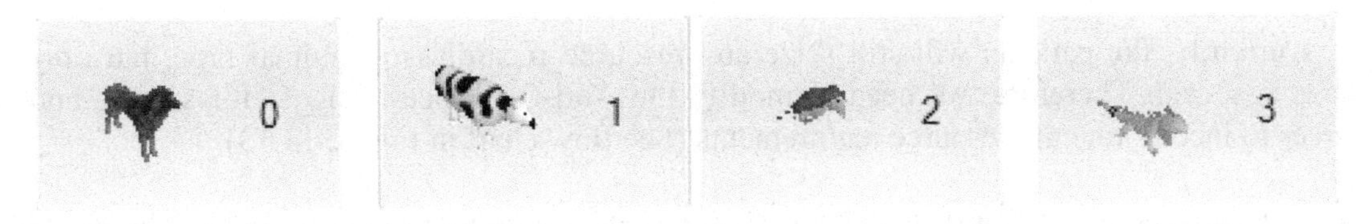

Figure 13.9: Trimble 3-D Warehouse Downloaded Dog, Cow, Bird, and Iguana Symbols

Step 4: The goal will be to randomly assign the animal types and pictures (color) based on the historic percentages. The assignment must be done after the entity has been created using the "*Created*" add-on process trigger in the SOURCE. Insert a new "*Created*" Add-On Process Trigger[178] which will be executed once the entity has been created by the **SrcPatients**.

- One of the features of SIMIO is the ability to associate one or more tables and/or rows of a table with an entity. Our newly created patient needs to be associated with a specific row of the **TableAnimals** data table and then assigned the correct picture symbol based on the "Animal Type" column. Insert a `Set Row` and an `Assign` process steps into the process as in Figure 13.10.[179]

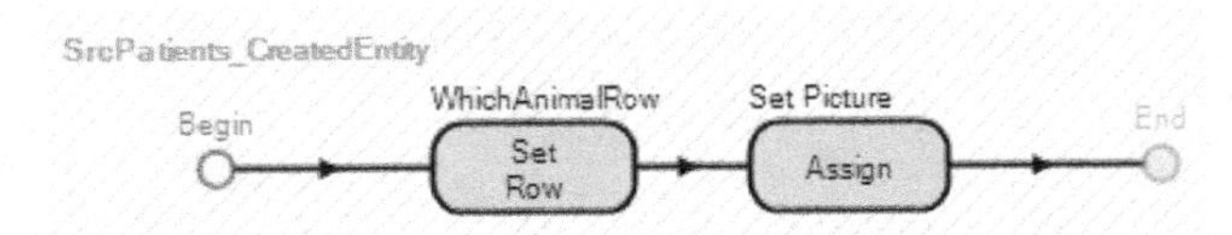

Figure 13.10: Determining Patient Type by Setting a Row of Datatable to Entity

[178] To invoke the "*Created*" add-on process, expand the "*Add-On Process Triggers*" of the SOURCE in the properties window and double click on the label to invoke/create a process.

[179] In this example, the *Table Reference Assignments* (i.e., On Created Entity) could be used to assign a table. Without knowledge of the object you cannot be sure that the picture assignment via *State Assignments* happens before or after the table assignment but with the processes approach you can control the order.

- Specify a specific row of the associated table randomly using the `RandomRow` method based on the "PercentType" column as seen in Figure 13.11.[180] Also make sure the *Object Type* under the *Advanced Options* is set to the "AssociatedObject".[181]

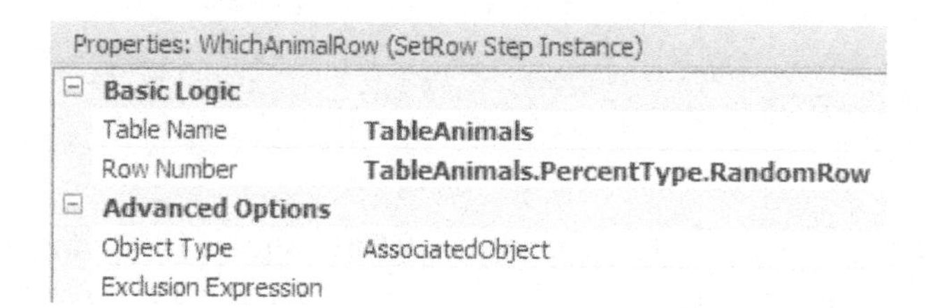

Figure 13.11: Specifying the Properties of the `Set Row` Process Step

- Once an animal type has been determined, Figure 13.12 shows the properties of the *Assign* process step used to assign the associated "Type" column from the table to the picture symbol state of the MODELENTITY.[182]

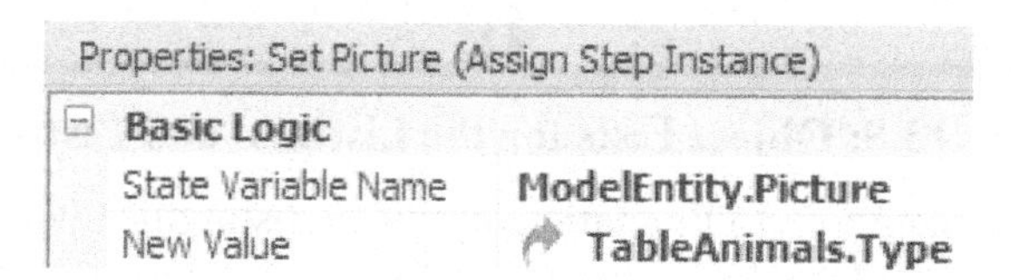

Figure 13.12: Changing the Picture of Entity to Reflect a Column in an Assigned Datatable

Step 5: Save and run the model to make sure the different animal types arrive the way they were expected. One may need to adjust the animated processing queues to allow the large animals to be visible.

Step 6: Currently, the patients will still seize any resource regardless of animal type, but only one staff member is seized. Therefore, we need to modify the Add-On Process trigger for seizing and releasing resources to incorporate the resource requirements (See flow chart in Figure 13.13),

- If the type is cat/dog or horse/cow then we can chose any resource, but for horse/cow patients two resources are needed.
- If the type is bird we need to choose a resource from the OBJECT LIST **ListBird**
- Finally if the type is iguana we will need to choose from the OBJECT LIST **ListIguana**

---

[180]Any numeric column can be used to specify a random row of table. The numbers do not need to add up to 1 or 100. SIMIO will take the sum of the column and then generate a random row based on the normalized percentages.

[181] The "ASSOCIATED OBJECT" is the object the current token is represents. In this case, it is the entity or patient. The other options are "TOKEN" for the current token executing the process, the "PARENTOBJECT" which would be the SOURCE since it owns the Add-on Process Trigger, and "SpecificObjectOrElement" which allows you to set the table on a particular object or element.

[182] Notice, the animal types were specified on purpose to be zero to three to represent the symbol numbering for the pictures. Also, the green arrow lets the user know this refers to a referenced property.

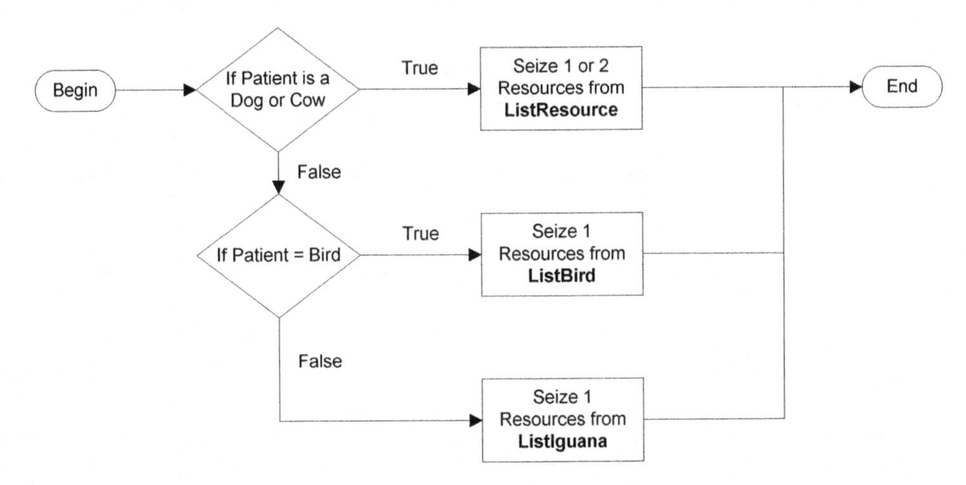

Figure 13.13: Seizing the Appropriate Veterinary Staff Based on Patient Type

- To perform the logic we need to make some decisions based on the animal type as seen in the above flow chart. The logic is performed by a SIMIO `Decide` process step (see Figure 13.14) which is basically an If/Else statement that has a **True** and **False** branch. In Figure 13.13, the `Decide` step's two major properties of interest are the *Decide Type* and the decision *Expression.*

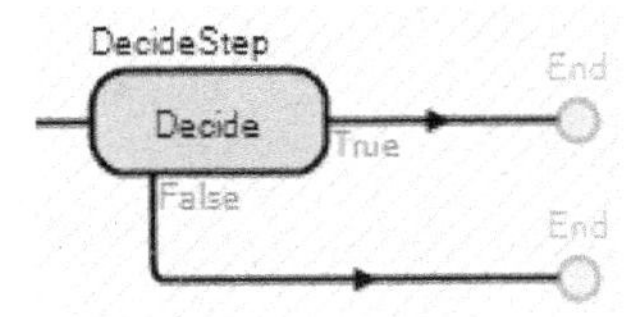

Figure 13.14: SIMIO `Decide` Process Step

Figure 13.15 shows the modified "*Before Processing*" add-on process trigger that will seize the appropriate staff members based on animal type.[183]

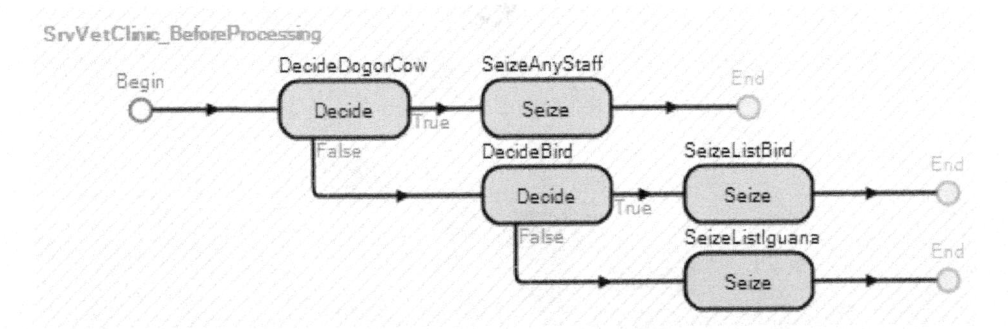

Figure 13.15: Seizing the Appropriate Resource Based on Animal Type

- Insert a `Decide` step before the current `Seize` step which will determine if the patient is a dog/cat or cow/horse. Based on which is true, the patient will seize from the resource object `LIST` **ListResource** as before. Specify the *Decide Type* to be condition based with an expression that checks to see if the animal type is a 0 or 1 (see Figure 13.16). [184]

---

[183] The *Processing* add-on process trigger occurs after the entities are transferred into the processing station so the animals will appear in the processing station even though they have not seized a vet. Therefore, use the *Before Processing* add-on process trigger to eliminate the animation issue.

[184] The expression `(TableAnimals.Type < 2)` determines if the type is a zero or one could have been specified `(TableAnimals.Type == 0 || TableAnimals.Type ==1)` where `||` equals the Or logical operator.

Basic Logic	
Decide Type	ConditionBased
Expression	**TableAnimals.Type< 2**

Figure 13.16: Specifying the CondtionBased Expression of a Decide Step

- The second `Decide` step will check to see if it is a bird.(Hint, just copy the first `Decide` block and paste it onto the False branch and change the expression to (`TableAnimals.Type==2`)).
- Copy the `Seize` step to both the *True* and *False* branches of the second `Decide` step. You will need to change which resource list the entity will seize from (i.e., **ListBird** and **ListIguana** respectively).
- The first `Seize` step is already setup to seize from the all the resources (**ListResources**). However, by default, it only seizes one object. But two resources are needed for the large animals. There are several ways to model the logic. Since the animal type of cats/dogs is zero and cows/horses is one, the type can be added to one to give a result of one or two. In the *Number of Objects*[185] property to seize, the following Figure 13.17 illustrates the expression to seize one or two objects based on the animal type.

Basic Logic	
Object Type	**FromList**
Object List Name	**ListResources**
Request Move	None
Advanced Options	
Number Of Objects	**1+TableAnimals.Type**
Units Per Object	1

Number of Resource Objects

Figure 13.17: Specifying an Expression to Determine the Number of RESOURCES to `Seize`

Step 7: Save and Run your model. You may need to experiment with the percentages to make sure the animals are being serviced by the correct staff members. Figure 13.18 shows a cow being serviced by both **ResBilly** and **ResJim**.

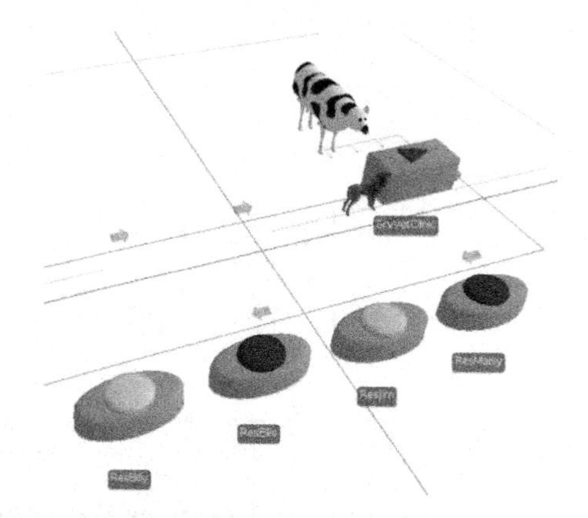

Figure 13.18: Using Two Resources to Service a Large Animal.

Question 4: Did the staff get released and why?

____

[185] The *Number of Objects* property specifies the unique number of objects to select out of a list while the *Units Per Object* specifies the number of each object to seize. For example, if your list had doctors and nurses and you set the *Number of Objects* to "1" and *Units Per Object* "2", then it would seize either a doctor or nurse object but it would require two doctors or two nurses. Resources can be given capacity of more than one (e.g., you could have ten nurses but you do not want to have ten different nurses (i.e., Billy, Ellie, etc.).

Step 8: The only thing left to do is to release the veterinary staff once they have finished processing the patient. It is similar to the *"Before Processing"* trigger but uses `Release` instead of `Seize` steps. It is important that you release the correct number of objects back to the **ListResources** pool.

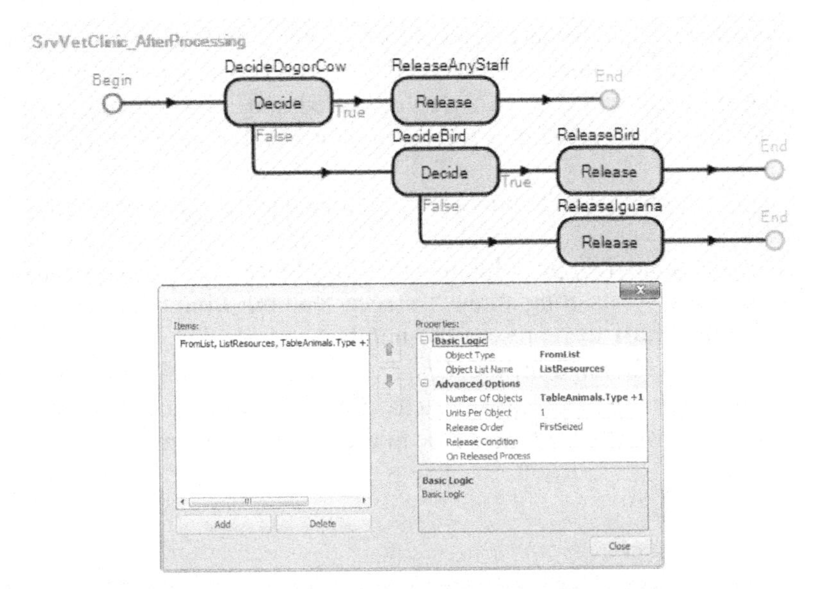

Figure 13.19: Releasing the Correct Vet to Appropriate Resource Set

Step 9: Save and Run your model and observe what happens (see Figure 13.20).

Figure 13.20: Figuring Showing Animals being Serviced and Waiting

Question 5: What is the overall average utilization of each of the RESOURCES?

___

Question 6: What is the total time a patient has to wait for a staff person as well as the average number in queue for the input buffer of the clinic?

___

Question 7: What is the total cycle time for the patient?

___

Part 13.3: Resource Decision Making

One of the most prominent features of SIMIO is that you can model very complex logic (e.g., process steps) or expressions, and supplement it by associating information via data tables on entities. In an earlier chapter, the processing time and routing information for different part types was linked directly to the part itself. When parts arrived at the SERVERS, the server would ask the entity how long to process it. Many simulation languages, including SIMIO, have the ability to specify attributes (i.e., entity state variables in SIMIO) which can be assigned the actual processing time value when the entity is created. However, SIMIO information in a data table can be dynamically changed for any object to reference to it. In the current example, there are four different animal types. What if the number of types increases to ten with eight or nine different lists to seize the correct vet? The decision logic would get very complicated and long. Therefore, we have the ability to make the objects more intelligent (i.e., they know the list and the number required for service).

Step 1: Create or open the model from the previous section and save it as a new name.

Step 2: Recall that only the animal **Type** column was utilized previously to make decisions. Now modify the data table **TableAnimals** by inserting a new "Integer" column from the *Standard Property* dropdown. Name this column **NumberStaff** to represent the number of vets that are required by the animal type. Next add an OBJECT LIST column from the *Object Reference* dropdown named **ListStaff** which will represent the list the resources that can be selected. Set the rows values according the ones in Figure 13.21. Now the patients know which group of people can service them as well as the number that are needed.

Table Anim...

	Animal	Type	Percent Type	Number Staff	List Staff
▸ 1	CatDogs	0	47	1	ListReources
2	CowsHorses	1	25	2	ListReources
3	Birds	2	20	1	ListBird
4	Iguanas	3	8	1	ListIguana

Figure 13.21: Including the Number of Staff Needed in the TableAnimals

Step 3: In the "*Processes*" tab remove all the steps except one `Release` and `Seize` step in the *"Before Processing"* and *"After Processing"* add-on triggers respectively as seen in Figure 13.22.[186]

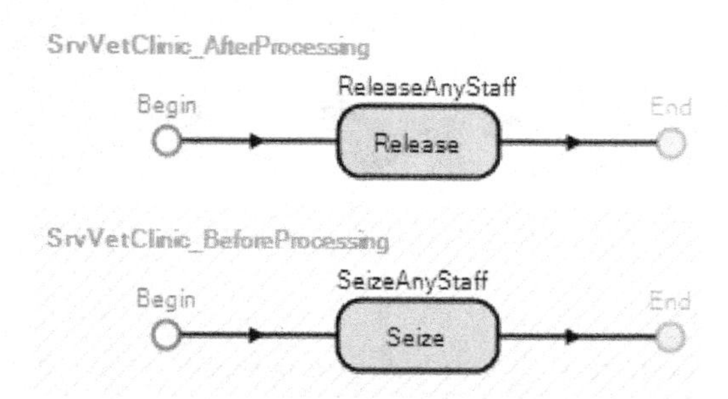

Figure 13.22: Improved Process Logic for Seizing and Releasing Vet Staff Members

Step 4: Modify the properties in both the `Seize` and `Release` steps to utilize the referenced/associated table properties. On the *Object List Name* and *Number of Objects*, right click and set the *Referenced Property* to the correct column of the table respectively. Now when a patient arrives to seize a staff member, it will dynamically seize the appropriate number from the correct group of resources.

---

[186] Modeling the logic in this fashion has removed the two decision steps which will speed up the simulation model as well as simplify the logic when, for example, more animal types are added. Now that information is contained in the data table, eliminating the need to add additional process logic.

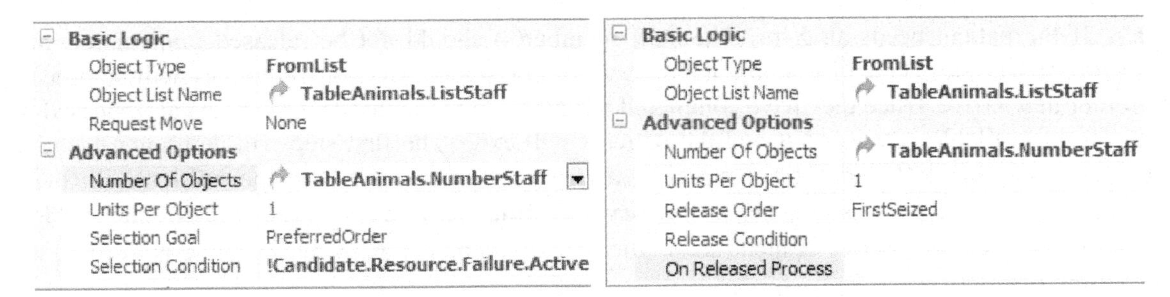

Figure 13.23: Seize and Release Properties using Referenced Properties

Step 5: Save and run your model and compare the results to previous model's results.

Part 13.4: Adding an Additional Process

Sometimes, the patients need to have an X-ray taken which is located in a different room. For example, 25% of cats and dogs typically need X-rays while only 15%, 0% and 2% of cows and horses, birds, and iguanas respectively do. The patients will leave the patient care rooms and be moved to the X-ray machine room. The veterinarians will accompany the patients to perform the X-ray. Afterwards the veterinarians will be released to serve other patients.

Step 1: Add an additional SERVER named **SrvXrayMachine as seen in** Figure 13.24.

- Set the capacity of the SERVER to one since there is only one machine
- The processing time to perform and read an X-ray follows a Pert distribution with a minimum, most likely, and maximum values of 20, 25, and 35 minutes respectively.
- Connect the output of the X-ray machine to **SnkExit** via a 30 meter logical PATH.
- Connect a 10 meter logical PATH from the **SrvVetClinic** output to the input of the **SrvXrayMachine**.

Step 2: Since each animal type has a different chance of needing an X-ray, add an additional Real column named **XrayLabPercentage** to the **TableAnimals** as seen in Figure 13.25.

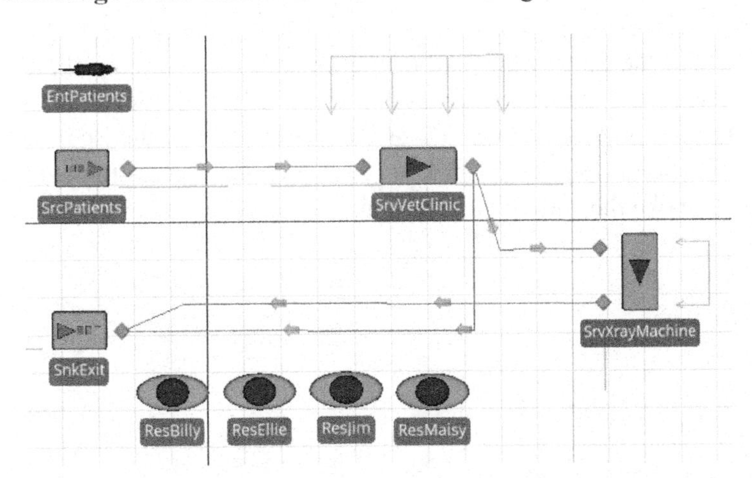

Figure 13.24: Adding an Additional Process that uses Resources Already Seized

Table Anim...

	Animal	Type	Percent Type	Number Staff	List Staff	Xray Lab Percentage
1	CatsDogs	0	47	1	ListResour...	0.25
2	CowsHor...	1	25	2	ListResour...	0.15
3	Birds	2	20	1	ListBird	0
4	Iguanas	3	8	1	ListIguana	0.02

Figure 13.25: Animal Percentages of Needing an X-ray

Step 3: If the patient needs an X-ray, the staff member(s) should not be released immediately after the patient care service. The entity should move to the X-ray station SERVER where they will be processed in the order they arrive. Once they have completed the X-ray process, the staff members are released back to the pool of available veterinarians and the patients will exit. The first step will determine if the patient should release the staff members or not (i.e., a `Decide` process step will be needed). In the "*After Processing*" add-on process trigger of the **SrvVetClinic** insert a `Decide` process step. Move the `Release` step to the "False" branch (see Figure 13.26) so that if an X-ray machine is needed, then the staff is not released.

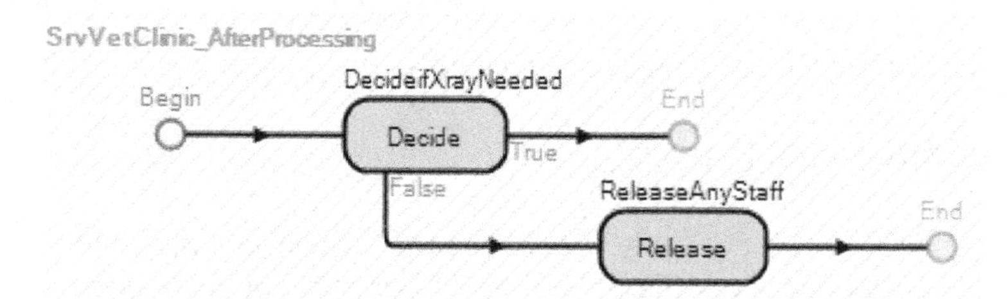

Figure 13.26: Modified Processed Process Trigger

- The `Decide` *step is shown in* Figure 13.27.

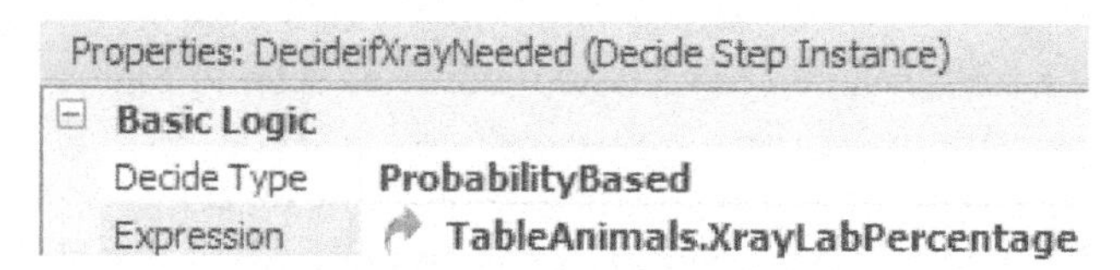

Figure 13.27: The Decide Step

Step 4: If the patient needs an X-ray, they release the staff members after the X-ray has finished processing them. Insert a new process for the *After Processing* add-on process trigger for the **SrvXrayMachine.** Copy and paste the `Release` step block from the **SrvVetClinic_AfterProcessing** process to the new process.

Step 5: Now the patients who need X-rays need to be routed to the new SERVER. By default, the patients will randomly go 50% of the time to the sink or the input at the X-ray machine owing to the equal link weights of one.[187] There are numerous functions (i.e., `f(x)` designation in the expression editor) associated with the various objects. Table 13.2 gives a description of the `SeizedResources` family of functions. The base function of `SeizedResources` which gets access to the objects that any intelligent object has seized will invoke the underlying function `SeizedResources.NumberItems`.

---

[187] Note, we could have specified the link weights to be `1-TableAnimals.XrayLabPercentage` for the link to the sink and `TableAnimals.XrayLabPercentage` for the link to the X-ray machine and the patients would have been routed correctly. However, this logic would only work if the patients did not need the same staff members to accompany them to the X-ray machine (i.e., they could be released before exiting the **SrvVetClinic**).

Table 13.2: Description of all the `SeizedResources` Functions

Function	Description
`SeizedResources.NumberItems`	This function will return the current number of objects (i.e., RESOURCES, SERVER and LINK capacities, etc.) seized and owned by the object using the function.
`SeizedResources.FirstItem`	All object instances in a SIMIO model have a unique numeric identifier (ID) (e.g., NODES, SERVERS, RESOURCES, etc.) This function will return the reference of the first object in the list of objects currently seized and owned by this object.
`SeizedResources.LastItem`	This function will return the reference of the last object in the list of objects currently seized and owned by this object.
`SeizedResources.ItemAtIndex (index)`	This function will return the reference of the object in the list of objects currently seized and owned by this object at the specified index.
`SeizedResources.IndexOfItem (resource)`	This function will return the index of a specified resource in the list of resources currently seized and owned by this object. It will return zero if the resource has not been seized.
`SeizedResources. CapacityOwnedOf(resource)`	This function will allow you to determine the number of units of capacity for a specified `resource` that are currently owned by this object (i.e., it would return two if they had seized two hammers).
`SeizedResources. Contains(resource)`	This function returns True(1) if the list of resources currently seized and owned by this object includes the specified resource. Otherwise the value False(0) is returned.

- Specify the link weight of the path leaving the **SrvVetClinic** and going to the exit (**SnkExit**) to be `ModelEntity.SeizedResources.NumberItems == 0` which means the patient has released the staff members and can exit the system.
- Specify the link weight of the path leaving the **SrvVetClinic** and going to the X-ray machine (**SrvXrayMachine**) to be `ModelEntity.SeizedResources.NumberItems > 0` which means the patient has still seized staff members because it needs to have an X-ray done.
- Note the link weights will evaluate to one or zero thus forcing the patients to take the path only if the link weight was a one.

Step 6: Save and run the model and observe what happens.

Part 13.5: Changing Processing Based on Animal Type and Vet Servicing

At this point, all of the animal types take the same amount of time to be seen by one of the vets (i.e., processing time is sampled from the same distribution). However, depending on which animal type as well as which vet is seeing the animal the time to be seen can vary. This is a common situation where the time is dependent on the resource that was seized (i.e., doctor or nurse, etc.). Recall that relational data tables were used previously to specify a different processing time for each of the part types for each of the operations in the sequence. This method works well for distinguishing processing times among different part types (i.e., animal types). Therefore, a child table containing the processing times will be created to allow the processing time to become also dependent on the resource being employed.

Step 1: First, select the **Type** column in the **TableAnimals** and make it the primary key of the table as seen in Figure 13.28. This table will become the parent table from which the child table will inherit the values.

Table Animals

	Animal	Type	Percent Type	Number Staff	List Staff	Xray Lab Percentage
1	CatsDogs	0	47	1	ListResources	0.25
2	CowsHorses	1	25	2	ListResources	0.15
3	Birds	2	20	1	ListBird	0
4	Iguanas	3	8	1	ListIguana	0.02

Figure 13.28: Specifying the TableAnimals Type Column to be a Primary Key.

Step 2: Next add a new data table named **TableProcessingTimes** via the *Data→Tables→Add Data Table* button as seen in Figure 13.31.

- Add a new Standard *Expression Property* column named **ProcessingTime** as seen in Figure 13.29 **making sure to set the properties** *Unit Type* **to "Time" and** *Default Units* **to "Minutes."**[188]

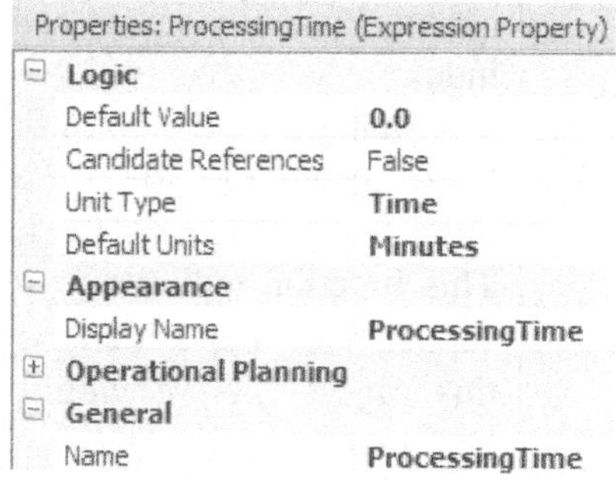

Figure 13.29: Specifying the Processing Time Column

- Next, add a new *Foreign Key* column named **Type** with the *Table Key* property set to `TableAnimals.Type` as seen in Figure 13.30.

Properties: Type (Foreign Key Property)

Logic	
Default Value	
Table Key	**TableAnimals.Type**
Appearance	
Display Name	**Type**
Operational Planning	
General	
Name	**Type**

Figure 13.30: Specifying the Foreign Key to inherit Column from Parent Table

- At first we will utilize constant times for each of the animal services. Set the time to 30, 60, 20, and 40 minutes for dogs, cows, birds and iguanas respectively to correspond to Figure 13.31.

Table Animals | Table Processing Ti...

	ProcessingTime (Minutes)	Type
1	30	0
2	60	1
3	20	2
4	40	3

Figure 13.31: Processing Time Table for the Different Animal Types

Step 3: Next we need to utilize the new processing time information into the model. Unlike prior cases where we were using the next sequence of the TRANSFERNODE to set the particular row in the child table,

[188] It has to be an Expression property to allow distributions to be specified.

we will need to specify it directly (i.e., set the *Processing Time* property for row one by `TableProcessingTimes[1].ProcessingTime` of **SrvVetClinic**). Without the relational table, the expression would set it to the first entry in the whole table always. Since this is a related table, it will set it to the first entry of the inherited rows which currently only has one row as seen in Figure 13.33 which shows for child table for an entity which has been assigned a bird. [189]

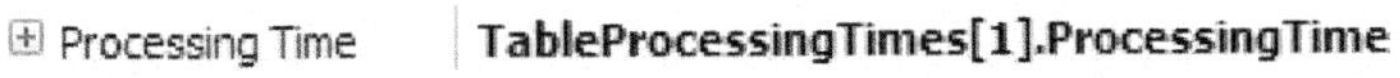

⊞ Processing Time **TableProcessingTimes[1].ProcessingTime**

Figure 13.32: Setting the Processing Time to be the First Row of the Child Table

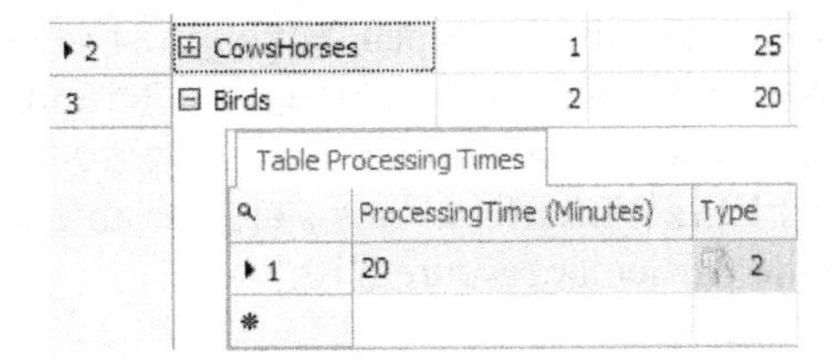

Figure 13.33: Showing the Child Table Associated with the Bird Row

Step 4: At this point we could run the model but it may be difficult to determine if the processing time is being assigned correctly.[190] Therefore, we will set the priority[191] state variable of each entity to reflect their processing time. In the *SrcVetClinic_Processing* process, insert an `Assign` step after the `Seize` step to set the patients PRIORITY state variable as seen in Figure 13.34.[192]

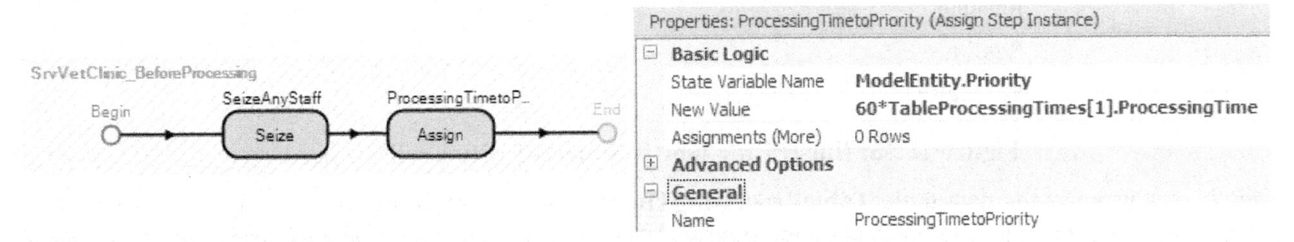

Figure 13.34: Specifying the Processing Time for Display Purposes

Step 5: In order to see the set priority value, we will add a status label to the animal that displays the processing time for each animal.[193] Select the MODELENTITY **EntPatients** and then insert a new *Animation→Status Label* which associated/attaches the label to the entity. Specify the expression of the label to be the entity's **Priority** state variable.

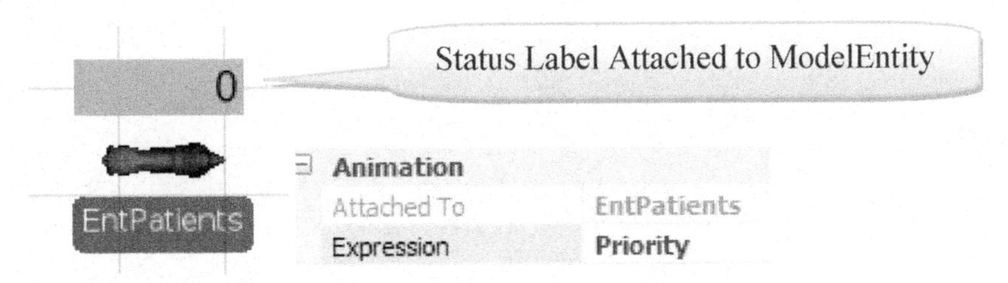

Figure 13.35: Inserting a Status Label Attached to the Entity

---

[189] Notice, an error would occur if we just specify `TableProcessingTimes.ProcessingTime` as was done in a previous chapter without first setting the row via a *Set Row* step or a *By Sequence* from the TRANSFERNODE.

[190] We could turn on the trace and watch what the processing time delay was when the animal went into service.

[191] Recall that "priority" is a SIMIO-defined state variable for entities.

[192] We are multiplying by 60 because SIMIO has converted all times to hours internally.

[193] SIMIO now provides a *Dynamic Label Text* property under the *Animation* section which can be used to a similar thing but the label appears below the MODELENTITY.

Question 8: Why do we have to multiply the processing time by 60?

Step 6: Save and run the model observing the status label times.

Question 9: Are the correct times being assigned to the different animals?

Step 7: Now the model can handle the situation where the processing time is based on the animal type. In many situations the processing time may also be dependent on the resource (i.e., vet) you have seized. The notion of position within the list of resources will be used to determine the processing time. First, select the MODELENTITY in the [*Navigation*] panel and add a new DISCRETE INTEGER STATE variable from the *Definitions→State* section named **EStaWhichRes**. This state variable will be assigned the position of the resource that has been seized in the particular resource list.

For example, if a current patient who is a dog has seized Ellie from the **ListResource** list the patient's **EStaWhichRes** variable would be assigned a value of "2" to correspond that **ResEllie** is in the second position of this list as seen in Figure 13.36.

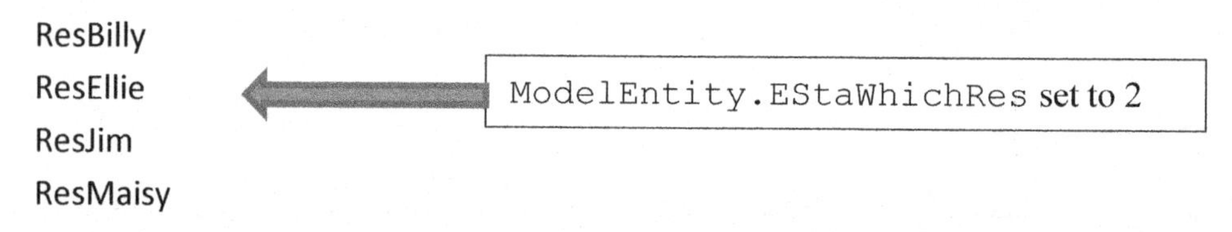

Figure 13.36: Illustrating how Resource Position will be used

Step 8: Currently the data table **TableProcessingTimes** contains only time for each animal type. Modify the table to contain a different processing time for each vet (i.e., resource) and each animal type as seen in Figure 13.37.[194] Since four different resources are capable of processing dogs, there are four different processing times (i.e., 31 minutes for Billy, 32 minutes for Ellie, 33 minutes for Jim, and 34 minutes for Maisy) for animal type "0" (i.e., cats and dogs). For birds and iguanas only two vets can see each of those types of patients, therefore only two different processing times are specified for types "2" and "3". As Figure 13.37 illustrates, Jim will process Iguanas in 42 minutes while it takes Ellie 43 minutes.

---

[194] One of the easiest ways to create the table is to copy it from an Excel worksheet where the table can be created easily. Tables can be copied to and from Excel. To copy from Excel, first copy the two dimensional table in Excel and then highlight all the current values in the SIMIO data table before pasting the new values which will overwrite any existing one and then create the any new rows as needed.

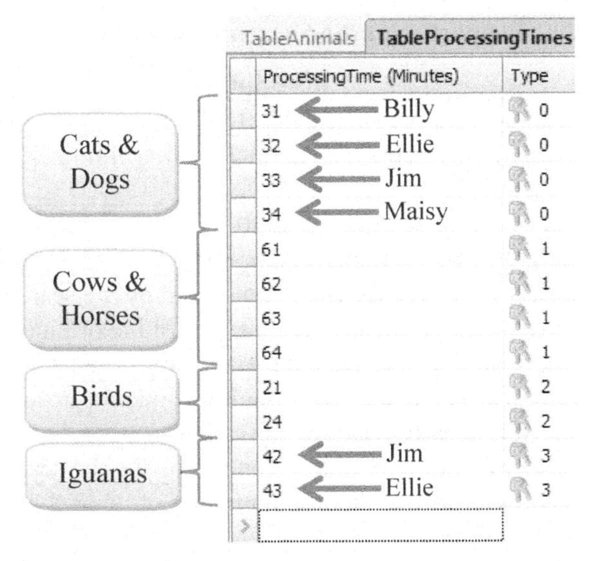

Figure 13.37: Add more Processing Times for Each Resource

Figure 13.38 shows the child related table for animal type zero (i.e., cats and dogs) when expanding out the type "0" from the parent table **TableAnimals.** As you can see, only four processing times are related to type zero. Now going back to the example in Figure 13.36, Ellie would corresponds to the second position in the list which translates to row two of the related table (i.e., 32 minutes).

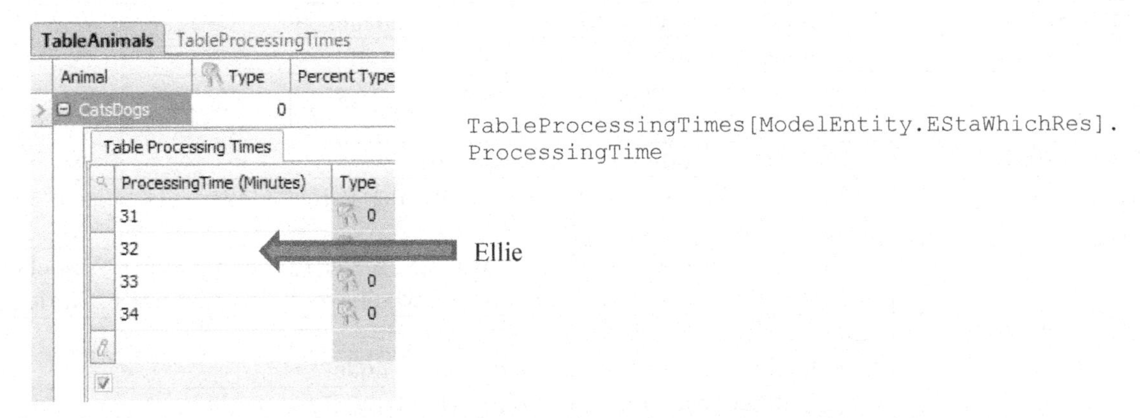

Figure 13.38: Showing the Child Sub Table for Cats and Dogs Animal Type

Step 9: The only step left is to appropriately assign the **EStaWhichRes** variable based on the resource that was seized. To accomplish this task, we will employ a `Search` process step where Table 13.3 explains all of the properties associated with the step.[195] The `Search` process step can search a collection of items (see Table 13.4) and return TOKENS associated with the items in the collection.

[195] Recall the TOKEN executing the process steps is typically associated with an entity. Therefore any steps (i.e., Transferring, Setting Nodes, etc.) are only applied to the MODELENTITY associated with the TOKEN. Often we may want to affect other objects (resources seized, etc.) besides the current entity, the search step can be used to create new TOKENS associated with other objects. In some process steps, the action can be applied to the parent object of the TOKEN'S associated object (e.g., the SERVER the logic resides).

Table 13.3: The Properties of the Search Process Step

Property	Description
Collection Type	Type of the collection to search (See Table 13.4 for more information).
Search Type	The type of search to perform (Forward, Backward, MinimizeReturnValue, and MaximizeReturnValue)[196]
Match Condition	Optional match condition that can be used to filter the items in the collection before the search begins. Don't forget to use the keyword CANDIDATE in the condition.
Search Expression	An expression that is evaluated for each item found by the search. The total sum for this expression is returned in the ReturnValue of the original TOKEN. One could save the particular Resource and Seize the exact same one later.
Starting Index	One based index to start searching. By default it is one for Forward searches while the number in the collection for Backward searches.
Ending Index	One based index to end searching to assist in narrowing the search within collection.
Limit	An expression that specifies the maximum number of items that can be found and returned (e.g., Infinity or one)
Save Index Found	Specify an optional discrete state variable to save the last item found
Save Number Found	Specify an optional discrete state variable to save the number of items found

Table 13.4: Different Types of Collections that Can Be Searched

Collection Type	Description
EntityPopulation	Search a particular entity population (i.e., MODELENTITY, WORKER, or VEHICLES)
QueueState	Search the objects currently in one of the queues
TableRows	Search the rows in a Data table
ObjectList	Search the objects in a list (e.g., ResourceList, etc.)
NodeList	Search all nodes in a particular list.
TransporterList	Search all transporters in a particular list.
SeizedResources	Search the objects currently seized by the Parent Object, Associated Object or a Specific Object
ResourceOwners	Search the objects that currently own (i.e., seized) a Parent Object, Associated Object or a Specific Object. Search all animals who have seized the **SrvVetClinic** capacity.
NetworkLinks	Search all links of a particular network.
NodeInboundLinks	Search all the inbound links of a particular node.
NodeOutboundLinks	Search all the out bound links of a particular node.
ObjectInstance	Search all objects of a particular class (e.g., all entities that are currently created).

Insert a *Search* Step between the *Seize* and the *Assign* (see Figure 13.39). The "Original" branch is associated with the original TOKEN (i.e., customer) while the "Found" branch will be invoked for a TOKEN associated with each item found.[197] The *collection type* will be an "object list" with the list name based on the animal type (***TableAnimals.ListStaff***). We only want to match the last resource seized by the patient with a resource from the resource list.[198] If it finds a match, it will save the index (i.e., row position) in the MODELENTITY's **StaWhichRes** state variable.

---

[196] *MinimizeReturnValue* and *MaximizeReturnValue* search values will search the collection and return up to the limit of items specified that minimizes or maximizes the *Return Value* expression property.

[197] The original token will not move forward until all the searching is done.

[198] Note the use of the CANDIDATE keyword to represent each member of the collection being searched (i.e., like a wildcard *). Also, in the case of a cow or horse which requires two vets, the second vet seized will be lower in the list and our assumption would reflect less seniority and therefore slower.

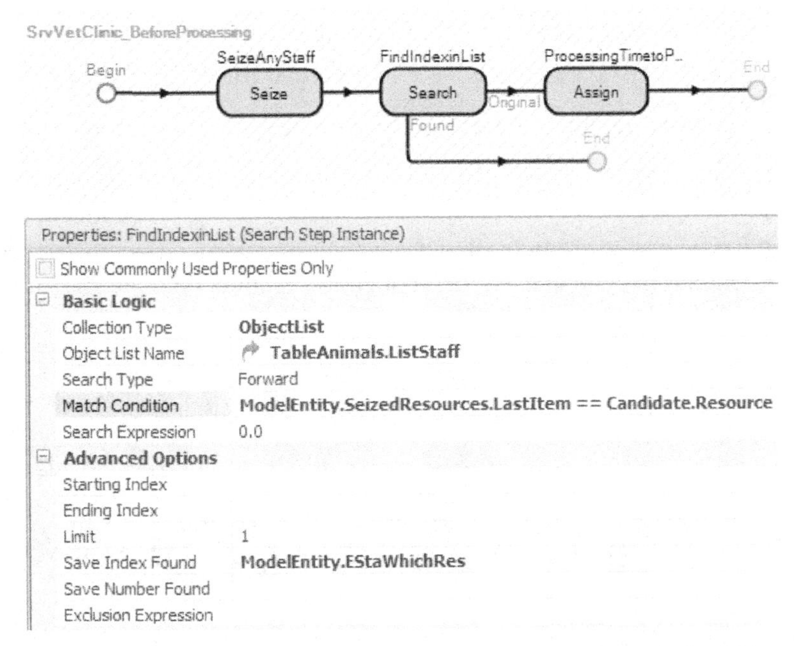

Figure 13.39: Adding a Search Step to Find the Index Position of the Seized Resource in the List

Step 10: Now that the **EStaWhichRes** has been set, you need to modify the *Processing Time* property of the **SrvVetClinic** as well as the `Assign` step to utilize the following expression (i.e., change the row index of "`1`" to `ModelEntity.EStaWhichRes`).

```
TableProcessingTimes[ModelEntity.EStaWhichRes].ProcessingTime
```

Step 11: Save and run the model observing the status label times to see if different processing times are being assigned based on animal type (i.e., in the 30, 60, 20, and 40 minute ranges) as well as which vet (i.e., resource is serving the patient).

Part 13.6: Changing the Resource Allocation Selection

In the original description, Jim was the preferred staff member to see Iguanas while Ellie could see them if he was busy. The clinic is considering a new policy that when Jim becomes available, he will check to see if an Iguana is waiting and if so, he will service these patients first before seeing the next patient. A "first-in and first-out" ranking scheme is currently being used by the **SrvVetClinic** to service the patients.

RESOURCES like SERVERS have a static ranking rule[199] which will order the requests waiting to be allocated capacity from this object. The SERVER and RESOURCE will process them in the order the entities are sorted.[200]

Step 1: Change the static ranking rule of the RESOURCE **ResJim** to be the "largest value first". The ranking expression should be the referenced property animal type since iguanas will always be serviced first if they have requested service.[201] The static ranking rule works because the type of the animal does not change, while the patient is waiting for a vet (see Figure 13.40).

---

[199] The static ranking rules are First-in First-out (Default), Last-in Last-Out, Smallest Value, and Largest Value, where an expression provides the value for the last two rules (e.g., `ModelEntity.Priority`)

[200] Also there is a dynamic selection rule, which will be considered in a later chapter.

[201] If **ResJim** is busy processing a patient and an iguana is at the front of the server's queue then **ResEllie** may still see them, which is acceptable. The goal is to maximize **ResJim's** chances of seeing Iguana patients but not to eliminate her's completely. If only **ResJim** could see iguanas then the **ListIguana** could be changed to only have Jim as a resource.

Ranking Rule	Largest Value First
Ranking Expression	TableAnimals.Type

Figure 13.40: The Largest Value Static Ranking Rule

Step 2: To visualize the ranking of **ResJim**, select **ResJim** and insert an animated queue (i.e., ALLOCATIONQUEUE) from the *Draw Queue* section of the *Symbols* tab. The allocation queue is the queue that contains the requests for allocation of capacity of an object (i.e., RESOURCE, SERVER, WORKER, VEHICLE, COMBINER, etc.). Make the queue long enough to accommodate the size of the animals as seen in Figure 13.41 which is drawn from right to left.

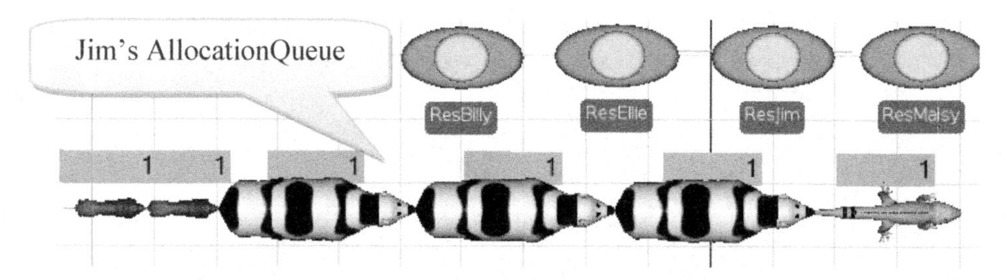

Figure 13.41: Allocation Queue of ResJim

Step 3: Save and run the model and observe what happens. The percentage of time Iguanas arrive may need to be altered to assist in verifying the model works the way it is intended.

Question 10: Are there any issues with ranking the queues in this manner?

____

Question 11: Did you observe cows jumping ahead of dogs in the queue?

____

Step 4: Jim should service the Iguanas first and then service the remaining two animals in order of their arrival. One way to accomplish this ranking is to modify the *Ranking Expression* property to the value in Figure 13.42. The logical expression will make all iguana patients have a value of one; while all other patient types would be zero and therefore preserve the arrival order to break the ties among remaining animals (arrival order is the tie-breaking for static ranking rule).

Ranking Rule	Largest Value First
Ranking Expression	TableAnimals.Type==3

Figure 13.42: Modifying the Ranking Rule to not Service Cows before Dogs

Step 5: Save and run the model and observe what happens now.

Part 13.7: Commentary

- A very common modeling technique is to add additional state variables to the general entity to hold values or states. In the second part, an ***EStaXray*** discrete state variable could have been added to the entity. In the True branch of the *"After Processing"* add-on trigger as seen in Figure 13.43, an `Assign` step could have been added to assign the value of one to the new ***EStaXray*** state variable to show that an X-ray is needed.

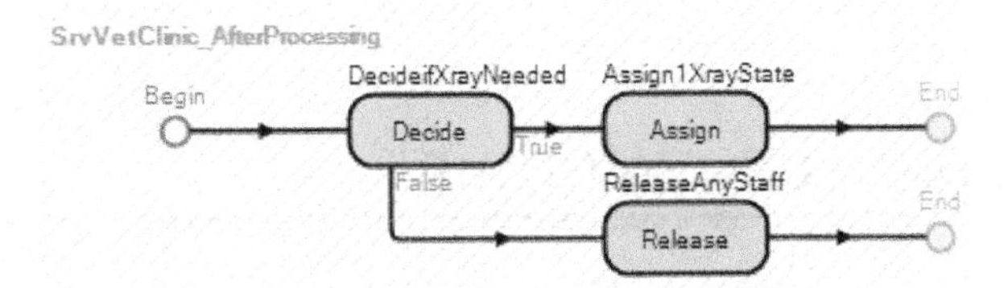

Figure 13.43: Using a State Variable to Assign Whether an X-ray is needed or Not

- The link weights could have been specified as (`1-ModelEntity.EStaXray`) and (`ModelEntity.EStaXray`) for the links to the exit SINK and the X-ray SERVER machine, respectively. Again the link weights will take on zero or one and will never be the same. If the patient does need an X-ray, then (`1-ModelEntity.EStaXray`) would evaluate to one while (`ModelEntity.EStaXray`) would evaluate to zero.

- To some extent, one could have used the *Secondary Resources* to seize and release the vets. One could of easily used *Before Processing* under *Other Resource Seizes* and *After Processing* under *Other Resource Releases* with the same input as Figure 13.23. However, the ability to release or not release if the animal needed a vet would require a little more modeling if you were trying to avoid processes all together. It is easy not to release the Vets at the clinic and then only release them under the *After Processing* under *Other Resource Releases* of the **SrvXrayMachine** if the animal needed an X-ray. However, there is no mechanism to release the vets for those who do not. You would have to add an additional SERVER between the **SrvVetClinic** and the **SnkExit** that would release them. As has been seen in other chapters, one has more control and flexibility in using processes.

Chapter 14
A Mobile Resource: The Worker

The previous chapter illustrated the flexibility of RESOURCES and RESOURCE LISTS to model very complicated relationships. However, these types of resources were fixed, in the sense that they do not move dynamically throughout the system model. When the patients need an X-ray, the resources were not released and the patient moved to the X-ray machine room, but the animation did not show the resources moving along with the patient. Also, in the DMV problem, resources were used in multiple locations but the travel time between the two locations was not considered. SIMIO has the ability to have multiple types of the objects (i.e., entities, workers, and vehicles) that can act as resources (i.e., objects that can be seized and released) and move throughout the system, similar to an entity.

Part 14.1: Routing Patients

Again, we return to the veterinary clinic. However this time the clinic will behave more consistently with a doctor's office where patients arrive to the clinic, check in at the front desk, and then wait in the waiting room for an available room. The patients will move to one of four examination rooms once one becomes available. Then one veterinarian will start from their office and will dynamically move to the rooms as needed and service the patients as requested. The veterinarian will be modeled as a movable resource that needs to be seized and released.

Step 1: Create a new model that contains one SOURCE (**SrcPatients**), a SINK (**SnkExit**), and five SERVERS (**SrvVetWaiting**, **SrvRoom1**, **SrvRoom2**, **SrvRoom3**, and **SrvRoom4**).

Step 2: Insert a new MODELENTITY named **EntPatients** into the model. The patients travel at a constant rate of 4.5 km per hour or 4500 meters per hour.

Step 3: Insert the rooms into a two hall configuration with the waiting room serving the two hallways as seen in Figure 14.1.[202]

- The capacities of all the servers should be one.
- The check-in process at the **SrvVetWaiting** uniformly takes between one and four minutes before they are placed into the waiting room.
- Each patient takes between 10 and 20 minutes to be seen with most of them taking 12 minutes.
- It has been observed that patients arrive exponentially with an interarrival time of 6.5 minutes.

Step 4: Insert a wall in the middle of the rooms to add a more visual cue for the hallway separation as seen in Figure 14.1. Make sure to specify a height of 3.5 meters meters[203] as well as give the wall a texture. You will notice we inserted a different line (i.e.., window texture) around the waiting room plus added a 3-D door as the entrance.

[202] To be more efficient, create one room named **SrvRoom** with all the correct properties and then copy it three times changing the name of the original one to **SrvRoom4** accordingly.

[203] After you have selected the line or rectangle, the *Object Height* and *Line Width* text edit area can be found under the "*Object*" section of the "*Drawing*" Tab.

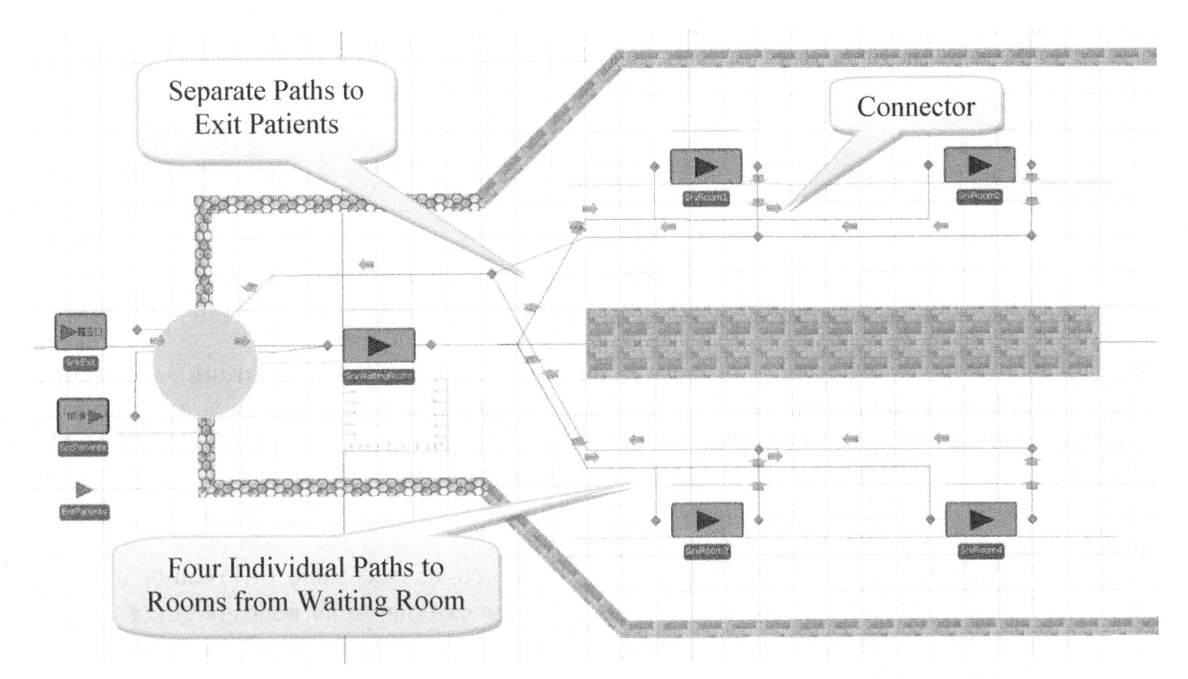

Figure 14.1: New Veterinary Clinic Model for Moveable Resources

Step 5: Connect the output of the **SrvVetWaiting** to each of the inputs to the four rooms using four PATHS named **PathRoom1**, **PathRoom2**, **PathRoom3**, and **PathRoom4** respectively.

- The distance between waiting room and rooms one and three is 15 meters and to the other rooms it is 30 meters.
- Select all four PATHS and set the traveler capacity to one and the speed to 4.5 kilometers per hour.

Step 6: Next, provide a means for the patients to leave the examining rooms and exit the clinic using paths and the same distances from the waiting room to the rooms for the rooms to the exit. TRANSFERNODES can be used to assist in directing the patients out as seen in Figure 14.2.[204] We can use common paths (i.e., paths between room one and room two) for the output as we will need to constrain the paths to only have one person traveling on the path as we will the input paths.

Step 7: Connect the SOURCE **(SrcPatients)** via a CONNECTOR to the **SrvVetWaiting** and use a CONNECTOR to the **SnkExit** from the TRANSFERNODE

Step 8: Modify the OUTPUTBUFFER.CONTENTS queue of the **SrvVetWaiting** server to have an *Oriented Point* alignment with several locations with the *Keep in Place* option checked which will improve the animation.

Step 9: As before, patients should select a room that is currently available. Create a new node list via the *Definitions* tab named **ListRooms** which will allow the patients to choose a room that is available. Figure 14.2 shows the list, noting the order of preference is one, three, two and then four (i.e., fill up the closest rooms first if more than one room is available).

---

[204] CONNECTORS are used to connect the outputs of the rooms to the TRANSFERNODES while paths with appropriate logical lengths are used to connect the nodes.

Figure 14.2: Creating a Node List of Examination Rooms

Step 10: For the output TRANSFERNODE of the waiting room (**Output@SrvVetWaiting**), specify the *Entity Destination Type* to select a node "*From a list*" using the **ListRooms** as the *Node List Name* property. The goal is to route patients to an available room. Therefore, specify the *Selection Goal* as the "Smallest Value". In previous chapters we used an expression to determine the total number of entities routing to the input node, waiting in the input buffer, plus those currently being processed by the server.

Step 11: However, a series of `AssociatedStation`[205] functions (see Table 14.1) can be used as part of the dynamic routing selection for all fixed objects that have external input nodes (e.g., SERVERS). These external input nodes are associated with an internal STATION (e.g., **Input@SrvRoom1** is associated with the input buffer station of **SrvRoom1**).

Table 14.1: Explanation of the `AssociatedStation` Functions

Function	Description
`AssociatedStation.Capacity`	This function returns the current capacity of the location inside the associated object (e.g., the capacity of the input buffer for the server). [206]
`AssociatedStation.Capacity.Remaining`	This function returns the current available unused capacity of the input buffer of the server.
`AssociatedStation.Contents`	This function will return the current number of entities in the location (e.g., number in the input buffer of a server (`InputBuffer.Contents`).[207]
`AssociatedStation.EntryQueue.NumberWaiting`	This function will return the current number of entities waiting to enter the location (`InputBuffer.EntryQueue`).
`AssociatedStationLoad`	This function will return the "load" which is the sum of entities on their way to the node, waiting to enter the object, and now in the object (i.e., input buffer and processing stations).
`AssociatedStationOverload`	This function returns how much a location is overloaded which equals the "Load" minus "Capacity" which can be both positive (overloaded) or negative (under loaded).

Use the expression (`Candidate.Node.AssociatedStationload`) which returns the total number of entities heading to the room, waiting to enter the room, waiting at the room plus the current number of entities being processed by the room as the *Selection Expression* property as seen in Figure 14.3.

---

[205] When you access the `AssociatedStation` functions for an external node, you are given access to the actual station and all of its properties, states and functions.

[206] The capacity of the SERVER (i.e., the **Processing** STATION) is not included unless the **InputBuffer** capacity is zero then all the functions report values associated with the **Processing** STATION instead of the **InputBuffer** STATION. Note that an infinite capacity is represented by the maximum integer value of 2147483647.

[207] The `AssociatedStation.Contents` for the input buffer does not include the entities that are currently being processed by the SERVER or those in the `OutputBuffer` station. Also, the `AssociatedStationLoad` does not take into account entities waiting in the `OutputBuffer` STATION.

⊟ Entity Destination Type	**Select From List**
Node List Name	**ListRooms**
⊟ Selection Goal	**Smallest Value**
Selection Expression	**Candidate.Node.AssociatedStationload**
Selection Condition	
Blocked Destination Rule	Select Available Only

Figure 14.3: Specifying the *Selection Expression* to Choose the Room with Smallest Number of Patients

Step 12: Save and run the model to observe what happens.

Question 1: Does the model behave the way it was intended?

Step 13: Patients seem to wait at the examining rooms after being checked-in instead of at the waiting room, which was not what was intended. Therefore, we need to force the patients to wait in the waiting room until a room is available in a similar fashion as was done in the Chapter X. Select each of the rooms and change the *Input Buffer*[208] capacity to "0". Select the output node of the **SrvVetWaiting (Output@SrvVetWaiting)** and make sure the *Blocked Destination Rule* is set to "Select Available Only."

Step 14: Run model to observe what happens. You might need to increase the arrival rate to see any issues.

Question 2: Does the model behave the way it was intended or are there animation issues still present?

The patients will now wait properly in the waiting room. However Figure 14.4 shows an animation issue of a patient waiting for a vet potentially in two places which did not occur in the previous models. In those models, the SERVERS were the only constraining resource but suppose now we need a room (the SERVER) and a vet (a RESOURCE). Since the input buffer is zero and we try to seize the vet (a resource) in the *Before Processing* add-on process trigger. The entity can become stuck in a limbo state if the room (SERVER) is available but the vet is not immediately available. Thus the entity cannot enter the input buffer because the vet is unavailable, but yet it has seized capacity of the SERVER). The entity has not been completely transferred to the processing station. An entity waiting for a station capacity (in this case the input buffer) to become available will wait in an "Entry Queue". Entry queues are used whenever an entity cannot enter a station because there is no capacity available. The animation anomaly occurs because the entity is both waiting in the entry queue and yet is in the station (due to the need for secondary resources).

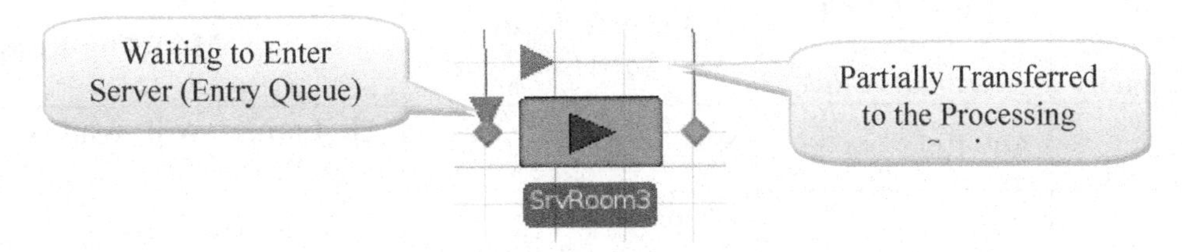

Figure 14.4: Animation Anomalies

Step 15: We could change the `Seize` of the vet to occur in the "*Processing*" add-on process trigger which happens after the entity has been completely transferred into the processing station. If you want to try it, select **SrvRoom3**, remove the *Before Processing*[209] add-on process trigger and then select the same the process from the dropdown of the *Processing* add-on process trigger. Run the model and observe what happens for the room.

---

[208] You can select all four rooms and then set the *Input Buffer* to "0" under the *Buffer Capacities* properties at once. Here the spreadsheet view is also useful.

[209] You can remove an add-on process trigger either by using the right click reset menu option or selecting the trigger name and then deleting it.

Step 16: The animation anomaly has been removed but the patient now waits in the processing station for the vet and then is processed when the vet arrives. However, this can be confusing as well as cause issues with statistics since both waiting and processing are combined now. Table 14.2 describes the options for choosing a MODELENTITY's destination node from a list. The *Selection Condition* and *Blocked Destination Rule* properties can be used to control which nodes are available for possible selection as the destination node.

Table 14.2: Choosing Entity's Destination via Select from List

Routing Logic	Description
Selection Goal	The goal determines the logic to utilize when selecting a destination node for an Entity. This can be random or cyclic (i.e., will select the next node in the list), preferred (i.e., will always select the nodes that are first in the list unless a *Selection Condition* eliminates a node), smallest and largest distance, and smallest and largest values.
Selection Expression	An expression evaluated for each candidate node destination when the *Selection Goal* property is set to the "Smallest or Largest Value."
Selection Condition	An optional condition that is evaluated for each candidate destination node. This expression must evaluate to be true for the node to be considered under the *Selection Goal.*
Blocked Destination Rule	Determines what happens if the destination nodes are blocked. The default value "**Select Available Only**," will only evaluate nodes to be selected that are currently not blocked while "**Select Any**" will select any node regardless of whether it is blocked or not based on the *Selection Goal*. The "**Preferred Available**", will first select among all non-blocked nodes and if all the nodes blocked will then revert back to select any. A node is considered blocked if the node is an input node of an object and the number of entities assigned to the station is greater than or equal to the capacity of the object (i.e., `AssociatedStationLoad ≥ AssociatedStation.Capacity`).

Table 14.3 looks at all possible scenarios associated with the server's input buffer size and the values of the *Selection Condition* and *Blocked Destination Rule* of the TRANSFERNODE. The "Modeling Result" column represents what happens to the additional entities when the rooms all have one patient currently. The *Selection Condition* will eliminate all nodes (i.e., filter the list) that do not meet the criteria before the selection goal is used while the *Blocked Destination Rule* may not allow the node from being selected if the node is currently blocked even though it was not filtered out.

Table 14.3: Understanding Blocking, Selection Condition, and Input Buffer Size has on Routing

Input Buffer Cap.	**Path Travel Capacity**	**TransferNode's** *Blocked Destination Rule*	**TransferNode's** *Selection Condition*	**Modeling Result**
Infinity	1, Infinity	Any Rule	None	All Entities will Wait at the SERVER's Input Buffer.
Infinity	1, Infinity	Any Rule	Candidate.Node. AssociatedStationload ==0	All Entities will wait at the TRANSFERNODE (i.e., Waiting Room)
0	1, Infinity	Select Available Only	None	All Entities will wait at the TRANSFERNODE
0	Infinity	Prefer Available, Select Any	None	All Entities will wait at the SERVER's Input Node waiting to go into SERVER.
0	1	Prefer Available, Select Any	None	One additional Entity will wait at the SERVER's Input Node waiting to go in all others will wait at the TRANSFERNODE.

Step 17: Set all the *Input Buffer* capacities back to "Infinity" or at least one to allow one entity to wait in the room. Also making sure to reset **SrvRoom3's** *Before Processing* and *Processing* add-on process triggers back to **SrvRoom3_BeforeProcessing** and nothing respectively.

Step 18: From Table 14.3, just relying on the blocking of the node's associated station maybe problematic. Therefore, we will set the **Output@SrvWaitingRoom**'s *Selection Condition* property to be `Candidate.Node.AssociatedStationLoad == 0` which says only allow nodes who currently do not have any entities routing to the node or currently in the server to be eligible to be selected.

⊟ Entity Destination Type	**Select From List**
Node List Name	**ListRooms**
⊟ Selection Goal	**Smallest Value**
Selection Expression	**Candidate.Node.AssociatedStationload**
Selection Condition	**Candidate.Node.AssociatedStationLoad==0**
Blocked Destination Rule	Select Available Only

Figure 14.5: Forcing Patients to wait until a Room is Available

Question 3: For 40 hours, what is the average time in the system and the average wait in the room?

____

Question 4: What assumption(s) does the current model make?

____

Part 14.2: Using a WORKER as a Moveable Resource

The current model addresses the exam rooms, but does not model the veterinarians. A fixed RESOURCE object similar to previous chapters could be used to constrain the processing to only one staff member, but the object is stationary. Since the vet needs to travel between rooms, a dynamic moving resource would better reflect the real system. WORKERS (and VEHICLES) in the system can be specified to be resources (i.e., they can be seized and released like RESOURCES and have capacities like STATIONS) but they are visibly mobile.

Step 1: From the [*Project Library*], insert a new WORKER named **WkrVet**. Change the symbol of the worker to an appropriate figure if you would like. Modify the properties as specified in Figure 14.6.

Properties: WkrVet (Worker)		
Show Commonly Used Properties Only		Change the *Park While Busy* property to **True** which will force the worker to park when it is busy service a customer. This will help minimize blocking issues on paths
⊟ **Resource Logic**		
Capacity Type	Fixed	Change the *Initial Desired Speed* to 4.5 kilometers per hour.
Ranking Rule	First In First Out	
Dynamic Selection Rule	None	
Park While Busy	**True**	
⊟ **Travel Logic**		
⊟ Initial Desired Speed	**4.5**	Specify that the *Initial Node (Home)* will be the office (i.e., **TOffice**) after performing Step 2 below.
Units	**Kilometers per Hour**	
Initial Network	Global	
Network Turnaround Method	Exit & Re-enter	
⊟ **Routing Logic**		Specify the *Idle Action* property to be "**Park At Home**"
Initial Priority	1.0	
Initial Node (Home)	**TOffice**	
Idle Action	**Park At Home**	
Off Shift Action	Park At Node	

Figure 14.6: Specifying the Worker Properties

Step 2: Next, build a separate network that the Veterinary staff will use to move throughout the system. Add six TRANSFERNODES named **TRoom1**, **TRoom2**, **TRoom3**, **TRoom4**, **TOffice**, and **TWaitRoom**. The **TRoom** nodes should be located near the same room with the office in the back as seen in Figure 14.7.

- Connect the nodes together using bi-directional PATHS since the staff can move in either direction to service the next patient. Use logical distances of 15 meters for each link.[210] The staff will travel on the inner loop.

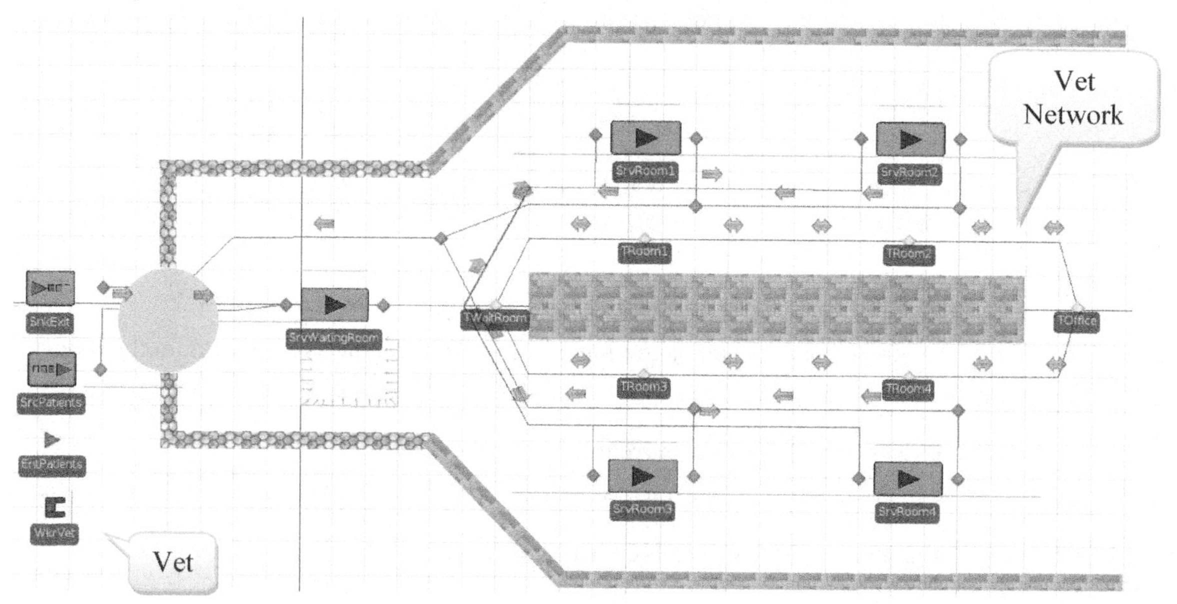

Figure 14.7: Modified Vet Clinic with Vet Network

Step 3: Since the WORKER entity acts like a VEHICLE, the vets will utilize the parking queues when it stops at each of the rooms and the office.[211] By default, a parking queue is automatically placed. However, you are unable to orient the queue or place it in a different location. If you wish to improve the animation, then select one of the transfer nodes associated with the Vet network (**TRooms** or the **TOffice** node) and toggle the *Parking Queue* button off. Next, select the `ParkingStation.Contents` Queue from the "Draw Queue" section of "*Attached Animation*" ribbon and draw the animated queue very near the node selected choosing the orientation. Repeat the process for the remaining nodes in the Vet network.

Step 4: Save and run the model and observe the dynamic worker.

Step 5: At this point our **WkrVet** stays in their office and does not actively seek any patients. The patients will need to request the **WkrVet** in order to be processed. When the patient goes into processing they need to seize the **WkrVet** and then release the **WkrVet** once they have finished processing.[212]

- From the *Processes* tab, create a new process named **SiezeVet** and insert a *Seize* step that will request a **WkrVet** to service a patient as seen in Figure 14.8. Select all four room servers and select the new process **SeizeVet** as the *Before Processing* Add-on process trigger.

---

[210] Select all the paths and specify the direction and the logical length at one time. Note, careful consideration has to be followed in using bi-directional paths because they can cause the system to become blocked and really should only be used when a very small number of entities, workers or vehicles are being used. Otherwise drawing two paths (one forward and one backward) from each node is the preferred method.

[211] This is for animation purposes since the Vets would disappear when they arrive at their destination and begin service.

[212] Again we will utilize the option that gives you the most control (i.e., processes). However, in this case *Secondary Resources* could be used and is demonstrated in the commentary.

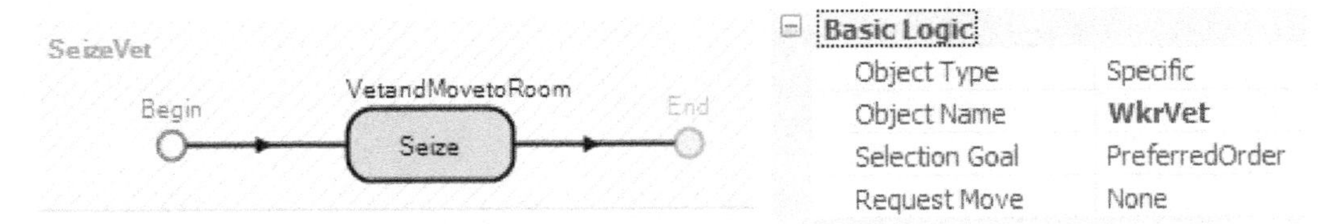

Figure 14.8: Seizing a Moveable Worker and Forcing a Visit

- Create a new Process called **ReleaseVet**. It will be triggered in the Add-on process trigger *"After Processing*". Insert a `Release` step that will release the Vet to service a new patient as seen in Figure 14.9.

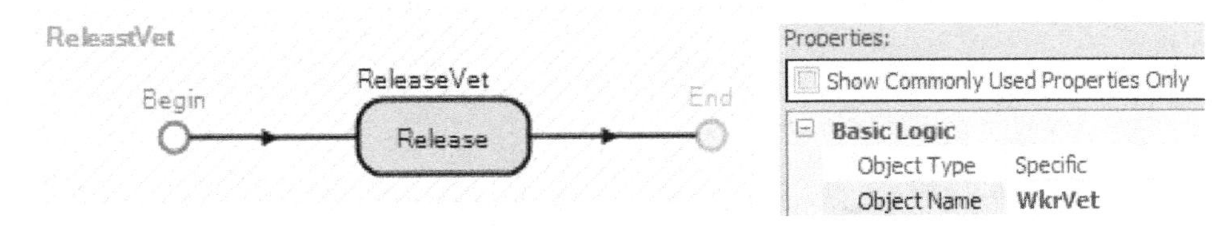

Figure 14.9: Releasing the Moveable Worker

- In the dropdown list for the *"After Processing"* Add-on process trigger for all four rooms, just select the **ReleaseVet** to release the veterinarian after the processing has finished.

Step 6: Save and run the model and observe the veterinarian.

Question 5: Does the vet get seized and release correctly as well as move to the appropriate room?

____

Step 7: Our vet is now being seized and released constraining the processing patients rather than assuming four stationary vets at each room. However, the vet seems to be servicing the patients from the **TOffice** node. We really do not want to seize the **WkrVet** worker until the vet is located at the room. From Figure 14.8, you will notice the Request Move property which can be change to be **"ToNode"** and then specify the appropriate destination node (e.g., **TRoom1** for room one). This will force the **WkrVet** to arrive at the **TRoom1** transfer node before the processing can occur. The *Move Priority* property can be used to specify the priority if multiple patients have requested the worker. Since this will be the same for all four rooms except the node will change (i.e., **TRoom1** to **TRoom2**), a tokenized process will be used.

- From the *Definitions* tab, add a new TOKEN named **TknVet** that will have a *Node Object Reference* state variable named **TStaWhichNode** that will allow us to pass in the correct node as seen in Figure 14.10.

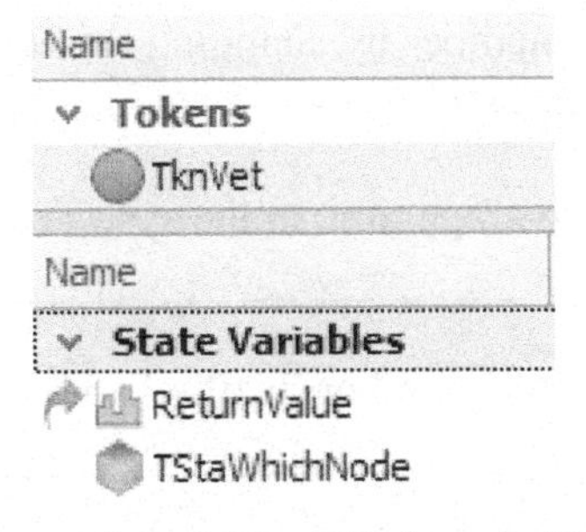

Figure 14.10: Creating a Specialized Token to Pass in the Move Node

- To create the Tokenized process, select the **SeizeVet** process from the *Processes* tab and specify the **TknVet** as the *Token Class Name* property value as seen in Figure 14.11. Specify one input argument that allow the user to pass into the process which room needs the service.

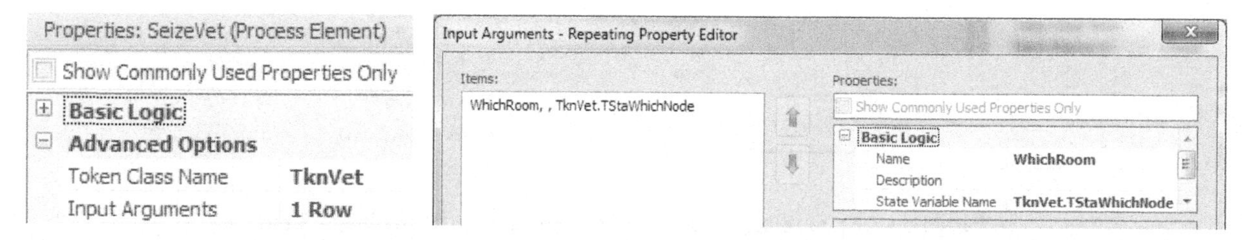

Figure 14.11: Making the SeizeVet Process Tokenized by Passing in one Input Argument

- Next, modify the `Seize` step to specify the *Request Move* to a particular node and the *Destination Node* will be the `TknVet.TStaWhichNode`. This will force the vet to move to a particular node before processing can begin on the patient taking into account the travel time to the room.

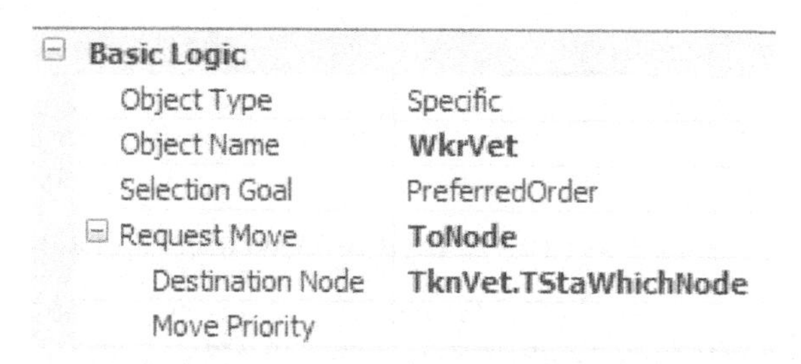

Figure 14.12: Modify the Seize to force Vet to Move to the Specified Node

- Finally, on each of the rooms specify the appropriate Which Room input argument property for the Before Processing add-on process trigger as seen in Figure 14.13 for **SrvRoom1**.

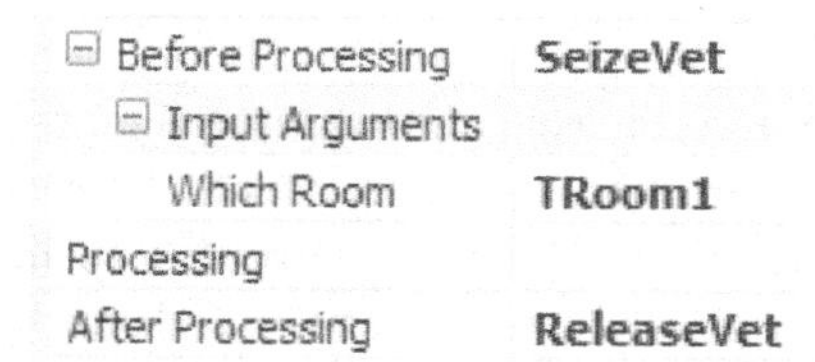

Figure 14.13: Specify the Correct Node Associated with each Room

Step 8: Save and run the model observing how the veterinarian dynamically moves to each room.

Question 6: For 40 hours, what is the average time in the system and the average wait in the waiting room?

____

Step 9: Add a second veterinary staff member by changing the *Initial Number in the System* property under the *Population* property category to be two.[213]

Question 7: For 40 hours what is the average time in the system and the average wait in the waiting room?

____

Question 8: Is the second veterinary staff member justified and why?

____

[213] Another option is to insert a new worker named **VetB**. Then a resource list would have to be created to allow entities to seize one from the pool of resources. This modeling approach may be necessary if the two staff members can service different types of patients, different processing times, have a different ordering or work schedule, etc.

Part 14.3: Returning to the Office between Patients

One additional modeling concern is the veterinarians need to return to their office after each visit to complete paper work on the current patient and look over the chart of the next patient. All this needs to be done before returning to one of the rooms to service the next patient. So we need to force them to go back to the office and then delay for a particular time before going back to service the next patient. Since resources respond to requests to be allocated (i.e., patients trying to seize them), one can model the situation where an entity requests the **WkrVet** at the office so they will return back to the office and process that entity. In the office, the paperwork uniformly takes between four and fourteen minutes.

Step 1: Save the current model as a new model and set the number of available veterinaries back to one.

Step 2: Insert a SERVER named **SrvOffice** near the **TOffice** transfer node and connect it to a SINK called **SnkOfficeWork** via a connector as seen in Figure 14.14. Specify the processing time in the **SrvOffice** to be a Uniform(4,14) minutes.

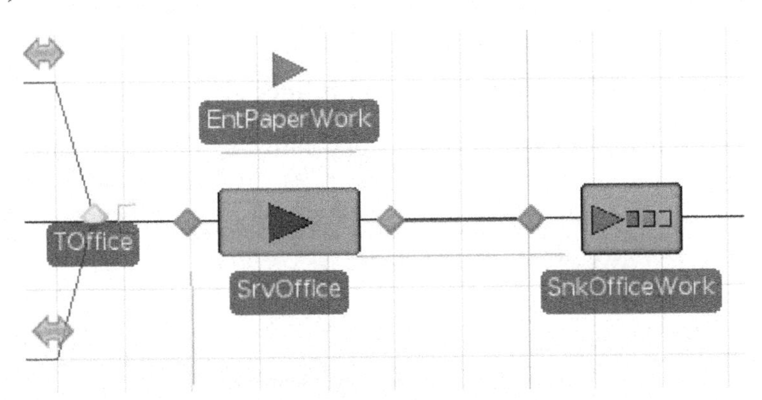

Figure 14.14: Modeling the Paperwork

Step 3: Insert a new MODELENTITY called **EntPaperWork**. It will be this entity that is processed in the office. Since we want the Vet to do the paperwork before seeing any other patient, let's give the *Initial Priority* of the **EntPaperWork** a value of "2" (recall that the default value is "1"). Then to insure that this paperwork is given priority, change the *Dynamic Selection Rule* property of the **WkrVet** to "Largest Value First", with the *Value Expression* as `Candidate.Entity.Priority`.[214]

Dynamic Selection Rule	**Largest Value First**
Value Expression	Candidate.Entity.Priority

Figure 14.15: Changing the Dynamic Selection Rule of the Worker

Step 4: When the vet finishes at a room, paperwork needs to be created and sent to the office. This paperwork needs to `Seize` the **WkrVet** before the vet is seized by another room. Also as a part of this process, the vet needs to be released from the room to visit the office. Therefore, we need to modify the **ReleaseVet process** to accomplish all this, as shown in Figure 14.16. Recall that this process is used by all the rooms when patients exit.

[214] The *Dynamic Selection Rule* is used here only to illustrate its use and is strictly not necessary. The only case when the dynamic selection rule is needed is when the Value may be changed while the entity is waiting, such as when a due date is reached.

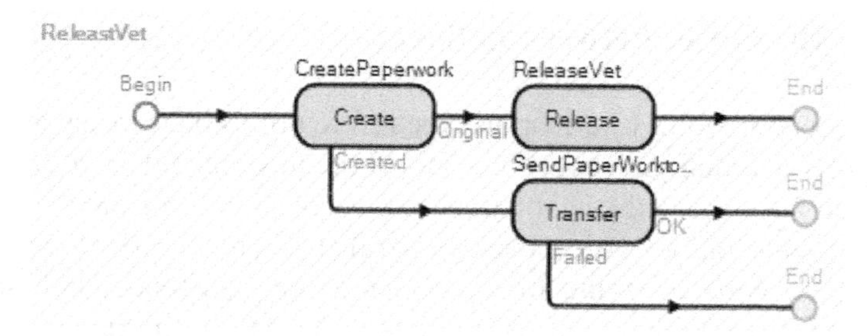

Figure 14.16: Creating and Transferring the Paperwork

- Use the `Create` step to generate a paperwork entity a new TOKEN associated with the new entity.

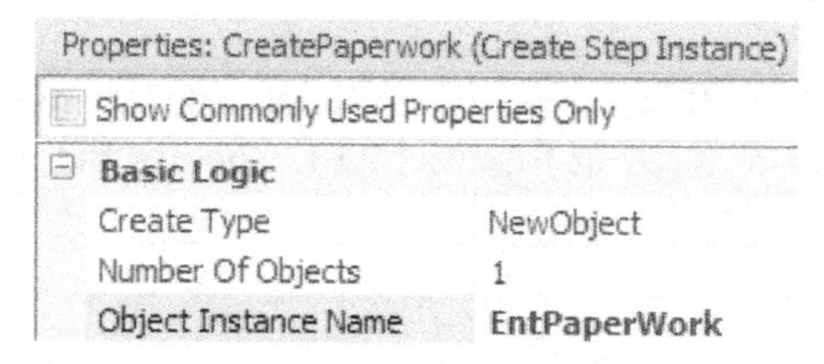

Properties: CreatePaperwork (Create Step Instance)

Show Commonly Used Properties Only

Basic Logic	
Create Type	NewObject
Number Of Objects	1
Object Instance Name	**EntPaperWork**

Figure 14.17: `Create` Step to Produce an MODELENTITY

- Use the `Transfer` step to send the paperwork entity to the input buffer station of the office SERVER.[215]

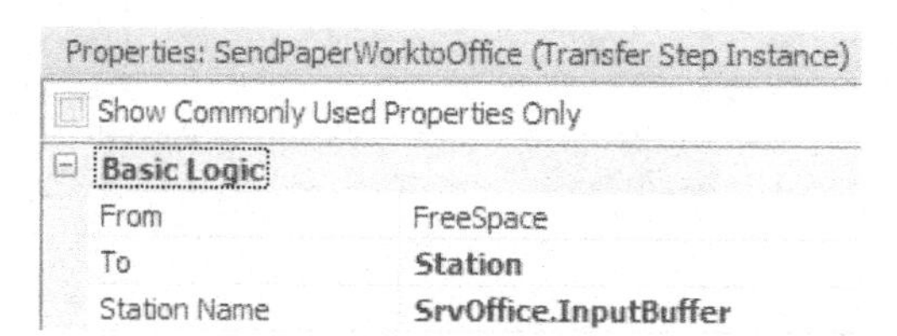

Properties: SendPaperWorktoOffice (Transfer Step Instance)

Show Commonly Used Properties Only

Basic Logic	
From	FreeSpace
To	**Station**
Station Name	**SrvOffice.InputBuffer**

Figure 14.18: Using the `Transfer` Step to move the New Paper ENTITY.

Step 5: For the **SrvOffice**, create two new add-on process triggers processes for *Before Processing* and *After Processing*.

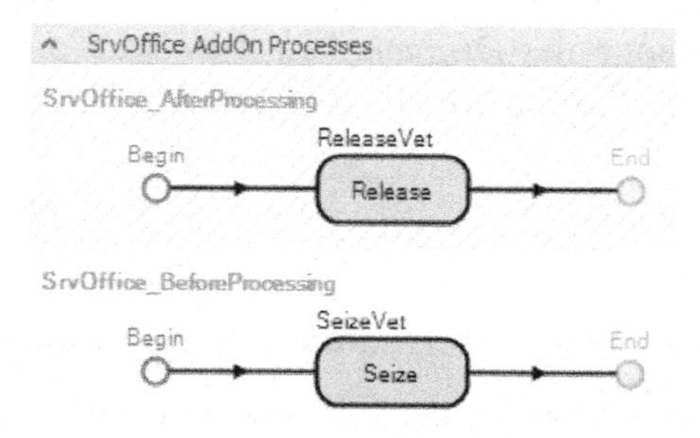

Figure 14.19: Requesting and Freeing the Vet from Paper Work at the Office

- Copy the `Seize` step from one of the four rooms to request the **WkrVet** for the paperwork. Remember to change the node to visit to the **Toffice** node in the `Seize` step.
- Copy the `Release` step from **ReleaseVet** to release the Vet after the paperwork is finished.

Step 6: Save and run the model and observe what happens.

Question 9: Is the model working as expected or do you observe any issues?

____

[215] Note this time we are sending the entity directly to the input buffer station rather than the Input node at the server which will force the paper work to seize the vet.

Step 7: The issue seems that the vet will process a second patient before heading back to the office. Essentially, the vet processes the previous patient's paper work but it is difficult to determine which paper work went with which patient since all the patients and paper work entities are green. To alleviate the situation, add three additional symbols for both the **EntPatients** and **EntPaperWork** giving each one of the additional symbols a different color making sure the symbols for each entity type are the same color (i.e., 0th symbol is green for both patient and paperwork, 1st symbol is magenta for both, 2nd symbol is blue,and the 3rd symbol is red).

Step 8: Patients will be assigned a color based on the room they are serviced too (i.e., patients will be green in Room 1, magenta in Room 2, blue in Room 3, and finally red in Room 4). Since this will be done when the patients goes into processing, we need to add an additional input argument to the **SeizeVet** process to specify which picture to use. First, add an `INTEGER STATE` variable named **TstaRoomNumber** to our **TknVet** and then specify the second input argument of the **SeizeVet** process to be the room number as seen in Figure 14.20.

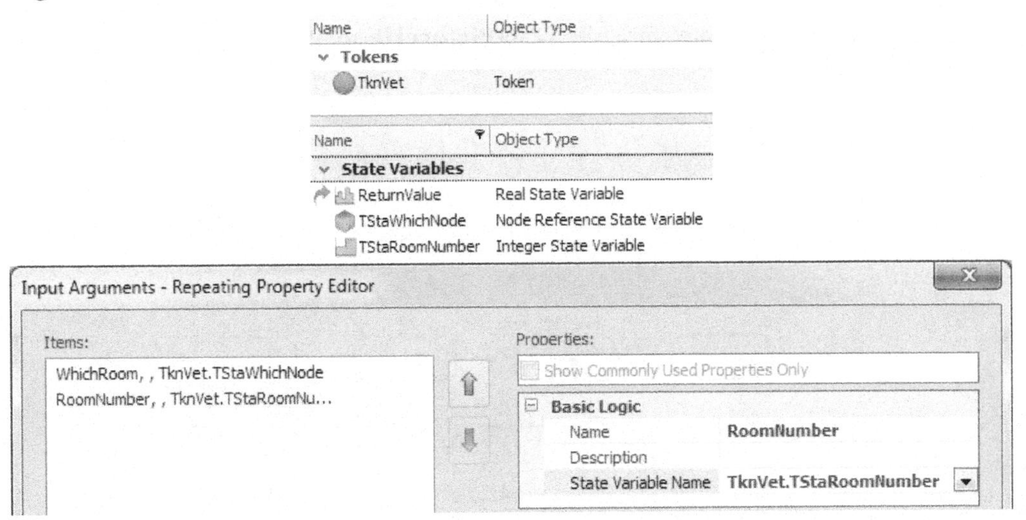

Figure 14.20: Adding a new Input Argument to the Tokenized SeizeVet Process

Step 9: Next, add an `Assign` step to **SrvRoom2_BeforeProcessing** before the `Seize` step that assigns a new value of "1" to the `ModelEntity.Picture` state variable.

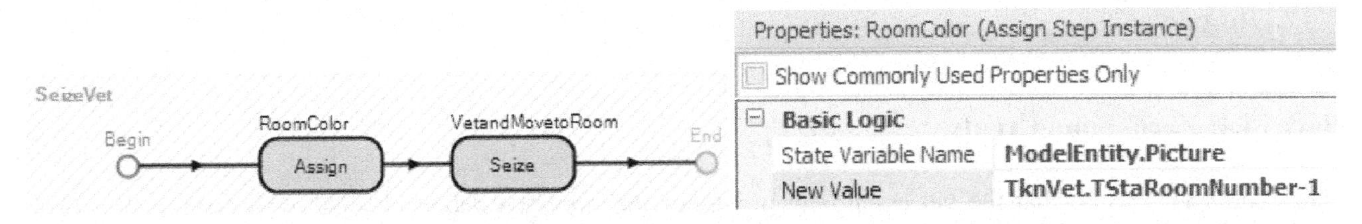

Figure 14.21: Changing the Patient Color based on the Room

Step 10: Modify the ***Before_Processing*** for each of the rooms and specify "1", "2", "3", and "4" for the Room Number input argument for **SrvRoom1**, **SrvRoom2**, **SrvRoom3**, and **SrvRoom4** respectively.

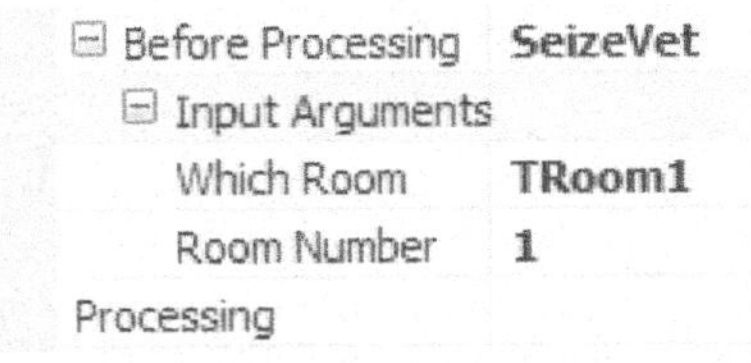

Figure 14.22: Passing the Room Number for SrvRoom1

Question 10: Why are we subtracting one from the room number in the `Assign` step?

____

Step 11: Save and run the model observing the entities as they enter the various rooms.

Step 12: Now, the patient entities are changing color based on the room they enter. The paperwork entity now needs to reflect the patient color that created it. Process steps like *Create* and *Search* cause new tokens to be created that are now associated with either the new entity created in the *Create* step or the objects found in the *Search* step. Therefore, state variables associated with the original token's object (i.e., the ModelEntity.Picture) are not accessible to the new objects. One common approach to copy states from one entity to another is to create a global (i.e., MODEL) state variable that is accessible by all objects of the model.[216] From the *Definitions* tab, add a new INTEGER DISCRETE STATE variable named **GStaPictureID**.

Step 13: Next add an *Assign* step before the *Create* step in the **ReleaseVet** process that assigns the **GStaPictureID** the ModelEntity.Picture of the patient entity that is creating the paper work. Then insert another *Assign* step before the *Transfer* which is associated with the newly created paper work entity that assigns the model state variable **GStaPictureID** to the ModelEntity.Picture state variable as seen in Figure 14.23.

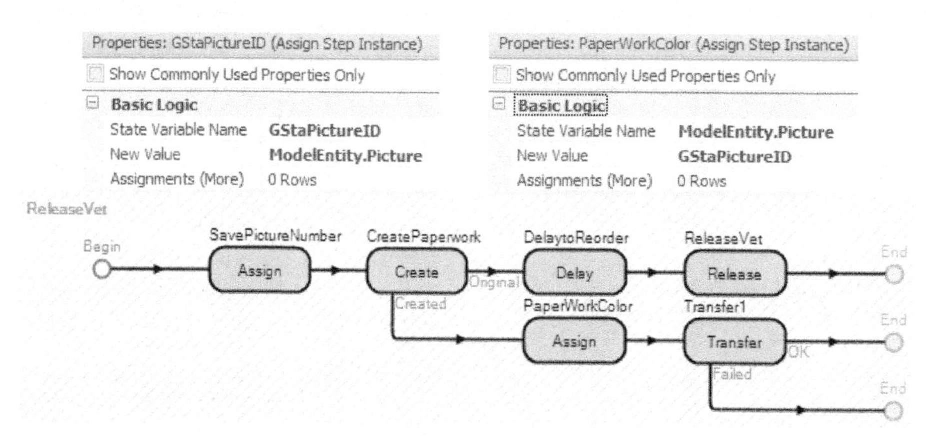

Figure 14.23: Copying the Value from One ModelEntity to another ModelEntity

Step 14: Save and run the model observing the entities as they enter the various rooms and the paper work.

Question 11: Can you see the different paperwork entities being created and how the vet always lags behind by one?

Part 14.4: Zero-Time Events

The issue seems to be that the vet essentially processes the previous patient's seize even though the paper work has a higher priority. A discrete-event simulation executes the simulation through an event calendar that is ordered by the time of the event. Zero time events are events that occur at the exact same time.[217] The *Seize* step for the paperwork entity happens at the same time the *Release* step occurs for the room. If the release step occurs before the seize step occurs, then the vet may choose to service a patient in a different

---

[216] Instead of creating a new Object in the *Create* step, a copy of the associated object could be created which would inherit all the properties and states of the creating object. This approach could have been used in this case but the statistics associated with **EntPatients** could not be used since this would include the paper work entities as well since they would be the same entity type. Also we could have created a tokenized **ReleaseVet** process that would take in the room number and therefore would know the color that needed to be set.

[217] There are dozens of events that can occur that same time and can become an issue especially when dealing with allocations of resources, workers, and vehicles.

room. Ordering of zero-time events can become critical in the execution of the model and the modeler won't know which is done first without some detective work.

A great tool for understanding what is happening during your simulation is through the use of the "Trace". The "*Model Trace*," initiated from the "*Run*" tab, is the most complete description of what is happening. However, in this case you want to trace the behavior of a process. You do this by selecting the step in the *Processes* tab and clicking the breakpoint button in the ribbon, which will place the break point symbol adjacent to the step[218]. Now in the "*Facility*" tab, select the *Model Trace* button which displays the trace in its own window.

Step 1: Click on the `Create` step in the **ReleaseVet** process and select the breakpoint button. The `Create` step should now have the breakpoint symbol beside it.

Step 2: Click the *Model Trace* button on the *Run* tab and then run the model via the *Fast-Forward* button.

Step 3: After the simulation "breaks", click the *Step* button a couple of times. (You can recognize the break because the line colored red describes the breakpoint.) When you click the *Step* button, a series of actions are shown, colored in yellow, representing that simulation step. You may need to scroll back up to see the breakpoint statement.

Step 4: Figure 14.24[219] is a printout of all the events that have occurred around the breakpoint.

Trace - Tracing to D:\MyStuff\Dropbox\Dropbox\SimioBook\NewBook\Chapter14\Chap 14-4new_Model_trace.csv

Time (Hours)	Entity	Object	Process	Token	Step	Action
0.285292362711871	EntPatients.61	Model	ReleaseVet	9	[Create] CreatePaperwork	Stopped at breakpoint on ReleaseVet.CreatePaperwork
						Creating '1' object(s) of entity type 'EntPaperWork' using creation method 'NewObject'.
	-	-	-	0	-	Object 'EntPaperWork.64' created.
	EntPaperWork.64	Model	ReleaseVet	6	[Assign] RoomColor	Assigning state variable 'EntPaperWork.64.Picture' the value '0'.
					[Transfer] Transfer1	Entity 'EntPaperWork.64' initiating transfer from '[FreeSpace] Model' into station location 'SrvOffice.InputBuffer'.
	EntPatients.61	Model	ReleaseVet	9	[Release] ReleaseVet	Releasing resource(s) owned by object 'EntPatients.61'.
						Object 'EntPatients.61' released '1' capacity unit(s) of resource 'WkrVet[1]'.
		WkrVet[1]	OnCapacityReleased	1	[Begin]	Process 'WkrVet[1].OnCapacityReleased' started.
					[Assign] OnCapacityRelease...	Assigning state variable 'WkrVet[1].ResourceState' the value '0'.
						Assigning state variable 'WkrVet[1].OnEvaluatingRiderReservation.Enabled' the value '1'.
						Assigning state variable 'WkrVet[1].OnEvaluatingRiderReservation.ReturnValue' the value '1'.
						Assigning state variable 'WkrVet[1].OnEvaluatingSeizeRequest.Enabled' the value '1'.
						Assigning state variable 'WkrVet[1].OnEvaluatingSeizeRequest.ReturnValue' the value '1'.
						Assigning state variable 'WkrVet[1].OnEvaluatingRiderAtPickup.Enabled' the value '1'.
						Assigning state variable 'WkrVet[1].OnEvaluatingRiderAtPickup.ReturnValue' the value '1'.
					[SetNode] ToClearDestination	Destination node of entity 'WkrVet[1]' set to value 'No Destination'.
					[Fire] RemainInPlaceEndedEv...	Firing event 'WkrVet[1].RemainInPlaceEnded'.
					[Fire] ReleasedEvent	Firing event 'WkrVet[1].Released'.
					[End]	Process 'WkrVet[1].OnCapacityReleased' ended.
		Model	SeizeVet	5	[Seize] VetandMovetoRoom	Object 'EntPatients.62' seized '1' capacity unit(s) of resource 'WkrVet[1]'.
					[Begin]	Process 'WkrVet[1].OnCapacityAllocated' started.

Figure 14.24: Sample Trace showing the Ordering of the Zero-Time Events

Step 5: Now you should read down the trace, noting the entity, the object, the step, and the action. Here we see that the paperwork entity is created (**EntPaperwork.63**), the picture variable is assigned the value of 0 and that entity is transferred to the "*SrvOffice.InputBuffer*." Next, the patient entity (**Entpatients.61**) releases one unit of capacity of the vet. Scrolling down through the other zero time events, you will see that the vet will then respond to a request from another room first because the paper work entity does not request one unit of capacity from the vet until near the very end of the current time.[220] All this is done at the same simulation time but the ordering is important.

Question 12: Can you now see that the release step is done before the seize step of the office?

____

[218] You can also right-click the step and select the "breakpoint".

[219] The trace is automatically saved to a CSV file as the model is running and can be imported into Excel for further analysis and manipulation if necessary. You can also filter each column by selecting a row and the right clicking on the column heading.

[220] It may be necessary to click the *Step* button again to get the next set of zero time events owing to the number of events that occur.

Step 6: We need the release to happen after the `Seize` step which implies the concurrent time events need to be reordered. To accomplish a reordering of events you can insert a `Delay` step before the release of the vet (see Figure 14.25) that will delay the current token (representing the patient) a very small amount of time, by specifying the delay time to be `Math.Epsilon`.[221]

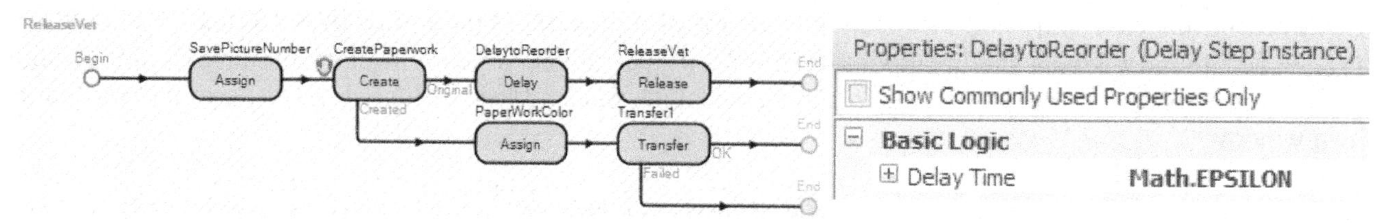

Figure 14.25: Reordering Zero Time Events by Delaying an Epsilon Amount of Time

Step 7: Save and run the model again preforming a few "Steps" after the breakpoint which yields the trace in Figure 14.26. As you can see the delay causes all the steps following it to become late priority events meaning they will be placed at the end of the current time allowing the paper work to seize the vet first which then allows the vet to select it based on its priority.

Time (Hours)	Entity	Object	Process	Token	Step	Action
	EntPatients.61	Model	SrvRoom1_AfterProcessing	7	[Create] CreatePaperwork	Stopped at breakpoint on SrvRoom1_AfterProcessing.CreatePaperwork
						Creating '1' object(s) of type 'EntPaperWork' using creation method 'NewObject'.
	-	-	-	0	-	Object 'EntPaperWork.63' created.
	EntPaperWork.63	Model	SrvRoom1_AfterProcessing	6	[Assign] PaperWorkColor	Assigning state variable 'EntPaperWork.63.Picture' the value '0' .
					[Transfer] SendPaperWork...	Entity 'EntPaperWork.63' initiating transfer from '[FreeSpace] Model' into station location 'SrvOffice.InputBuf
	EntPatients.61	Model	SrvRoom1_AfterProcessing	7	[Delay] DelaytoReorder	Delaying token by 'Epsilon' until time '0.243481765778934' Hours. Token exit scheduled as late priority event
	EntPaperWork.63	SrvOffice	OnEnteredInputBuffer	3	[Begin]	Process 'OnEnteredInputBuffer' initiated.
					[Delay] TransferInTime	Delaying token for '0' Hours until time '0.243481765778934' Hours.
					[EndTransfer] IntoInputBu...	Entity 'EntPaperWork.63' ending transfer into station 'SrvOffice.InputBuffer'.
					[Seize] Server	Requesting seize of resource(s) in behalf of owner 'EntPaperWork.63'.
	-	-	-	0	-	Object 'EntPaperWork.63' seized '1' capacity unit(s) of resource 'SrvOffice'.
	EntPaperWork.63	SrvOffice	OnEnteredInputBuffer	3	[Execute] BeforeProcessin...	Executing process 'SrvOffice_BeforeProcessing'.
		Model	SrvOffice_BeforeProcessing	8	[Begin]	Process 'SrvOffice_BeforeProcessing' initiated.
					[Seize] SeizeVet	Requesting seize of resource(s) in behalf of owner 'EntPaperWork.63'.
						Could not immediately complete resource seize of type Specific 'WkrVet'.
						Token waiting at Seize step for completion of resource seizes or move requests.
0.243481765778934			SrvRoom1_AfterProcessing	6	[End]	Process 'SrvRoom1_AfterProcessing' ended.

Figure 14.26: Reordered zero-time events

Step 8: Remove the break point by selecting the `Create` step and toggling the Breakpoint button. Now run the model and observe what happens.

Question 13: Is the model working as expected?

Part 14.5: Handling Multiple Vets

In the previous chapter, there were multiple vets that could process patients. Like other dynamic objects MODELENTITIES and VEHICLES, multiple WORKERS can be created at the start of the simulation or during the simulation run. Let's add two more vets to help in the clinic.

Step 1: Select the **WkrVet** and change the *Initial Number in System* property to "3" under the *Population* category.

Step 2: Change the **SrvOffice** capacity to "Infinity" so each of the vets can service their own paperwork.

Step 3: Save and run the model observing the vets and paperwork.

[221] `Math.Epsilon` represents the smallest real value number greater than zero. `Delay` steps will actually interpret the number as zero, but it forces the event to be put at the end of the current event calendar. If the delay time is specified as zero, SIMIO will not place the event on the calendar thus not changing the order at all. In the commentary, another solution to the ordering is presented.

Question 14: Did the vet that processed the patient always process the correct paper work?

____

Step 4: The difficulty is the paper work entity just requested a vet and the closest one that was not busy went to the office to perform paperwork (or was already sitting at the office) which was not what was intended. We wanted the vet who performed the service to also complete the paper work. Therefore, the paper work entity needs to request the same vet as the patient.[222] First, select the MODELENTITY in the [*Navigation*] panel and insert a new MODELENTITY DISCRETE INTEGER STATE variable named **EStaResourceID** which will be used to store the ID of the resource to request for every paper work entity.

Step 5: Transition back to the **Model** and insert a new model DISCRETE INTEGER STATE variable named **GStaModelResID** which will be used to pass the resource ID that was seized by the patient to the paper work entity in a similar fashion as the picture number was passed.

Step 6: Next we need to set the global (i.e., model) variable (**GStaModelResID**) with the vet the current patient has seized before we create the paper work entity and then set the paper work entity's **EStaResourceID** variable the value of model variable. Inside the **ReleaseVet**, modify the first *Assign* step (i.e. **GStaPictureID**) to include another assignment which sets **GStaModelResID** a value of `ModelEntity.SeizedResources.LastItem.ID`.[223] Then, modify the second *Assign* step on the "Created" path (i.e., PaperWorkColor) to set the **ModelEntity.EStaResourceID** variable a value of **GStaModelResID** as seen in Figure 14.27.

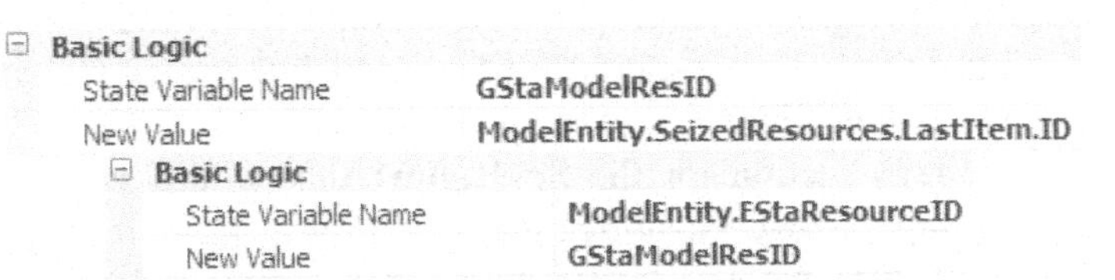

Figure 14.27: Passing Seized Resource ID to the Newly Created PaperWork Entity

Step 7: Modify the *Seize* step in the **SrvOffice_BeforeProcessing** process to request a particular Vet that needs to perform the paperwork based on the state variable **ModelEntity.EStaResourceID**. Under the *Advanced Options*, specify the *Selection Condition* property to be `Candidate.Worker.ID == ModelEntity.EStaResourceID` as seen in Figure 14.28. Recall the CANDIDATE is a wild card designator and will match all the **WkrVets** that are currently in the system. The *Selection Condition* property has to evaluate to "True" for the **WkrVet** to be in the pool of potential ones to be seized and in this example all **WkrVets** will be eliminated for consideration except the one whose ID matches the **EStaResourceID** value.

---

[222] All objects created in the simulation receive a unique identifier (ID) which can be used. For example **EntPatients.61** is a patient whose `ID` is 61.

[223] The vet will be the last resource seized since the patient had to first seize the server before it was allowed to go into processing and then request the vet. See Table 13.2 for more information on the `SeizedResources` property of the entity.

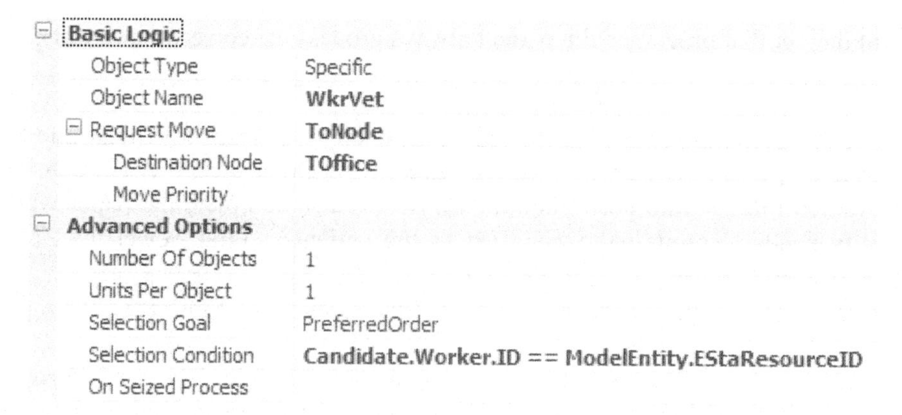

Figure 14.28: Selecting Only the Correct Vet for the Paper Work

Step 8: Save and run the model observing the vets and paperwork.

Question 15: Did the vet that processed the patient always process the correct paper work?

____

Part 14.6: Commentary

- Dynamic resources offer considerable modeling capability. In this chapter, we utilized processes to seize and release the work stating it offers the most control over the process. However, in this case we could have utilized secondary resources to accomplish the same task of the two processes used to seize and release the vet. Figure 14.29 displays *Resource for Processing* property values under the *Secondary Resources* section for the **SrvRoom1** that would seize the vet and wait for it to arrive before processing. However, notice the *Selection Condition* property used in the previous section is not exposed. It is exposed for the *Other Resource Seizes* properties but using processes was even easier.

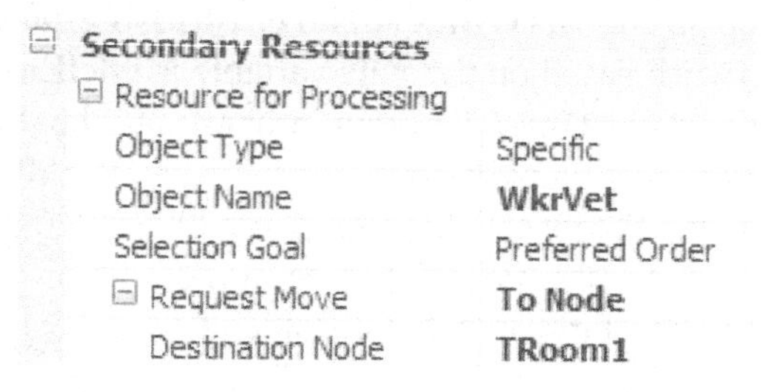

Figure 14.29: Specifying the Seizing and Release of the Vet Using Secondary Resources

- We fixed the zero-time event by reordering the concurrent events using a delay step. Another solution would have been to seize the vet before transferring as seen in Figure 14.30. The "Original" token waits until the "Created" token reaches end of the current process. Therefore, the seize will now happen before the release does in the zero-time events.

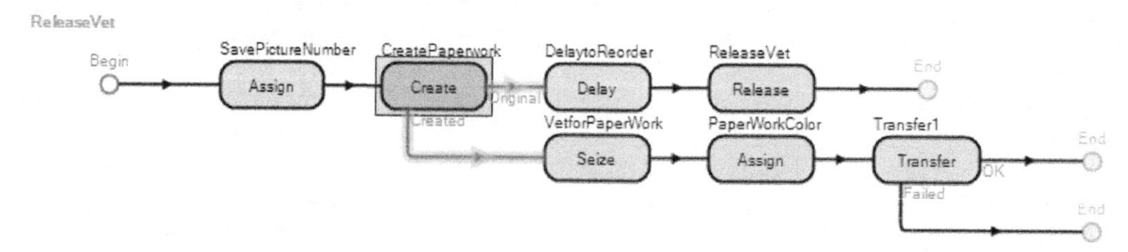

Figure 14.30: Seizing the Vet before Transferring the Paperwork

- Decision-making objects offer a number of modeling options especially as they relate to resources, transporters, and workers. Because there is so much flexibility, it can confound the choices.

However, by working with these objects, you begin to develop some modeling patterns that seem more appropriate than others.

- While it is our intent to use as many modeling approaches as we can, it will still be necessary for you to supplement our efforts. Features such as the breakpoints, trace, and watch can be extremely helpful in learning how SIMIO is behaving. It's hard to modify a behavior unless you understand the existing behavior.
- Unfortunately, the behavior of workers and vehicles is quite complex. Because of the complex behavior, these objects could be referred to as " heavyweight." In some object-oriented designs, heavyweight objects are composed of "lightweight" objects which are less complex. But that is not the case with SIMSO, so we will continue to investigate and employ the complexity of these objects.

Chapter 15
Adding Detail to Service: A Bank Example

We have previously modeled a moveable resource as a worker. Entities would request the resource to visit a specific node in the *Seize* step, which forces the worker to move to the node before allowing the entity to go into service. VEHICLES can also be seized and used as resources in the exact same way. The WORKER object is a specialized VEHICLE that can transport entities as well as being seized and released as a resource.[224] The modeling concepts relative to moveable resources used in this chapter will apply to vehicles or workers.

A small bank has al loan officer that assists customers with various loans and other account information as seen in Figure 15.1. Customers arrive and wait for the available loan officer. The loan officer goes to the waiting room and escorts (i.e., picks up) the customer back to the office to help the customer. After this process is complete, some of the customers require a visit to their safety deposit boxes which then requires the loan officer to escort them to the vault area and assist the customer again. The loan officer can be modeled using a worker or a vehicle.

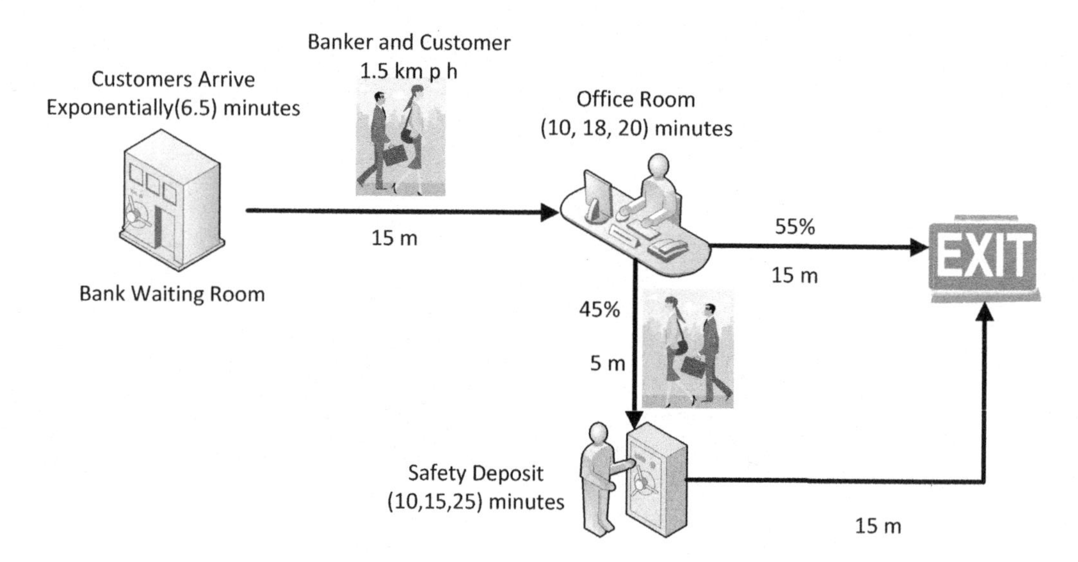

Figure 15.1: Bank Example where Worker Picks up Customers as well as Services Them

Part 15.1: Using a Worker as a Resource and a Vehicle

One advantage of using a worker as a moveable resource is the worker can also be used as a vehicle. For example, the vet or a technician may go to the waiting room and escort (i.e., picks up) the patient to the room where the vet then sees the patient (i.e., vet is seized). That kind of transportation is being done in the bank.

Step 1: Create a new model.

Step 2: Insert a SOURCE named **SrcWaitingRoom**, a SERVER named **SrvRoom**, and a SINK named **SnkExit**.

- Connect the output of the **SrcWaitingRoom** to the input of the **SrvRoom** via a 15 meter path and then the output of the room to the SINK (**SnkExit**) by another 15 meter path.[225]
- Customers arrive exponentially with an interarrval time of 6.5 minutes.

---

[224] A WORKER object must work "on demand" and has no routing type, meaning they don't follow a specific sequence.

[225] Select both paths and change the *Drawn To Scale* property to **False** and the *Logical Length* to be 15 meters simultaneously.

- The capacity of the SERVER should be one.
- Each customer takes between 10 and 20 minutes to be seen with most of them taking 18 minutes

Step 3: Insert a new MODELENTITY named **EntCustomer** and a WORKER[226] named **WkrBanker** into the model. The customers and banker travel at a constant rate of 1.5 km per hour or 1500 meters per hour.

- For the **EntCustomer** entity, add four additional symbols and change each of the five symbols to be represented by one of the *People\Animated* symbols in the *Library*[227]. If you utilize just the "static" "people" symbols, the people will "skate" across the paths. The animated symbols will allow the people to physically walk, run, and/or skip. For the **EntCustomer**, set the *Animation→Random Symbol* to "True" so it will randomly select one of the five people.
- Change the symbol of the **WkrBanker** to one that is more representative using one of the animated people as well. To simplify the presentation, we delete all ten of the alternative symbol states.
- For the WORKER, move the animated ride queue (Ride Station Queue)[228] so it is beside the **WkrBanker** instead on top for a normal VEHICLE as seen in Figure 15.2. Notice that the ride station queue is automatically given the option of "Match Attached Animation Speed" which means the "rider" will walk at the same speed as the worker.
- Also for the worker, set the *Initial Node (Home)* to the **Output@SrcWaitingRoom** and set the *IdleAction* to ***Park at Home.***
- Now for the **WkrBanker** set the *Park While Busy* property under the *Resource Logic* section to "True" to force the banker to park when it is utilized as resource.
- Make sure the SOURCE creates the **EntCustomer** MODELENTITIES.

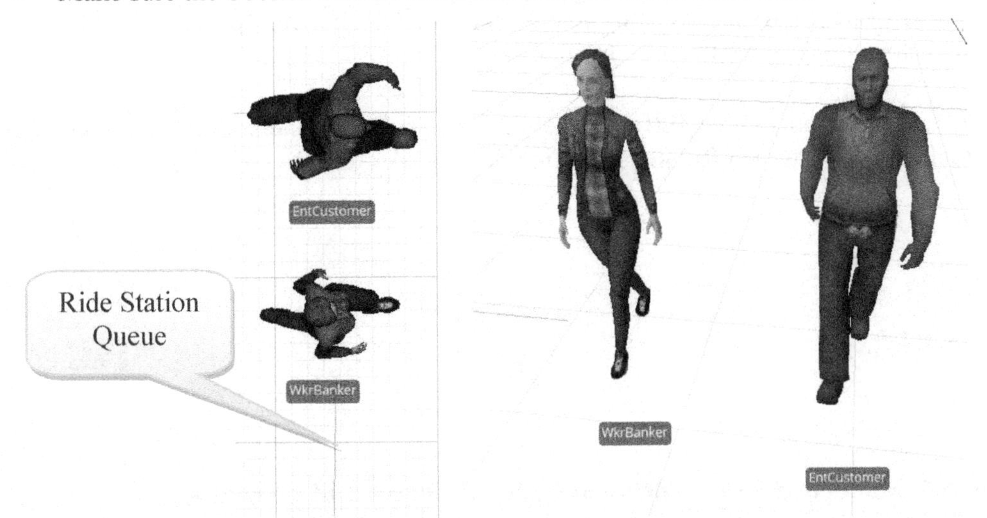

Figure 15.2: Two and Three-Dimensional Versions of a Banker and Customer

Step 4: Save and run the model, making sure the entities are flowing correctly.

Question 1: Does it generate random people symbols arriving to the bank which are physically walking and what is the banker currently doing?

____

Step 5: In the next step, the **WkrBanker** should pick up the customers from the waiting room and deliver them to the room. Change the outgoing transfer node **(Output@SrcWaitingRoom)** at the

[226] One can also use a Vehicle as a transporter and resource.

[227] Symbols in the Library can be selected based on "Domain", "Type", and "Action".

[228] By holding down the *Shift* key while in 3-D mode will allow you to move the queue up and down. Once you have it on the floor, switch back to the 2-D view to position it beside the banker.

SrcWaitingRoom to require a transporter as seen in Figure 15.3. Change the *Ride On Transporter* property to "True" and choose a specific *Transporter Type* and specify the **WkrBanker** as the vehicle requested.

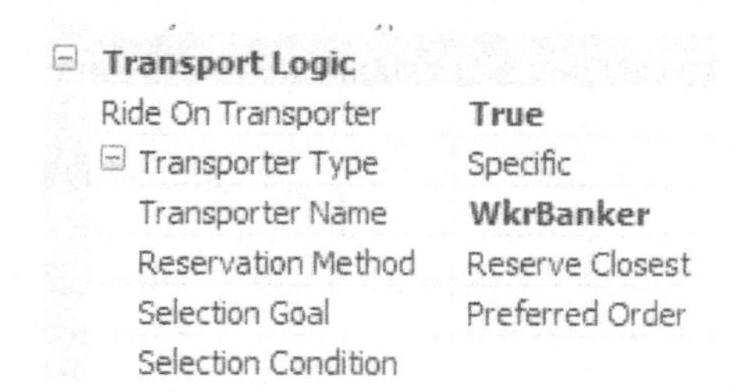

Figure 15.3: Changing the Node Logic to Request a Transporter

Step 6: Make the path between the **SrcWaitingRoom** and the **SrvRoom** bidirectional, so the banker can move between the two locations. Generally two unidirectional paths are better to prevent blockage.

Step 7: Save and run the model making sure the entities are moved by the **WkrBanker** to the room.

Question 2: What did you observe (i.e., does the **WkrBanker** stay with the **EntCustomer** while they are being processed)?

____

The **WkrBanker** is currently operating just as a VEHICLE which transports the customers from the waiting room to the room. It does not necessarily stay with the customer while it is being processed, but continues back to pick up the next customer and move them to the room.

Step 8: Insert a new TRANSFERNODE named **TRoom**[229] above the **SrvRoom** and connect it via a one meter bidirectional path to the input of the **SrvRoom** as seen in Figure 15.4.[230]

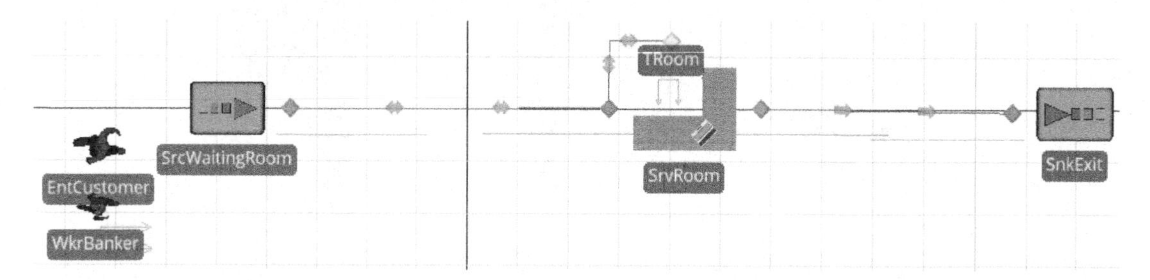

Figure 15.4: Model that Uses Vehicles as both a Resource and a Transporter

Step 9: In the *Before Processing* add-on process trigger of the **SrvRoom** SERVER, use a `Seize` step that will seize the **WkrBanker** and request a visit to the **TRoom** node (see Figure 15.5 for more information).[231]

[229] This node will be used as the node to visit in the `Seize` block. Note, the node that a worker or vehicle is dropping off entities cannot be the same node they need to visit. If the same node is specified, the model will result in a deadlock situation with the `Worker/Vehicle` stuck at the node and the `Entity` frozen in the **Inputbuffer** and not able to proceed into processing.

[230] This bidirectional path will not consume time because its speed is "Infinity".

[231] Allocation as a `Resource` (i.e., responding to seizures) takes priority over handling ride requests from entities. If there is an entity that has requested a ride and one that has requested its service as resource, the vehicle will handle the resource request first.

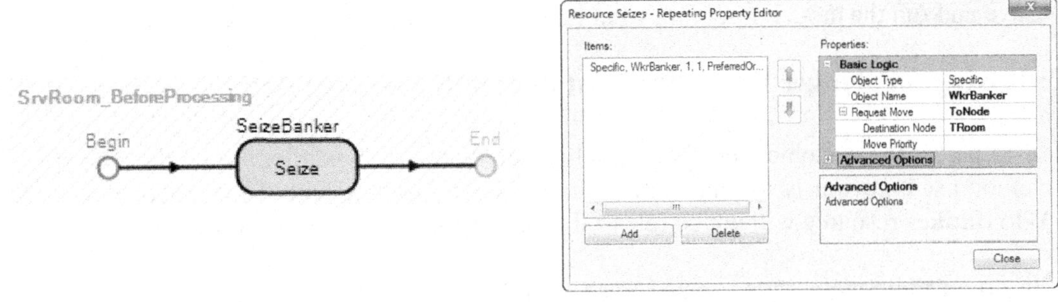

Figure 15.5: Requesting a Banker for a Customer to be seen at the TRoom Node.

Step 10: In the *"After Processing"* add-on process trigger of the **SrvRoom** SERVER, use a `Release` step that releases the **WkrBanker** once it has seen the customer.[232]

Step 11: Save and run the model observing what happens.

Question 3: Did the **WkrBanker** transport the customer to the room and then move into place to service the entity before heading back to get the next customer?

____

Step 12: When the banker is seized by the customer at the office, it is automatically shown in the parking queue of the **TRoom** TRANSFERNODE and the customer is shown automatically in the processing station contents queue above the server **SrvRoom**. We will improve the animation to have the customer and banker face each other across a desk as seen in Figure 15.6.

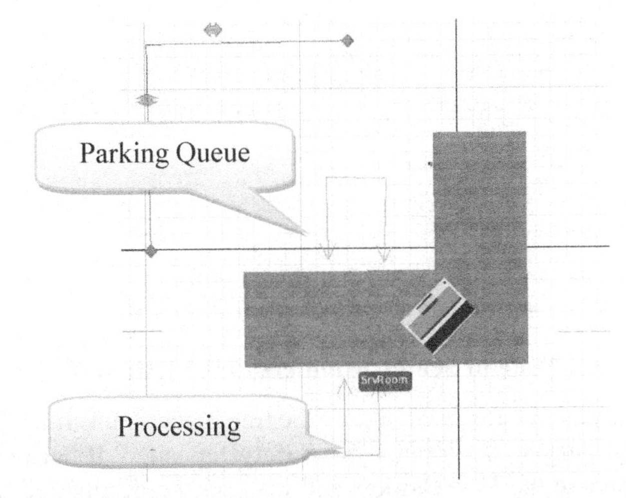

Figure 15.6: Modifying the Animation of the Banker's Office

- Select the **SrvRoom** server and the change the symbol to a desk by selecting the *"Desk1"* symbol in the *Library/Furniture* section.
- Move the processing contents queue below the desk and change its alignment to *Oriented Point*. Make sure the oriented points are facing the desk as seen in the figure.
- Next select the **TRoom** transfer node and turn off the automatic parking queue by toggling the *Parking Queue* button in the *Appearance→Attached Animation* section. From the *Draw Queue* dropdown, insert a *Parking Station.Contents* with *Oriented Point* alignment.[233]

---

[232] We could have used the Secondary Resources section *Resource for Processing* that will seize and release the worker. However, in the next sections we will make modifications that cannot be done with Secondary Resources.

[233] You will need to orient the desk by selecting the desk. While holding down the *Ctrl* key, grab one of the corners and rotate the desk around to the correct orientation.

Step 13: Save and run the model observing what happens when they arrive at the office.

Part 15.2: Having the Banker Escort the Customer to the Deposit Box

In this bank, 45 percent of customers need to visit their safety deposit box after seeing the banker.[234] A similar modeling method used previously to route patients to an X-ray machine with the assistance of the Vet will be used. The **WkrBanker** resource will not be released if the customer needs to visit the safety box.

Step 1: Insert an additional SERVER named **SrvSafetyDeposit**.

- Set the capacity of the SERVER to one since there is only one vault area.
- The processing time for a customer to retrieve their safety deposit box and perform any actions follows a Pert distribution with a min, most likely, and maximum values of 10, 15, and 25 minutes respectively.
- Connect the output of the **SrvRoom** to the input of the **SrvSafetyDeposit** via a five meter PATH.
- Connect the output of the **SrvSafetyDeposit** to the **SnkExit** via a 15 meter logical PATH.

Step 2: Modify the **SrvRoom_AfterProcessing** add-on process trigger to model the banker behavior when the customer needs to be escorted to the safety deposit box 45% of the time. From the *Definitions* tab, add a new REAL PROPERTY named **GProbSafetyDeposit** and set the default to 0.45. Insert a `Decide` step (as seen in Figure 15.7) that is probabilistic based with `GProbSafetyDeposit` as the expression.

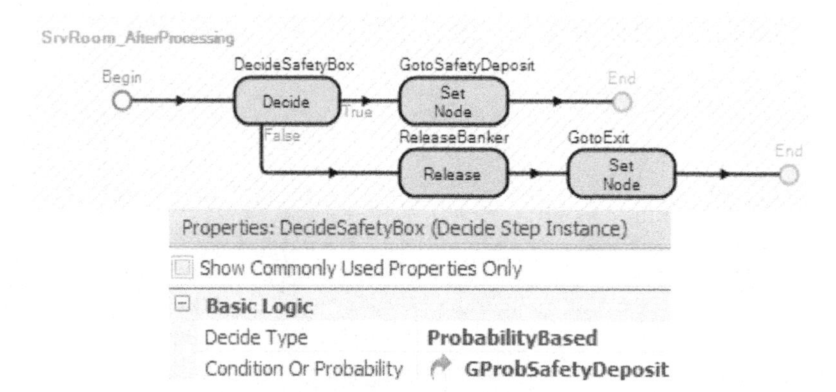

Figure 15.7: Using Set Node to Send Customers to the Exit or Vault Machine

Step 3: If a visit to the safety deposit box is needed (i.e., the true branch), insert a `Set Node` step that will set the destination node of the entity to the input at **SrvSafetyDeposit**.[235] If not visiting the safety deposit box (i.e., the false branch), release the **WkrBanker** and then set the destination node to the input at the **SnkExit**.[236] See Figure 15.8 for the property details of the two Set Node process steps. When the entity enters the TRANSFERNODE of the **SrvRoom** where the *Entity Destination Type* has been set to continue, the destination node that was set using the `Set Node` step will be used to route the entities.

---

[234] The percentage is deliberately high to determine if the model is behaving correctly.

[235] The `Set Node` step is identical to setting the *Entity Destination* property of a `TransferNode` (i.e., "Continue", "Specific", "By Sequence", or "Select from List"). The *Destination Type* can be a "Specific Node", choose the node in the next sequence (i.e., "By Sequence") or a complicated expression.

[236] In the previous chapter, the `SeizedResources` functions were used in the link weights to route the entities to either the exit or the X-ray machine.

Properties: GotoSafetyDeposit (SetNode Step Instance)	
Show Commonly Used Properties Only	
Basic Logic	
Destination Type	Specific
Node Name	**Input@SrvSafetyDeposit**

Properties: GotoExit (SetNode Step Instance)	
Show Commonly Used Properties Only	
Basic Logic	
Destination Type	Specific
Node Name	**Input@SnkExit**

Figure 15.8: Property Settings for the Set Node Property to Send the Entity to the Vault or the Exit

Step 4: Insert the *"After_Processing"* add-on process trigger for the **SrvSafetyDeposit** that will release the **WkrBanker** once the service is completed as seen in Figure 15.9.[237]

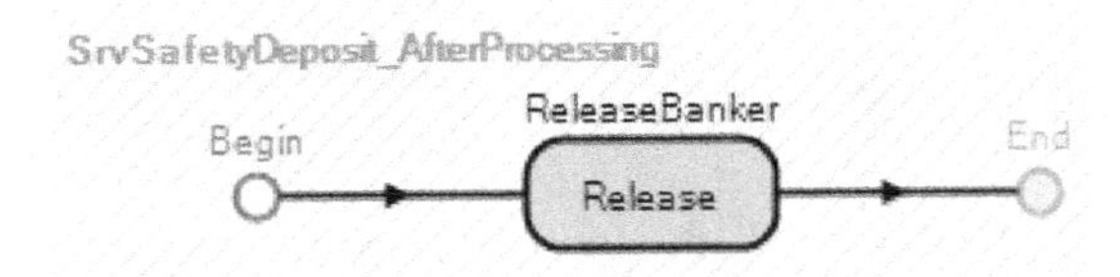

Figure 15.9: Releasing the Banker after the Customer is finished at the Deposit Box

Step 5: Save and run the model.

Question 4: What is the advantage of defining a property instead of specifying 0.45 directly in the `Decide`?

____

Question 5: Do some of the customers flow to the safety deposit vault server?

____

Question 6: Does the banker resource wait until the customer is completely done visiting the vault before heading back to see another customer?

____

Question 7: If the previous question is true, where does the Banker wait while the customer is visiting?

____

Step 6: The **WkrBanker** does not move with (i.e., escort) the customer from the room to the safety deposit machine. To facilitate the moving of the **WkrBanker** to the **SrvSafetyDeposit**, paths have to be added to allow the **WkrBanker** to move to the vault room as seen Figure 15.10.

- Insert a new TRANSFERNODE named **TSafetyDeposit** near the **SrvSafetyDeposit** which will be the node the **WkrBanker** will transfer to for processing the customer at the safety deposit machine.
- Insert a one meter path from the **TRoom** transfer node to the output node of the **SrvRoom** so the **WkrBanker** can use the same path to escort the customer to the safety deposit box.
- Connect the input of the **SrvSafetyDeposit** to the **TSafetyDeposit** transfer node via a one meter PATH and then connect the **TSafetyDeposit** node back to the output of the **SrcWaitingRoom** via a 15 meter PATH which allows the **WkrBanker** to travel back to the waiting room to get the next customer.

---

[237] Just copy the `Release` step from the **SrvRoom_AfterProcessing** to the **SrvSafetyDeposit_ AfterProcessing.**

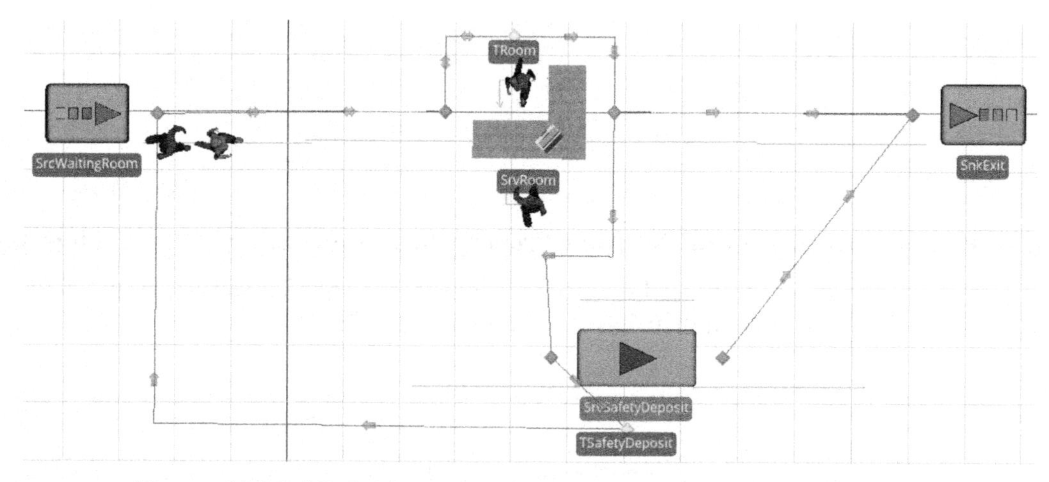

Figure 15.10: Model for Transferring Customers and the Banker

If the customer needs to see his safety deposit box, the **WkrBanker** needs to travel to accompany them to the room. The process logic in the *After_Processing* add-on trigger in Figure 15.7 moves the entities to either the safety box or the exit based on a probabilistic *Decide* branch using a *Set Node* process step. When this process trigger runs, the TOKEN that is executing the process steps is associated with the entity. Therefore, any steps (i.e., Transferring, Setting Nodes, etc.) are only applied to the entity[238] associated with the TOKEN.

Step 7: Insert a *Move* step which will request a move of one or more moveable resources that have been seized by the MODELENTITY (i.e., the associated object) as seen in Figure 15.11.

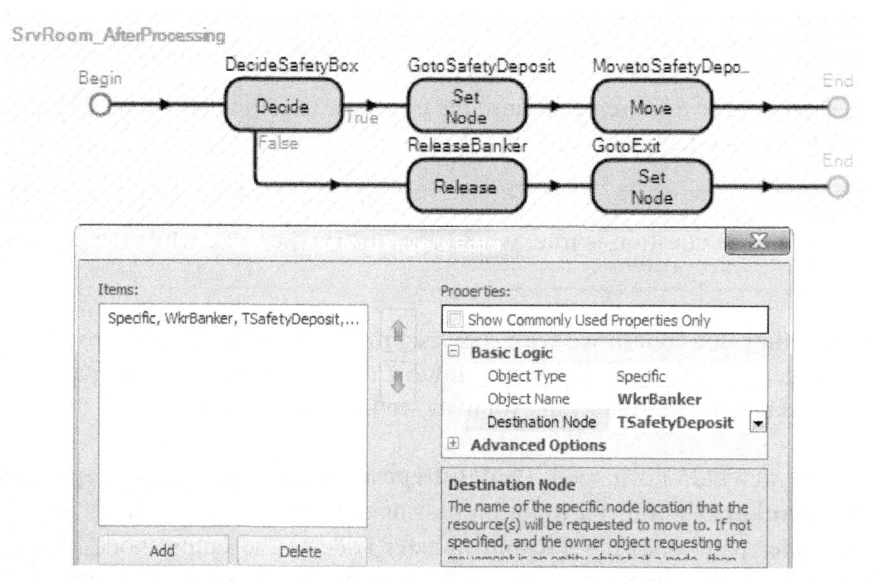

Figure 15.11: Using the *Move* Step to cause the Banker to Move to the Safety Deposit Box

Step 8: Save and run the model observing what happens with the banker and customer.

Question 8: Does the banker now move to the safety deposit box with the customer?

____

Step 9: It may appear to move with the customer based on the speed of the animation. However, the banker will actually move first to the **TSafetyBox** node before the customer does. To see this phenomenon, place a break point on the path connecting the **TRoom** and **Output@SrvRoom** nodes. Fast forward the simulation till it stops and then slow down the animation speed under the "*Run*" tab.

[238] In some process steps, the action can be applied to the parent object of the Token's associated object (e.g., the Server the logic resides).

Question 9: Does the banker move first now?

____

Step 10: When the customer (i.e., MODELENTITY) requests a dynamic resource to be moved using the *Move* step, the token associated with the entity will wait until the movement occurs before proceeding forward (i.e., move itself). Therefore, we would like the banker and customer to move together. To manipulate the banker, it will be necessary to identify the banker and then influence it. To accomplish that in this model we will employ a *Search* process step. The *Search* process step can search a collection of items and return Tokens associated with the items as specified in Table 15.1 and Table 15.2.

Table 15.1: The Properties of the Search Process Step

Property	Description
Collection Type	Type of the collection to search (See Table 15.2 for more information).
Search Type	The type of search to perform (Forward, Backward, MinimizeReturnValue, and MaximizeReturnValue).[239]
Match Condition	Optional match condition that can be used to filter the items in the collection before the search begins. Don't forget to use the keyword CANDIDATE in the condition.
Search Expression	An expression that is evaluated for each item found by the search. The total sum for this expression is returned in the ReturnValue of the original TOKEN. One could save the particular Resource and Seize the exact same one later.
Starting Index	One based index to start searching. By default it is one for Forward searches while the number in the collection for Backward searches.
Ending Index	One based index to end searching to assist in narrowing the search within collection.
Limit	An expression that specifies the maximum number of items that can be found and returned.
Save Index Found	Specify an optional discrete state variable to save the last item found.

Table 15.2: Different Types of Collections that Can Be Searched

Collection Type	Description
ObjectInstance	Search all objects of a particular class (e.g., search all the entities that are currently created)
EntityPopulation	Search a particular entity population (i.e., MODELENTITY, WORKER, or TRANSPORTERS)
ObjectList	Search the objects in a list (e.g., ResourceList, etc.)
NodeList	Search all nodes in a particular list.
TransporterList	Search all transporters in a particular list.
SeizedResources	Search the objects currently seized by the Parent Object, Associated Object or a Specific Object
QueueState	Search the objects currently in one of the queues
TableRows	Search the rows in a Data table
NetworkLinks	Search all links of a particular network.
NodeInboundLinks	Search all the inbound links of a particular node.
NodeOutboundLinks	Search all the out bound links of a particular node.

Step 11: If the customer needs to be escorted to the safety deposit box, insert a *Search* Step after the *Set Node* as seen Figure 15.12. The "Original" branch is associated with the original TOKEN (i.e., customer) while the "Found" branch will be invoked for a Token associated with each item found.

---

[239] **MinimizeReturnValue** and **MaximizeReturnValue** search values will search the collection and return up to the limit of items specified that minimizes or maximizes the Return Value expression property.

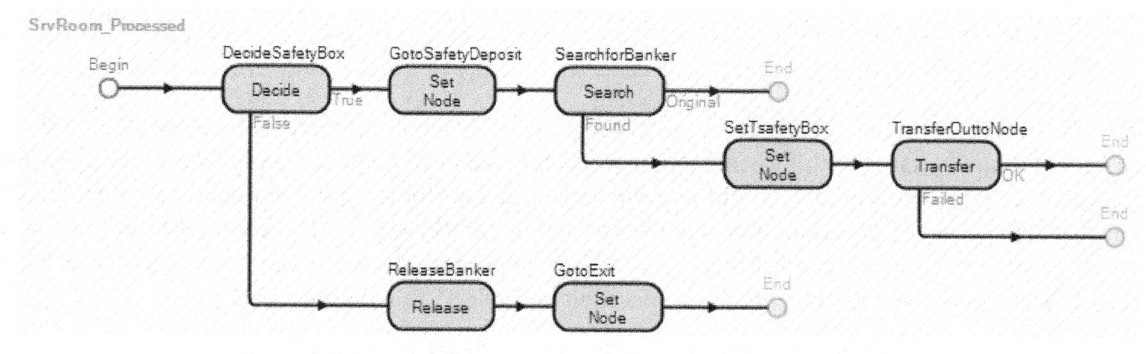

Figure 15.12: Searching for and Transferring the Banker

- Set the *Collection Type* property to `SeizedResources` and make sure the *Owner Type* is "AssociatedObject" which will search all the objects that are seized by the customer.
- Set the *Match Condition* to `Candidate.Object.Is.WkrBanker` to only search for the banker that has been seized.[240] This condition will insure that only the seized banker is.
- Set the *Search Expression* property to return the "ID" of the WORKER seized, which can be used for a further enhancement.

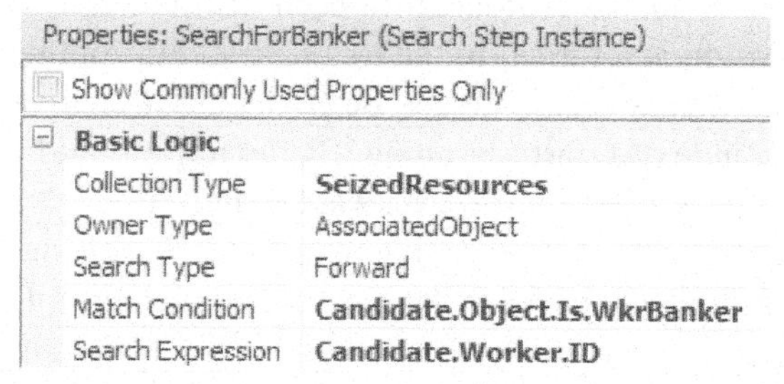

Figure 15.13: Properties of the `Search` Process Step

Step 12: Once the **WkrBanker** has been found, we need to set the node to the **TSafetyBox** and then using a *`Transfer`*[241] process step transfer out of the parking station (i.e., the current stations) and into the **TRoom** node as seen in Figure 15.12 with the properties as in Figure 15.14.[242]

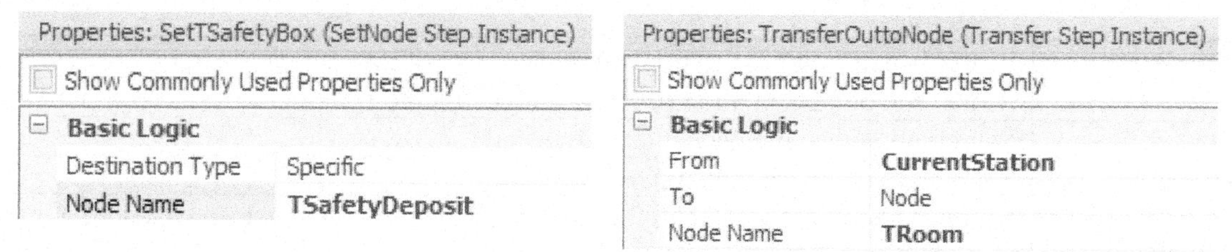

Figure 15.14: Transfer Process Step Properties to Move Banker to the Vault

Step 13: Save and run the model.

Question 10: Does the banker now move with the customer when it needs to visit the safety deposit box?

____

Question 11: Does it behave the way you envisioned?

____

[240] The customer entity has also seized one unit of capacity from the `Server` and we do not want to return the `Server` object.

[241] The *`Transfer`* step can be used to transfer dynamic objects (`Agents`, `Entities`, `Vehicles`, etc.) from Current Node, Current Station, Free Space, or Link to a specific Node, another station, Parents External Node, or outbound link.

[242] If you transfer it directly to the **TSafetyBox** it will beam the banker there like Star Trek and not route it as this will take it out of the station and place in the **Troom** node which then cause the banker to move to the **TSafetyBox** node.

Part 15.3: Using the Transport Function of the Worker

The difficulty of the previous model is that the customer and the banker worker moved independently of one another. Instead, the banker should pick up the customer (i.e., act as a vehicle) and then transport the customer to the safety deposit vault where the customer would then `Seize` it again similar to moving the customer to the banker's office. Here the banker will both serve the customer and provide transportation.

Step 1: Save the current project.

Step 2: For the **SrvSafetyDeposit** SERVER, insert a *Before_Processing* add-on process trigger with a `Seize` step (or you can use the *Secondary Resources)* to seize the **WkrBanker** once the customer is dropped off. Use the `Seize` step in the *Before_Processing* that will seize the **WkrBanker** and request a move to the **TSafetyDeposit** node as seen in Figure 15.15.

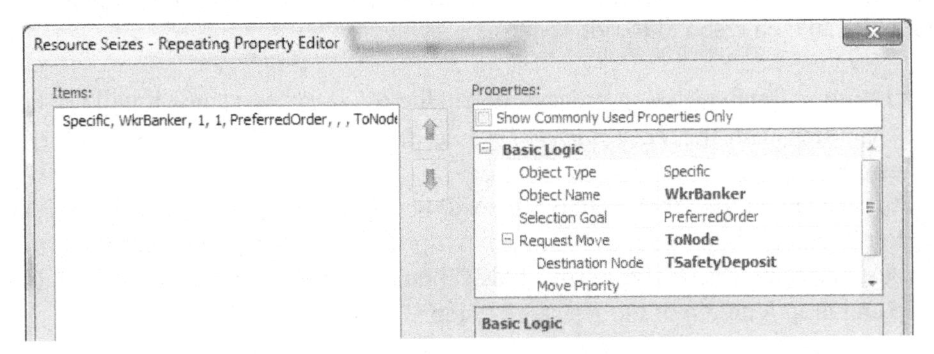

Figure 15.15: Seize the Banker at the Safety Deposit Vault

Step 3: However, only certain entities need to ride on a worker/vehicle (i.e., the ones heading to the safety deposit box) while the other customers will walk to the exit. Therefore, the output node at the **SrvRoom** cannot be specified to "Ride on Transporter" like the output node for the SOURCE since this would force all customers to ride. A `Ride` process step can be used to force an "Entity to Ride" on a vehicle. However, the `Ride` process step can only be utilized in a process trigger inside a `TransferNode` (e.g., **Output@SrvRoom**). Therefore, all of the logic used to determine where the entity will proceed after the processing will need to be removed.

- Remove the *After Processing* add-on process trigger from the **SrvRoom** by choosing "null" from the drop down[243] or using "reset" on the right-click, since this logic will be needed in a different object's add-on triggers as seen in Figure 15.16.

Figure 15.16: Removing an Add-on Process Trigger from an Object

- Select the output TRANSFERNODE for the **SrvRoom** and then choose the **SrvRoom_AfterProcessing**[244] process as the *"Entered"* add-on process trigger as seen in Figure 15.17. The *Entered* add-on process will be fired when an entity, worker, or vehicle enters the TRANSFERNODE.

---

[243] You can also remove the process trigger by selecting it and then deleting it. Removing the process trigger only removes the process from being invoked and does physically remove it

[244] The **SrvRoom_Processed** is just the name of the process and can be easily changed. You could have created a new process trigger and then copied all the logic into the new process.

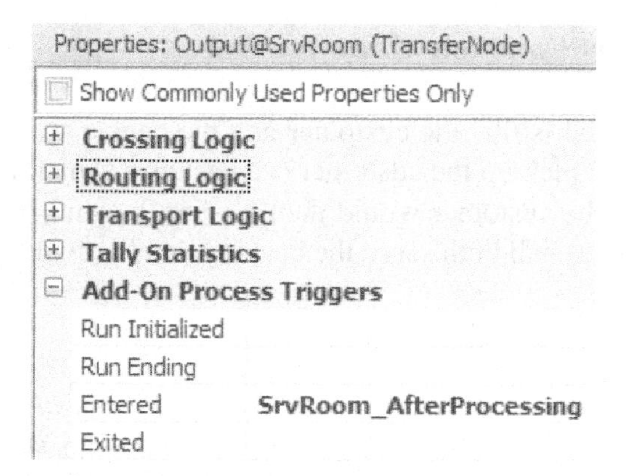

Figure 15.17: Specifying the Entered Add-On Process Trigger

Step 4: If the customer needs to visit the vault, the customer should request a ride and then release the **WkrBanker** so it can then respond to ride requests.

- In the **SrvRoom_AfterProcessing** process, delete the `Search` step since it will not be needed.
- Insert a `Ride` step with the *Transporter Type* set to "Specific" and the specific *Transporter Name* specified as a **WkrBanker** (see Figure 15.18).
- If the entity needs to see their safety deposit box and the customer has requested a ride, the **WkrBanker** needs to be released so it can pick up the customer to transport it to the safety deposit box. To reuse the same `Release` in the "False" branch, drag the end of the branch (-○) associated the `Ride` step and drop it on top of the `Release` step.[245]

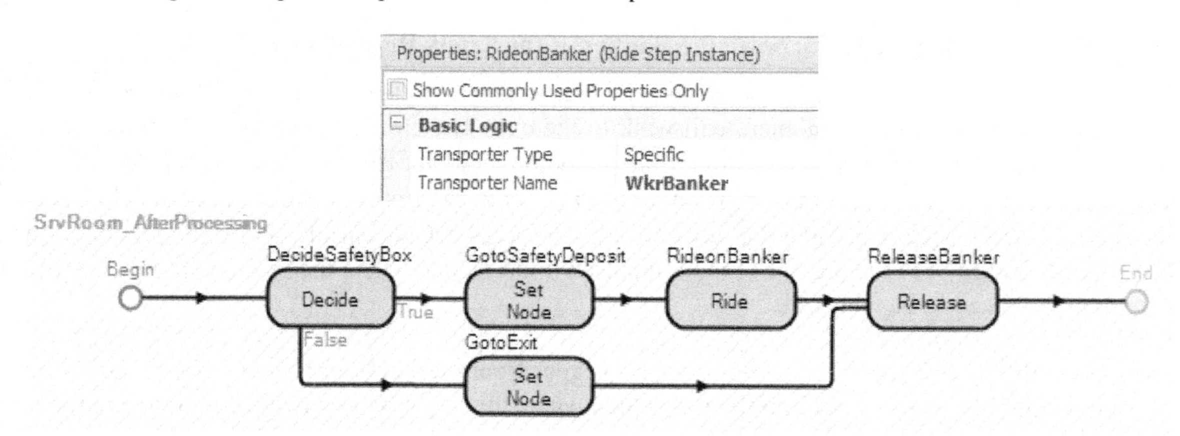

Figure 15.18: Adding the Ride Step and then Releasing the Banker

Step 5: Save and run the model.

Question 12: What do you observe when the customer needs to visit safety box now?

____

Step 6: The banker is stalled at the desk while the customer waits for a ride to the safety deposit box. More specifically the TOKEN associated with the entity in the process is waiting at the `Ride` step owing to the *Token Wait Action* property (within the *Advanced Options*) which specifies, by default, to "*WaitUntilTranferredEvent*" (i.e., until the entity has finished being transferred into the vehicle). Now change the *Token Wait Action* property to "None (Continue)" as shown in Figure 15.19.

---

[245] You could have copied and pasted another `Release` step after the `Ride` instead

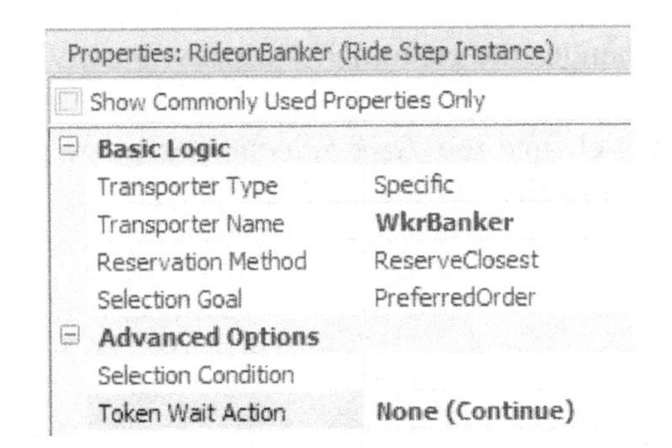

Figure 15.19: Changing the Token Wait Action

Step 7: Save and run the model.

Question 13: What do you observe when the customer needs to visit the safety deposit box and the **WkrBanker** moves to the output node of the **SrvRoom**?

____

Question 14: Does it behave the way you envisioned?

____

Step 8: When the **WkrBanker,** which is a VEHICLE, reaches the output node of **SrvRoom** to pick up the customer, it actually enters the node thus running the same logic in Figure 15.18 which tells it to exit or get picked up by a banker. This then creates the system deadlock. Only customers should run the logic.

Step 9: Insert another `Decide` step (see Figure 15.20) before the current `Decide` step and that checks to see if the object associated with the Token is a MODELENTITY. The `Is` operator will check to see if the object is a particular object which in this case `Is.ModelEntity` will return "True" if it is a customer and "False" if it is Banker (i.e., WORKER/VEHICLE) as seen in Figure 15.20.

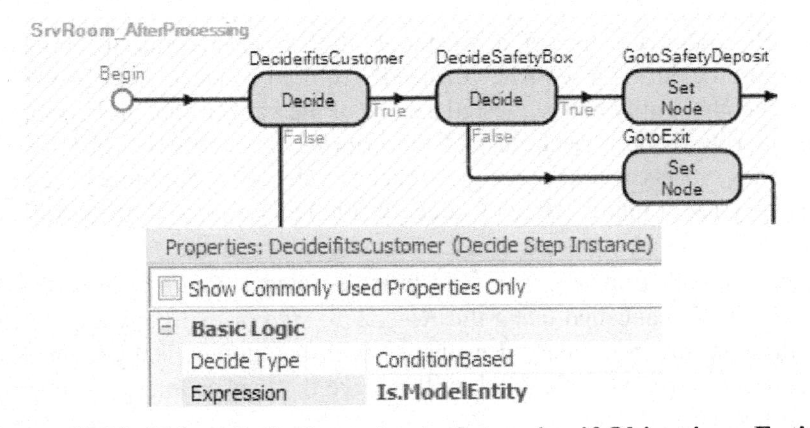

Figure 15.20: Using the Is Operator to determine if Object is an Entity

Step 10: Save and run the model.

Question 15: Does the model behave the way you envisioned?

____

Step 11: The problem is the **WkrBanker** responds to ride requests in the order they arrive. The customers going to the safety deposit machine should have priority.

- Add an `Assign` step that changes the customer's priority (`ModelEntity.Priority`) to two (see Figure 15.21).
- Select **WkrBanker** and then change the *Task Selection Strategy* property under the *Transport Logic* properties to be the "Largest Priority" as seen in Figure 15.22.[246]

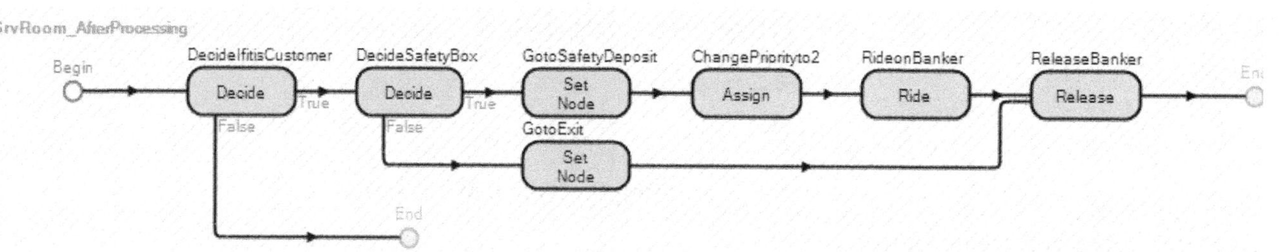

Figure 15.21: Using the Entered Add-On Trigger to Specify a Ride on Vehicle

⊟ **Transport Logic**	
Initial Ride Capacity	1
Task Selection Strategy	**Largest Priority**
⊞ Load Time	0.0
⊞ Unload Time	0.0
Park to Load/Unload	False
Minimum Dwell Time Type	No Requirement

Figure 15.22: Changing the Task Selection Strategy of a Vehicle

Step 12: Save and run the model.

Question 16: Does the model work as intended now?

Part 15.4: Resource Reservations

The previous section illustrated the use of a worker both for service and for transportation. However continued use of this resource during a sequence of needs required a careful set of seizes, rides, and releases. You should note that if another entity had requested a seize of the banker, that entity request would be handled before responding to the ride request, regardless of the priority. In these cases, it is easier to use a resource reservation. An entity may "reserve" a resource or worker or vehicle for continued use at different stages. By reserving it, the entity may use it immediately or later. While being reserved, the resource cannot be used by another entity. A resource may be reserved at a TRANSFERNODE or in the "*Resource for Processing*" in the "*Secondary Resources*" section of an object or in a process using a `Reserve` step or the `Release` step. Reservations may be implicitly cancelled using the `Release` step or explicitly using the `Unreserve` step. Reservations made through the Secondary Resources will be automatically cancelled when service is complete. So each customer arriving to the bank will reserve the banker for use throughout the process until the banker is no longer needed.

Step 1: At the **Output@SrcWaitingRoom**, modify the "Transport Logic" to cause the transporter to be reserved as shown in Figure 15.23. Adding this specification causes the banker to be reserved for this customer and unavailable to any other customer for any type of service until the reservation is cancelled.[247]

---

[246] Changing the *Task Selection Strategy* reorders the Rider Pickup Queue of the Vehicle.

[247] The act of cancelling a reservation is called unreserving.

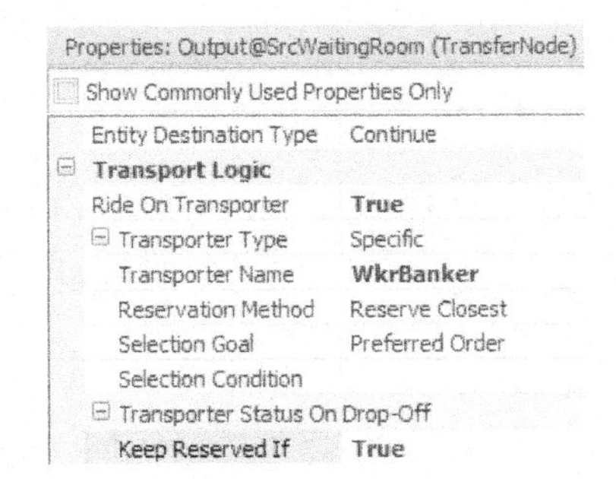

Figure 15.23: Creating a Reservation at Transport

Step 2: The customer requests service from the reserved banker at the office. After the banker serves at **SrvRoom**, the customer will either release the banker and exit or keep the banker to be escorted to the safety deposit. Change the **SrvRoom_AfterProcessing** to have a separate `Release` as shown in Figure 15.24 and the `Assign` step is no longer needed as we will continue reserving the **WkrBanker**.

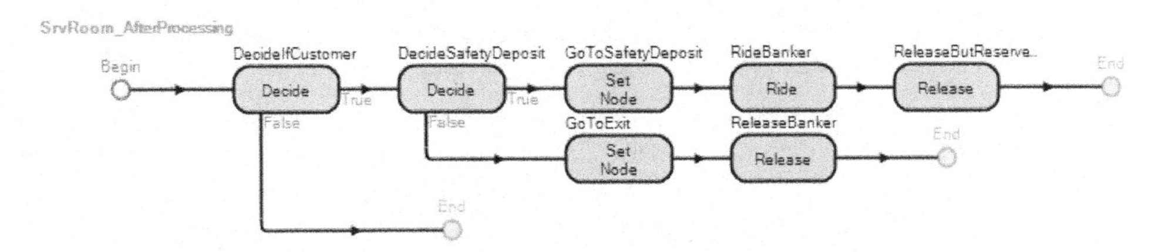

Figure 15.24: Modifying the Process

Step 3: When the customer exits, the banker is released. Unless the `Release` step specifies keeping the reservation, the reservation is automatically cancelled. However, those customers going to the safety deposit box keeping the reservation in the `Release` step in the process going to the safety deposit box, as shown in Figure 15.25. Note the *Reservation Timeout* expression property allows you specify logic that will cancel the reservation if the entity still has reserved the object when the time expires.

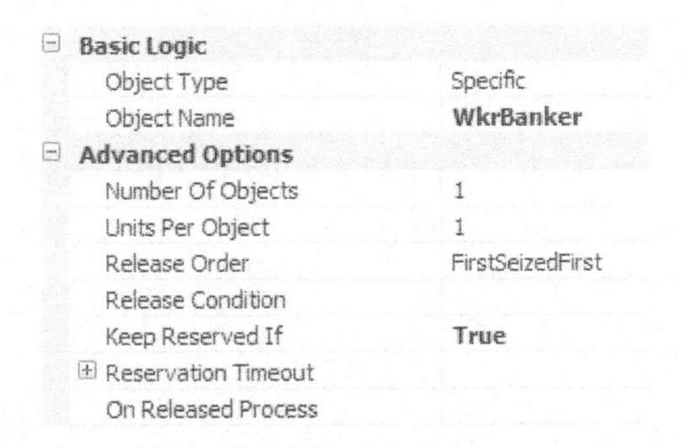

Figure 15.25: Keeping Reservation on Release

Step 4: Rerun the model, looking closely at the behavior of the banker.

Question 17: Does the model work in a similar fashion as before?

____

Part 15.5: Animated Entities

The "animated people" that have been added to SIMIO can enhance the visual appeal of a simulation. Each animated people symbols (e.g., female, male, children, elderly, soldier, and cartoon people) has a list of animation options. You can see the options by right-clicking an animated *ModelEntity/Worker/Vehicle* and select "*List Animations of Active Symbol*". Table 15.3 shows the partial lists for male and female animated people. You can invoke an animation by its "Index" or its name "Name". Unfortunately the animation lists are not identical across people symbol groups as can be seen in the partial list as male animated people can perform 41 different animations while female animated people can perform 26 animations and not the same ones as the males.

Table 15.3: Partial List of Animated Symbols for Male and Female Animated People

Male		Female	
Index	**Name**	**Index**	**Name**
1	Walk	1	Walk
2	Run	2	Run
3	Jump	3	Jump
4	Attack_Kick	4	Scared
⋮		⋮	
18	Walk_Stealth	18	Sit_Chair_Loop
⋮		⋮	
41	Sit_Chair_Loop	26	Walk_Push

Step 1: To illustrate the use of animated entities, let's reconsider the model from the first section of this chapter. Add two chairs at the desk, one for the banker and one for the customer, using the *Drawing → Place Symbol,* as shown in Figure 15.26. You may need to adjust the placements of the chairs, the parking station queue and the processing station queue as you see the animation perform.

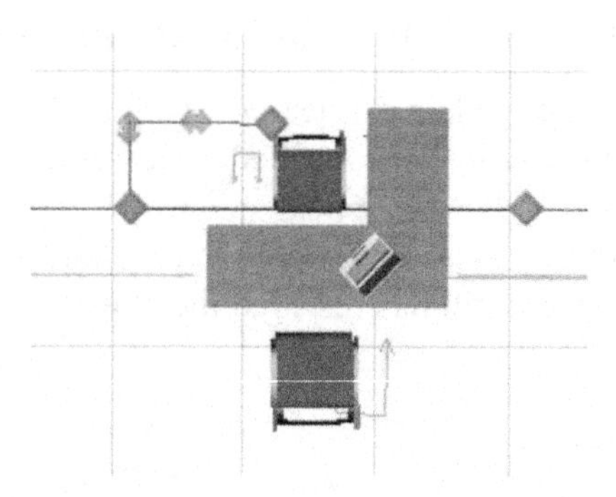

Figure 15.26: Add Chairs around Desk

Step 2: One of the MODELENTITY state variables that are defined by SIMIO in *Definitions→ States* is "Animation." This "string" state variable can be used to hold current entity animation and it can be used to assign a new animated behavior. Specify the "Animation" properties section of the **EntCustomer** as shown in Figure 15.27.

Animation	
Current Symbol Index	ModelEntity.Picture
Random Symbol	**True**
Current Animation Index	ModelEntity.Animation
Default Animation Action	**Moving**
Dynamic Label Text	

Figure 15.27: Animation Properties of the Customer

Step 3: WORKERS and VEHICLES do not have individual state variables like MODELENTITIES. So we need to define a new Model DISCRETE STRING STATE variable named **GStringStateWorker** for the banker which can be changed. So now specify the "Animation" properties of the **WkrBanker** as shown in Figure 15.28.

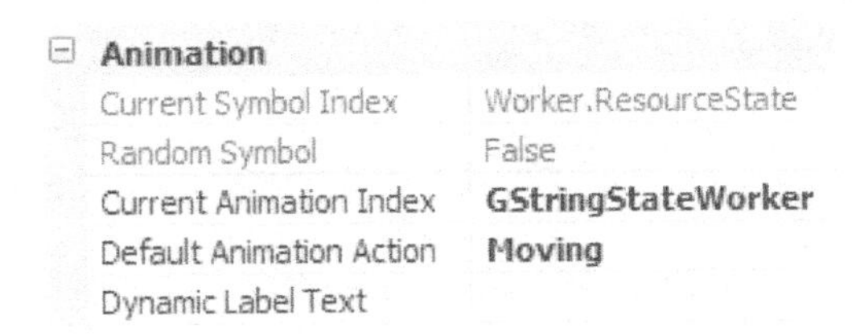

Figure 15.28: Animation Properties of the Banker

Step 4: The customer should walk to the banker's desk and then sit in the chair. So in the *Before Processing* add-on process trigger of the **SrvRoom** SERVER, add the assignment shown in Figure 15.29.

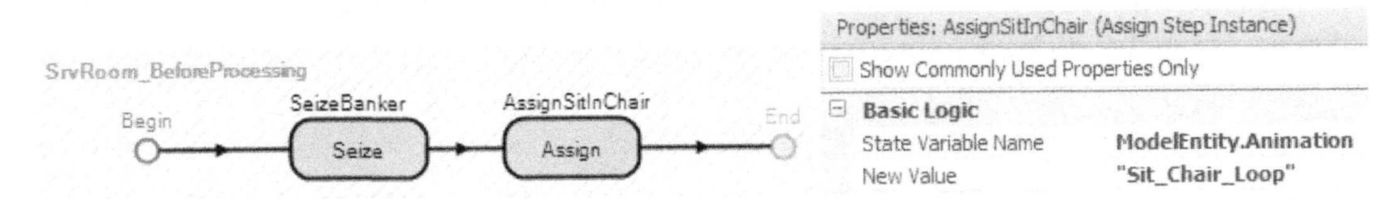

Figure 15.29: Customer Sits at Banker's Desk

Step 5: Then the customer should walk to the exit or the safety deposit room. So in the In the *After Processing* add-on process trigger of the **SrvRoom** SERVER, add the assignment shown in Figure 15.30.

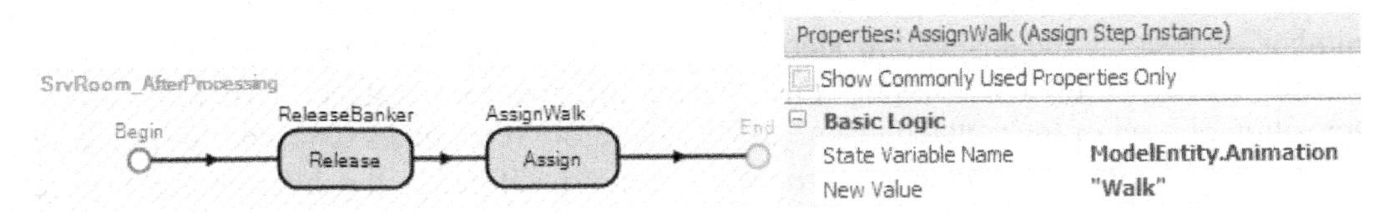

Figure 15.30: Customer Walks to Exit

Step 6: The banker needs to sit at the desk. So in the in the *"Entered"* add-on process trigger of the **TRoom** TRANSFERNODE, add the assignment shown in Figure 15.31.

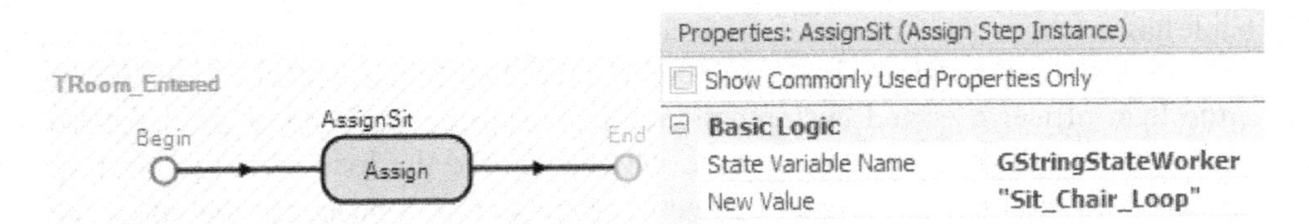

Figure 15.31: Banker Sits

Step 7: The banker needs to walk again when they leave the desk to pick up the next customer or take the current customer to the safety deposit box. So in the *Exited* add-on process trigger, add the `Decide` and `Assign` step as shown in Figure 15.32. Recall we specified that the **WrkBanker** will park when they are seized. When the WORKER object goes into the parking station, it exits the node and therefore would run the *Exited* add-on process trigger forcing the banker to walk after they were told to sit. The `IsParked` function returns true once the worker enters parking station before exiting the node and therefore will not set the variable to "Walk."[248]

[248] The worker will enter the node when they leave the parking station as well executing the "*Entered*" add-on process trigger. However, we do not need to add a `Decide` in the entered process because the *Exited* runs after the *Entered* making them walk.

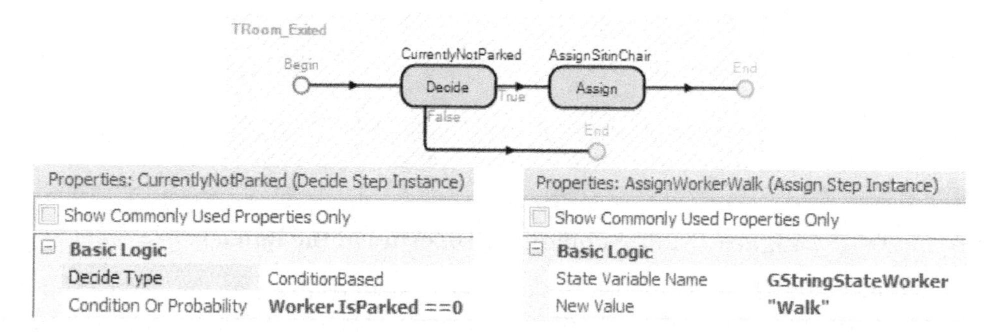

Figure 15.32: Banker Walks

Step 8: Save and run the model, looking carefully at the animation[249].

Question 18: Are the animations working as expected?

Part 15.6: Detailed Service: Tasks and Task Sequences

A service activity may be composed of several tasks rather than a simple processing time. For example, the kitting of an electronic product may employ a sequence of tasks to complete the kit. Or an inspection may require a number of tests at the inspection station. Often the detail in these types of operations can be conveniently summarized by their processing time. There are, however, instances when the processing time needs a composition of a set of interrelated tasks. SIMIO therefore recognizes two types of processing stations within a SERVER, which are referred to as *Process Types*. Specifically, the alternative specifications are "*Specific Time*" or "*Task Sequence*". Currently, we have only modeled using the *Specific Time* which is the default value. When more detail about a process needs to be included in the model, the model was extended with more objects (i.e., additional servers were inserted). Using a *Task Sequence* process type, detail about processing can be incorporated into a set of task sequences within a single SERVER.

Currently, the bank has a loan officer that assists customers with various loans, account information, special deposits, and safety deposit box usage. A single processing time distribution was used to describe that service. In the example, the loan officer escorted customers back to the office for general service. A portion of the customers required a visit to their safety deposit box, which required the loan office to accompany them. In that example we employed the WORKER for service and for transportation broke up the tasks into two separate servers and flow. We want to re-consider the bank service and employ a different modeling approach to incorporating details of the office service. More specifically, we want to recognize that the service being provided by the loan officer is composed of a set of "tasks." Each of the tasks has its own task properties. Some tasks must be performed before others while some tasks may be performed in parallel at the same time.

Step 1: First, change the "*Process Type*" property of the **SrvRoom** to *Task Sequence* as seen in Figure 15.33.

---

[249] Note that the **TRoom** and **Input@SrvRoom** nodes have two-way traffic and so the animation instructions are invoked in both directions. In this case, that is not a problem, but it could be in other instances.

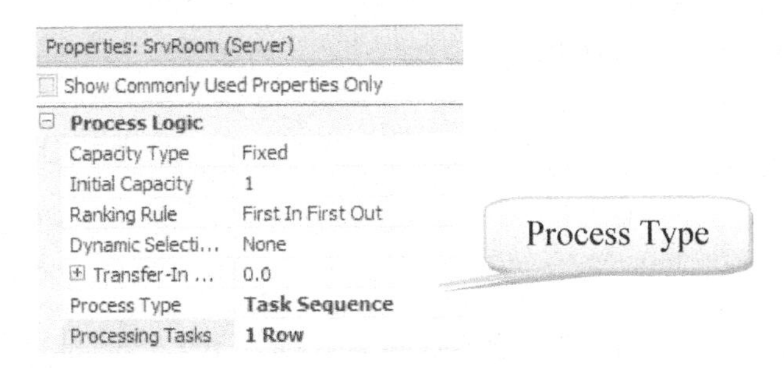

Figure 15.33: Changing Process Type

Step 2: This change reveals a *Processing Tasks* property which is specified through the *REPEATING PROPERTY EDITOR* that allows you to list the tasks. Add a task as shown in Figure 15.34. It is important to note that: (1) the *Processing Time* for the task is the same as before, (2) we added the *Name* of the task, and (3) we specified that this task needs the **WkrBanker** and it should be moved to **TRoom** node.

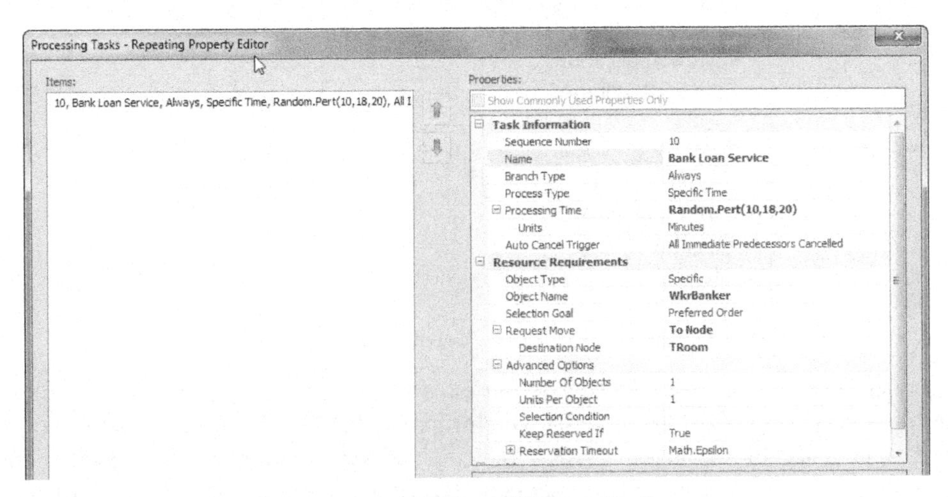

Figure 15.34: Adding a Task

Step 3: By default a *Sequence Number* of 10 is provided. We accepted the *Branch Type* of *Always* and specified the *Process Type* as *Specific Time*. We will modify these specifications in later sections.

Step 4: Since we are specifying the use of the **WkrBanker** in the task, we can remove the **SrvRoom_BeforeProcessing** process and the `Release` steps in the **SrvRoom_AfterProcessing** process. Note that in Figure 15.34 the resource is *Reserved* by default specification of the task. This specification of a *Reservation Timeout* of `Math.Epsilon` insures this resource is held until the end of this time step event so continued use of the resource is given priority.

Question 19: Why is the resource reservation important in this problem? (consider the need to accompany the customer to the safety deposit box)

___

Step 5: Save and execute the model.

Question 20: Is the model executing as it did previously?

___

Question 21: Did the use of a task make it easier to create this model?

___

Part 15.7: Using Task Sequences

The previous example doesn't illustrate the unique task capability especially since we handled it with processes earlier or could have used secondary resources. A task sequence refers to a network of tasks ordered by precedence. For example, you need a nurse to take vitals and information, then you need a doctor, and potentially followed by more nurse visits. This would be harder to handle using one SERVER. Let's consider the banking service example which has more complications. Suppose the Loan Officer service is really composed of three distinct tasks: (1) Task1: a review of the application, which takes Pert(3,5,8) minutes, (2) Task2: a request for records, which only happens in 20% of the cases, and (3) Task3: a final assessment of the loan, which takes Pert(5,7,10) minutes. The request for records doesn't require time from the Loan Officer but is an automatic search that takes a Uniform(2,6) minutes. In other words Task1 and Task2 are done in parallel but only Task1 requires time from the loan officer while Task2 only requires time. Task3 cannot start until both Task1 and Task2 are completed. A task sequence diagram is shown in Figure 15.35.

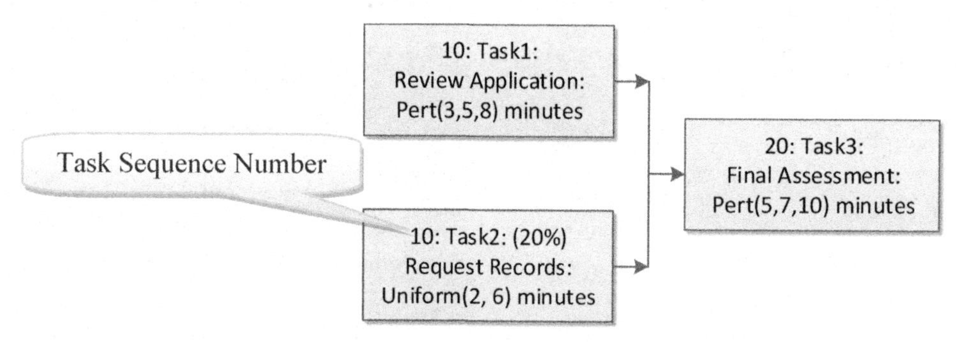

Figure 15.35: Task Sequence Diagram

The general sequence number formation is X.Y.Z. and so forth where X,Y, Z, etc. are whole numbers. One only needs to define the X value (e.g., 10 and 20) which represents the root number of the sequence and helps to determine the precedence of the task. The Y's, Z's, and other numbers, are optional integer suffixes separated by periods that are added to the root sequence to define subsequences (e.g., 10.1 and 20.2). Table 15.4 discusses the rules associated of how the task sequence numbers define the precedence of tasks.

Table 15.4: Rules of Precedence Associated with Task Sequence Numbers

Rule	Description
Rule 1	If the task sequence number has only a root number (e.g., 20), then all tasks which have lower task root numbers regardless of suffixes must precede that task (e.g., 10, 10.1, 10.2, 10.1.1 all precede task with sequence number 20).
Rule 2	If a *task A* has a sequence number consisting of the required root number and one or more suffixes (e.g., 20.1, 20.2, or 20.1.1), then tasks with sequence root numbers that are less, must precede *task A* provided they either have no suffixes (i.e., 10 would precede 20.1 or 20.1.1) or match all the suffixes (e.g., 10.1 precedes 20.1, 10.2 precedes 20.2, 10.1.1 precedes 20.1.1, etc.).
Rule 3	If a *task A* has a sequence number that has a root number followed by at least one suffix (e.g. 10.1, 10.2 or 10.1.1 and 10.1.2), then all tasks with the same root number without any suffixes must precede *task A* (e.g., 10 would precede both 10.1 and 10.2). However, a task with sequence 10.1 does not precede 10.1.1 or 10.1.2 as these will all be considered starting at the same time.
Rule 4	Note, one or more tasks may have the same root sequence number with or without suffixes. For example, *task A* could have sequence number 20 and *task B* could have sequence number 20 which means these two tasks will start in parallel once the preceding tasks have finished. Starting at the same time is true for tasks with the same root sequence numbers followed by any number of suffixes (e.g., 10.1, 10.1, 10.1.1, or 10.1.2 sequence numbers all start at the same time).

In our problem there are two task sequence numbers 10, so they must be done before the task sequence number 20. Identical task sequence numbers are done in parallel as specified by Rule 4 in Table 15.4 (i.e., these tasks begin at the same time) and we will start task associated with sequence number 20 until both sequence number 10 tasks are complete as stated by Rule 1.

Step 1: For the **SrvRoom**, will need to modify the one task that is defined and then add two additional tasks.

- Modify the current task to represent the application review process as seen in Figure 15.36 which requires the **WrkBanker** to be moved to **TRoom** and has a sequence number of 10.

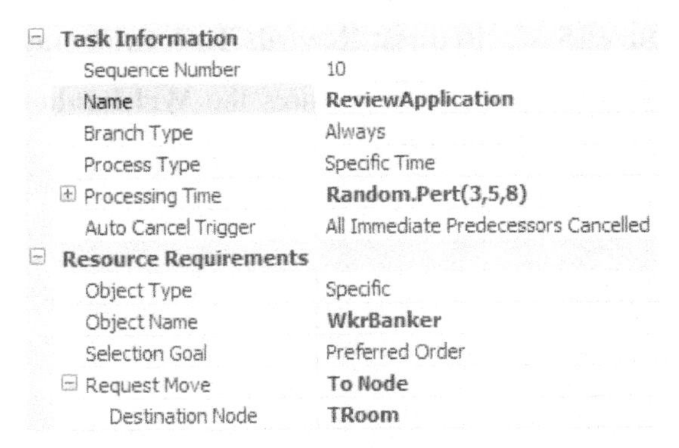

Figure 15.36: Review Application Task Information

- Recall the request records task only occurs 20% of the time as seen in Figure 15.37 and requires no resources in the current scope of the model. Change the *Branch Type* property to "Probabilistic" which is explained in Table 1.5. Again we will use a property to represent the 20% of time a request is needed. Right click on the *Condition or Probability* property and specify to create a new property named **GProbRecordCheck**.

Table 15.5: Task Branch Type in the Task Sequence Definitioin

Branch Type	Description
Always	Branch will be executed all of the time.
Conditional	Branch will conditionally be executed based on the expression independent of any other tasks (i.e., if the expression evaluates to true then the branch is executed otherwise it will be cancelled).
Probabilistic	Branch will be executed probabilistically based on the percentage value stated and is dependent on the other probabilistically branches at the same level. Probabilistic branching is mutual exclusive meaning only one branch or none of the probabilistic branches will be executed. For example, consider a sequence with two tasks (e.g., *task A* and *task B*) each with sequence number 20. If *Branch Type* is set to "Probabilistic" for both *task A* and *task B* with *Condition Or Probability* values 0.25 and 0.4 respectively. Both *task A* and *task B* cannot be executed at the same time. The system will execute *task A* 25% of the time, *task B* 40% of the time, and neither task 35% of the time.[250] Note the sum of all the probabilities at the same level cannot exceed 100%.

[250] If you wanted both *task A* and *task B* to be executed independent of each other, then set the *Branch Type* to "Conditional" with the *Condition or Property* value set to `Random.Uniform(0,1) < .25 and Random.Uniform(0,1) < 0.4` respectively.

Task Information	
Sequence Number	10
Name	**RequestRecords**
Branch Type	**Probabilistic**
Condition Or Probability	**GProbRecordCheck**
Process Type	Specific Time
Processing Time	**Random.uniform(2,6)**
Auto Cancel Trigger	All Immediate Predecessors Cancelled

Figure 15.37: Request Records Task Information

- For the last task, the loan assessment task requires the **WrkBanker** again as seen in Figure 15.38 making sure to set the Sequence Number to 20.

Task Information	
Sequence Number	**20**
Name	**AssessLoan**
Branch Type	Always
Process Type	Specific Time
Processing Time	**Random.Pert(5,7,10)**
Auto Cancel Trigger	All Immediate Predecessors Cancelled
Resource Requirements	
Object Type	Specific
Object Name	**WkrBanker**
Selection Goal	Preferred Order
Request Move	None

Figure 15.38: Assess Loan Task Information

- Once you close up the tasks, we need to make sure the **GProbRecordCheck** property is set to 0.2. Either right click on the MODEL in the [*Navigation*] panel and select *Properties* or select somewhere in the model to access the MODEL properties, under the *Controls→General* section set the value to 0.2.

Question 22: Why do we not have to request a move for the banker in this task?

____

Step 2: Run the model.

Question 23: Is the model executing as it did previously?

____

Part 15.8: Some Observations Concerning Tasks

The addition of tasks to SIMIO is rather new and their potential is still being explored. However when a service activity needs detailed modeling the task approach offers some interesting possibilities. For example, let's assume the banker may have two primary sets of tasks: loan application (75% of the time) and other service visits (25% of the time). Also, for the Assess Loan task, 20% of the customers have issues and take longer to process. For our problem, we would like to model the more complicated set of activities to be performed by the banker at his office as seen in the task sequence diagram in Figure 15.39.

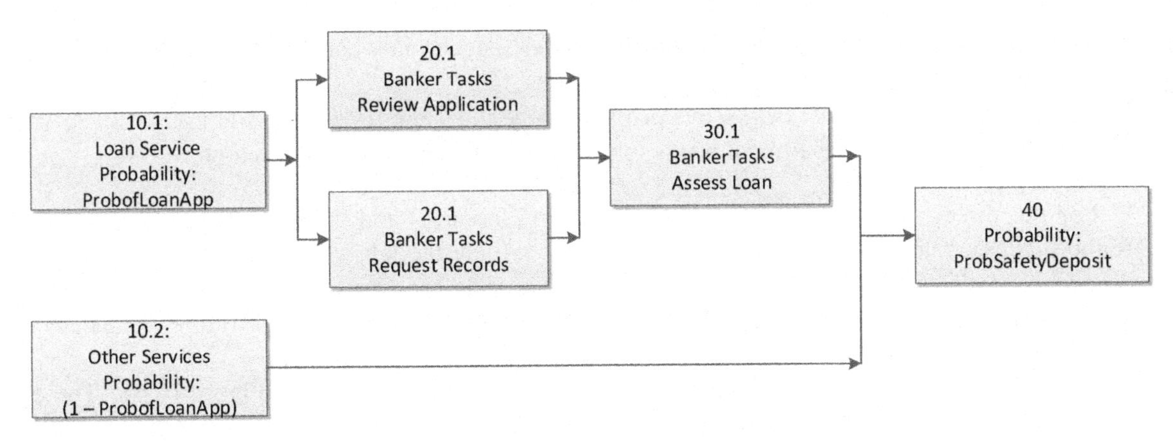

Figure 15.39: Single Task Sequence Diagram

In this task sequence, the sequence numbers will utilize a suffix (i.e., 10.1), to denote a subsequence. One subsequence performs the loan application tasks and a second subsequence consists of only one task that performs other services by the bank with both subsequences potentially utilizing the visit to the safety deposit box. Even though for this set of tasks some can be handled in a similar fashion as the last section (i.e., basic processing time and seizing of resources), we will demonstrate the use of the other process types as described in Table 15.6 to show the flexibility by using processes and sub-models. We will use processes to model the "Normal Service," "Assess Loan" application and the "Safety Deposit" tasks and utilize a sub-model for the "Request Records" task. For the "Review Application" task we will use the same *Specific Time* process type as well as for 10.1 which will be a zero-processing time task needed to complement the probability of performing normal services as explained in Table 15.6.

Table 15.6: Process Type Option Explanation

Process Type	Description
Specific Time	Expression representing the time needed for the task to complete.
Process Name	Specify a process that will be executed when the task starts. The task will finish once the process has finished executing.
Sub-Model	Allows you to create a different model within the current facility window. When the task starts, an entity of a type that is specified will be created and sent to the starting node of the sub-model. You can save a reference to the original entity to pass information to the created entity. The task finishes once the created entity is destroyed by the sub-model.

Step 1: From the *Definitions→Properties* tab, insert a new "model" *Expression* property named **GProbofLoanApp** with a default value of 0.75 for the probability that a customer needs a loan versus other services. Repeat the process for **GProbLoanAppIssues** for the percentage of customers who have issues with their loan application setting the default to be 0.20.

Step 2: As mentioned, we will utilize a sub-model for the request records to represent this as being done in a different department once it submitted by the banker. In a different portion of the model, insert a new MODELENTITY named **EntRecordSearch**, a SERVER named **SrvRecordSearch**, and a SINK named **SnkFinishRecSearch** as seen in Figure 15.40.

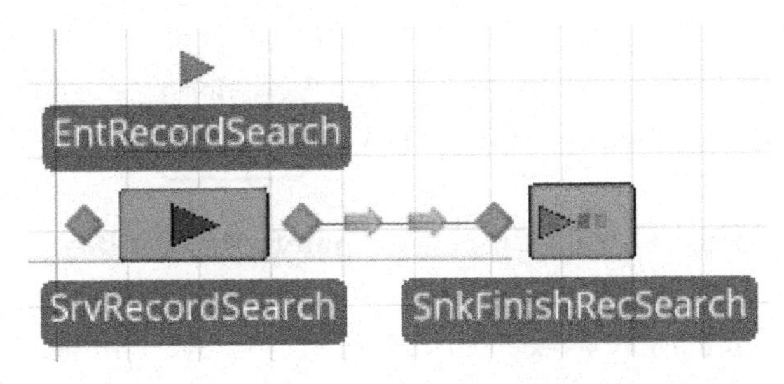

Figure 15.40: Submodel for Request Records Department

- Connect the **SrvRecordSearch** to the **SnkFinishRecSearch** via a connector as we are not concerned with travel time.
- For the SERVER, set the *Processing Time* property to be the same `Random.Uniform(2,6)` minutes to represent the time the department needs to process the request for record search.

Step 3: For the three tasks (i.e., "Assess Loan," "Other Service," and "Saftey Deposit Visit"), we will create the processes that are associated with each task. Even though we could easily use the *Resource Requirements* to request the **WkrBanker** as before and use the processes, we will `Seize` and `Release` the bank as part of the process because we can control when they occur, in case of future enhancements.

- Insert a new process named **TaskLoanApplication_AssessLoan** as seen in Figure 15.41. The `Seize` step should request the banker to the **TRoom** as seen in Figure 15.5. The `Decide` step should be use "ProbabilityBased" with the *Condition Or Probability* property expression sent to `1-GProbLoanAppIssues`. Use the same `Random.Pert(5,7,10)` minutes for the loan assessment with loans with no issues and `Random.Pert(10,12,15)` minutes for the assessment of loans with issues. Then release the banker making sure to reserve the resource as in Figure 15.25 and the set the destination node of the customer to **Input@SnkExit** so this customer will be routed to the exit.

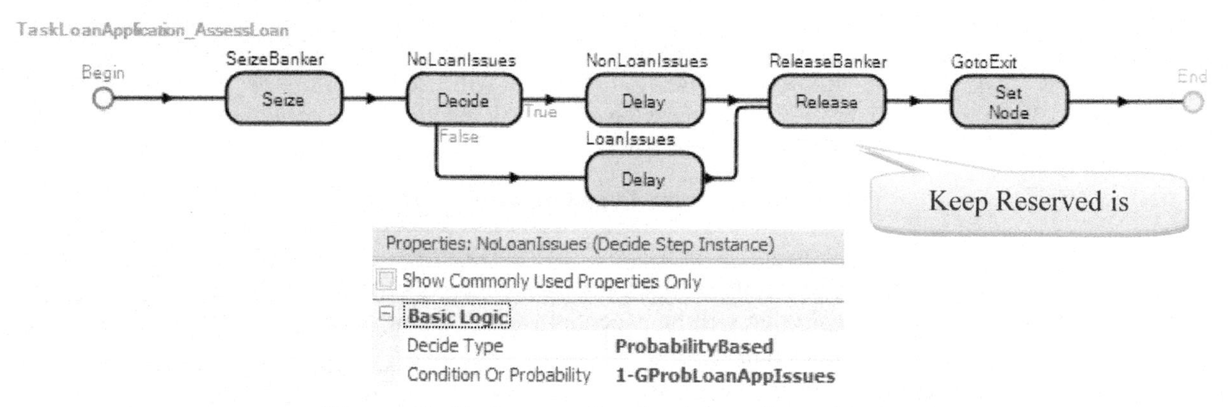

Figure 15.41: Process for the Assess Loan Task

- For the other services task, insert a process named **Task_OtherServices**. Copy all the steps from the previous process changing the delay time to `Random.Pert(4,6,8)` minutes.

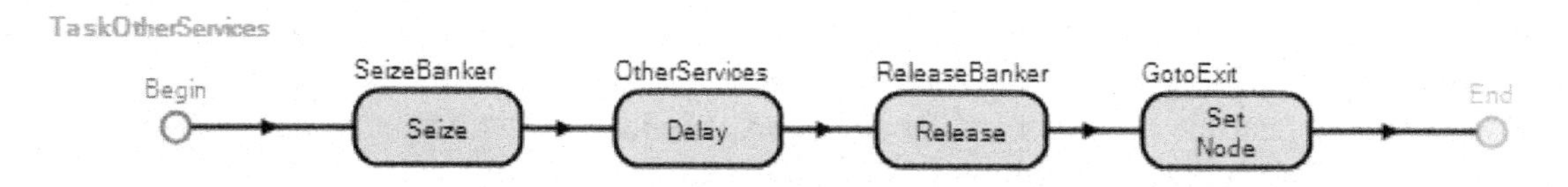

Figure 15.42: Process for Other Services Performed by Banker

- Finally, for the task associated with the safety deposit. Copy the **TaskOtherServices** process and rename it to **Task_SafetyDepositBox** changing the delay associated with getting the key for the safety deposit box takes `Random.Pert(2,4,6)` minutes.

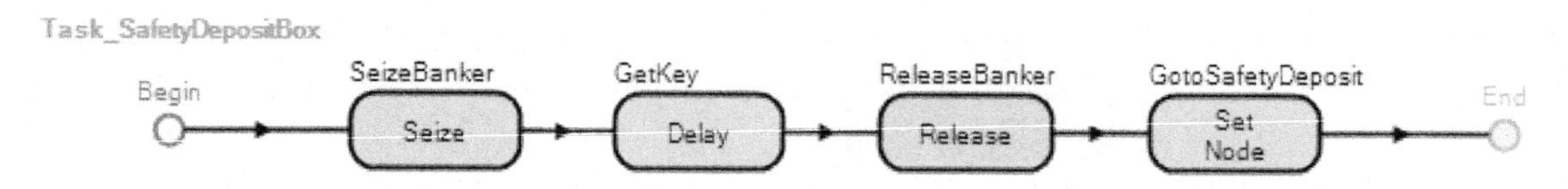

Figure 15.43: Process for Safety Deposit Box

Step 4: Since we have defined the task submodel and the task processes, now we need to implement the task sequence based on Figure 15.39 as seen in Figure 15.44. We will need to modify the three current task sequences.

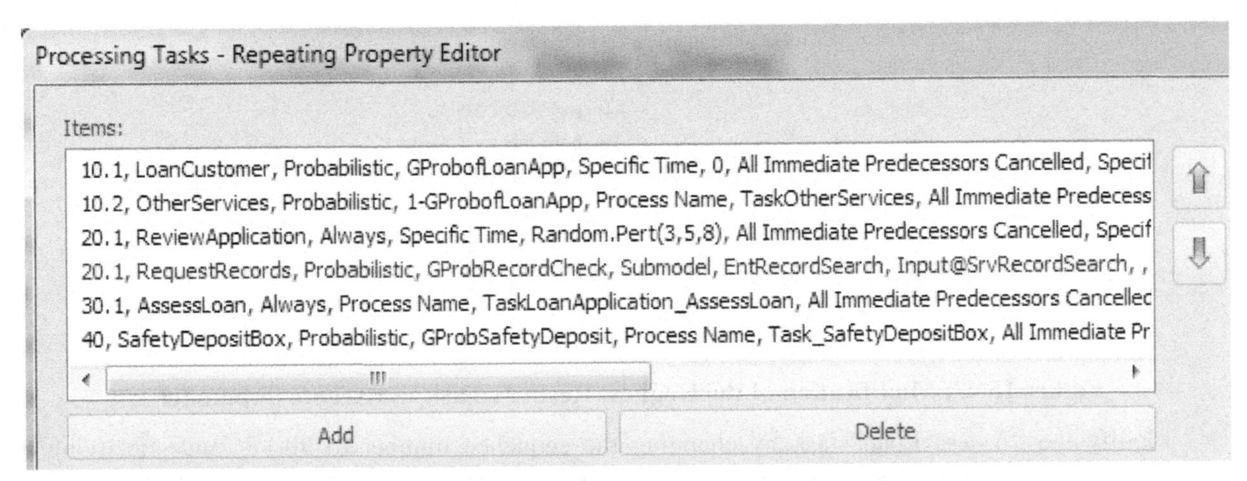

Figure 15.44: Implementation of Task Sequence

- From Figure 15.39, there are two potential sequence paths that customers can take (i.e., loan application or other services). A *Branch Type* of Probabilistic will be used to select one of these two paths. Therefore, we need to insert a new sequence task (10.1) named "LoanCustomer" that takes zero time and has a *Condition Or Probability* value equal to **GProbofLoanApp** that allows the selection of that sequence path.

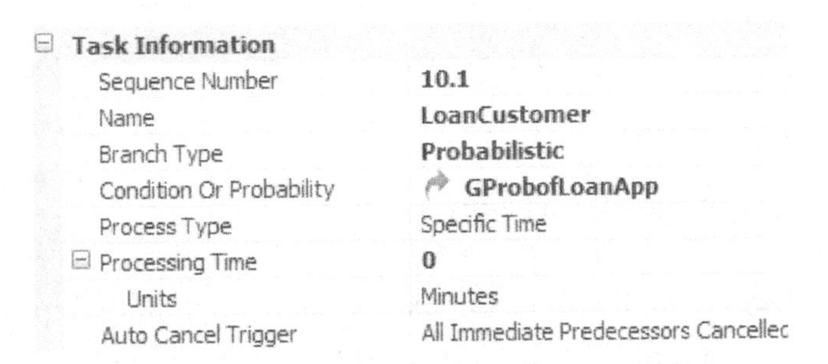

Figure 15.45: Task Select Loan Customer Path

- Insert a new task sequence (10.2) named "OtherServices" that has `1-GProbOfLoanApp` probability to ensure that the customer will select either the 10.1 or 10.2 sequence paths. Change the *Process Type* to "ProcessName" and set the *Process Name* property to "TaskOtherServices."

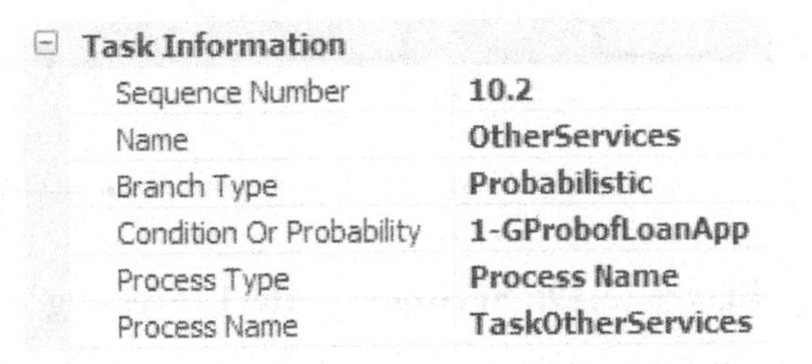

Figure 15.46: Task Associated with Other Services

- Modify the "Review Application" task by changing just the sequence number to 20.1 so the task will be executed only if the previous 10.1 is executed since the Auto Cancel Trigger property cancels successive sequential tasks when the earlier task is not done.
- Modify the "Request Records" task to execute the sub-model instead of using simple processing time as seen in Figure 15.47 as well as changing the sequence number to 20.1.

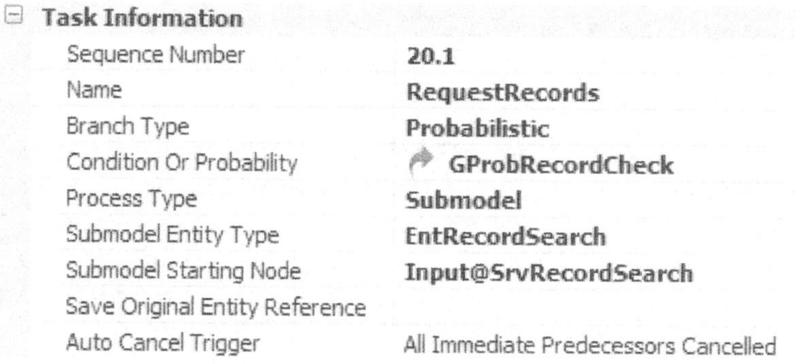

Task Information

Sequence Number	**20.1**
Name	**RequestRecords**
Branch Type	**Probabilistic**
Condition Or Probability	**GProbRecordCheck**
Process Type	**Submodel**
Submodel Entity Type	**EntRecordSearch**
Submodel Starting Node	**Input@SrvRecordSearch**
Save Original Entity Reference	
Auto Cancel Trigger	All Immediate Predecessors Cancelled

Figure 15.47: Modification of the Request Records Task to Execute Submodel

- Modify the "Assess Loan" task by changing the sequence number to 30.1[251] and specifying the *Process Type* be "Process Name" and specify the **TaskLoanApplication_AssessLoan** process. Also, make sure to remove the *Resource Requirements* of the **WkrBanker**.
- Finally, insert a new task sequence with a sequence number equal to 40 and use the **Task_SafetyDepositBox** as the process. Also make sure the *Branch Type* is "Probabilistic" with the *Condition Or Probability* set to **GProbSafetyDeposit** because only a percentage of customers perform this task.

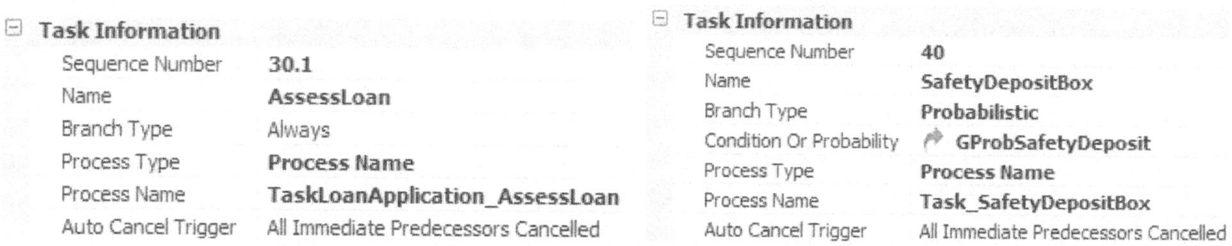

Task Information

Sequence Number	**30.1**
Name	**AssessLoan**
Branch Type	Always
Process Type	**Process Name**
Process Name	**TaskLoanApplication_AssessLoan**
Auto Cancel Trigger	All Immediate Predecessors Cancelled

Task Information

Sequence Number	**40**
Name	**SafetyDepositBox**
Branch Type	**Probabilistic**
Condition Or Probability	**GProbSafetyDeposit**
Process Type	**Process Name**
Process Name	**Task_SafetyDepositBox**
Auto Cancel Trigger	All Immediate Predecessors Cancelled

Figure 15.48: Task Sequence for Assess Loan and Safety Deposit Box.

Step 5: Modify the **SrvRoom_AfterProcessing** as shown in Figure 15.49. Remove the `Set Nodes` because those are done in the task sequences. Modify the second `Decide` step is Figure 15.49 based on the condition `ModelEntity.DestinationNode == Input@SrvSafetyDeposit.`

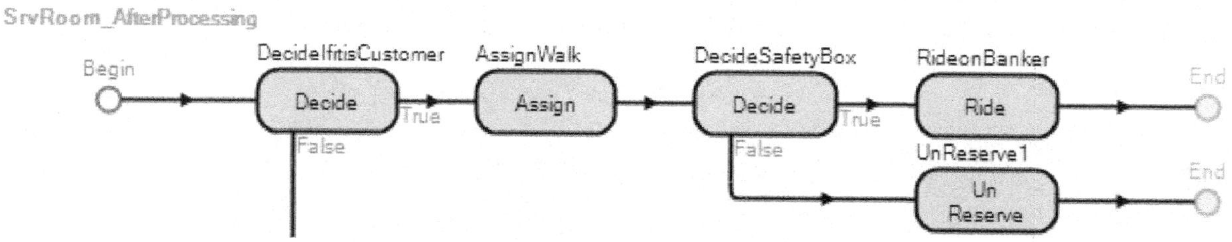

SrvRoom_AfterProcessing

Begin → DecideIfitisCustomer (Decide) —True→ AssignWalk (Assign) → DecideSafetyBox (Decide) —True→ RideonBanker (Ride) → End; DecideSafetyBox —False→ UnReserve1 (Un Reserve) → End; DecideIfitisCustomer —False

Figure 15.49: SrvRoom_AfterProcessing

Question 24: Why could we not put the `Ride` step in the task sequences?

____

Step 6: Save and run the model. You might want to place breakpoints on the **SrvRecordSearch** to be sure it is executed as expected.

Question 25: Does it run as you expected?

____

[251] A subsequence number is needed so this task is sequential to 20.1. A sequence number of 30 would have 10.2 as a prior task.

Question 26: Has the use of tasks and task sequences enhanced the modeling of the banker's service?

Part 15.9: Commentary

- Using the vehicle or the worker for both transportation and service causes the model to become very expressive of the real system.
- Resource reservations make continued use of a resource convenient.
- The animated people show promise to greatly enhance the visual appeal of the animation.
- Tasks and task sequence promote a detailed view of a service activity and can incorporate a wide range of behaviors.
- Tasks have two add-on process triggers (i.e., *Starting Task* and *Finished Task), which we did not illustrate.* These processes can be used to record when tasks start and when they end to help calculate critical paths, earliest finish, latest start, etc.

Chapter 16
Modeling of Call Centers

In previous chapters, workschedules, non-homogenous rate tables, and routing have already been extensively used. This chapter will reinforce those concepts while demonstrating how to handle two of the most common behavior of people in queues: "balking" and "reneging".

A call center receives support calls from customers who purchase kitchen appliances. The call center is open seven days a week from 8:00 am to 7:00 pm (i..e,11 hours). It is, however, staffed from 8:00 am to 8:00 pm, which leaves an hour to finish calls that are in-process at 7pm. There are two lines for technician service and customers are routed to each (see Figure 16.1) depending on whether or not the first line can answer the customer questions.

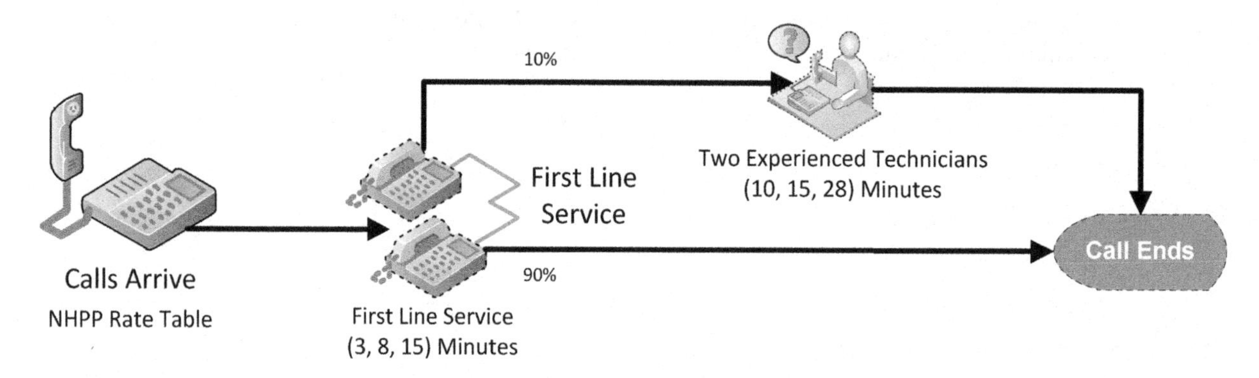

Figure 16.1: Model of a Simple Call Center

There is information about the arrival rate of calls during the day. Calls for support arrive at different rates (calls per hour) throughout the day and th following table (taken from historical data) displays how the rates change from hour to hour during the 12 hour day. Note that no calls are taken during the 11th hour.

Table 16.1: Hourly Rate of Support Calls to the Center

Time Interval	Hourly Rate
8am - 9am	35
9am - 10am	55
10am - 11am	70
11am -12pm	85
12pm - 1pm	95
1pm - 2pm	110
2pm - 3pm	90
3pm - 4pm	88
4pm - 5pm	60
5pm - 6pm	50
6pm - 7pm	30
7pm - 8pm	0

Calls are serviced in the order they are received at the call center and, we are told that ten support technicians answer the calls. Apparently the time it takes to answer a call generally takes 8 minutes to answer a support question, with the minimum being 3 minutes and the maximum being 15 minutes. The first line service technicians can completely answer ninety percent of the calls. So after working with the customer on the original question, ten percent of the calls must be routed to more experienced technicians, of which there are two technicians. The second line technicians end up taking about 15 minutes on the call, with a minimum of 10 and a maximum of 28 minutes.

Since the call center uses telephonic equipment, the calls are transferred automatically to the service centers. And, to the best of our knowledge, once a call comes in, the customer stays on the line until it is processed (i.e., currently there are no call abandonments).When we are able to talk with people at the call center, we find staffing is not constant throughout the 12 hour day, but has there are a number of different schedules for the workers, including part-time and full-time. In fact, the schedule of work is described by the following table.

Table 16.2: Number of First Line Staff per Time Period

Time Period	Number of First Line Technicians
8am - 9am	5
9am - 10am	5
10am - 11am	8
11am -12pm	10
12pm - 1pm	10
1pm - 2pm	16
2pm - 3pm	16
3pm - 4pm	12
4pm - 5pm	8
5pm - 6pm	8
6pm - 7pm	8
7pm - 8pm	2

Part 16.1: Building the Simple Model

Step 1: Create a new model that has one SOURCE named **SrcCallsArrive,** a SINK named **SnkCallsExit**, and two SERVERS named **SrvFirstLineTech** and **SrvAdvancedTech** as well as add a MODELENTITY named **EntCalls**.

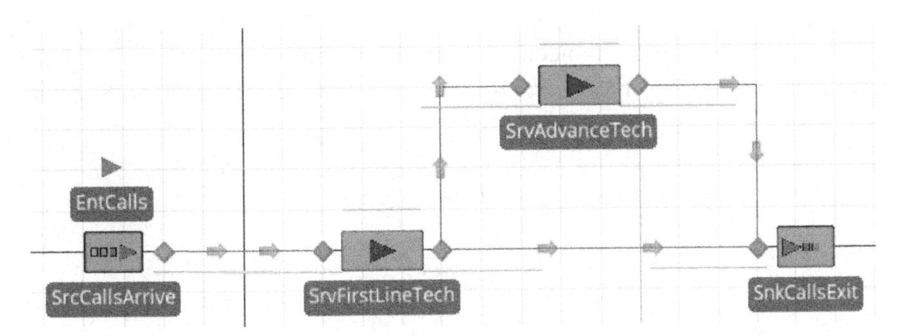

Figure 16.2: SIMIO Model of Simple Call Center

Step 2: Connect all of the objects via CONNECTORS since these are electronically switched calls. To handle the case that 10% of the calls have to be routed to the advanced technical support line, specify a selection weight of `10` for the connector linking **SrvFirstLineTech** to **SrvAdvanceTech** and a selection weight of `90` for the connector between the **SrvFirstLineTech** and **SnkCallsExit**.

Step 3: From the *Data→Rate Tables* section, create a new RATE TABLE name **RateCallTable** that has 12 intervals with an interval size of one hour.[252]

Rate Tables

RateCallTable — Rate Table

Starting Offset	Ending Offset	Rate (events per hour)
Day 1, 00:00:00	Day 1, 01:00:00	35
Day 1, 01:00:00	Day 1, 02:00:00	55
Day 1, 02:00:00	Day 1, 03:00:00	70
Day 1, 03:00:00	Day 1, 04:00:00	85
Day 1, 04:00:00	Day 1, 05:00:00	95
Day 1, 05:00:00	Day 1, 06:00:00	110
Day 1, 06:00:00	Day 1, 07:00:00	90
Day 1, 07:00:00	Day 1, 08:00:00	88
Day 1, 08:00:00	Day 1, 09:00:00	60
Day 1, 09:00:00	Day 1, 10:00:00	50
Day 1, 10:00:00	Day 1, 11:00:00	30
Day 1, 11:00:00	Day 1, 12:00:00	0

Figure 16.3: Arrival Rate Table for the Call Center

Step 4: Return to the model and change the *Arrival Mode* property of the **SrcCallsArrive** source to "Time Varying Arrival Rate" and specify the **RateCallTable** as the *Rate Table* property.

Entity Arrival Logic	
Entity Type	EntCalls
Arrival Mode	Time Varying Arrival Rate
Rate Table	RateCallTable

Figure 16.4: Specifying the *Arrival Mode* of the Call Arrivals SOURCE

Step 5: Select the **SrvAdvanceTech** server and change the *Initial Capacity* property to two and set the *Processing Time* property to `Random.Pert(10, 15, 28)` to simulate two advanced technicians.

Step 6: Next, the work schedule of the first line technicians needs to be created. Recall, SIMIO already defines a default work schedule and day pattern for you. Select the "*StandardDay*" pattern from the "*Day Patterns*" tab under the *Data→Schedules* section and modify it to match the data in Figure 16.5.[253] Let us use the WORKSCHEDULE on the left as it specifies every hour separately versus the one on the right which groups some time periods into one for the same values. We will need each time period in a subsequent section.

[252] We could have modeled the whole 24 hours with the other intervals values being set at zero. However, we changed the run length of our simulation to only be 12 hours.

[253] We find it easier to set the *Start Time* and then the *duration* which will automatically set the *End Time*.

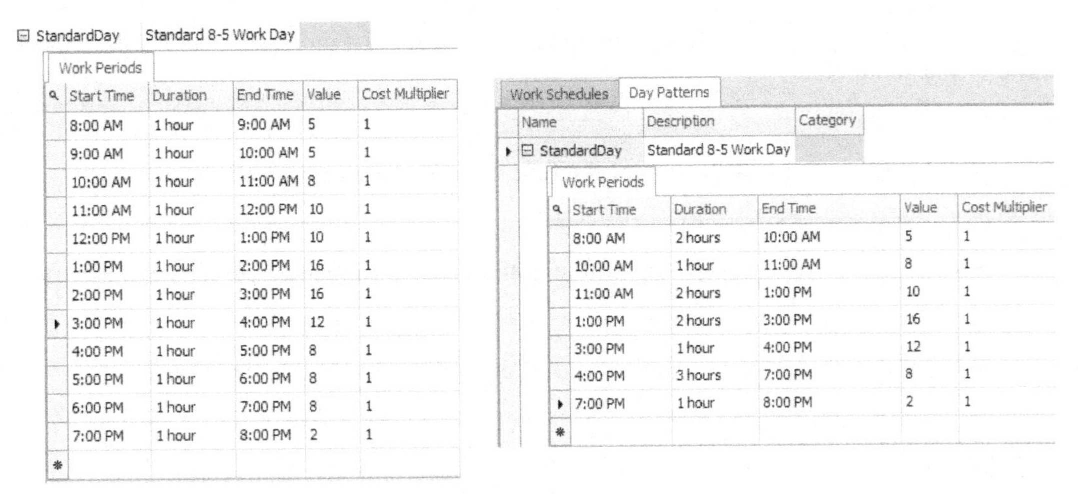

StandardDay — Standard 8-5 Work Day

Work Periods

Start Time	Duration	End Time	Value	Cost Multiplier
8:00 AM	1 hour	9:00 AM	5	1
9:00 AM	1 hour	10:00 AM	5	1
10:00 AM	1 hour	11:00 AM	8	1
11:00 AM	1 hour	12:00 PM	10	1
12:00 PM	1 hour	1:00 PM	10	1
1:00 PM	1 hour	2:00 PM	16	1
2:00 PM	1 hour	3:00 PM	16	1
3:00 PM	1 hour	4:00 PM	12	1
4:00 PM	1 hour	5:00 PM	8	1
5:00 PM	1 hour	6:00 PM	8	1
6:00 PM	1 hour	7:00 PM	8	1
7:00 PM	1 hour	8:00 PM	2	1

Work Schedules | Day Patterns

Name	Description	Category
StandardDay	Standard 8-5 Work Day	

Work Periods

Start Time	Duration	End Time	Value	Cost Multiplier
8:00 AM	2 hours	10:00 AM	5	1
10:00 AM	1 hour	11:00 AM	8	1
11:00 AM	2 hours	1:00 PM	10	1
1:00 PM	2 hours	3:00 PM	16	1
3:00 PM	1 hour	4:00 PM	12	1
4:00 PM	3 hours	7:00 PM	8	1
7:00 PM	1 hour	8:00 PM	2	1

Figure 16.5: Work Schedule for the First Line Technicians.

Step 7: Return to the **Model** and select the **SrvFirstLineTech** and specify the new work schedule and the processing time as seen in Figure 16.6.

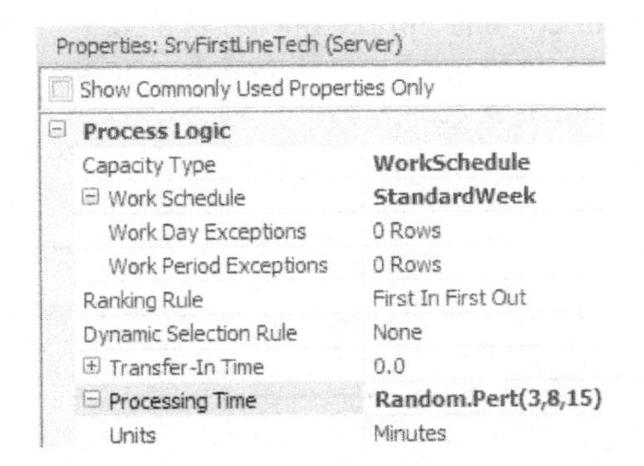

Figure 16.6: Setting the Properties of the First Line Technician Server

Step 8: Even though we do not allow any more calls to come after 7:00 pm, how many people on average will be in the queues at 8:00 pm when there are no more technicians to handle these calls. An OUTPUT STATISTIC can perform this calculation at the end of the simulation run. From the *Definitions→Elements*, insert a new OUTPUT STATISTIC named **OutStatNumberofCallsLeft** with the following Expression property: `EntCalls.Population.NumberInSystem`.

Step 9: In the *Run Setup*, change the *Starting Time* to be 8:00 AM and the *Ending Type* should be 12 Hours. Save and run the model.

Question 1: What is the average customer time in the system and the number of calls still in the system at the end of the simulation?

____

Question 2: What is the average utilization of the first line and the experienced technicians?

____

Question 3: What is the average holding time in minutes for both of the two server's input buffer?

____

Question 4: What is the average number of calls waiting in the two server's input buffer?

____

Step 10: Next insert a new EXPERIMENT and run 20 replications of the model.

Question 5: What is the average customer time in the system and the number of calls still in the system at the end of the simulation?

Question 6: What is the average utilization of the first line and the experienced technicians?

Question 7: What is the average holding time in minutes for both of the two server's input buffer?

Question 8: What is the average number of calls waiting in the two server's input buffer?

Part 16.2: Balking

Now you observe that people who may encounter waiting for service sometimes abandon that service. In queuing theory this is called "balking." Effectively a person who observes the queue to be "too long" simply will not join. In our case, the wait time is proportional to the number in the queue, so we will model the waiting time by using the number (or contents) in the queue. If there are 20 or fewer people waiting only 10 % will balk while 25% will balk when there are 21 to 30 people waiting. And finally, if there are more than 30 people waiting, the chance of balking is 50%.

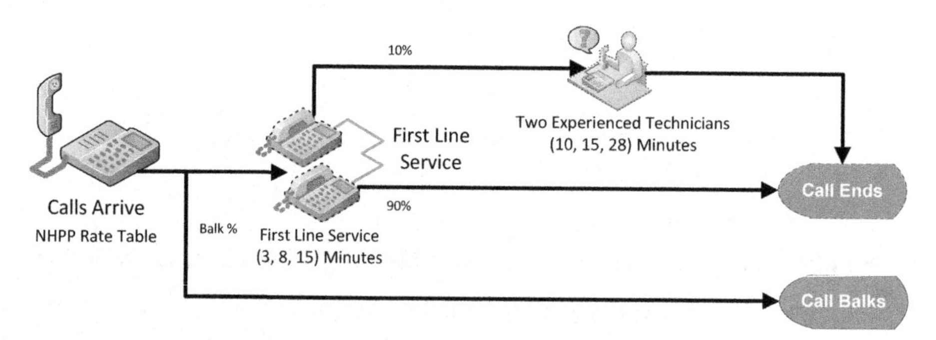

Figure 16.7: Call Center with Balking

Step 1: Insert a new SINK named **SnkBalk** below **SnkCallsExit** and connect the **SrcCallsArrive** to the new SINK via a CONNECTOR.

Step 2: To model the changing balking percentage based on the input queue of the first line service, a LOOKUP TABLE offers a great modeling mechanism.[254] From the ***Data*** tab, create a new LOOKUP TABLE named **FuncLookupBalkPercent** with the appropriate values as seen in Figure 16.8.

---

[254] There are many ways to model the balking procedure. One could add three TRANSFERNODES with the first transfer based on the number in the queue by setting the link weights (i.e., <= 10 , > 10 && <=20, >20 && <= 30, and > 30). Then set the appropriate link weights based on the percentages from the TRANSFERNODES to the **SrvFirstLineTech** or the **SnkBalk**.

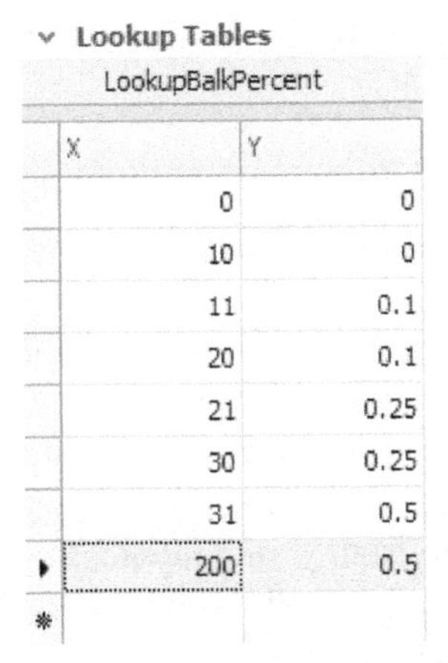

Lookup Tables

LookupBalkPercent

X	Y
0	0
10	0
11	0.1
20	0.1
21	0.25
30	0.25
31	0.5
200	0.5

Figure 16.8: Balking Percentage Lookup Table[255]

Step 3: Since the expression is very long, we will setup a function to make it easier as well as limit errors from occurring. From the *Definitions* tab, insert a new `FUNCTION` from the *Edit* section named **FuncBalkPercent**. Set the expression to return a value from the Lookup Table based on the number waiting at the SrvFirstLineTech input buffer (i.e., `LookupBalkPercent[SrvFirstLineTech.InputBuffer.Contents.NumberWaiting]`) as seen in Figure 16.9.

Properties: FuncBalkPercent (Function)

Show Commonly Used Properties Only

Value	
Expression	**LookupBalkPercent[SrvFirstLineTech.InputBuffer.Contents.NumberWaiting]**
Return Type	**Number**
Unit Type	Unspecified
General	
Name	**FuncBalkPercent**

Figure 16.9: Utilizing a Function

Step 4: For the connector linking the SOURCE to the SERVER (SrvFirstLineTech) set the set the *Selection Weight* property to `1-FuncBalkPercent` and the *Selection Weight* property of the link connecting the source to the balking sink to `FuncBalkPercent`.

Step 5: Save and rerun the experiment.

Question 9: What is the average customer time in the system and the number of calls still in the system?

___

Question 10: What is the average utilization of the first line and the experienced technicians?

___

Question 11: What is the average holding time in minutes for both of the two server's input buffer?

___

Question 12: What is the average number of calls waiting in the two server's input buffer?

___

[255] It is not strictly necessary to include the values for 0 and for 200, since the table assumes the last value if the lookup is outside the range of the table – a value less than or equal to 10 will be 0 and values equal to or greater than 31 will be 0.5.

Question 13: How many calls on average balked during the 12 hour period and do you feel the current system is handling enough of the calls?

Part 16.3: Modeling Reneging of Customer Calls

Now we allow customers to renege (i.e., leave) from the queue for standard calls when they have waited for a period of time. In other words, we will allow abandonment of calls to occur. More specifically, every customer who enters the call queue will wait for at least ten minutes. However if after waiting in the queue for ten minutes, the customer examines its place in the queue (the telephonic system informs them of this). If they are not within the first five places from the front of the queue, the customer reneges and leaves the system. If they are within the first five places of the front, then the customer will extend their stay in the queue for another four minutes in hopes of getting served. If after the extended stay, they are still in line, then they will also renege.

Step 1: Save the current model and then insert a new SINK named **SnkRenege** below the **SnkBalk** but do not connect it to anything. This SINK will be used to collect statistics on those customers that renege.

Step 2: The modeling methodology to be used is when a call comes into the service queue of the first line technician that a delay occurs for ten minutes and then we check to see if it is still in the queue. If it is in the queue and the position is greater than five you renege otherwise you wait another four minutes to see if you have been served. Select the **SrvFirstLineTech** SERVER and insert the *Entered Add-on Process Trigger*.

Question 14: What would happen if we placed the delay and checked for reneging directly in this process?

Step 3: The actual call MODELENTITY would wait ten minutes then perform the check before actually entering the server to be placed into the queue or be processed. Recall the discussion of how a token associated with the actual MODELENTITY invokes the process and executes the steps. In this case we need one TOKEN to continue on through the process (i.e., enter the queue as well as be processed) while a second TOKEN goes through the delay and potential removal of the customer from the queue. Figure 16.10 illustrates how an `Execute` step can be used to spawn a new TOKEN associated (i.e., blue one) with the entity to execute another process. When the first TOKEN associated with the "Call 21" entity runs the `Execute` step, a new 2nd TOKEN is created that will execute the steps in the new ***RenegeProcess***. The first token can be allowed to continue on allowing the call to enter the queue and be possibly be served.

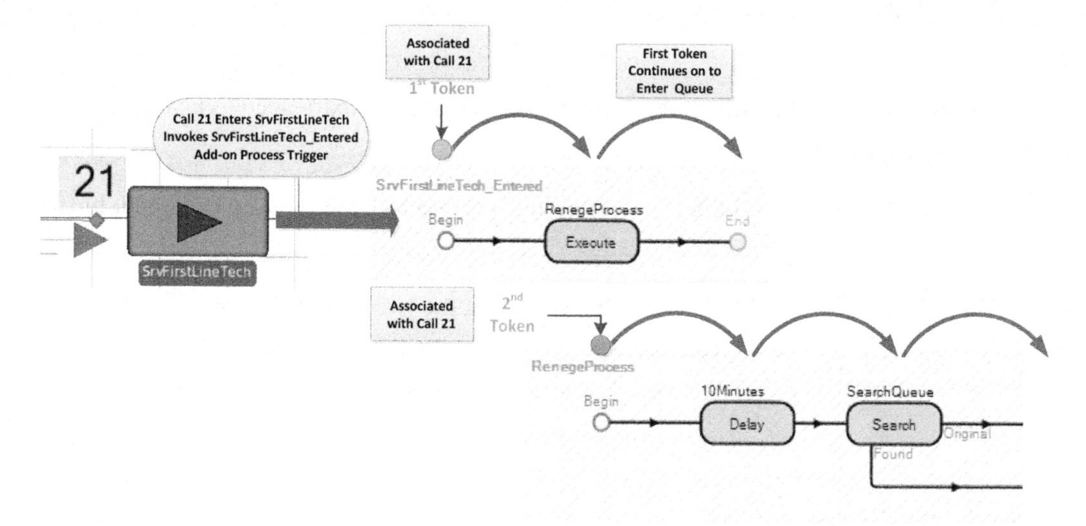

Figure 16.10: Show how Two TOKENS are Associated with the Same Call MODELENTITY

Step 4: From the *Processes* tab, create a new process named **RenegeProcess**. Under the *Advanced Options*, set the *On Associated Object Destroyed* to **EndProcess**. Refer to Figure 16.11. If the MODELENTITY that is executing the process is destroyed, we want the **RenegeProcess** to end.[256] This will be important has it will be executing in parallel while the actual entity moves through the network.

Properties: RenegeProcess (Process Element)	
Show Commonly Used Properties Only	
+ Basic Logic	
− **Advanced Options**	
Token Class Name	Token
Input Arguments	0 Rows
Return Values	0 Rows
Token Action On Associated Object Destroyed	**EndProcess**

Figure 16.11: Ending the Process if the Entity has been Destroyed

Step 5: Insert an `Execute` step into the **SrvFirstLineTech_Entered** process as seen in Figure 16.12 which executes the ***RenegeProcess*** process. Change the *Token Wait Action* property to "None (Continue)" which allows the TOKEN executing this step to continue on. By default it is set to wait until the ***RenegeProcess*** finishes, which would not allow the customer to continue to obtain service until the renege process ends (specify the `Execute` step as shown in Figure 16.12).

Properties: RenegingProcess (Execute Step Instance)	
Show Commonly Used Properties Only	
− **Basic Logic**	
Process Name	**RenegeProcess**
− **Advanced Options**	
Token Wait Action	**None (Continue)**

Figure 16.12: Executing the Renege Process but Allowing First Token to Continue

Step 6: From the *Definitions* tab, add an INTEGER DISCRETE STATE variable named **GStaPosition** which will be used to store the position of the caller in the queue.

Step 7: Return to the "*Processes*" tab and inset a `Delay` step into the ***RenegeProcess***"as seen in Figure 16.13 that delays the TOKEN for ten minutes.[257]

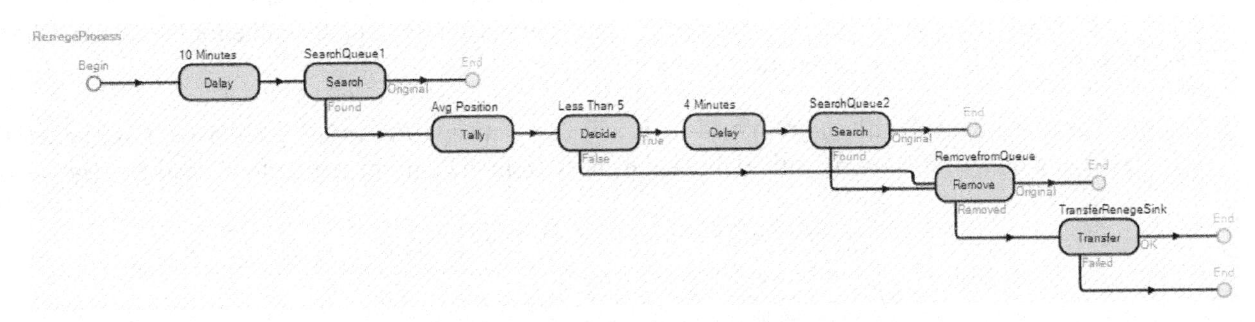

Figure 16.13: Process to Handle Reneging

Step 8: After the ten minute delay, we need check to see if the associated call is still in the queue. If it is in the queue determine its position in the queue. Insert a `Search` step after the `Delay` step which will be used to search the queue as well as determine its position.[258] Change the *Collection Type* property to

---

[256] Otherwise, the SERVER will be executing the process without the MODELENTITY and it will crash when performing the `Search` step.

[257] Typically it is a good idea to create properties for values rather than hardcoding values which makes it easier to change later.

[258] See Chapter 13 for a discussion on using the `Search` step to look through collections.

"QueueState" and the *Queue State Name* to **SrvFirstLineTech.AllocationQueue**. Specify the **GStaPosition** state variable as the *Save Index Found* property so it will be assigned the index of the item found. Since we want to search for the associated call in the queue, specify the *Match Condition* property to be `ModelEntity.ID == Candidate.ModelEntity.ID` which will search the queue and match the ID of the item in the queue with the associated entity ID as seen in Figure 16.14.

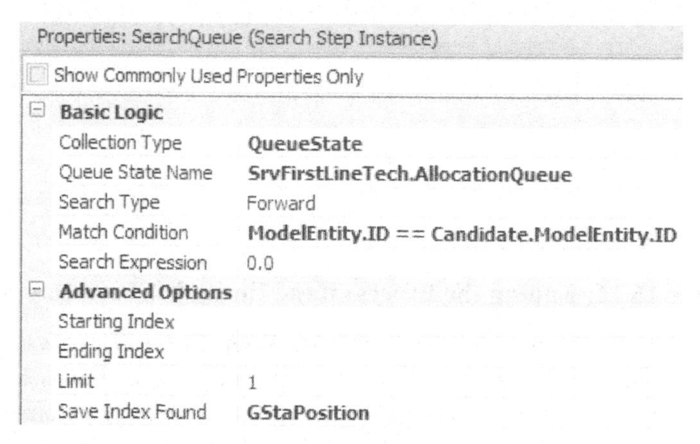

Figure 16.14: Specifying the Search Step Parameters to Search for the Call in the Service Queue

Step 9: If the TOKEN associated with the **EntCalls** MODELENTITY finds itself in the queue, then the **GStaPosition** state variable needs to be checked to see if it is less than or equal to five. Therefore, insert a `Decide` step that performs a conditional test.

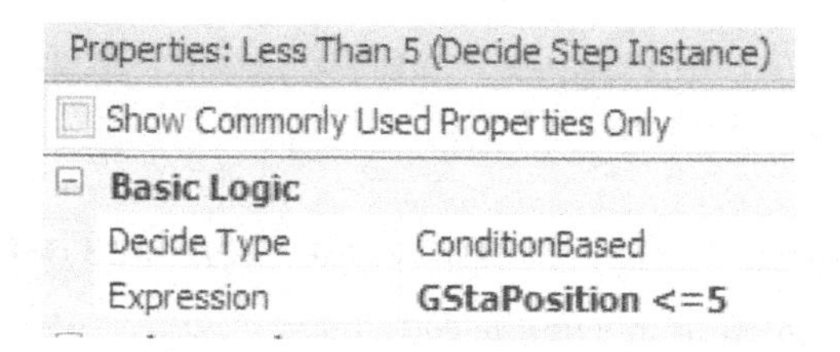

Figure 16.15: Checking the Index of the Entity in the Queue

Step 10: If the position of the entity in the queue is greater than five (i.e., the *False* branch), then we will remove the entity from the queue and transfer it to the SINK **SnkRenege**. Insert a `Remove` step on the "False" branch that will remove the current call associated with the "Found" TOKEN from the `SrvFirstLineTech.AllocationQueue`.[259] You will need to type in the value for the *Queue State Name* property directly if it does not appear in the list.

Step 11: A TOKEN associated with the removed item will exit out the "*Removed*" path. Therefore, insert a `Transfer` step on the "*Removed*" path to move the entity from the current station to the **SnkRenege**.

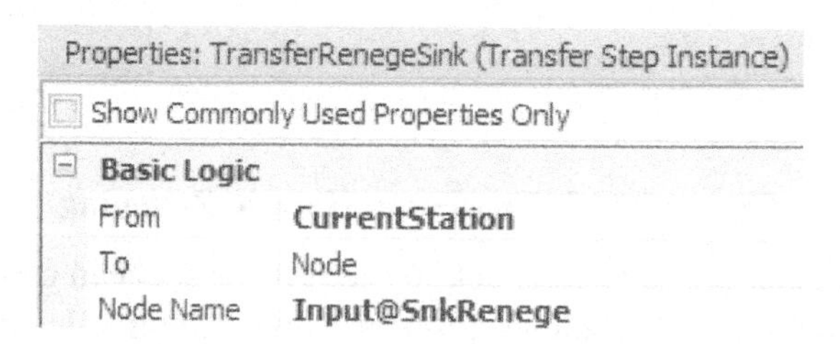

Figure 16.16: Transferring the Reneging Call to the Appropriate Queue

[259] Recall the `Remove` step can be used to remove entities from storage and allocation queues only.

Step 12: If the position in the queue is less than or equal to five then delay the TOKEN for an additional four minutes and search again. Insert a `Delay` step on the "True" path that delays for four minutes.

Step 13: Copy the first `Search` step and place it after the four minute delay to check to see if the entity is still in the queue.

Step 14: If the call is found in the queue (i.e., it is still waiting to be serviced), it will automatically renege. Grab the ○ (End) associated with "Found" path and drag it on top of the `Remove` step to utilize the same `Remove` and `Transfer` steps as seen in Figure 16.13.

Step 15: For those people who have been waiting for over ten minutes to talk with a technician, we would like to know on average what their position was after ten minutes. From the *Definitions→Elements* section insert a new TALLY STATISTIC named **TallyStatPosition**. In the **RenegeProcess**, insert a `Tally` step right before the `Decide` step.

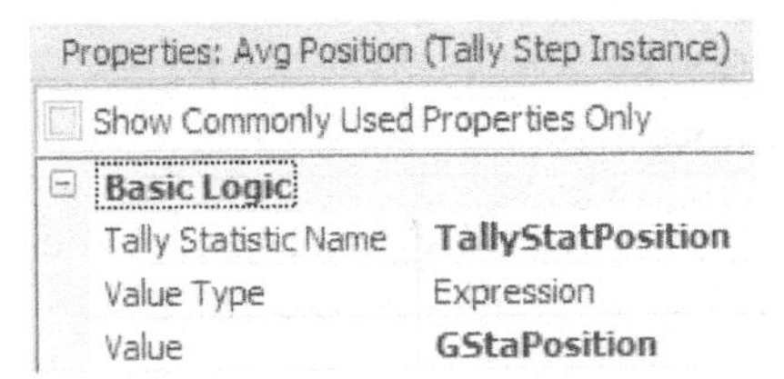

Figure 16.17: Tracking Average Line Position

Step 16: Save and rerun the Experiment.

Question 15: On average, how many calls renege during the 12 hour simulation run?

____

Question 16: What is the average position for people who have waited at least ten minutes?

____

Question 17: What is the average time in the system for calls that don't balk or renege?

____

Question 18: What is the average time in the system for calls that renege?

____

Part 16.4: Optimizing the Number of First Line Technicians

We would like to determine if we have enough first line technicians at the right times that minimize our costs and wait times by our customers. Optimization with OptQuest™ was introduced earlier and will be used again. There are times when the WORKSCHEDULES may not be flexible enough and you may need to resort to another technique. Nevertheless at the end of the section we return to specifying properties for the WORKSCHEDULE.

Step 1: Save the current model to a new name.

Step 2: From the *Definitions→Properties* section, insert 12 STANDARD EXPRESSION PROPERTIES with the names and values as seen Table 16.3.[260] These properties will be optimized to determine the best number to have during each time period where `TimePeriod1` would correspond to 8-9 AM.

---

[260] Create one property with the correct category, a name of **TimePeriod**, and then copy it changing the default value to the correct values.

Table 16.3: Expression Property for the Time Periods

Expression Property Name	Default Value	Category
TimePeriod1	5	Call Center
TimePeriod2	5	Call Center
TimePeriod3	8	Call Center
TimePeriod4	10	Call Center
TimePeriod5	10	Call Center
TimePeriod6	16	Call Center
TimePeriod7	16	Call Center
TimePeriod8	12	Call Center
TimePeriod9	8	Call Center
TimePeriod10	8	Call Center
TimePeriod11	8	Call Center
TimePeriod12	2	Call Center

Step 3: Next, insert a new DATA TABLE named **TableSchedule** under the ***Data→Tables*** section which will become our own manual schedule table as seen in Figure 16.18.

Step 4: Add a ***Standard Expression*** property column named **Capacity.**

Step 5: Insert 12 rows with the value of the expression for each row to be the corresponding property created above as seen in Figure 16.18.[261]

Table Sched...

	Capacity
1	TimePeriod1
2	TimePeriod2
3	TimePeriod3
4	TimePeriod4
5	TimePeriod5
6	TimePeriod6
7	TimePeriod7
8	TimePeriod8
9	TimePeriod9
10	TimePeriod10
11	TimePeriod11
▸ 12	TimePeriod12

Figure 16.18: Creating our own Work Schedule Table

Step 6: Since we are creating our own schedule, a TIMER will be needed to signal the time to change the capacity of the **SrvFirstLineTech** server at given intervals. From the *Definitions→Elements* section, insert a TIMER named **TimerSchedule** with a *Time Interval* of one hour and *Time Offset* equal to zero.

Step 7: Under the *Definitions→States* section, insert a new DISCRETE INTEGER State variable named **GStaTimePeriod** which will be used to set the current time period.

---

[261] The easiest way is to utilize Excel to create the 12 rows. First, add **TimePeriod1** and **TimePeriod2** into two different cells. Then select both of these cells and drag down letting excel create the other ten rows. Then copy and paste the Excel table into the **TableSchedule**.

Step 8: Next, navigate to the *Processes* tab and create a new process named **ChangeCapacity** that responds to the `TimerSchedule.Event` as its *Trigger Event* property as seen below.

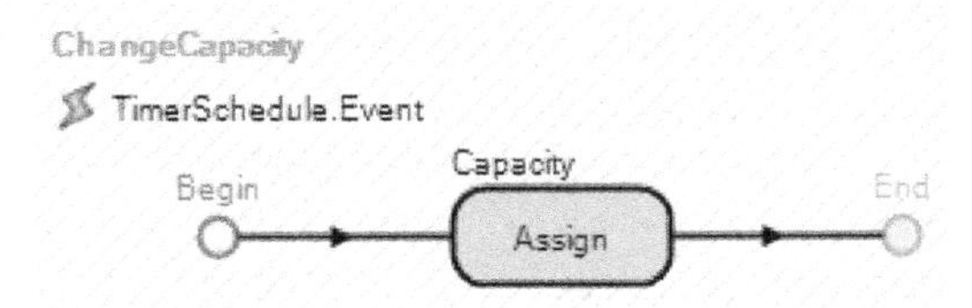

Figure 16.19: Changing the Capacity on the Timer Event

Step 9: Insert an `Assign` step that does the following two assignments. The first assignment will assign **GStaTimePeriod** the value of zero through eleven by using the `Math.Remainder` function.[262] The next assignment will assign the current capacity of the **SrvFirstLineTech** SERVER a value from the table **TableSchedule**.[263] The simulation will start at `Run.TimeNow` equal to zero and the reason we add one.

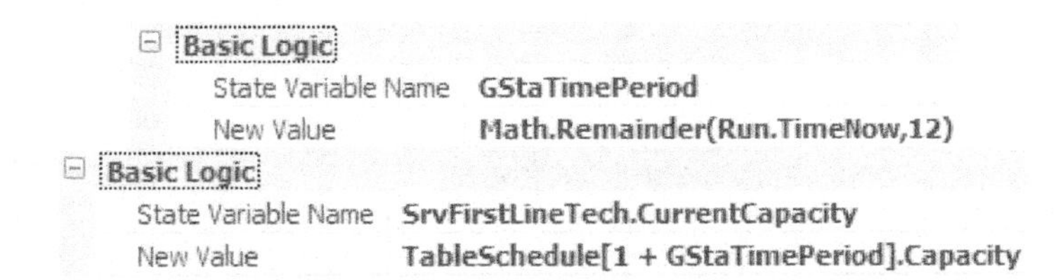

Figure 16.20: Assignments to Perform our Manual WorkSchedule

Step 10: Return the *Facility* tab and change the *Capacity Type* back to "Fixed" since we are performing the manual work schedule.

Step 11: In order to visually see if our new manual schedule is working, select the **SrvFirstLineTech** and from the *Animation* tab, insert a new STATUS LABEL with an expression of **CurrentCapacity**.

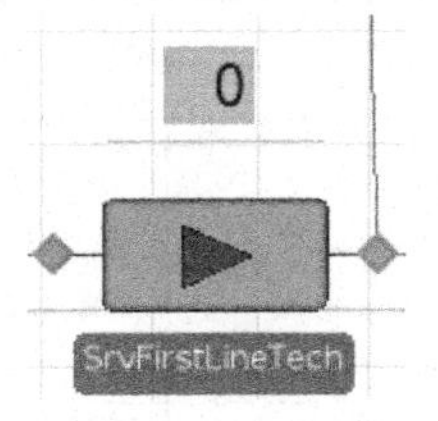

Figure 16.21: Inserting a Status to Watch the Current Capacity

Step 12: Save and run the model comparing the results to previous model.

Question 19: Does the capacity of the server change at each hour break?

___

Question 20: Compare the results of the following questions to the previous model.

What is the average customer time in the system and the number of calls still in the system at the end of the simulation?

___

What is the average utilization of the first line and the experienced technicians?

---

[262] In this example, the remainder function takes care of the fact that `Run.TimeNow` will reach 12 and therefore the remainder will be zero and start over. If you are running 24 hours for more than one day then using `Math.Remainder(StaTimePeriod, 24)` is the way to get the correct time period for each day.

[263] Row indexes start at one and therefore we need to add one to the **StaTimePeriod** to get the correct period.

What is the average holding time in minutes for both of the two servers' input buffer?

What is the average number of calls waiting in the two server's input buffer?

How many calls on average balked during the 12 hour period and do you feel the current system is handling enough of the calls?

Step 13: In the assignment, we manually converted the current simulation time into a row index (i.e., **GStaTimePeriod**) which if we ran the simulation for 14 hours would have started over at row one and two for the 13th and 14th hour. DATA TABLES in SIMIO can be specified as being *Time-Indexed* which allows one to access a column value based on the current simulation time. Select the table properties of the **TableSchedule** and specify under the *Advanced Options* that it is *Time-Indexed.* The starting time will tell SIMIO how to convert the current simulation time into which row while the *Interval Size* determines the length of the time bucket to map to each row in the data table.[264]

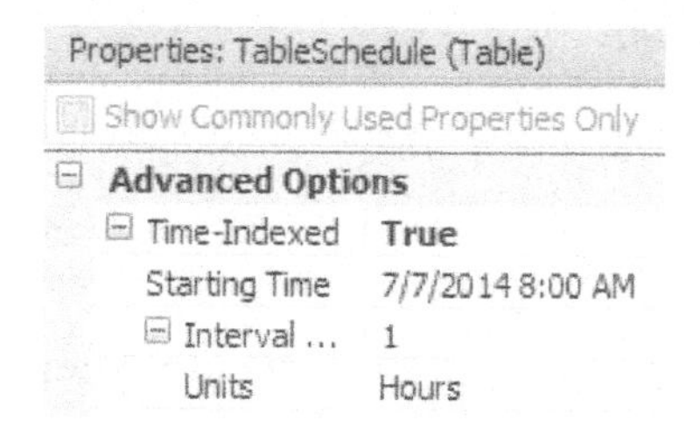

Figure 16.22: Making a Data Table Time-Indexed

There are two table functions associated with time-indexed data tables that can be used to retrieve column values or the current row index as described in Table 16.4.

Table 16.4: New Data Table Functions for Accessing Time-Indexed Columns

Function	Description
`TableName.ColumnName.TimeIndexedValue`	The `TimeIndexedValue` will return the value of the column or property that has been specified for the current simulation time (i.e., `TimedIndexedValue`)
`TableName.TimeIndexedRow`	This function will return the current row index based on the current simulation time.

Step 14: From the *Processes* tab, modify the `Assign` step by removing the first assignment which set the StaTimePeriod and then set the current capacity to `TableSchedule.Capacity.TimeIndexedValue` as seen in Figure 16.23.

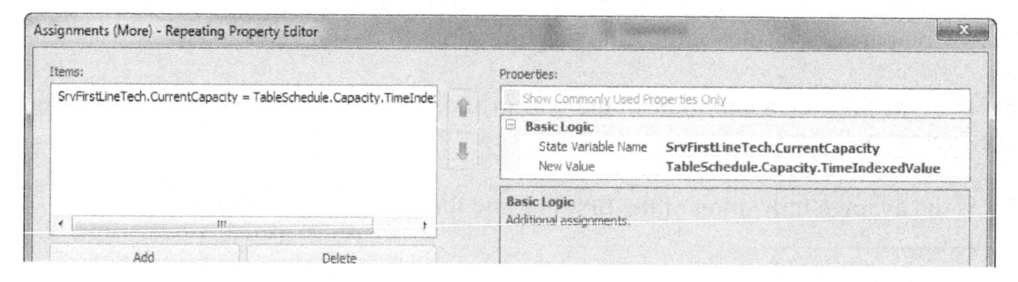

Figure 16.23: Using the TimeIndexedValue Function to set the Current Capacity

---

[264] SIMIO is internally doing what we did to calculate the row index.

Step 15: In the *Facility* window, insert two more status labels with expressions set to `TableSchedule.Capacity.TimeIndexedValue` and `TableSchedule.TimeIndexedRow` to see how these values change overtime.

Step 16: Save and run the model comparing the results to previous model.

Question 21: Does the capacity of the server change at each hour break in a similar fashion?

____

Question 22: Compare the results of the following questions to the previous model.

What is the average customer time in the system and the number of calls still in the system at the end of the simulation?

____

What is the average utilization of the first line and the experienced technicians?

____

What is the average holding time in minutes for both of the two servers' input buffer?

____

What is the average number of calls waiting in the two server's input buffer?

____

How many calls on average balked during the 12 hour and is this a good system now?

____

Step 17: Now that we have created a manual work schedule which can be optimized to demonstrate the use of time-indexed tables, we will utilize the built-in WORKSCHEDULE. From the *Data* tab, modify the WORKSCHEDULE to utilize all the time period properties for each of the values as seen in Figure 16.24.[265]

StandardDay — Standard 8-5 Work Day

Work Periods

Start Time	Duration	End Time	Value	Cost Multiplier
8:00 AM	1 hour	9:00 AM	TimePeriod1	1
9:00 AM	1 hour	10:00 AM	TimePeriod2	1
10:00 AM	1 hour	11:00 AM	TimePeriod3	1
11:00 AM	1 hour	12:00 PM	TimePeriod4	1
12:00 PM	1 hour	1:00 PM	TimePeriod5	1
1:00 PM	1 hour	2:00 PM	TimePeriod6	1

Figure 16.24: Using Properties for the Values of a WORKSCHEDULE

Step 18: Save and run the model comparing the results to previous model.

Part 16.5: Using the Financial Costs as the Optimizing Objective

Since the TABLE to perform the task of developing a manual WORKSCHEDULE has been set up, we can use OptQuest™ to optimize the 12 time period properties. Our goal is now to minimize the costs associated staffing the call center while balancing the wait time of our customers along with the amount of balking and reneging that occurs to reflect customer dissatisfaction. SIMIO introduced ABC costing for many of the objects under the "*Financials*" property section. This cost information can be associated with dynamic (i.e., MODELENTITIES, TRANSPORTERS, and WORKERS) as well as fixed objects (i.e., SERVERS, SINKS, RESOURCES, etc.). Therefore, you do not have to manually capture this information using state variables. The following tables explain the financial cost properties of fixed objects (see Table 16.6), vehicle and worker objects (Table 16.5), and model entities (see **Table 16.7**).

[265] The easiest way to set the values is to copy them again from Excel which can be used to create the column.

Table 16.5: Financial Properties Associated with Vehicle and Worker Objects

Property	Sub Property	Description
Parent Cost Center		See Table 16.6 for explanation.
Capital Cost Per Worker/Vehicle		One time occurrence cost to add this VEHICLE or WORKER.
Transport Costs	Cost Per Rider	Cost associated with loading/unloading and transporting an entity regardless of ride time.
	Transport Cost Rate	This is cost rate applied per entity based on the amount of time the entity waits in the buffer.
Resource Costs	Idle Cost Rate	The cost charged per unit time for any scheduled capacity of the server that is not utilized.
	Cost Per Use	This is a onetime use for every time the object is used regardless of how long the object is used.
	Usage Cost Rate	The cost charged per unit time for any scheduled capacity of the VEHICLE/WORKER that is utilized.

Table 16.6: Financial Properties Associated with Fixed Objects

Property	Sub Property	Description
Parent Cost Center		The cost center element that costs which are accrued to this server are summed up into. If the property is left blank, then the costs accrued are rolled up into the parent object containing the server (i.e., the MODEL).
Capital Cost		One time occurrence cost to build this fixed object which occurs at the beginning of the simulation
Buffer Costs (Input, MemberInput, MemberOutput, ParentInput, ParentOutput and Output Buffer)	Cost Per Use	Costs associated with the buffers. This is a onetime use for every entity that enters the buffer regardless of how long they wait in the buffer (i.e., all entities will enter the input and output buffer regardless if the only stay in zero minutes.
	Holding Cost Rate	This is cost rate applied per entity based on the amount of time the entity waits in the buffer.
Resource Costs (All fixed objects)	Idle Cost Rate	The cost charged per unit time for any scheduled capacity of the object that is not utilized.
	Cost Per Use	This is a onetime use for every time the object is used regardless of how long the object is used.
Resource Costs (Except WORKSTATIONS)	Usage Cost Rate	The cost charged per unit time for any scheduled capacity of the object that is utilized.
Resource Costs (WORKSTATIONS Only)	Setup Cost Rate$^{\tau}$	The cost charged per unit time for any setup activity at a WORKSTATION.
	Processing Cost Rate$^{\tau}$	The cost charged per unit time for any processing activity at a WORKSTATION.
	Teardown Cost Rate$^{\tau}$	The cost charged per unit time for any tear down activity at a WORKSTATION.

Table 16.7: Financial Properties Associated with ModelEntities Objects

Property	Description
Initial Cost	Initial cost for the creation of the MODELENTITY.
Initial Cost Rate	The cost charged per unit time for any scheduled capacity of the server that is not utilized.

Step 1: Select the **SrvFirstLineTech** SERVER and expand the "*Financials*" property category. The first line technicians receive $25 per hour regardless if they are helping a customer or not. Therefore set the *Idle*

and *Usage Cost Rate* properties to "25" as seen in Figure 16.25. Owing to the trunk line rentals, there is a fixed cost (i.e., $10) regardless of usage as well as $5 per hour per call usage rate that needs to be specified. We will leave the *Parent Cost Center* blank to utilize the MODEL's cost center.[266]

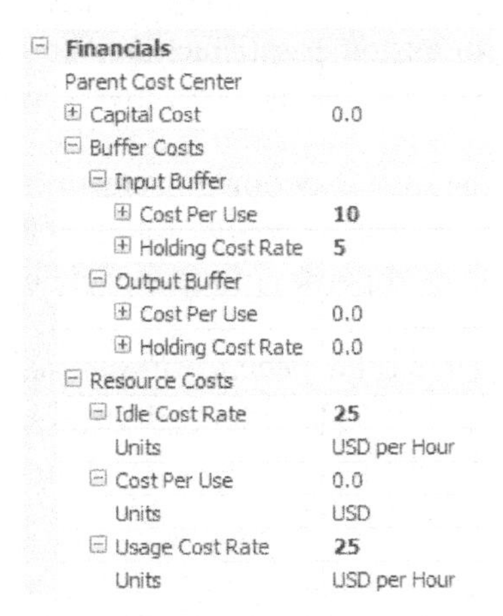

Figure 16.25: Specifying Cost Information on the SERVER

Question 23: Save and run the model observing the new cost values in the results section. What is the total cost of the system associated with the MODEL?

____

Step 2: Since our cost information is setup, we now need to optimize the number of technicians by minimizing the total costs associated with the system while balancing the number of unhappy customers. Create a "New Experiment" named **Optimization.**

Step 3: Add the following three responses to the experiment that will determine the total cost of the system as well as keep track of the number of people who balk and renege.

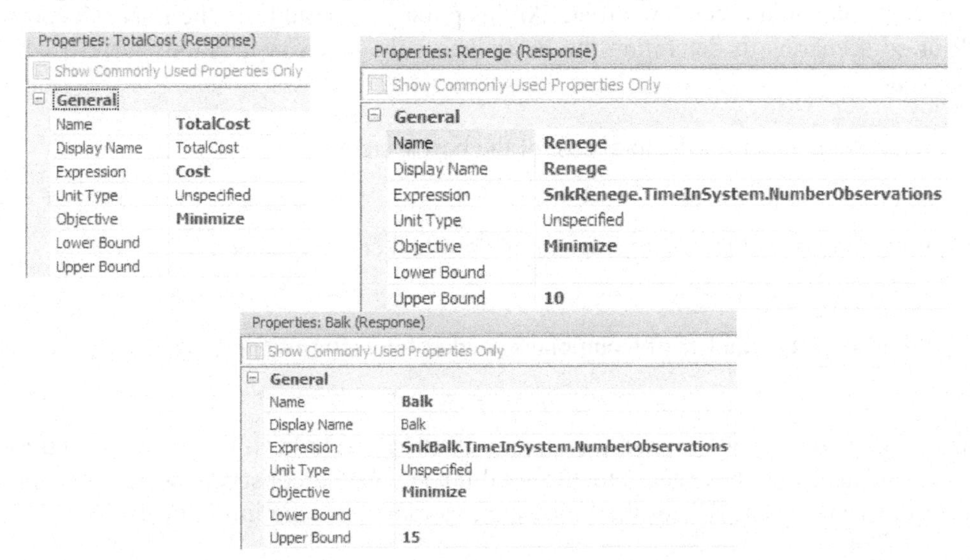

Figure 16.26: Add the Three Responses of Cost, Number Reneging, and Number Balking

Step 4: Like many problems, this is a multi-objective problem where the objectives often conflict (i.e., if you run the optimization it will most likely chose only one technician at each time period while customers

[266] A COST CENTER element can be specified to have the cost summed up. These cost centers can be specified in a hierarchical fashion (i.e., several object roll up to one center for total costing (e.g., department costing), several centers can roll up to higher level centers (e.g., departments to a plant, plants to organizations, etc.). You can also use Assign steps to assign costs to centers dynamically.

would like the maximum allowed at each period. As was done previously, we can convert the multi-objective by combining the objectives in a weighted fashion, which generally is not usable idea.[267] Instead, we will utilize thresholds or bounds on the number of balking and reneging as seen in . Having 25 people balk or renege represents on average approximately 3% of the callers that are seen in day which is what management feels is adequate for customer satisfaction. We set the upper bound of 15 and 10 for the number of people it is acceptable to balk and renege respectively as seen in Figure 16.26.

Step 5: From the *Add-Ins* section of the new experiment, select the OptQuest for SIMIO add-in. Next, for each of the 12 Controls variables, specify the *Minimum Value* to be "`1`", *Maximum Value* to be "`20`" and the *Increment* property value to be "`1`". Properties of the model become *Control* variables automatically.

Step 6: Specify the **TotalCost** response as the *Primary Response* as seen in Figure 16.27. Also, set the OptQuest™ parameter values based on the figure. We will need to run this optimization longer based on the number of variables and threshold constraints.

Primary Resp...	TotalCost
⊞ Advanced Options	
⊟ OptQuest for Simio - Parameters	
Min Replications	7
Max Replicati...	20
Max Scenarios	100
Confidence L...	95%
Relative Error	0.1
Objective Type	Single Objective

Figure 16.27: OptQuest™ Parameter Values

Step 7: Save and run the optimization experiment.

Step 8: Once the optimization is complete filter the scenarios such that only the ones that are checked are present which represents the scenarios that meet the number of balker and reneging thresholds.[268] Next sort the total cost column in ascending order. At this point you should run the K-N selection algorithm on the top 20 or so scenarios to determine the best since only seven replications were used. Use a $100 indifference zone.

Question 24: What did you get for the total cost of the best system?

____

Question 25: What did you get for the average number of balkers and renegers?

____

Question 26: What were the number of technicians needed for each time period?

____

Step 9: The optimization has determined that greater than 140 technician hours are needed based on costs. However, management only has access to 135 total hours. Add a constraint which is a constraint on the decision control variables that utilizes the following expression as seen in Figure 16.28.

```
TimePeriod1+TimePeriod2+TimePeriod3+TimePeriod4+TimePeriod5+TimePeriod6+
TimePeriod7+TimePeriod8+TimePeriod9+TimePeriod10+TimePeriod11+TimePeriod12
```

[267] We could arbitrarily put a large cost on the input buffers of both the balk and renege SINKS to try to force the system to minimize the number of balking and reneging that occurs. However, determining the right value of the cost can be difficult.

[268] Select the filter option in the column heading row above the check boxes.

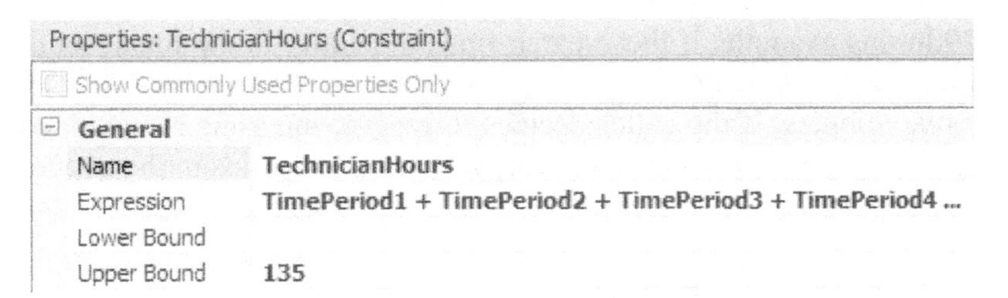

Figure 16.28: Specifying Control Variable Constraint

Step 10: Save and run the optimization experiment.

Step 11: Once the optimization is complete filter the scenarios such that only the ones that are checked are present which represents the scenarios that meet the number of balker and reneging thresholds. Next sort the total cost column in ascending order.

Question 27: What did you get for the total cost of the best system as well as the average number of balkers and renegers?

____

Question 28: What were the number of technicians needed for each time period as well as the total number of technicians needed?

____

Step 12: Next, run the KN Select Best Scenario algorithm with an indifference zone of $10 on the feasible solutions.

Question 29: What do you observe about the feasibility of the solutions now?

Step 13: Recall the KN selection algorithm does not take into account constraints on the non-primary objectives. Select the minimum one that meets the two objective thresholds.

Question 30: What did you get for the total cost of the best system as well as the average number of balkers and renegers?

____

Question 31: What were the number of technicians needed for each time period as well as the total number of technicians needed?

____

Part 16.6: Commentary

- In this chapter, the `Execute` step was used to create a new TOKEN that worked independently of the original TOKEN. There is an additional modeling concept that can cause the same phenomena to occur. The *Exited* add-on process triggers for objects can work in a similar fashion. The *Exited* add-on process triggers are executed when the trailing edge of the MODELENTITY has exited the node, link, station, or object depending on the trigger. Many people do not realize that logic placed in these add-on processes actually run in parallel while the entity continues on to the link, station, etc. To illustrate this fact, you can clear the *Entered* Add-on process trigger for the **SrvFirstLineTech** and then select the Input node of the server. Set the *Exited* add on-process trigger of the **Input@SrvFirstLineTech** BASICNODE to be the ***RenegeProcess*** and rerun the model comparing the results to the previous section.
- More complicated reneging processes can also be utilized. For example, every five minutes you are told how many people are still in line and the person could make a probabilistic decision based on some behavior (i.e., number of people and how long they have already been waiting) as seen in

Figure 16.29.In this example, if they search finds the entity in the queue as before, it will make a decision to renege or not. If they decide not to renege (i.e., "True" path), it loops back and delays for another five minutes. If the caller decides to renege, the same remove and transfer is utilized. The looping will continue until either the caller no longer remains in the queue or the caller reneges and leaves the system.

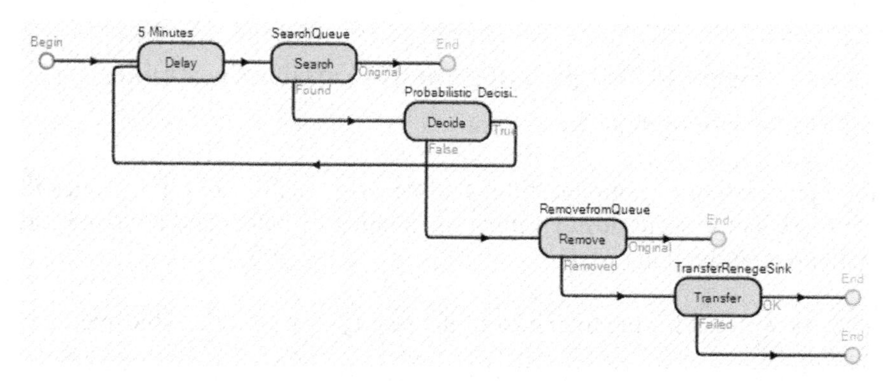

Figure 16.29: Reneging Process that Checks Every Five minutes

Chapter 17
Sub-Modeling: Cellular Manufacturing

Object-oriented simulation languages offer the ability for users to extend the base language by creating new objects through composition and extension/inheritance. Most object oriented languages offer the ability to create new objects out of existing ones which is referred to as composition. SIMIO provides the ability to create objects via composition as well as through extension. This chapter will explore the easier object creation (i.e., composition having facility models inside of other models (sub-models).

A manufacturing operation has ten multiple work cells where each work cell consists of two machines that are connected by a short conveyor as seen in Figure 17.1. A dedicated operator is needed to load parts into the machine for processing on the first machine and then to unload them from the second machine. The conveyor system will automatically transfer the parts from the first machine to the next machine. Some of the work centers (i.e., machines within the work cell) are capable of handling multiple parts in parallel. Also, priority is given to unloading parts from the second machine to ensure parts are processed through the cell quickly. One way to do this would be to insert ten Machine A servers, ten Machine B servers, ten conveyors, and ten fixed resources that are seized and released. Since the work cells are the same we can use composition to create a work cell object and then use it nine times to model the whole manufacturing system.

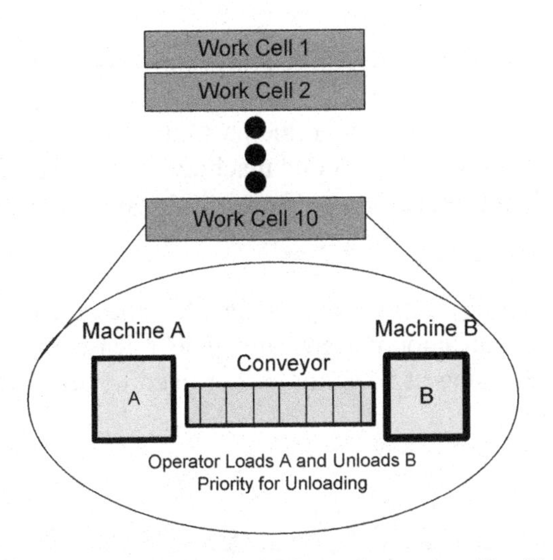

Figure 17.1: WorkCell Manufacturing Facility

Part 17.1: Model of One Work Cell

Before building the larger model, it is better to work initially with a single work cell. Eventually, this model will become a building block (i.e., composition object) and is referred to as a "sub-model" (see Figure 17.2).

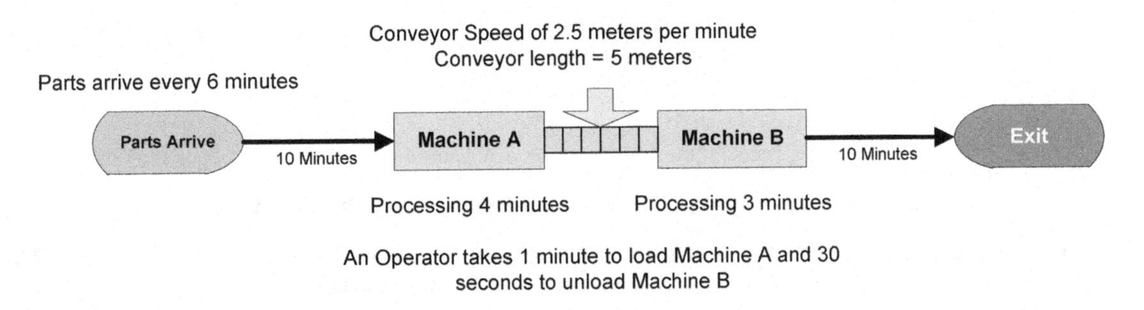

Figure 17.2: Individual Work Cell

Step 1: Create a new project SIMIO. Change the name of the model to **WorkCell** by changing the *Model Name* in the Model Properties.

Step 2: Construct the model according to Figure 17.3.

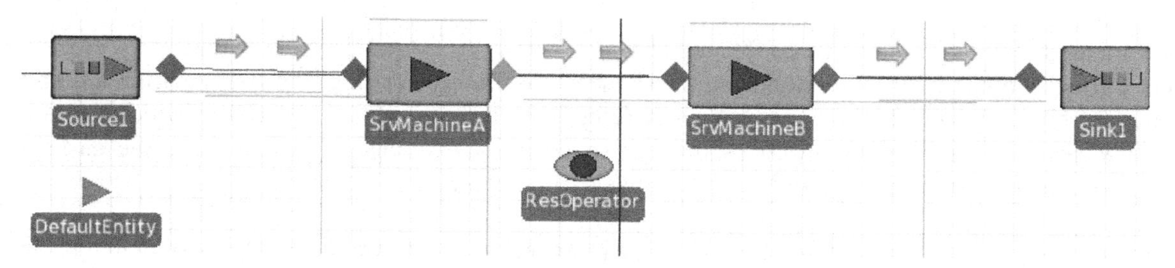

Figure 17.3: Initial Two-Machine Model

- Add two SERVERS named **SrvMachineA** and **SrvMachineB.** The processing times should be four and three minutes respectively for now.
- Connect these two machines by a five meter conveyor with a speed of 2.5 meters per minute.
- Add a MODELENTITY, SINK and SOURCE that generates arrivals every six minutes. Do not worry about naming since these are only temporary.
- Connect the SOURCE to **SrvMachineA's** input and the sink to **SrvMachineB's** output node via ten minute time paths.

Step 3: Save and run the model to make sure entities flow correctly.

Step 4: The machines require an operator to load the part into the first machine which generally takes one minute and then to unload the part from the second machine which takes approximately 30 seconds. Insert a new fixed RESOURCE named **ResOperator**. Add an additional symbol to indicate working and color the iris green.

Step 5: Add the *"Before_Processing"* add-on trigger for Machine A that will seize the "Specific" resource **ResOperator**, delay for a certain amount time, and then release the specific resource. This process represents parts that enter processing at Machine A but need to be first loaded into the machine by the operator before being processed.

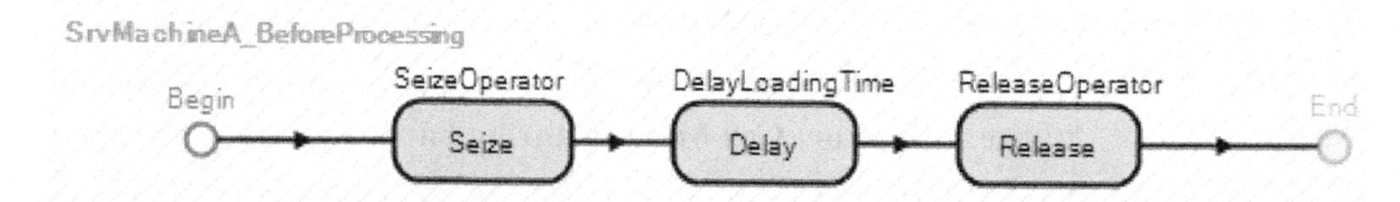

Figure 17.4: Process Logic to Seize and Release Operator to Load First Machine

Step 6: The *Delay Time* property for the `Delay` step should be a referenced property. Right click on the *Delay Time* and "Create a New Reference Property" named **LoadDelayTime**.

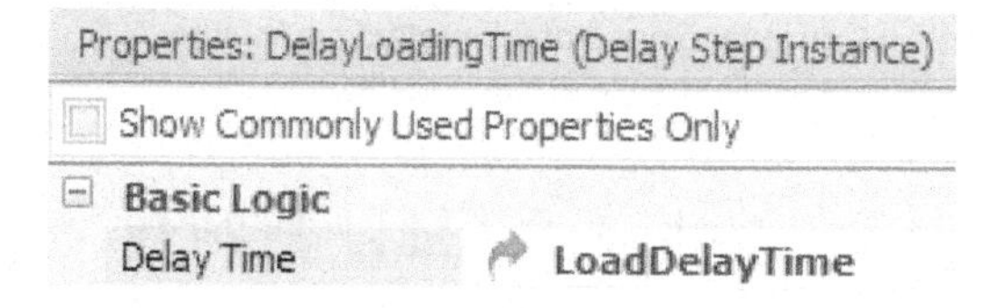

Figure 17.5: Specifying a Referenced Property as the Delay Time

Step 7: Repeat the procedure for the unloading of parts out the second server **SrvMachineB**. The only difference is the name of the referenced property should be **UnloadDelayTime** and the add-on trigger should be the *"Processed"* trigger (i.e., **SrvMachineB_Processed**). Once the process has been created, just copy and paste the process steps from the **SrvMachineA_Processing** and create the new reference property for the *Delay Time*.

Step 8: Switch to the "*Definitions*" tab, change the *Category* of the two new properties to "Process Logic" and set the *Default Units* property to "Minutes." The default value should be set to one minute for the **LoadDelayTime** property and 0.5 minutes for the **UnloadDelayTime** property.

Step 9: Save and run the model making sure it behaves as intended.

Question 1: For 24 hour replication, what is the average time in system (in minutes)?

Part 17.2: Creating the Sub-Model

Currently, we have the basic work cell model working correctly but we need to create the sub-facility model. In order to create the sub-model, an External view of the sub-model needs to be defined before you can use it as an object in other models.

Step 1: Delete the SOURCE, SINK, and MODEL ENTITY from this facility model since they were only needed to test the logic of the two machines connected via a conveyor.

Step 2: Since the **WorkCell** model will be used as a sub-model, none of the nodes in the model (i.e., **Input@SrvMachineA** or **Output@SrvMachineB**) are visible to external model (i.e., models using the **WORKCELL** object). Therefore, two external nodes need to be added to the external view of the object (i.e., *Definitions→External* section) that facilitates the flow of entities into and out of the sub-model.[269]

- Since entities entering the **WORKCELL** object need to enter the input node of machine A (i.e., **Input@SrvMachineA)**, right click the BASICNODE to bring up the sub menu seen in Figure 17.6 and select "*Bind to New External Input Node*" and then name the external node **Input** to be consistent.[270]
- Next, right click the output node of **SrvMachineB** but this time select "*Bind to New External Output Node*" and name it **Output**.
- By default, all objects (i.e., animated queues, objects, etc.) in the facility are visible to the external world as indicated by the "*Externally Visible*" button being toggled on as seen in Figure 17.6. Toggle the *externally visible* button off for both the **Input@SrvMachineA** and **Output@SrvMachineB** nodes. This will make it easier to see our new **Input** and **Output** nodes.

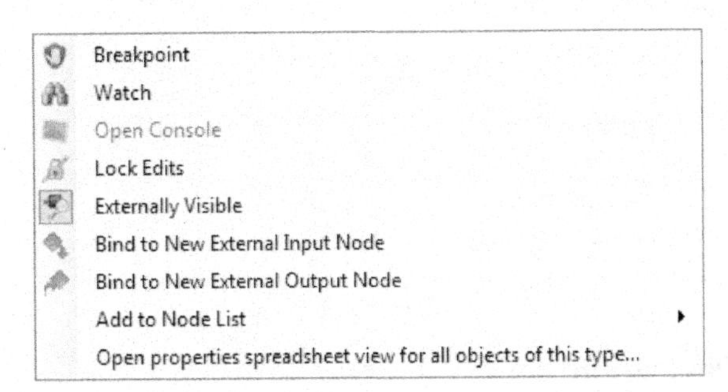

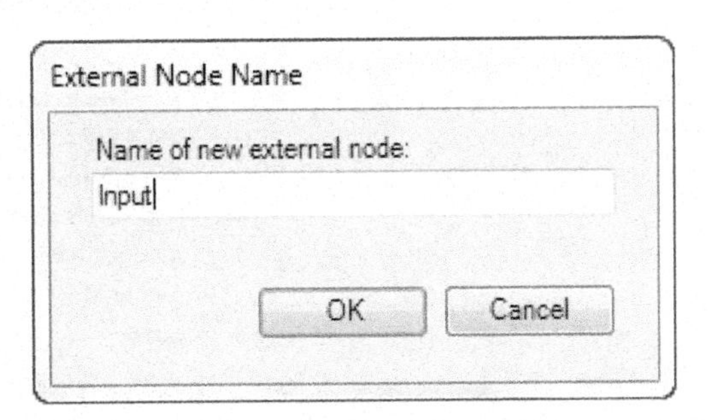

Figure 17.6: Automatically Adding External Nodes as Making Objects Visible to the External View

---

[269] All the basic objects (e.g., SOURCE, SERVER, WORKSTATION, etc.) have nodes in their external view that allow entities to flow into and/or out of the object.

[270] These can be added manually from the *External* panel underneath the "*Definitions*" table however all of the property settings also have to be done manually.

Step 3: Navigate to the *External* panel underneath the *Definitions* tab and survey the external view which includes the two external nodes named **Input** (a BASICNODE) and **Output** (a TRANSFERNODE) as well as a symbol representing the model as seen in Figure 17.7. Only the two nodes are accessible.

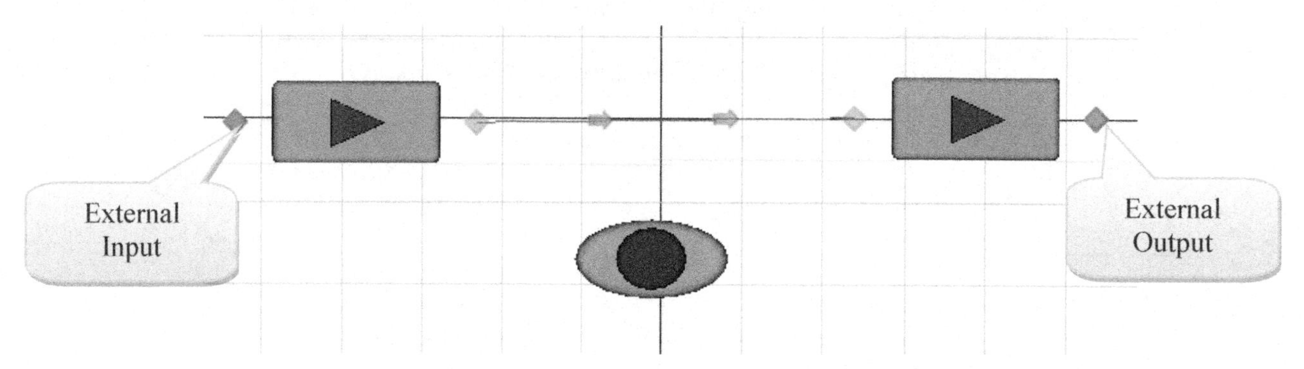

Figure 17.7: External View of the WorkCell Object

Step 4: Select the **Input** node which can be seen as an instance of the BASICNODE NODE *Class*. Generally entities are transferred from this input node into a station. However, the sub-facility model does not expose a station. Therefore, these entities will be transferred to the input of the **SrvMachineA**. The *Input Location Type* property is set to "FacilityNode" and the node is the **Input@SrvMachineA** as in Figure 17.8. As entities flow into the **Input** of the **WorkCell** they will be automatically routed to **SrvMachineA**.

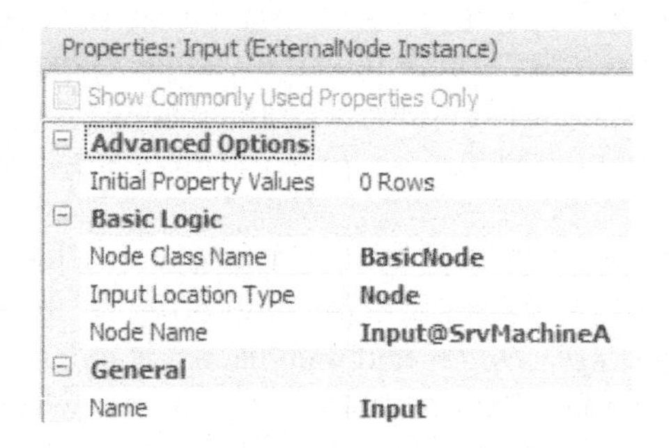

Figure 17.8: The Input Node

Step 5: After selecting the **Output** node, you can see this is an instance of the TRANSFERNODE class to allow entities to be routed to their next destination once they leave the **WorkCell** (see Figure 17.9).

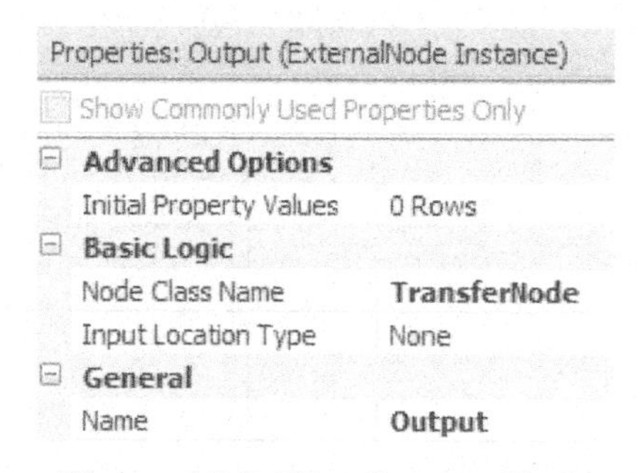

Figure 17.9: The Output Node

Step 6: Entities that enter the external input node of the **WORKCELL** object will automatically be transferred to the input of the **SrvMachineA** via the **Input** node. When entities reach the output node of the **SrvMachineB**, they need to be transferred to the external output node (**Output**) defined in the external view. Switch back to the "Facility" tab and select the `Output` node of **SrvMachineB** (i.e., **Output@SrvMachineB**). Under the *Advanced Options* category, you can see that SIMIO automatically changed the *Bound External Output* property to specify the **Output** as the external node name property as shown in Figure 17.10. Now entities will automatically be transferred to the external **Output** when they reach the output of the **SrvMachineB**.

Advanced Options	
Display Name	
Sequence Expected Operation Time	0.0
Random Number Stream	0
Bound External Output Node	**Output**

Figure 17.10: Connecting the External Node

Step 7: Save the **WorkCell** model.

Part 17.3: Creating a Model using the WorkCell Sub-model

The **WORKCELL** Sub-facility model has been built and can now be saved to a library and used as any other model that is part of SIMIO. We are going to build our manufacturing systems using the **WORKCELL**.

Step 1: From the *Project Home* tab, insert/create a new "Fixed Class" model.

Step 2: In the new model, insert a SOURCE named **SrcParts**, a SINK named **SnkExit**, and new work cell named **Cell1** which is part of the *Project Library* panel. Connect the SOURCE and the SINK to the **WORKCELL** via a ten minute time paths as seen in Figure 17.11.

Step 3: Make the interarrival time of parts be Exponential with a mean of 5 minutes.

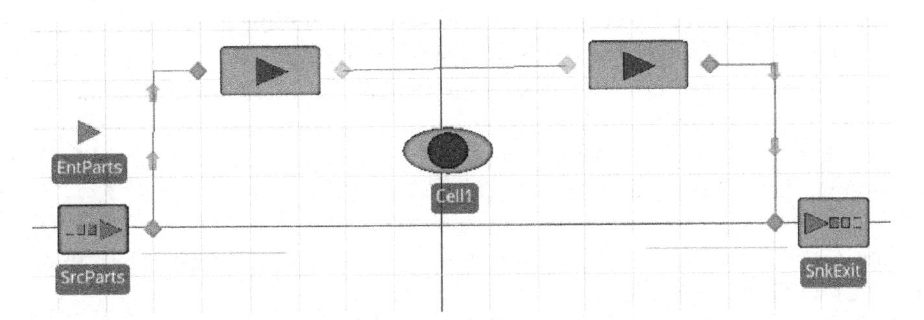

Figure 17.11: Simulation Model Using the New WorkCell Sub-Simulation Model

Step 4: For the **Cell1**, make sure the *Load Delay Time* property is one minute while the *UnloadDelay* time property is 0.5 minutes.

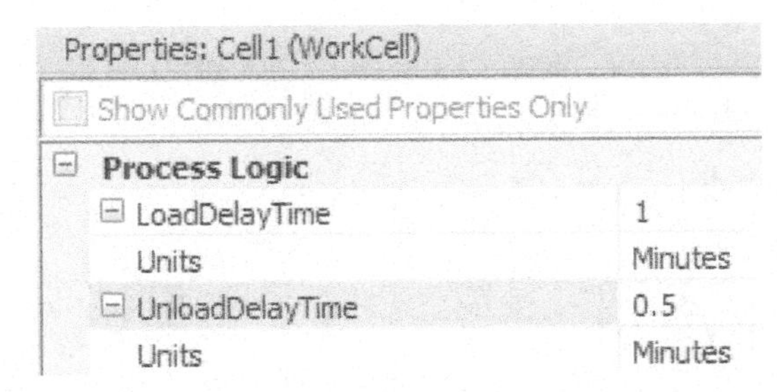

Figure 17.12: Setting the Properties of the WorkCell

Step 5: Save and run the model for 24 hours looking at the results. Notice that the statistics of all the pieces of the work cell are visible under the **WORKCELL** object type.

Question 2: For a single 24 hour replication, what is the average time in system (in minutes)?

____

Question 3: From an animation standpoint, what did you observe when the parts entered the workcell?

____

Step 6: Select the **WORKCELL Cell1** and choose a different symbol for animation purposes.[271]

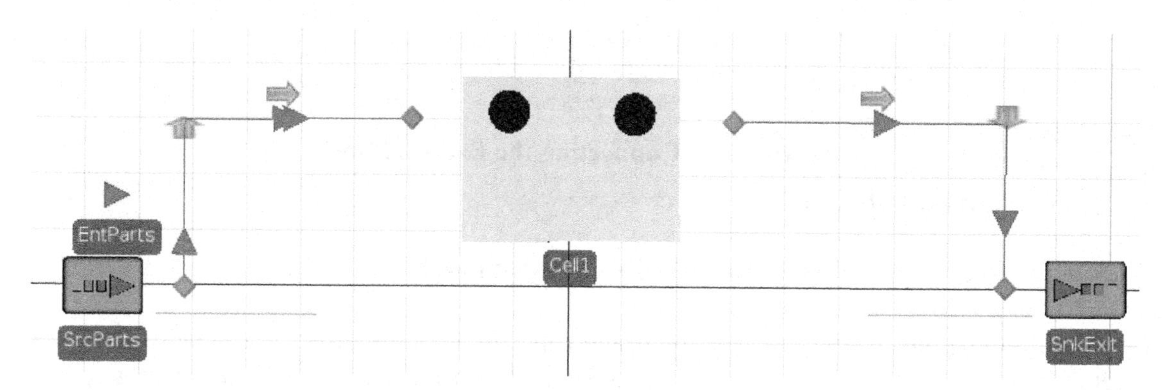

Figure 17.13: Changing the Symbol

Question 4: What did you observe when the parts entered the workcell now?

Part 17.4: Adding to the WorkCell Object

The internal view of the composition object (i.e., sub-model) is really the symbol of the object, which when replaced with a new symbol, the internal workings were removed. In this simple example, showing the internal workings may make sense but in more complicated sub-models, hiding the details of the sub-model maybe more beneficial (i.e., just having a plain symbol to represent the object). However, when using a different symbol, it would be helpful to see some indication of the locations of the entities (i.e., animated queues).

Step 1: Select the **WORKCELL** object in the [*Navigation*] panel and select the "*Facility*" tab. Select all of the animated queues and then right click to toggle off the *Externally Visible* option.

Step 2: Go to the *Definitions→Externa"* section; add four animation queues from the "*Animation*" tab on the ribbon for the new work cell object.

- One queue should animate the `SrvMachineA. InputBuffer.Contents` to represent entities waiting to be processed while both processing queues of each server (`SrvMachineA.Processing.Contents` and `SrvMachineB.Processing. Contents`) should be animated.
- Finally, animate the output buffer queue of the **SrvMachineB** to animate entities waiting to leave the work cell (`SrvMachineB.OutputBuffer.Contents`).[272]

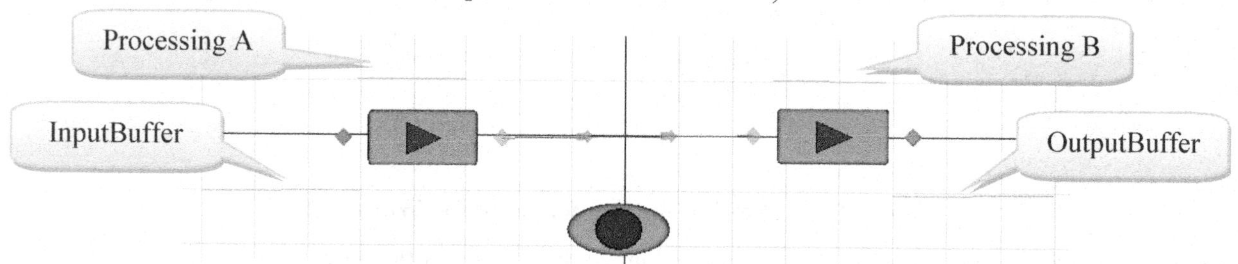

Figure 17.14: External View Showing the Four Animated Queues

Step 3: When the **WORKCELL's** animation symbol is changed, these four queues will still remain visible. Navigate back to the original model. The new animated queues will not be visible on objects already

---

[271] In this example, a factory symbol was chosen and it was resized and rotated.

[272] Draw the queues from right to left because the first point drawn will be the start/head of the queue.

declared in the model. You can either add these manually to the current **Cell1** object or you can delete it and add it back to the model making sure to specify the 10 minute paths, changing the symbol, etc.

Question 5: What did you observe when the parts entered the workcell now?

____

Step 4: Run the model observing the animation of the **WorkCell** when a different symbol has been used.

Currently, the entity will be hidden from animation view when it is on the conveyor as well as waiting for processing at machine B or any other place that does have an animated queue associated with it. Since the animation of the internal submodel is not shown, we can reveal some of the behavior by placing labels that display the number of entities at the various locations. However, it would be helpful to have a representation of the number of parts currently being processed by the work cell either by machine A or B as well as the number traveling on the conveyor as well as those waiting in the input buffer of the machine B.

Step 5: SIMIO has the ability to specify user defined storages (queues). Select the **WorkCell** model in the [*Navigation*] panel and navigate to the *Definitions* tab. Under the *Elements* section, insert a new STORAGE element named **StorageWorkCellInProcess**. We will accept the default ranking rule of "FirstInFirstOut".

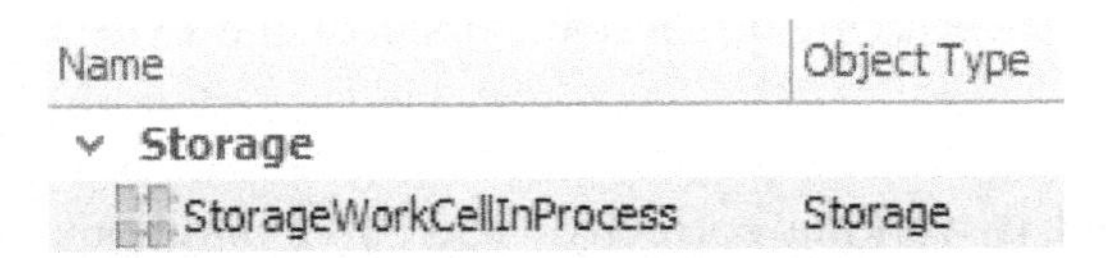

Figure 17.15: Adding a Storage

Step 6: Recall, there are two process steps (i.e., `Insert` and `Remove`) that can be used to manipulate user defined queues.[273] Once the part enters processing at **SRVMACHINEA**, they should be inserted into the **StorageWorkCellInProcess** queue to indicate they have gone into processing. Add an `Insert` step after the operator has been released in the processing with the *Queue State Name* property associated with the new user defined queue `WorkCellInprocess.Queue`. The *Object Type* should remain the "AssociatedObject" because the token is associated with the part.[274]

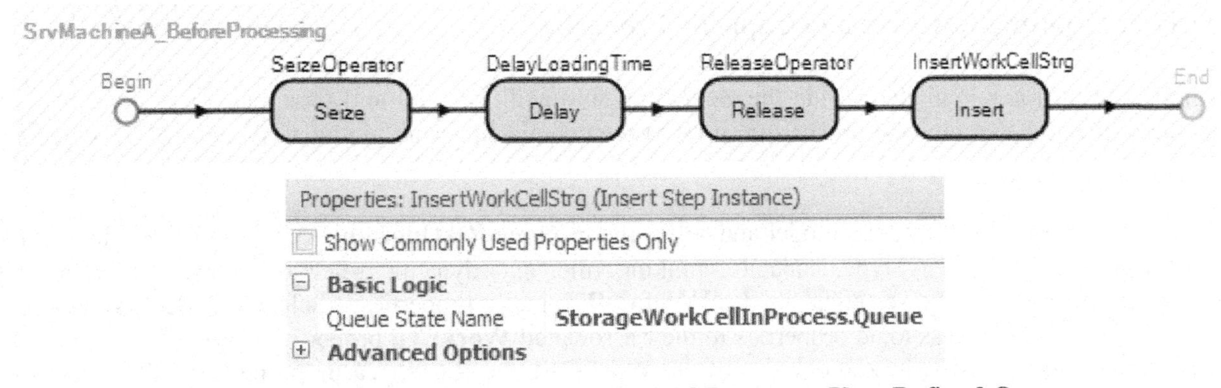

Figure 17.16: Using an `Insert` Step to Add Parts to a User Defined Queue

Step 7: Once the part has finished processing and removed from the machine, delete the part from the user defined queue using a `Remove` step as seen in Figure 17.17.

[273] These steps do not work on queues defined by SIMIO (which is quite a limitation).

[274] The *Rank Placement* property can be used to specify the placement of the entity in the queue ignoring the overall ranking of the queue.

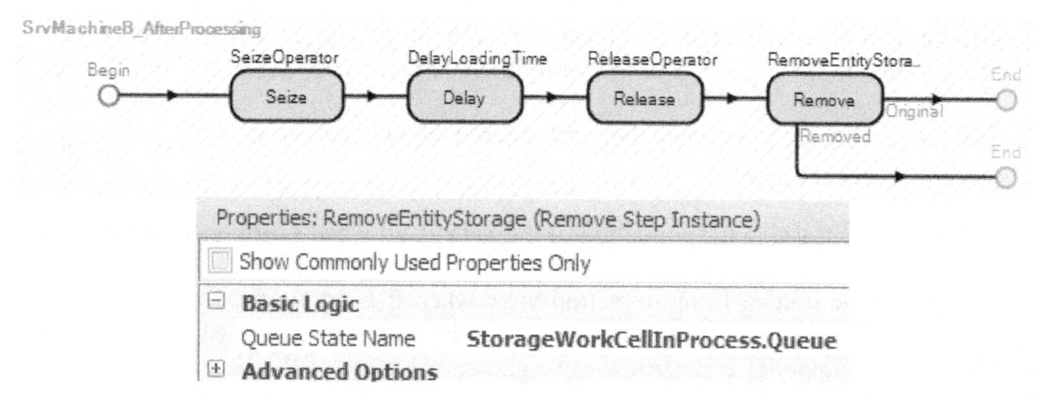

Figure 17.17: Using a `Remove` Step to Delete Parts to a User Defined Queue

Step 8: Go to the *Definitions→External* view section for the **WorkCell** Object and add an Animation Queue above the object that animates the `StorageWorkCellInProcess.Queue` (you may want to delete to the `Processing.Contents` queues for machine A and B inserted earlier).

Step 9: Return to the main model and manually insert an animated queue for **Cell1**, namely `Cell1.StorageWorkCellInProcess.Queue`. Recall, none of the external view changes will appear for objects already inserted.

Step 10: To see that the model is working correctly, insert five additional symbols for the Parts coloring each one with a different color. Set the *Animation→Random Symbol* property of the Parts to "True." Change the mean interarrival time to be Exponential(2) minutes.

Step 11: Save and run the model observing that the parts are now animated even if they currently are on the conveyor. The object now acts like a server with an input buffer, processing queue (i.e., **StorageWorkCellInProcess**), and an output buffer.

Question 6: For 24 hour replication, what is the average time in system (in minutes)?

Part 17.5: Exposing Resource and Capacity Properties

Currently the operator is hidden inside the **WORKCELL** sub-facility model and thus, cannot respond to work schedules. Also, the capacity and processing times of the two machines are hidden to the user of the objects as well.

Step 1: Return to the **WORKCELL** model and select the resource **ResOperator**. All fixed models have built-in properties of capacity type, capacity, ranking rule, and dynamic selection rules. Therefore, the **ResOperator** should be set equal to the **WorkCell** object's properties. Select the **ResOperator** and specify all of the process logic properties to their associated **WorkCell** properties as seen in Figure 17.18. When the user specifies these properties of the **WorkCell**, the **ResOperator** properties will be set to the same.

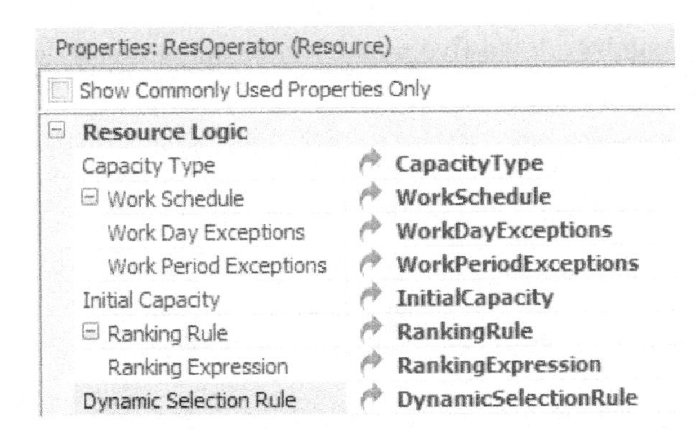

Figure 17.18: Exposing the Resource Capacity and Selection to the User

Step 2: Each of the fixed model inherited properties by default is hidden. Therefore, you will need to go to the *Definitions→Properties* section and expand down the *Properties (Inherited)* and set the *Visible* property to "True" for all the properties in Figure 17.18 as well as change the *Category* to "Process Logic".

Step 3: Create a new reference property for the *Processing Time* property for each of the two servers named **MachineAProcessingTime** and **MachineBProcessingTime** respectively and the *Initial Capacity* property named **MachineAInitialCapacity** and **MachineBInitialCapacity**.

Step 4: In the *Definitions→Properties* section, place all of these new properties into the "Process Logic" category and set the default units of the processing times to minutes.

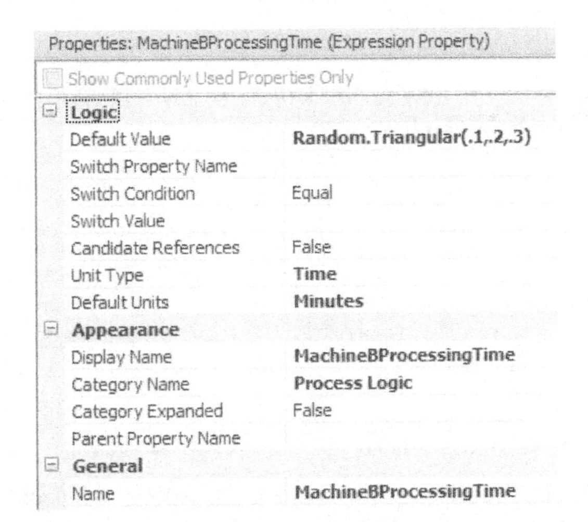

Figure 17.19: Machine B Processing Time Property Values

Step 5: Return the main model making sure to set the processing times of machines A and B to be equal to four and three minutes respectively.

Step 6: Save and run the model observing that nothing has changed.

Step 7: If you were only going to have one cell, then it does not make sense to take the time to create a sub-model object. However, the benefit of a sub-model object is in reusing the object over and over, without having to repeat the same process logic and properties. Another benefit is if the logic changes or additional capabilities are needed, the changes are made in only one place.

Step 8: Insert two new WORKCELL objects named **Cell2** and **Cell3** by copying **Cell1** object to include the same symbol with the same processing times of four and three minutes but make the capacities equal to two for the capacities for the two machines. Notice the new animated queue is automatically added.

Step 9: Connect the SOURCE and SINK via five minute time paths to these new workcells and increase the arrival rate to two per minute.

Step 10: Save and run the model observing the results.

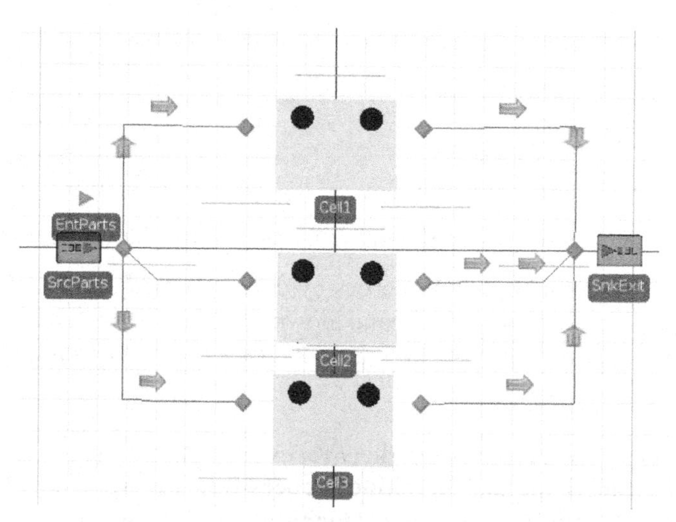

Figure 17.20: Demonstrating the Reuse of the WorkCell Object

Step 11: Next create a DATA SCHEDULE named **SchedCell.** Change the Standard Day such that there is a repeating schedule with three hours being "Off Shift" and three hours being "On Shift". "On shift" should have a capacity of "4" as seen in Figure 17.21.

Step 12: Select all three work cells and specify the *Capacity Type* property to follow a WORKSCHEDULE with the *Work Schedule* property the **SchedCell** as seen below.

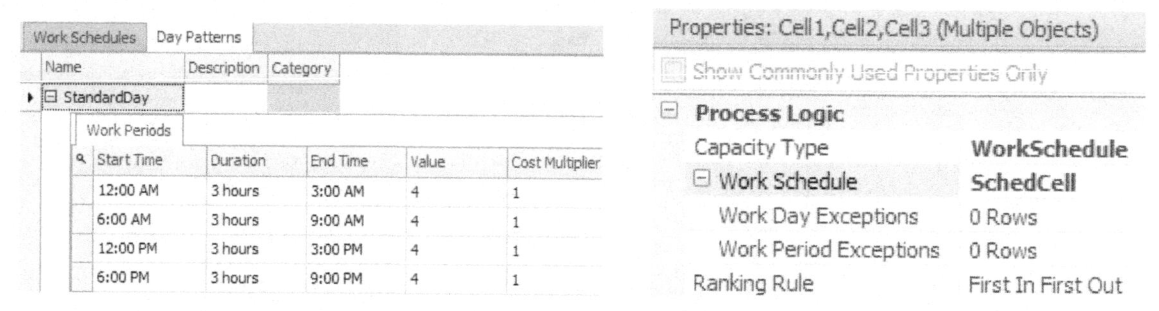

Start Time	Duration	End Time	Value	Cost Multiplier
12:00 AM	3 hours	3:00 AM	4	1
6:00 AM	3 hours	9:00 AM	4	1
12:00 PM	3 hours	3:00 PM	4	1
6:00 PM	3 hours	9:00 PM	4	1

Figure 17.21: Specifying the WorkCells follow a WorkSchedule

Question 7: What do you notice about the parts in-process in the cell when the shift changes?

___

Question 8: For 24 hour replication, what is the average time in system (in minutes)?

___

Part 17.6: Passing Information between the Model and its Sub-Models

Sometimes the sub-model must be constructed using information from the model, but before the model has itself been constructed. For example, suppose the **WorkCell** needs an operator for **MachineA**, but the operator is supplied from the model and not known to the sub-model. The sub-model needs to be created knowing that an operator will be available, but it doesn't know its exact specification. Here a property "reference" is useful for the sub-model, expecting that the property instance will be supplied in the model.

Step 1: Using the model from the previous section, select the **WorkCell** from the [*Navigation*] panel. From the *Defintions→Lists* add a new Object List Property from the Object Reference dropdown named

ResourceList. Set the *Required Value* property to False, since the main model will be supplying the actual list.

Step 2: *Click on the* **SrvMachineA** server and use the *Secondary Resources* section using the *Resource for Processing* specification to select a resource randomly from the **ResourceList** as shown in Figure 17.22.

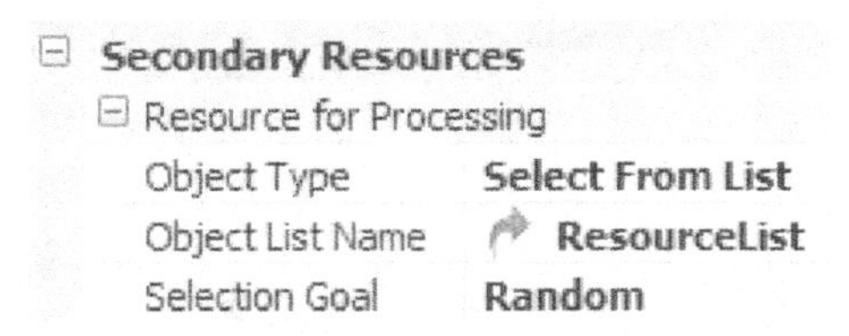

Figure 17.22: Specifying the Resource

Step 3: Select the MODEL and add two resources named **ResOperator1** and **ResOperator2**. Within the *Definitions* tab and the *Lists* option, add a new list of objects. Call it **GResourceList** and have it contain the two operators.

Step 4: By clicking on each cell, specify the **GResourceList** for the individual cell's **ResoruceList**. Doing this will link the model list to each of the sub-model lists.

Step 5: Run the model to see if it appears to be executing appropriately.

Step 6: Currently, we do not know where the operators are working since they are working at the sub-model level. At the model level, add an INTEGER STATE VARIABLE to the MODELENTITY called **EStaCellNumber**, which will be used to identify the cell.

Step 7: In the *State Assignments* section of the TIMEPATHS between the **SrcParts** and the cells, use the ***Before Exiting*** specification to assign the cell number to the **EStaCellNumber.** Figure 17.23 shows the assignment on the path to Cell 3. Repeat for the other two cells.[275]

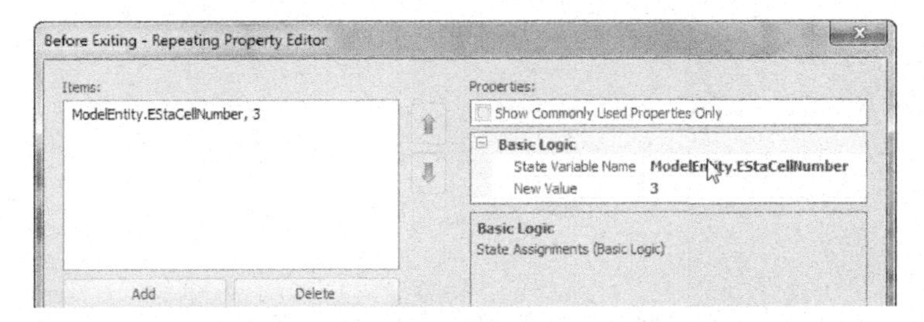

Figure 17.23: Assigning the Cell Number

Step 8: Add *Status Labels* under each resource to dispaly which cell the resource is working (refer to Figure 17.24). Show the status of each of the resources. The following expression is for **ResOperator1** which first checks to make sure the RESOURCE has been seized and if it has will display the entities cell number.

```
Math.If(ResOperator1.ResourceOwner > 0,
     ResOperator1.ResourceOwners.FirstItem.ModelEntity.EStaCellNumber, 0)
```

275 This is another great place to use a Tokenized process to set the cell number instead of using *Before Exiting* assignments especially if the number of work cells increased way beyond three.

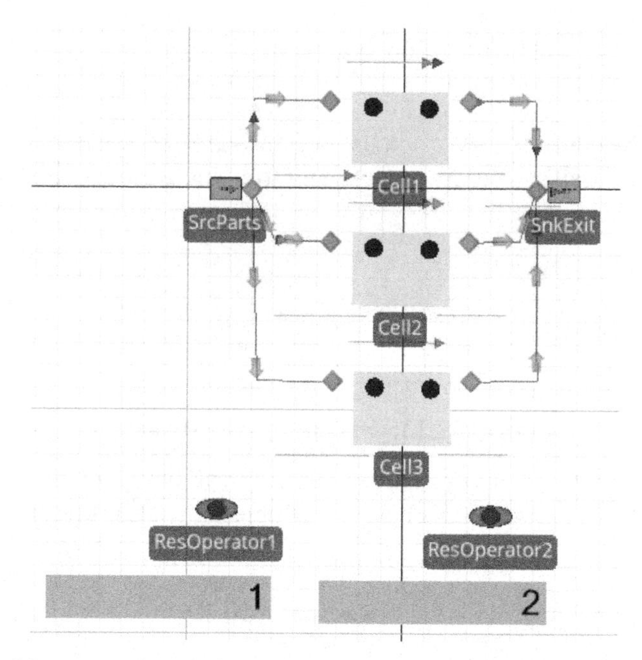

Figure 17.24: Communicating With the Sub-Model

Question 9: Is the model executing at expected?

Part 17.7: Commentary

- The idea behind the sub-model is that you can create your own object class by "composing" a model and then using it as an object. Once that object class is created/defined, you can use the object from that class multiple times in the same project or in different projects.

Chapter 18
The Anatomy of Objects: Server

Many simulation languages do not have the ability to modify the behavior of the objects and thus require another object to perform supporting activities (reneging is an example). Unlike most simulation languages, if you do not like the way the designers of SIMIO implemented the SERVER object (or any other SIMIO object) you can modify it or create your own object.

Part 18.1: A Simple Resource Model: Warehouse Pickup

A warehouse store has a pick-up area where customers come for items that are not stored on shelves. The pick-up area is serviced by an associate who waits on those customers. Because the pick-up area does not require the full attention of the associate, the associate is also asked to do some specialty stocking. Quite appropriately, the store wants the customers serviced before doing any stocking.

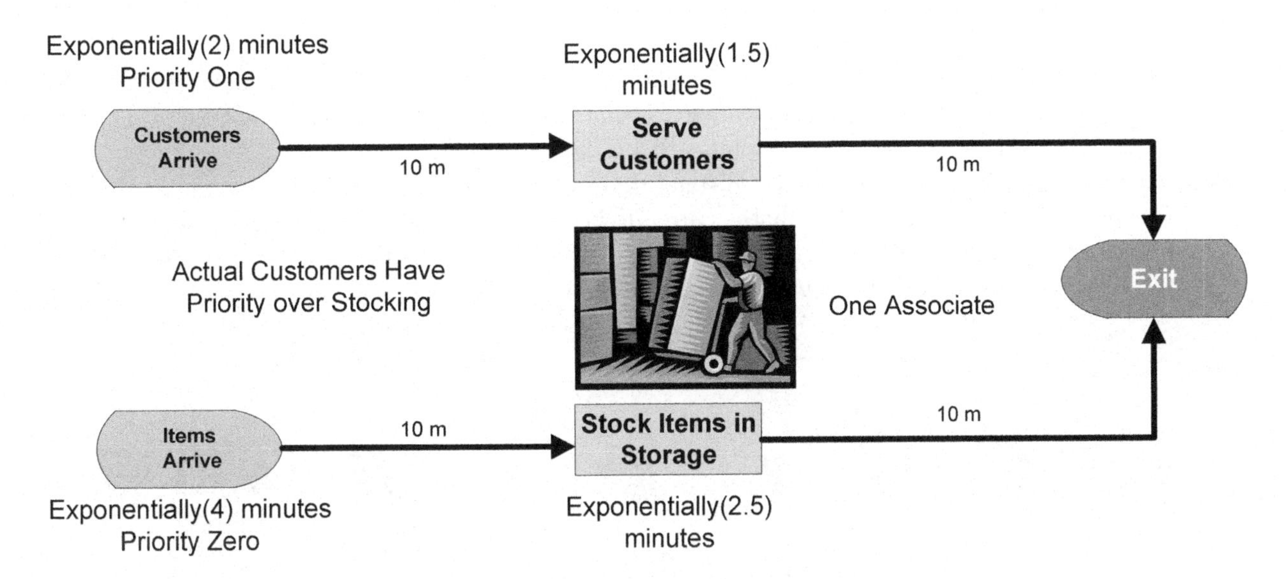

Figure 18.1: Warehouse Operations

Step 1: Create a new model.

Step 2: Insert two SOURCES named **SrcCustomers** and **SrcItems** and two MODELENTITIES named **EntCustomers** and **EntItems**.

- The **SrcCustomers** should create **EntCustomers** while the **SrcItems** creates **EntItems**.
- Customers arrive exponentially every two minutes while items arrive exponentially every four minutes.
- Color the **EntCustomers** red.
- Set the *Initial Priority* to zero for **EntItems** since by default MODELENTITIES have a priority of one.

Step 3: Insert two SERVERS named **SrvCustomer** and **SrvStorage** and one SINK named **SnkExit**.

- Each of the SERVERS should have a capacity of one.
- The *Processing Time* property for **SrvCustomer** should be set to Exponential (1.5) minutes and Exponential (2.5) for **SrvStorage**.
- Connect the two SOURCES to their appropriate SERVER via 10 meter logical paths and the two SERVERS to the SINK via ten meter paths as seen in Figure 18.2.

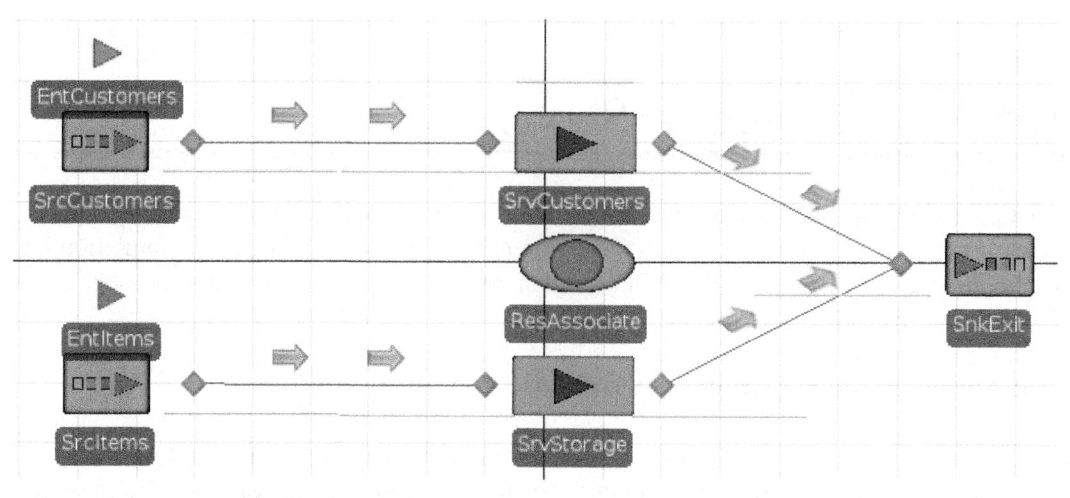

Figure 18.2: Model of a Warehouse Pick-up Stations

Step 4: Insert a fixed RESOURCE named **ResAssociate** between the two SERVERS as seen in Figure 18.2 to constrain the processing of entities.

- Change the **ResAssociate's** *Dynamic Selection Rule* property to be "Largest Value First" with the *Value Expression* left as the default property (i.e., `Candidate.Entity.Priority`). This should force the associate to service the customers first if there is a choice.
- Recall the RESOURCE has seven defined states defined by the `ResourceState` variable. In order to visually indicate the type of part that is currently being served, add an additional symbol coloring the round circle red and the outside green for the 8th symbol to represent busy for red parts. Then select the 1st symbol and color the iris green as seen in Figure 18.3.

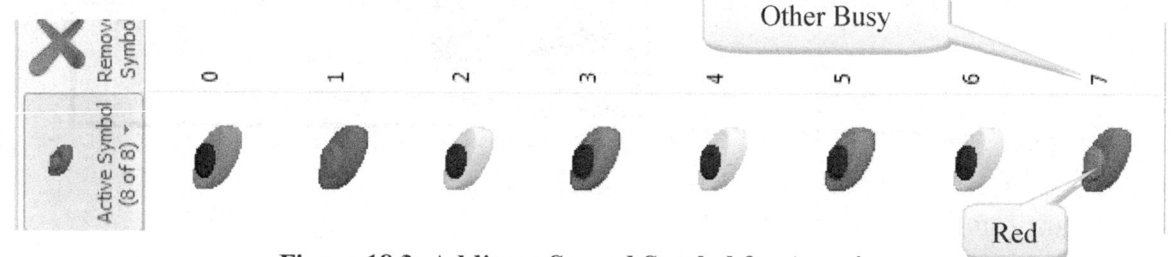

Figure 18.3: Adding a Second Symbol for Associate

Step 5: From the *Definitions→States* section, add a new DISCRETE INTEGER STATE variable named **GStaAssociatePicture** to be used to specify the symbol associated with the customer type since the RESOURCE picture cannot be specified directly.

Step 6: Change the *Current Symbol Index* of the property of the **ResAssociate** to use the new state variable as shown in Figure 18.4. If the state is busy and the entity is a customer then display the 7th symbol otherwise show the default state symbol.

Animation	
Current Symbol Index	**Math.If(Resource.ResourceState==1 && GStaAssociatePicture == 1,7, Resource.ResourceState)**
Random Symbol	False

Figure 18.4: Animation Symbol Index

Step 7: From the *Processes* tab, create two new processes named **SeizeAssociate** and **ReleaseAssociate**. The `ReleaseAssociate` process should release the associate while the `SeizeAssociate` should seize the associate as well assign the **GStaAssociatePicture** the value of the `ModelEntity.Priority` as seen in Figure 18.5.

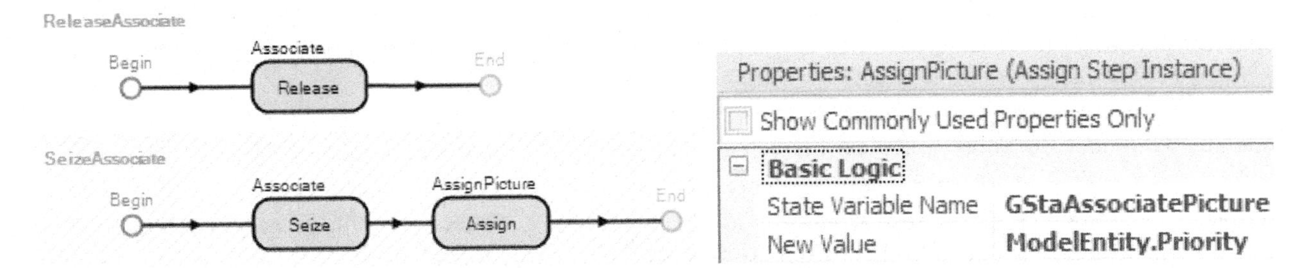

Figure 18.5: Processes for Seizing and Releasing the Associate

Step 8: For both the **SrvCustomer** and **SrvStorage**, specify the *Before_Processing* add-on process trigger to be the **SeizeAssociate** process and the *After_Processing* trigger to be the **ReleaseAssociate** in similar fashion.[276]

Step 9: Save and run the model.

Question 1: When the associate finishes with a customer, does he always attend to another customer if they are waiting?

___

Question 2: Why does this happen, since the *Dynamic Selection* criteria of the **ResAssociate** was specified correctly?

___

Step 10: Select the **ResAssociate** and from the *Attached Animation* section, insert the *Allocation Queue* drawing it from left to right somewhere in the model. The allocation queue represents all the requests for the **ResAssociate** and will be ordered by priority. Save and rerun the model.

Question 3: What do you notice about the arrival of entities with regard to the queue?

___

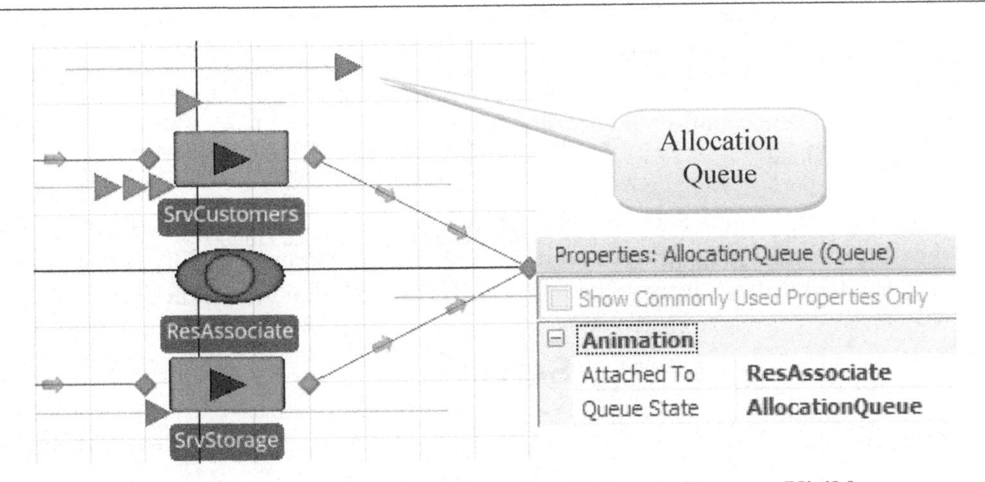

Figure 18.6: Causing the Allocation Queue to Become Visible

Part 18.2: Taking an Object Apart to Figure out how it Works

For some reason, a customer waiting at the **SrvCustomer** does not register a request for service when the associate finishes with another customer. At this point, the problem may not be evident. Unlike other

[276] You can select both of them at the same time and specify the add-on process triggers.

simulation languages, SIMIO objects can be examined by taking them apart, and possibly modifying the behavior. Existing objects can examined as well as modified in SIMIO via "subclassing."

To try to understand why the queue has not registered the other customers, the SERVER will be sub-classed to understand its inner workings. Subclassing provides the ability to extend and modify an existing object.

Step 1: To subclass any of the standard SIMIO objects, right mouse click on the object in the standard library and choose subclass in the menu option as seen in Figure 18.7.

Figure 18.7: Subclassing the Server

Step 2: After subclassing the SERVER, select the **MYSERVER** object in the [*Navigation*] panel. Now all the characteristics of the SERVER become visible. Look at the *Definitions→Element*'s section. The SERVER object defines ten different elements: one COSTCENTER, one FAILURE, three STATIONS, one TASKSEQUENCE, and four TIMER events.

- The COSTCENTER element is used to define the parent activity-based costing for the object.
- The FAILURE element is used to define a failure mode. Failure downtime occurrences are started and ended using the `Fail` and `Repair` steps.
- The STATION element is used to define a capacity constrained location within an object where one or more visiting entity objects can reside. Each STATION defines an `EntryQueue` and `Contents` Queue. The `Contents` Queue contains the entities which have seized capacity of the station and are being processed while the `EntryQueue` will hold the entities waiting for capacity.
- The TASKSEQUENCE element is used to reference the set of tasks that compose the processing time when the *Process Type* is *Task Sequence.*
- The TIMER element fires a stream of events according to a specified *IntervalType* (i.e., calendar time, processing count, event count, or processing time). These timers determine when failures occur if a failure has been defined.

When entities arrive at the input node they are automatically transferred to the "**InputBuffer**" Station if it has capacity. From the "**InputBuffer**" the entities are transferred to the "**Processing**" station and once they have finished processing they will be transferred to the "**OutputBuffer**" station as seen in Figure 18.8.

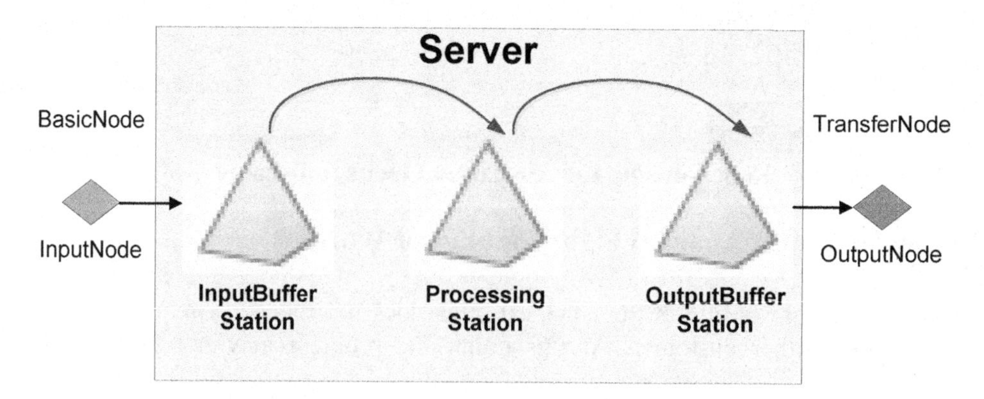

Figure 18.8: Showing the Entity Flow through the Server Object

Step 3: Next, select the "*Processes*" tab to look at the behaviors. The SERVER object defines 13 processes that govern how the SERVER reacts to failures, capacity changes, entity arrivals and exits, etc. Some of the processes respond directly to events (e.g., entities entering the Processing station, entities entering the Inputbuffer, etc.). The two processes of interest (see Figure 18.9 and Figure 18.10) are: the `OnEntered InputBuffer` which is executed when entities enter the **InputBuffer** station from the input node if the input buffer has capacity, and the `OnEnteredProcessing` which is executed when entities enter the processing station from either the input buffer or the input node.

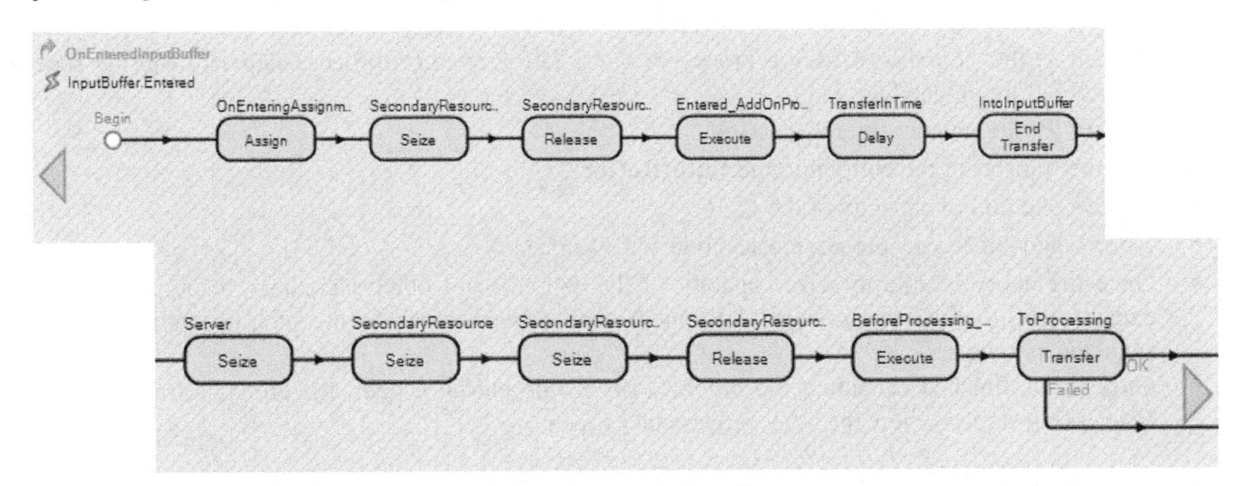

Figure 18.9: OnEntered Processes for InputBuffer Station

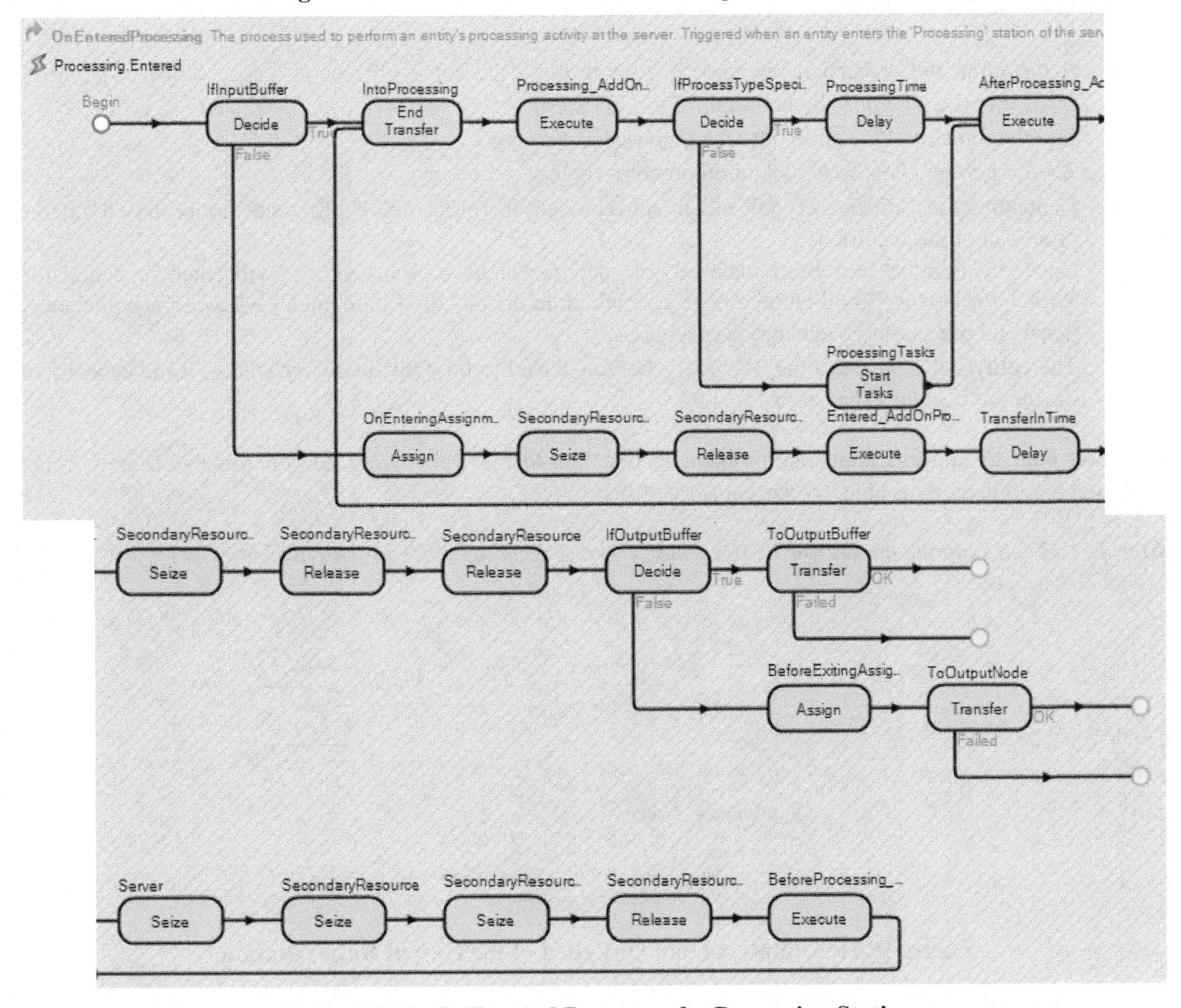

Figure 18.10: OnEntered Processes for Processing Station

When entities arrive at the SERVER they are transferred to the **InputBuffer** which automatically executes the `OnEnteredInputBuffer` process which has the following steps. Examination of these steps will yield some insight into our problem.

- Perform any On Entering State Assignments
- *Seize* and *Release* On Entering Secondary Resources
- *Execute* the *"Entered"* Add-on process trigger. The *Execute* process step will run the specified process before proceeding.
- *Delay* for the transfer time if specified.
- End the transfer of the entity into the **InputBuffer** STATION.
- *Seize* one unit of capacity of the SERVER.
- *Seize* and *Release* processing secondary Resources.
- Once the entity is able to seize capacity of the SERVER and other secondary resources it will then execute the *"Before_Processing"* Add-on process trigger which in the pick-up station model will seize the associate.
- Once it has finished executing the *Before_Processing* add-on trigger, the entity is transferred to the **Processing** station where the other process takes over.

Once entities are transferred to the processing station the `OnEnteredProcessing` is invoked where the actual processing is performed.

- If the input buffer capacity is zero, it goes through the same process as the `OnEnteredInput Buffer` (i.e., the false branch of the *Decide* step).
- End the transfer of the entity into the **Processing** STATION.
- *Execute* the *"Processing"* Add-on process trigger.
- Depending on "Process Type" either start processing tasks or *Delay* the entity based on the processing time specified.
- Once the entity has been delayed or the tasks have finished, it will then execute the *"After_Processing"* Add-on process trigger which in the pickup station model releases the associate.
- Seize and release any Secondary Resources.
- The entity will either exit the SERVER or be transferred to the **OutputBuffer** station if the capacity of the output is greater than zero.

The first clue to understanding our problem is that the *Before_Processing* add-on process trigger is not executed until the entity is able to seize capacity of the SERVER.

Step 4: Next examine the **OutputBuffer** *Entered* and *Exited* processes as seen in Figure 18.11.

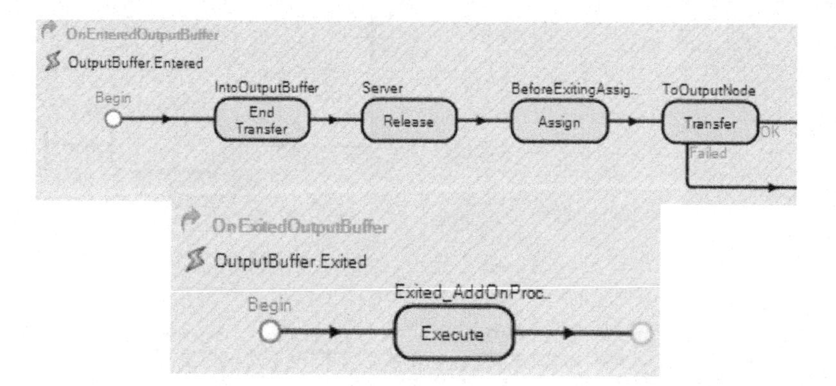

Figure 18.11: OnEntered and OnExited of the Output Buffer Station.

The `OnEnteredOutputBuffer` process will be invoked when the entity is transferred to the **OutputBuffer** STATION and goes through the following steps.

- Ends the transfer into the output buffer station.
- After ending the transfer, the entity releases capacity of the SERVER at this point.
- It then performs the before exiting state assignments.
- The entities are then transferred to the output node which invokes the `OnExitedOutputBuffer` process which executes the *"Exited"* add-on process trigger.

Step 5: Now it can be seen that the *After_Processing* add-on process trigger which releases the associate happens before the SERVER capacity is released. Therefore, the customer waiting in the input buffer cannot request the associate before the associate evaluates the current requests since it first has to seize the server before the `SeizeAssociate` process is executed.

Step 6: To fix the problem, set the capacity of the two SERVERS to "Infinity" which will eliminate the SERVER constraints. Therefore, every entity entering the two servers will immediately execute the *After_Processing* add-on process trigger and request the associate.

Step 7: Save and run the new model.

Question 4: After observing the model for a while, does the associate always service customers first?

___

Step 8: Setting the capacities to "Infinity" eliminates the problem. However there are situations where the SERVER capacity should not be infinite because available space in the server may also limit the number of entities being served. For instance, there may be more than one associate but no room to serve more customers. Notice from Figure 18.11, the *Exited* add-on process trigger happens after the `Release` step and can be used to release the associate.

- Change the capacity of the SERVERS back to one.
- Remove the *After_Processing* add-on process trigger from each SEVER by setting the value to null, removing it manually, or right clicking and then select *Reset.*
- Choose the **ReleaseAssociate** as the *Exited* add-on process trigger for both SERVERS.

Step 9: Save and run the new model. You may want to increase the item arrival rate to one or two per minute.

Question 5: After observing the model for a while, does the associate always service customers first?

___

In other instances, having the correct order may have resulted in the same phenomena owing to zero time events which can be checked by looking at the "Model Trace". Recall one can reorder the zero time events by placing a `Delay` step right before a step you want to force to the end of the zero time by delaying it for zero time using `Math.Epsilon` as seen in **Figure 18.12**. When SIMIO encounters a delay of `Math.Epsilon` it places the steps after at the end of the current event time.

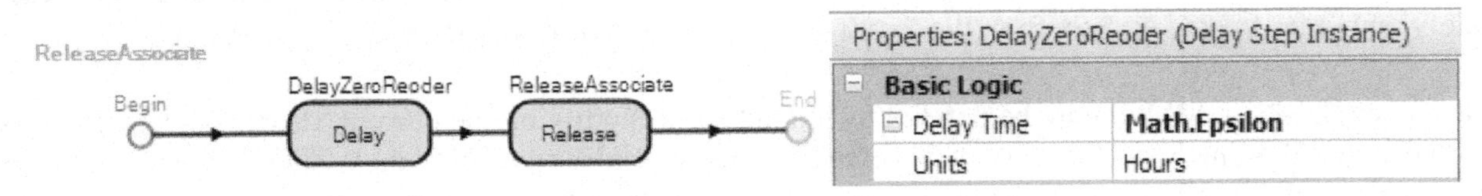

Figure 18.12: Adding the Zero Time Delay

Part 18.3: SIMIO Objects and Class Heirarchy

Objects are created (instantiated) from classes. The act of clicking on a SIMIO object in the [*Standard Library*] panel and dropping it into a position on the modeling canvas causes an object or model of that type to be created (instantiated). The object will possess all the characteristics defined for that class. Once on the modeling canvas, you can change (even delete) its "Default" characteristics such as initial property values, processing time, capacity, add-on processes, and so forth. You can also add your own properties, events, states, and elements. Each SIMIO object will have its own collection of processes, elements, properties, states, and events. Objects can be saved in a library and used in other objects. SIMIO makes no distinction between a model and an object.

Figure 18.13 shows the base object classes in SIMIO as well as the [*Standard Library*] model classes. New classes can be created from base classes and new classes can be obtain from the [*Standard Library*] (or your library) by "sub-classing".

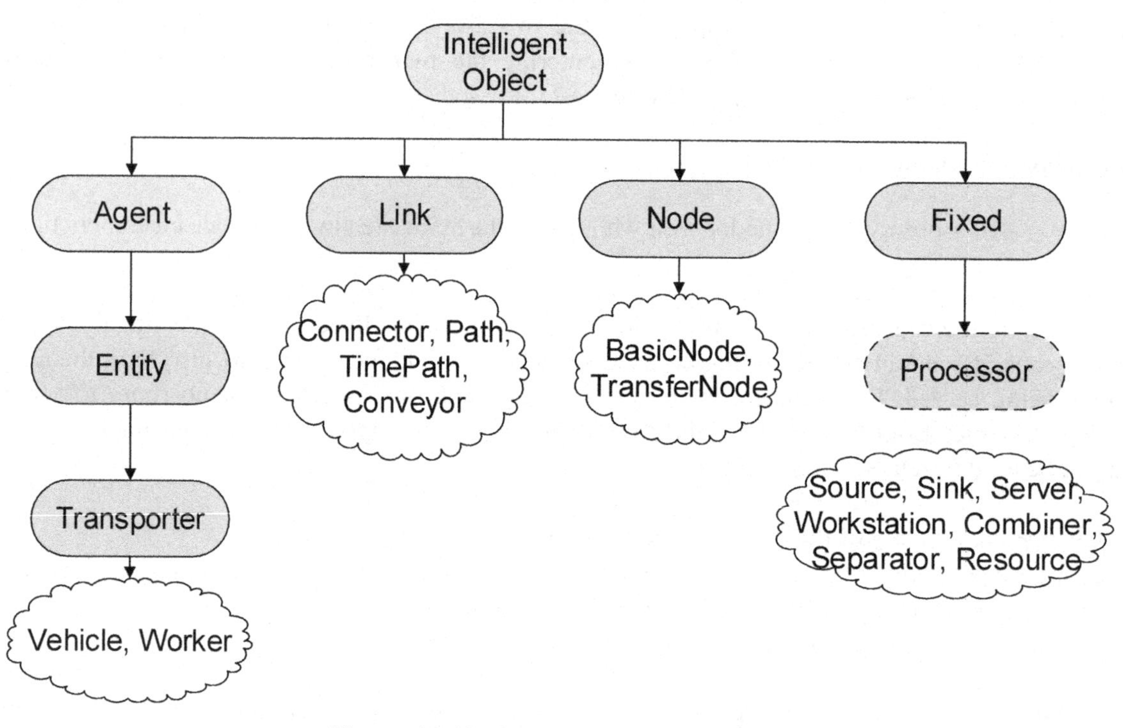

Figure 18.13: SIMIO Class Hierarchy

The "origin" object class is the INTELLIGENT OBJECT (IO) class which gives objects the ability to be seized and released as well as respond to schedules. The AGENT class inherits the properties and characteristics of IO class but adds additional characteristics and behaviors (i.e., objects that can be dynamically created and destroyed as well as move around in continuous and discrete space (grid)). . The LINK adds the ability (i.e., properties and behaviors) for ENTITIES to enter, exit and move along. ENTITIES are specialized agents that have the ability to move on links (paths, conveyors, and connectors) as well as move in and out of FIXED objects (combiners, servers, workstations). FIXED objects represent intelligent objects that are fixed and don't move in the space. The TRANSPORTER is a more specialized entity that has the additional capability of being able to pick up and drop off entities. They can do this by moving in free space or along links. At present models cannot be created directly from the INTELLIGENT OBJECT or from the ENTITY classes.

New object classes in SIMIO may be defined by extending or modifying the current SIMIO objects. In object oriented terms this method of extension is referred to as specialization or "subclassing." When a model is subclassed, it inherits all of the characteristics (i.e., processes, elements, properties, states, and events) but has the capability to modify those inheritable features as well as add new ones.

The object classes in the SIMIO [*Standard Library*] should be viewed as model classes which have been specialized from other classes. Part of their characteristics are derived from the base classes, but some characteristics are added or changed. For example, a WORKER object is a specialization of an entity that can act as a RESOURCE and a VEHICLE. WORKERS and VEHICLES are specialized TRANSPORTERS. Several properties (e.g., **Initial Node**, **Idle Condition**, etc.) can be modified and added as well as adding new process logic to force the entity to behave similar to resources and vehicles. The LINK, NODE and SERVER objects have the same properties as the FIXED class, but add more functionality. Likewise the PATH, CONNECTOR, and CONVEYOR objects are more specialized versions of the LINK class. The SERVER class is a fairly general object used to constrain and delay entities in the system. The PROCESSOR class is a specialization of the FIXED class

The ability to extend and add objects is an important feature of SIMIO that needs to be studied carefully. Thus if you are only going to need the sewing machine class for one instance of one model, you may not want to take the time to create an object class. However, if you're going to use this class in multiple places, then having a reusable object will be important and its utility will be demonstrated over the next few chapters.

SIMIO uses a three tier object hierarchy composed of the "object definition", the "object instance", and the "object realization". The object definition is simply the specification of the object class through its processes, states, elements, properties, and events. The definition may be determined by the SIMIO developers for the base classes and [*Standard Library*] objects, however a modeler may create their own object definitions (as we will see in the next several chapters). The object definition is shared by all instances from that class. While all objects from the same class share the same definition, each object instance has its own characteristics such as their own add-on process, state variable value, capacity, etc. The object instance may have its characteristics changed during the simulation.

Object realizations hold the state values for the instances and these occupy a fixed location in the model. However ENTITIES, VEHICLES, and WORKERS, are dynamic and are created from the instance definition during the execution of the simulation. So SIMIO creates static representation for that part of these object definitions that are common, but then creates a dynamic representation for the instances that move throughout the simulation. You can see this in the model with a vehicle, whose object realization shows on the modeling canvas, while the individual vehicles move around the model.

Chapter 19
Building New Objects via Sub-Classing: A Delay Object

In the prior chapter we demonstrated how specializing/subclassing existing objects allows the characteristics and behaviors of an object to be exposed. The subclassed object may have characteristics and behaviors added or existing ones to be modified. To illustrate we will create a new object through subclassing an existing objects and change its character. In particular, we address the need for a delay node. Sometime entities need to be delayed for a certain amount of time before continuing to their next destination. The SERVER object can be used to do this, but it is really more complicated than is needed to perform a simple delay. We will extend the TRANSFERNODE by adding the ability to delay entities that pass through it.

Part 19.1: Sub-Classing the TRANSFERNODE to Create a DELAYNODE

The TRANSFERNODE is an object that is used to transfer entities out of standard objects (e.g., SERVER, SOURCE, etc.) or as transfer locations in a network of links. It provides an excellent choice to delay an entity before continuing to its next destination. The logic to ride on transporters, continue by sequence, and take the shortest path is already built into the object and these properties and behaviors do not need to be reinvented in a new object. Therefore, the TRANSFERNODE will be specialized/sub-classed. This approach illustrates how an existing object is extended and modified. Objects created from existing objects are referred to as a "derived" objects.

Step 1: First create a new project and model in SIMIO.

Step 2: To subclass or "derive" any of the standard objects[277], right mouse click on the object and choose subclass in the menu option as seen in **Figure 19.1**.

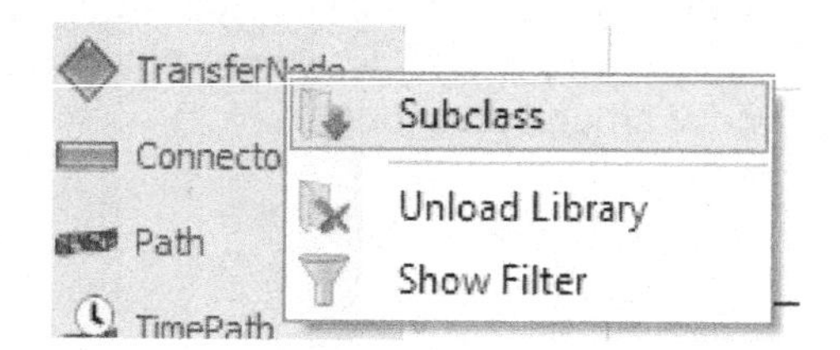

Figure 19.1: Sub-Classing the TRANSFERNODE

Step 3: A "derived" version of the TRANSFERNODE class is placed in the [*Navigation*] panel as **MYTRANSFERNODE**. Right click the new class and select the *Model Properties* option to display all of the properties.

- Change the *Model Name* property to **DELAYTRANSFERNODE**.
- Change the *Icon* property to any appropriate icon (i.e., a png or bitmap file).[278] The icon is the picture that is seen in the project window that is used to select the object. For example, the *Delay.png* file seen in Figure 19.2 will be used in the chapter.

[277] You can also subclass any user created object as well.

[278] Icons are typically 32 pixels wide by 32 pixels high. Many image editing software can create these icons (e.g., Photoshop, Paint or freeware GIMP program. (http://www.gimp.org)).

Figure 19.2: Icon for the DELAYTRANSFERNODE

- Under the "General" category, change the color of the new **DELAYTRANSFERNODE** to something like maroon.[279]
- Select the new **DELAYTRANSFERNODE** in the [*Navigation*] panel and let's take a moment to look at the processes and characteristics that were inherited. Navigate to the *Processes* tab and see the processes (i.e., behaviors) in **Table 19.1** that have been inherited from the parent TRANSFERNODE class.

Table 19.1: Processes Inherited from the Parent TransferNode

Process Event	Process Description
OnEntered	This process runs after the **leading edge** of an object (ENTITY, TRANSPORTER, or AGENT) has entered the node.
OnEnteredFromAssociatedObject	This process runs after an ENTITY has entered the node from another object (i.e., dropped off by a VEHICLE).
OnEnteredParking	This process runs after the **leading edge** of an object (ENTITY, TRANSPORTER, or AGENT) has entered the parking station.
OnEnteredToAssociatedObject	This process runs after an ENTITY has entered the node and is heading to an associated object. An object uses the node as its input node (e.g., entering the SERVER object.).
OnExited	This process runs after the **trailing edge** of an object has exited the node.
OnRunEnding	This process runs when the simulation ends.
OnRunInitialized	This process runs when the simulation starts initially.
OnEnteredParking	This process runs when the ENTITY or TRANSPORTER enters the parking station which only ends the transfer.
RoutingOutLogic	A process used to handle the logic of routing objects out of the node (i.e., ride on a TRANSPORTER, by sequence, etc.)
TransferFailureLogic	If an object fails to transfer to an outbound link it will destroy the object or park it if it fails (i.e., Transporters that cannot be destroyed).

- Select the *Definitions* tab to access the characteristics of the node. Elements represent objects in a process that change state over time (i.e., statistics, timers, monitors, failures, etc.). There are two elements defined: a parking STATION and a ROUTING GROUP.
  - The PARKINGSTATION station is used to house entities, operators, and transporters parked at the node. The station element will be discussed in more detail in the next chapter.
  - The ROUTINGGROUP element is used with a `Route` step to route an entity object to a destination selected from a list of candidate nodes. The TRANSFERNODE object uses a `RoutingGroup` element in its internal logic to determine where the entities will travel when they leave the node.
  - The new sub-classed node inherits several properties and states as well as four events from the parent TRANSFERNODE as described in Table 19.2.[280]

[279] You cannot change the picture (i.e., diamond) because nodes are special objects that take up zero space in the actual system.

[280] To access Inherited items, click on the ˅ to make the items visible.

Table 19.2: Events Inherited from the Parent TransferNode

Event	Event Description
`CapacityChanged`	The node's capacity has changed.
`Entered`	An object's leading edge has entered the node.
`Exited`	An object's trailing edge has left the node.
`RiderWaiting`	An ENTITY is waiting for transport at this node.

- Two string lists (i.e., **TransferNodeDestinationType** and **TransferNodeTallyConditionType**) have been defined which are used in property dropdowns. The **TransferNodeDestinationType** list is used to specify the property of how to route the entity to its next destination node (i.e., Continue, Specific, BySequence, and SelectFromList) while the **TransferNodeTallyConditionType** is the condition type (NoCondition, IsEntity, IsTransporter, CustomCondition) used to compute Tally statistics at this node.

Part 19.2: Modifying Processes and Adding a Properties for the New Node

The new delay node sub-classed from the standard TRANSFERNODE will operate just like the original node since nothing was added or modified. The node needs to delay the objects that enter the node either individually or via a transporter.

Step 1: Under the *Definitions* tab in the delay node, add an *Expression* property[281] named **DelayTime** that can be used by a user to specify the delay time for the entity. Specify the *Default Value* property to be zero while the *Unit Type* should be "Time" with a *Default Units* of "minutes" as seen in **Figure 19.3**. Place the new property underneath the "Process Logic" category.

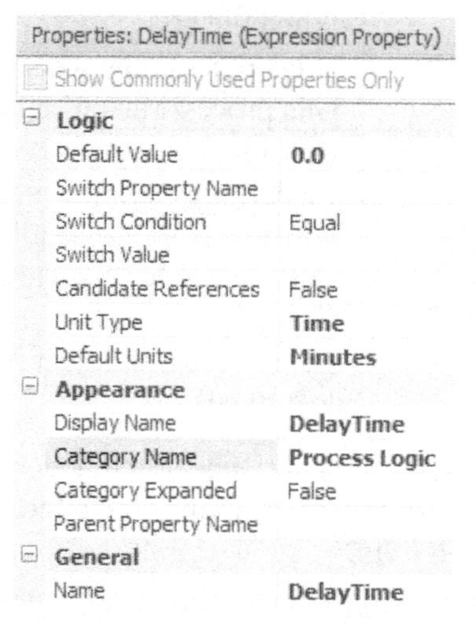

Figure 19.3: Inserting a Delay Expression Property

Every time, an entity enters this transfer node either on its own or via an associated object, it should be delayed by the delay time expression property. Therefore, the `OnEnteredFromAssociatedObject`[282] and the `OnEntered` processes will need to be modified to reflect this consideration. Currently the processes are not editable, which can be seen by selecting the `OnEntered` process and noticing the steps and properties are grayed out. After selecting the process,[283] click the *Override* button ( )which will allow changes to be

[281] The *Expression* Property allows the user to specify any mathematical expression similar to a *Processing Time* of the SERVER.

[282] For example, the object is entering via transporter.

[283] Select the *Processes* tab to access all the processes.

made. Once the process has been overridden the *Override* symbol is placed by the process name to indicate it has been modified. Note, the *Restore* button ( ) can be used to return the process back to its original state.

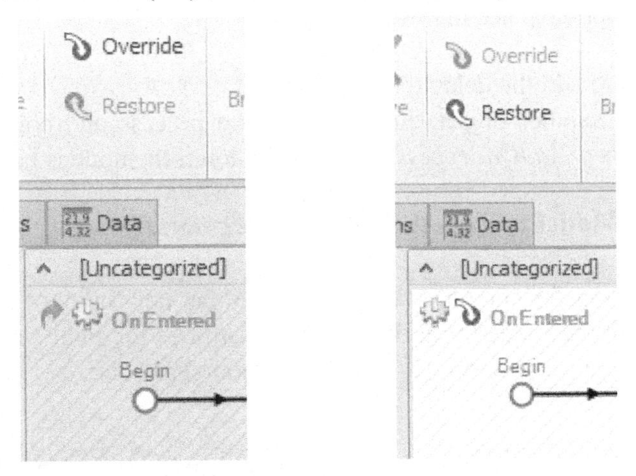

Figure 19.4: Overriding and Restoring Inherited Processes

The `OnEntered` process is fairly complex with the following basic steps.

- On Entering Tally statistics to be collected if specified.
- `Fires` the "*Entered*" Event.
- Then `Executes` the "EnteredAddOnProcess" that a user may have specified.
- Checks to see if node is "Bound to an External Node" and if so transfers it to the specified parent external node.
- The `Visit Node` step is used to invoke the entity's `OnVisitingNode` process to indicate the entity has arrived at the destination node.
- If this entity is a transporter then transfer it to the outbound link since it does not go through the same logic and if there is no outbound link, the transporter is sent to free space or processes a transfer error.
- If the entity is not a transporter, then it executes the `RoutingOutLogic` process to transfer the entity out of the node with the correct logic (i.e., uses the "`Route`" step in conjunction with the `ROUTINGGROUP` element). If not riding on transporter in routing out, then the transfer to the outbound link occurs and the prior logic applies.

Question 1: Where should the delay step be inserted into the process?

____

Step 2: In some cases, users may want change the delay time based on some condition when they enter the node. Therefore, insert a `Delay` step after the *Entered Add On Process* is executed to allow the processing time to be potentially set in the add on process before the delay occurs. The *Delay Time* property should be set to the **DelayTime** reference set in the previous step as seen in Figure 19.5.[284] Under the *General* Section, the *Name* property can be specified to be "**DelayTime**" for a better description.

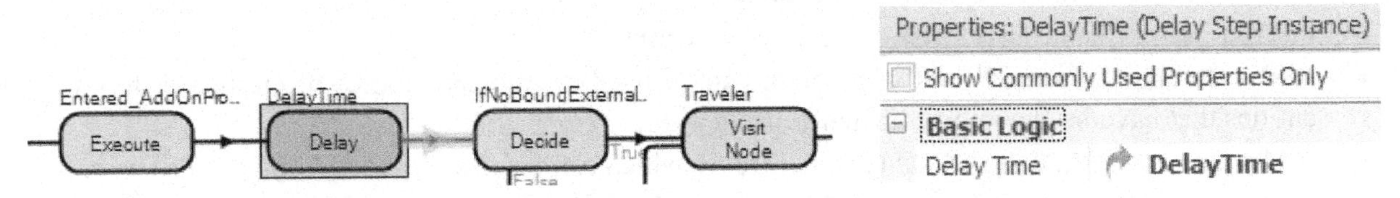

Figure 19.5: Modifying the OnEntered Process to Include the Delay

[284] Right click on the property label to specify a referenced property.

Step 3: Repeat the procedure for the `OnEnteredFromAssociatedObject` process by adding the `Delay` step in the same location. Once the process has been overridden, just copy the delay step and paste the step into the appropriate place in this process.

Step 4: We may want to add the delay to the `OnEnteredToAssociatedObject` if you were to use as an external node for another object. Just override the process and copy the `Delay` step between the execution of the *Entered_AddOnProcess* and the decision if the node is bound to an external node.

Part 19.3: Creating a Model to Test the New DelayTransferNode

Now that the new delay transfer node has been created, it can now be used in any model where the original TransferNode is used with the added benefit of delaying entities for some time as they enter the node. Figure 19.6 shows a network where we will test our new **DelayTransferNode**.

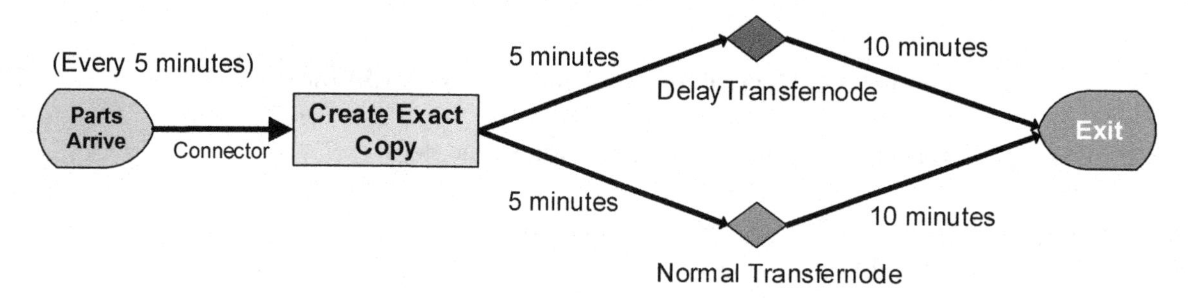

Figure 19.6: Building an Example to Test our DelayTransferNode

Step 1: Go back to the original Model by selecting it in the Navigation panel.

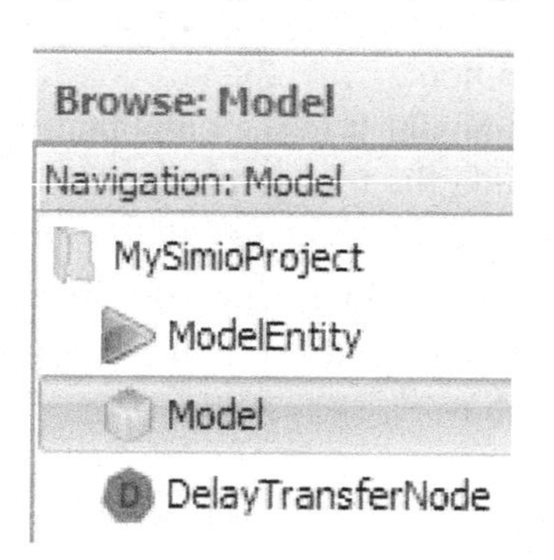

Figure 19.7: Navigating Back to the Original Fixed Model

The model as seen in Figure 19.8 will create an entity and then duplicate it sending the original through a **DelayTransferNode** and the duplicate through a standard TransferNode. The time in the system will be used to illustrate the delaying behavior.

- Insert a new ModelEntity named **EntParts**.
- Insert a Source named **SrcParts** that creates **EntParts**.
- Add a Separator named **SepCopy** which can be used or to create copies of entities or to un-batch entities that have been combined/batched using a `Combiner`.
- Add a new DelayTransferNode named **TDelay** which can be accessed via [*Project Library*] panel.
- Add a normal TransferNode named **TNormal**
- Add a sink named **SnkExit**.
- Connect the **SrcParts** to the Separator by a connector.
- Connect the parent output to your new **TDelay** and the member output to the **TNormal** node via TimePaths that each takes five minutes.

- Connect both **TNormal** and **TDelay** to the SINK via ten minute time paths.

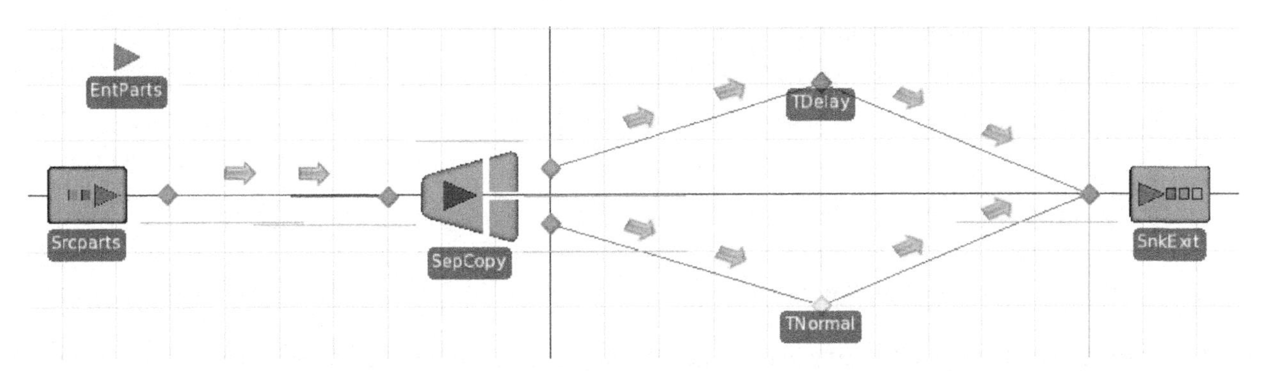

Figure 19.8: Model Used to Demonstrate the New Delay TransferNode

Step 2: Configure all the fixed objects with the following parameters.

- SOURCE: Parts should arrive at a constant interarrival time of five minutes.
- SEPARATOR: *Separation Mode* property should be specified to "Make Copies" and the *Copy Quantity* property should be one, as seen in Figure 19.9. This will make one copy of the current entity. All properties of the entity will be copied. However, the entity's creation time is the current time and not necessarily the same time as the original entity. The original entity will flow out of the **ParentOutput** node while the copy will flow out of the **MemberOutput** node.

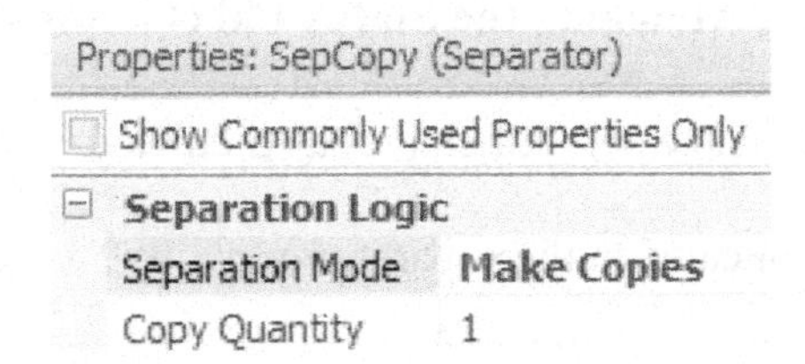

Figure 19.9: Separator Properties

Step 3: A new characteristic needs to be added to the **EntParts** MODEL ENTITY to track the current time at certain points in the system. Select the `ModelEntity` in the [*Navigation*] panel and add a new Discrete Real State variable named **EStaTimeClock** underneath the *Definitions* tab. A real state variable is needed since the value will change throughout the run of the system. Make the *Unit Type* **Time** and the *Units* **Minutes**.

Step 4: Let's add a status label to the ENTITY that displays its associated **EStaTimeClock** state variable. Select the **EntParts** object in your model and then choose the *Status Label* under the *Symbols→Attached Animation* section.[285] Draw a slender status label near the entity with the expression equal to `String.Format("{0:0.###}",EStaTimeClock)`[286] seen in **Figure 19.10**.

---

[285] We could put the expression in the *Dynamic Label Text* under the *Animation* section of the entity as well but this is more visible.

[286] This will format the time clock to have three digits of precision.

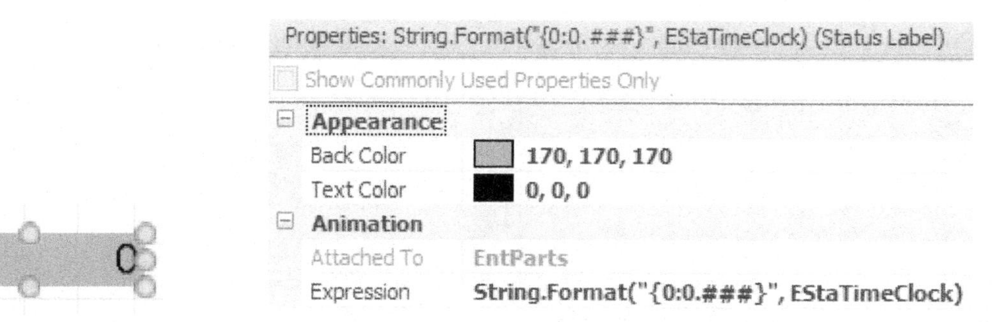

Figure 19.10: Creating a Status Label to Track the Part's EStaTimeClock State

Step 5: Since the status label is attached to the MODELENTITY it will travel with it. Therefore, rotate (via *Ctrl→Mouse*), size and move the label so it sits right on top of the entity as seen in the Figure 19.11 below.

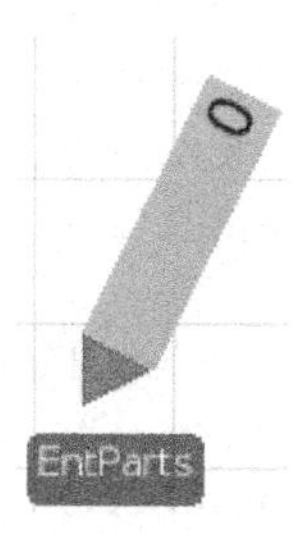

Figure 19.11: Attaching the Status Label to the Part Entity

Step 6: Save and run the model to make sure it behaves as predicted.

Question 2: Does the part get duplicated at the separator correctly?

___

Question 3: Does the status label travel with the **EntParts** and what is the value of all the labels?

___

Step 7: The value of the **EStaTimeClock** state variable has not been updated at this point. Navigate to the "*Processes*" tab and create a new process named **EStaTimeClock**. Insert an `Assign` step that sets the **ModelEntity.EStaTimeClock** to a new value of `60*Run.TimeNow`[287] which set the variable to the current simulation time in minutes.

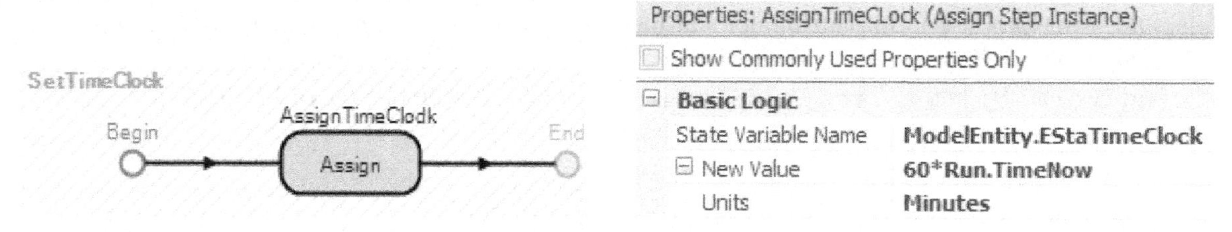

Figure 19.12: Creating a Generic Process to Set the Model Entity's StaTimeClock

Step 8: When the **Parts** exit the SEPARATOR either through the Parent or Member output, the **Parts EStaTimeClock** variable needs to be set to the current time. Select the **ParentOutput@SepCopy** transfer node and specify the **SetTimeClock** process for the *Entered* add-on process trigger which will assign the

___

[287] `Run.TimeNow` is a function that returns the current time of the simulation in hours.

current simulation time to the **EntPart's EStaTimeClock**. Perform the same process for the **MemberOutput@SepCopy** *Entered* add-on process trigger.

Step 9: Repeat the process for the *Entered* and *Exited* add-on process triggers for both the **TNormal** and **TDelay** transfer nodes.

Step 10: Save and run the model. Observe the time stamps as the entities leave the separator as well as the times to enter and exit the transfer nodes. You would think that the time stamps should be the same since the delay time is zero and whole numbers are used for the arrival and time paths. You may want to set a breakpoint on one of the nodes and then single step until the entity leaves the transfer node.

Question 4: Are the entering and exiting times different for the parts and if so why?

It seems there is a small amount time being delayed in one of the transfer nodes as compared to one another. They enter the nodes at the same time (i.e., at time five in the left side of Figure 19.13) but appear to leave at slightly different times as seen in the right side of Figure 19.13. However, transfer nodes take up zero space and therefore should not have caused any delay (i.e., discrepancy). The time an entity enters the node should be equal to time it left.

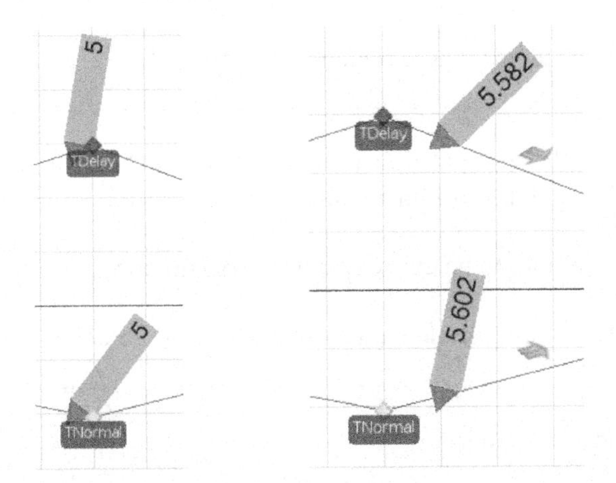

Figure 19.13: Showing Discrepancies Between Exited and Entered Times

The discrepancy deals with when the *Exited* process event is fired. This event will not fire until the trailing edge of the entity has left the node, which depends on the length of the entity and how fast the entity is transferred onto the link (i.e., function of the entity's speed and the speed of the path its leaving and the one its entering). The *Entered* process event is fired once the leading edge has entered the node. Therefore, when setting time stamps the add-on process trigger selected should be based on the leading edge of an ENTITY. Figure 19.14 demonstrates that both entities enter the node at the same time. However, the smaller entity will exit the node first as shown.

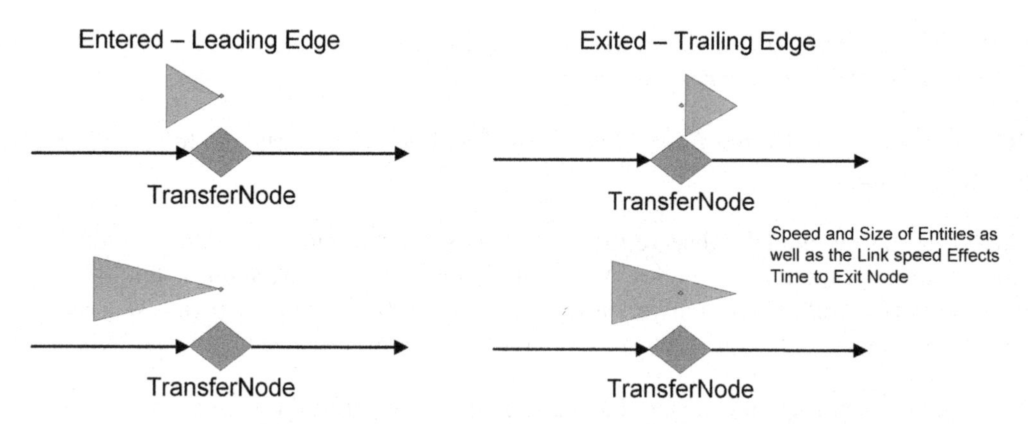

Figure 19.14: Demonstrating Entering and Exiting from Nodes

Step 11: Reset (remove) the *Exited* add-on processes of both transfer nodes (i.e., **TDelay** and **TNormal)** and reset the *Entered* the two outputs of the **SepCopy** to remove the add-on process trigger.[288] Specify the **SetTimeClock** process for the *Entered* add-on process for each of the four time paths leaving the two transfer nodes and the SEPARATOR.

Step 12: Save and run the model to demonstrate that the entities are correctly synced.

Question 5: Are the time stamps whole numbers and synced?

____

Step 13: Next, specify a delay time of four minutes for the **TDelay** transfer node and rerun the model. The part entities are now being delayed four minutes before proceeding into the rest of the network.

Step 14: Change the arrival rate to be every two minutes and observe what happens.

Question 6: What happens to multiple entities that arrive at the delay node?

____

Part 19.4: Commentary

- Derived objects are created by subclassing in SIMIO. Derived objects add or change characteristics and behaviors of existing objects. In this chapter, delay features were added to the transfer node and the processing modified accordingly.

[288] Recall to reset a process trigger, right click the process and select the "*Reset*" item.

Chapter 20 Creating New Objects

Traditionally an object is created (instantiated) from an object class. The class describes an object's composition in terms of the attributes an object has and the methods (procedures) it performs. In SIMIO the model itself is an object (from the FIXED CLASS). When SIMIO is initiated, a MODEL and a MODELENTITY is automatically created within a new PROJECT LIBRARY. The MODELENTITY defines the DEFAULTENTITY as part of the project and instances of this run-time class can be created throughout the simulation run. The MODEL is formed by combining instances/objects from the libraries available, usually including the [*Standard Library*]. It is not possible in SIMIO to create completely independent simulation objects as one might do in a programming language. Models must be formed from either the base classes or by "subclassing" an existing object class. The five base classes are: FIXED CLASS, ENTITY CLASS, TRANSPORTER CLASS, NODE CLASS, and LINK CLASS. There is a PROCESSOR class that is derived from the FIXED CLASS to help in creating processing objects.

When developing new objects, the user must adopt an "architect's view" in addition to a "user's view" since few object class are developed to create only a single object. Instead, new object classes, like a SERVER or TRANSFER NODE in SIMIO, is expected to be used many times in one or more simulation projects. Therefore the purpose or scope of an object's use becomes central to its design. SIMIO object classes have been developed for many purposes and with diverse users in mind.

Consider the SERVER object class which is basically designed as a single queue (input buffer) with fixed or scheduled multiple identical servers (at processing). However, it allows for many specifications including static and dynamic ranking, transfer-in time, flexible processing time with the option of tasks, buffer capacities, material handling, reliability of servers, activity-based accounting, secondary resources, state assignments, etc. In other words, this object class is designed for a wide variety of uses. While state assignments, the acquisition of additional resources, and many of the other features could be user-implemented in SIMIO processes, they are made "convenient" specifications for users not wanting to write processes, making the object class more accessible to casual or learning users. As you review the design of the SERVER you can begin to understand the architectural "paradox" – namely, you want your objects to be "powerful" with lots of features for widespread application, and yet you want them to be "simple" so they can be quickly understood and used. It's not an easy compromise. Generally the SIMIO objects tend to favor the powerful side of the paradox while in our development, we will want to explore the simple side.

Typically, object classes with lots of features like those in SIMIO are called "heavyweight" because each object requires a lot of information and has numerous procedures.[289] Lightweight objects may have limited use but are easy to understand and use, and, if designed properly, can become easy to extend through object inheritance and/or composition. This chapter will focus on the creation of a lightweight specialized object, with limited purpose (as compared to SIMIO objects). But in this creation, significant reference is made to the SIMIO objects as guides for development and the new object class will become a part of our custom library. Also we use the words "object" and "model" interchangeably which SIMIO also uses.

We begin with a simple delay object and then embellish this object so that it provides a simple queuing capability. It will eventually be able to create multi-server, single queue objects.

Part 20.1: Creating a Simple Delay Object

In the previous chapter, a new delay transfer object was created by subclassing the standard TRANSFERNODE to delay entities that enter the node. However, the entities will queue up at the transfer node right on top of one another. Also, there is no opportunity to add a different symbol (i.e., an oven, etc.) to represent the delay

[289] The Vehicle/Worker is clearly the most complicated of the SIMIO standard object classes.

location since the transfer node takes up zero space. Initially, a new fixed object (i.e., a simple delay object) will be built that only delays an entity in its path, similar to any of the standard objects. Of course, such a delay can be accomplished by an infinite capacity SERVER or even just a TIMEPATH, but creating this simple delay object will be instructive both in its creation and as a base for extension.

Step 1: Use the previous chapter model and save it as Chapter20-1.spfx.

Step 2: Insert a "New Model" icon by selecting the "Fixed Class" option to produce a new type of object class and name it **DelayObject**. In the "*Advanced Options*" properties of this object class, be sure that its *Resource Object* property is "*False*" since we don't plan to use these objects as resources.[290]

Step 3: Within this new object class, add an expression property named ***DelayTime***. Its *Default Value* is `Random.Pert(1,2,4,1)`.[291] The *Unit Type* should be *Time* and the *Default Units* would be *Minutes*. Finally specify "*Process Logic" as* the *Category Name*.

Step 4: Objects that contain entities (SOURCES, SERVERS, COMBINERS, SEPARATORS, WORKSTATIONS, etc.) utilize STATION elements to house these entities. The STATION element may be used to define a capacity constrained location within an object, where one or more visiting entity objects can reside. From the *Definitions→Element*, insert a new STATION named **StationDelay**. Its *Initial Capacity* should be *Infinity* because we don't want to restrict the number of entities using the station.

The station element has two queues associated with each station: `Contents` and `EntryQueue`. The `EntryQueue` is the queue where entities wait until there is available capacity to allow the entities to move into the `Contents` queue. If you specify that the input buffer is ten for a server and the eleventh entity arrives to the server, the entity will be placed into the `EntryQueue` and not be allowed to enter the server (i.e., it will stay at the input node of the server).

Step 5: The delay STATION element provides a "place" for entities to reside while they are being delayed. However, entities need a way to enter and exit the **DelayObject**. The external view is used to create the "visual" representation as well as provide enter and exit nodes to the object. Go to the "External" section underneath the *Definitions* tab to create the visual interface.

- First, we need to define the "symbol" that will represent objects from this class. You can use a symbol from the existing SIMIO symbol library, create your own using the *Drawing* tab, download a symbol from the Trimble 3D Warehouse, or import from several cad programs (e.g., SketchUp © 2015 from Trimble Navigation Limited). In this case, we created the symbol using the drawing shown in Figure 20.1.
- From the *Drawing* section, insert an Ellipse centering it on the grid. In the *Object* section, change the *Object Height* to 0.3 and the *Line Width* to 0.02. From the 3D view, color the top grey and the side and line black.
- Next insert a half diamond to represent the delay. Using the Polygon tool just define the top, the right and bottom points. Place the shape so it is positioned halfway on the ellipse. Switch into 3-D view, select the shape and while holding the *Shift* key down, raise the label to its just barely above the ellipse. Switch back to the 2-D view, to center in the in Figure 20.1.

[290] Most SIMIO objects have capacity.

[291] Since the expression editor is not active, type the expression directly into the *Default Value*. We will tend to use the random number stream specification throughout to aid comparisons.

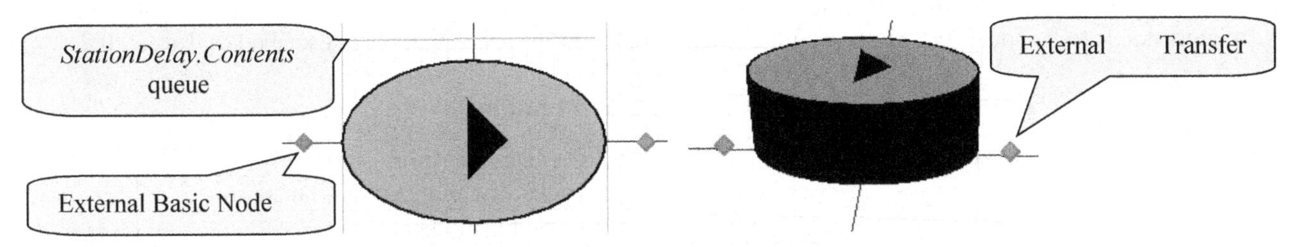

Figure 20.1: Delay Symbol

Step 6: Add the **Input** and **Output** nodes which provide the mechanism for objects to arrive and leave the delay object. Click the *External Node* from the *Drawing→Transfers* section to add two external nodes and position them to edges of the object as seen in Figure 20.1. External nodes have specifications explained in Table 20.1 that need to be defined.

Table 20.1: External Node Specifications

Specification	Description
Node Class Name	The type of node this external node should be (BASICNODE, TRANSFERNODE, **DELAYTRANSFERNODE**, etc.). Transfer nodes are typically used as exit/output nodes since they contain the logic to route entities to the next destination while basic nodes are used as entrance/input nodes.
Input Location Type	Location type in the object that an entity transfers into using this external node (e.g., Station, Node, or Container). If you have added objects to the facility window and you want to transfer the entities to one of the nodes in the fixed model of the new object, select "**Node**."
Station Name	Which station the entity will transfer into from this node if "**Station**" is the *Input Location Type*.
Node Name	Which node in the facility of the object should entities be transferred into from this external node if "**Node**" is the *Input Location Type*.
Container Name	Which container in the facility of the object should entities be transferred into from this external node if the *Input Location Type* is "**Container**."

- The **Input** node should be a BASICNODE as specified in Figure 20.2. When entities enter the node, we want them to be transferred automatically to the STATION **StationDelay** by SIMIO.
- The **Output** node will be a TRANSFERNODE to allow entities to be routed and/or ride on transporter.

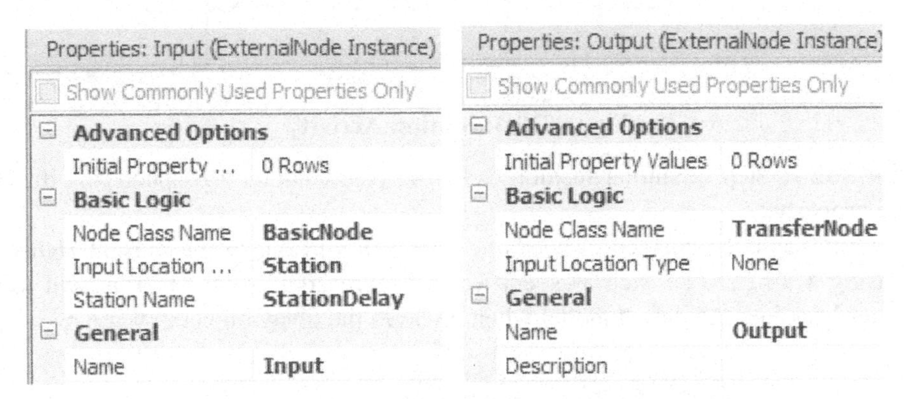

Figure 20.2: Specification of the External Nodes

Step 7: As entities enter the delay object and are delayed, the entities need to be visually seen in the delay (like the `Server.Processing.Contents`). Insert the queue animation above the symbol by selecting the *Queue* under the "*Animation*" tab as shown in Figure 20.1. The `QueueState` should be the `Contents` Queue of the **StationDelay** (i.e., `StationDelay.Contents`).

Step 8: Since we have defined the default external view, the logic to process the entities entering the objects needs to be specified. Stations automatically respond to three events as described in Table 20.2.

Table 20.2: Events at a Station

Event	Description
`Entered`	Fired when entities leading edge enters the station
`Exited`	Fired when entities exit the station
`CapacityChanged`	Fired when capacity of the station changes

Step 9: Create a new process named **OnEnteredDelay** which will be executed every time an entity enters the STATION according to the *Triggering Event Name* `StationDelay.Entered` as seen in Figure 20.3.

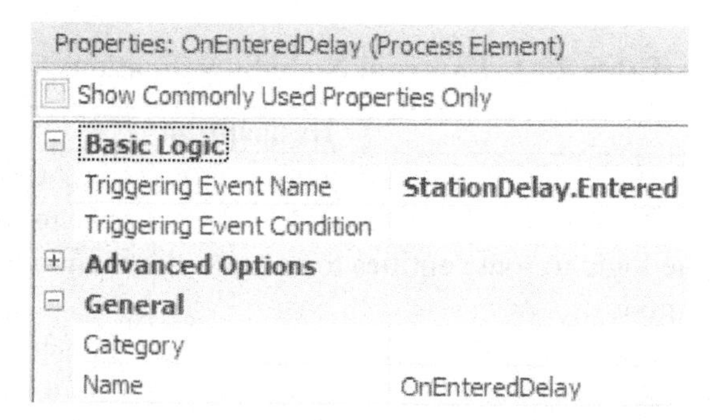

Figure 20.3: Specifying the Entered Triggering Event of the Delay Process

Step 10: The first step for the entity token is to end the transfer into the station using an *`End Transfer`* step which indicates the ENTITY has left its present location and finished the transfer into this location. This step also fires the `Transferred` event for the entity and alerts the input node that this transfer process has completed.

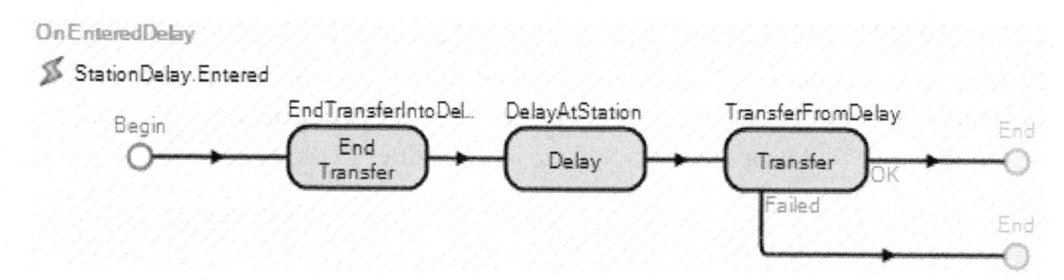

Figure 20.4: Station Activity

- Next add a *`Delay`* step in similar fashion as before (see Chapter 19) specifying the **DelayTime** as a referenced property.
- Once the entity has been delayed, it needs to be transferred out of the current **DelayObject** location (station) using a *`Transfer`* step as specified in Figure 20.5. The entity is sent to external output node to enter back into the parent model when it leaves the delay object:

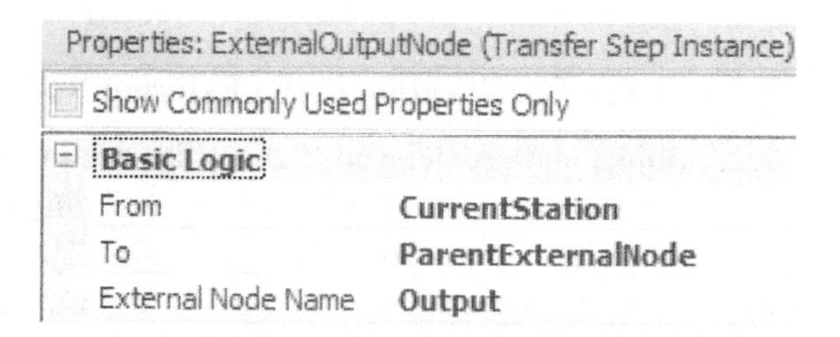

Figure 20.5: Transfer Step Specifications

Step 11: Next, we need to use the new DelayObject in our model as seen in Figure 20.6. Switch back to the main create **Model** (within the *Navigation* window). To facilitate multiple changes as we modify the object we will insert the delay between two nodes so we can avoid having setting the paths each time.

- Delete the TIMEPATH connecting the **TNormal** to the **SnkExit**. Insert a new TRANSFERNODE and connect it to the **SnkExit** via a 10 minute TIMEPATH. Also, you will need to specify the **SetTimeClock** as the *Entered* add-on process trigger.
- Insert a **SimpleDelay** object from the [*Project Library*] and call it **DelOperation**. Specify the delay time to be four minutes to match the **TDelay** time. Use connectors to link the input and output nodes.

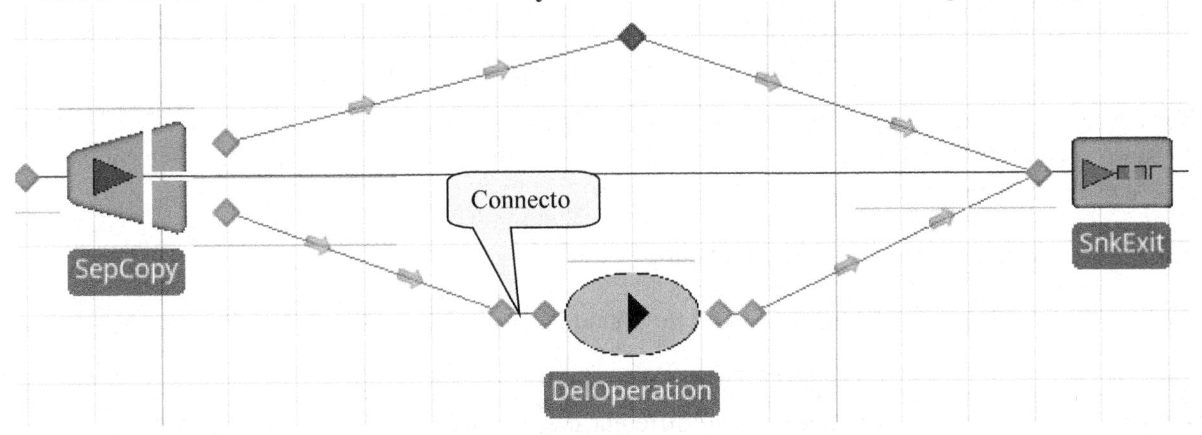

Figure 20.6: SIMO Model with Simple Delay

Step 12: Save and run the model for 24 hours observing the differences between the delay object and the delay transfer node. Set the arrival rate to be two minutes to see the differences. Note, if your entities are upside down in the queue, then the animated queue in the "*External*" view was drawn in the wrong direction (i.e., should be right to left).

Question 1: What is the visual difference between the **DELAYTRANSFERNODE** and the **SIMPLEDELAY** with regard to flow of entities?

____

Step 13: Our new **SIMPLEDELAY** object acts more similar to the standard SERVER object now with entities shown delaying in the station Contents queue. To compare our new object to the SERVER, replace the **DELAYTRANSFERNODE** with a SERVER named **SrvComparison**

- Set the *Processing Time* and *Delay Time* to Random.Pert(1,2,4,1) minutes for both the **SrvComparison** and the **DelOperation** respectively.
- Set the **SrvComparison's** *Initial Capacity* to "Infinity" as seen in Figure 20.7.
- Connect the **ParentOutput@SepCopy** to the **SrvComparison** via a five minute TIMEPATH.
- Connect the **SrvComparison** to the **SnkExit** via a ten minute TIMEPATH.
- Also, connect the **SrcParts** directly to both the **SrvComparison** and **DelOperation** via CONNECTORS.

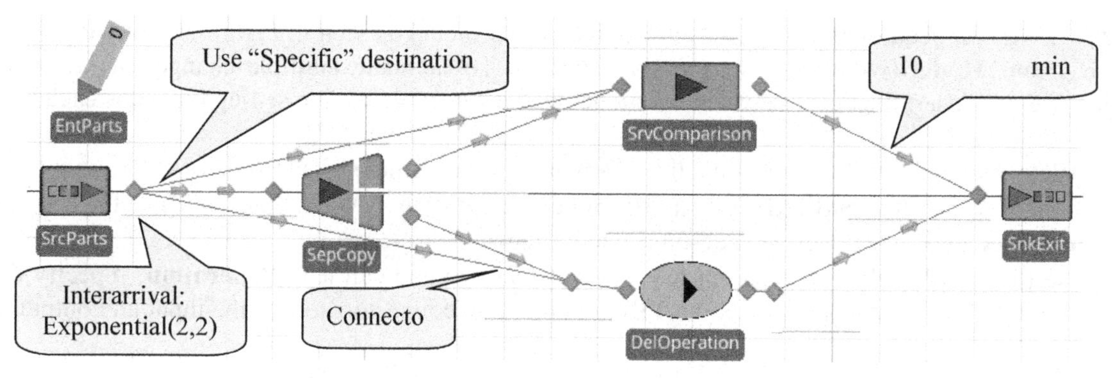

Figure 20.7: Comparing the SimpleDelay with the Server

Step 14: Change the *Interarrival Time* of the **SrcParts** to `Random.Exponential(2, 2)`[292] To allow us to do a comparison of the system, use a *"Specific"* as the *Entity Destination Type* for routing from the SOURCE output and specify the `Input@SepCopy`.

Step 15: Save and run the model for 100 hours.

Question 2: When you execute the model for 100 hours, can you see any difference in behavior or results?

____

Step 16: So we can run experiments with different destinations, define a *Node Object Reference* property called ***DestinationNode*** whose *Default Value* is `Input@DelOperation`. Change the Entity Desitination Type from the **Output@SrcParts** is based on a specific destination using the node reference property.[293]

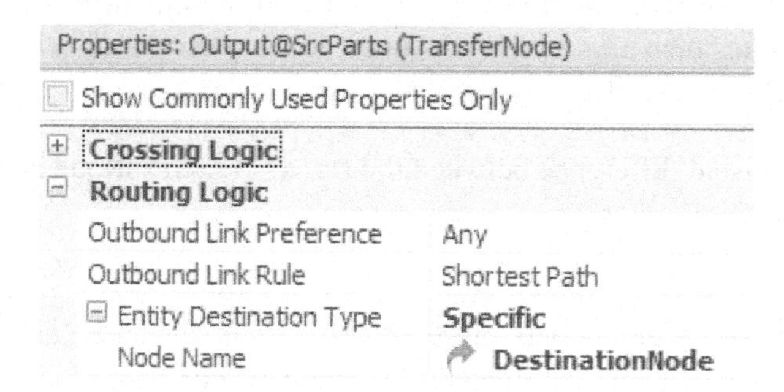
Properties: Output@SrcParts (TransferNode)
Show Commonly Used Properties Only
Crossing Logic
Routing Logic

Outbound Link Preference	Any
Outbound Link Rule	Shortest Path
Entity Destination Type	Specific
Node Name	DestinationNode

Figure 20.8: Specifying the Use of DestinationNode Property

Step 17: Remove the *Entered Add-on process triggers* for setting the time clock from the time path heading to the **DelOperation**, the time path leaving the **DelOperation** and the **TNormal** TRANSFERNODE.

Step 18: Create a new Experiment and set the parameters to one replication of 1000 hours. Run the experiment with the ***DestinationNode*** control set to new delay node (Input@DelOperation) and then rerun it again with the destination being the infinite capacity server.

Question 3: What did you get for the execution time for each of the experiments? Do you think this difference in time is important?

____

Question 4: What other factors influence this execution time comparison?

____

[292] Note we are using the second parameter of the Exponential to specify the specific random number stream.

[293] Again you will need to type in the reference since it is not shown in the drop-down list.

Question 5: Although execution time is better for a **SimpleDelay** object versus a SERVER object, is execution time the most important consideration when creating a simulation object?

Part 20.2: Adding Color and States

One of the features of the SIMIO objects (e.g. SERVER) is they change color based on their status (state). With some work we can mimic this behavior in our delay object, so it turns a different color when entities are being delayed versus when the delay object is idle. Furthermore, we can also update their state and automatically generate statistics on the object's use. We begin with the state determination.

Step 1: The SERVER keeps track of its states using a LIST STATE variable named "**ResourceState**". The LIST STATE VARIABLE defines a discrete integer variable with a list of possible values from zero to *N*. Select the **SIMPLEDELAY** from the [*Navigation*] panel. Within the new object, first define a string list with the names. From the *Definitions→Lists*, insert a new string list named **LstDelayStateNames** with "Idle" and "Delaying" as the two potential states.

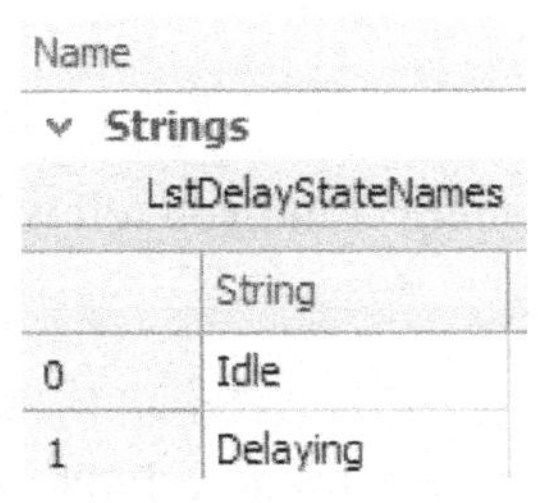

Figure 20.9: Specifying the Names of the Potential States

Step 2: Insert a new LIST STATE variable named **StaDelayState** which uses the **LstDelayStateNames** list. Since SIMIO will automatically track statistics on each state, specify the *Data Source* and *Category* property as seen in Figure 20.10.

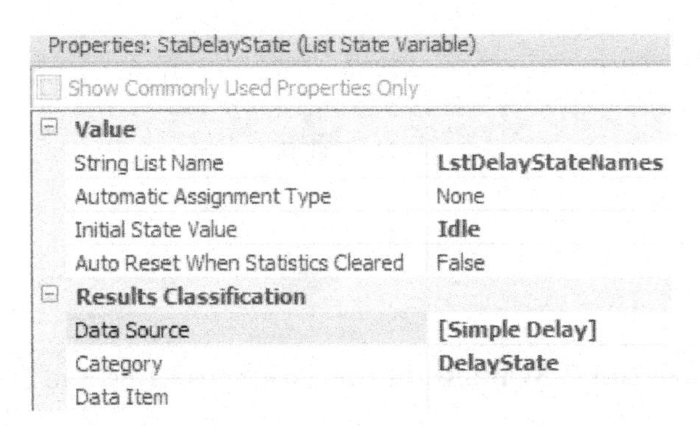

Figure 20.10: Setting up the List State Variable.

Step 3: Once the entity enters the delay station, the delay state will be delaying and will potentially become idle when it leaves the simple delay object as seen in Figure 20.11.

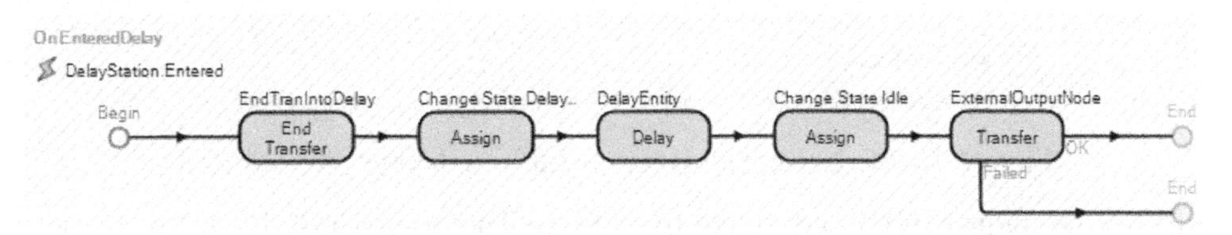

Figure 20.11: Modifying the OnEnteredDelay to Change the Delay State

- Insert an *Assign* step that sets the **StaDelayState** to "1" immediately after the *End Transfer*.

- Right before the entity leaves (i.e., `Transfer` step), we need to update the status to idle if this is the only entity in the delay station. Insert another `Assign` step that sets the **StaDelayState** variable to `StationDelay.Contents.NumberWaiting > 1` which will evaluate to zero if this is the last entity in the delay station or one as seen in Figure 20.12.

Figure 20.12: Changing the State of the SIMPLEDELAY **Object**

Step 4: Return back to the main **Model** and set the arrival destination node to `Input@DelOperation`.

Step 5: Save and run the model for 24 hours looking at the results.

Question 6: What percentage of the time was the delay object idle and delaying?

____

Question 7: While the simulation was running could you tell when the **DelOperation** was delaying?

____

Step 6: Currently there is only one symbol defined. To add additional symbols to our simple delay, click first on the "*External View*" for the **SIMPLEDELAY** object and select the *Additional Symbols* tab. Now add an additional symbol and color the ellipse of the new one "green" (i.e., the green color will represent the status "Delaying", meaning there are entities in the delay). See the symbols in Figure 20.13.

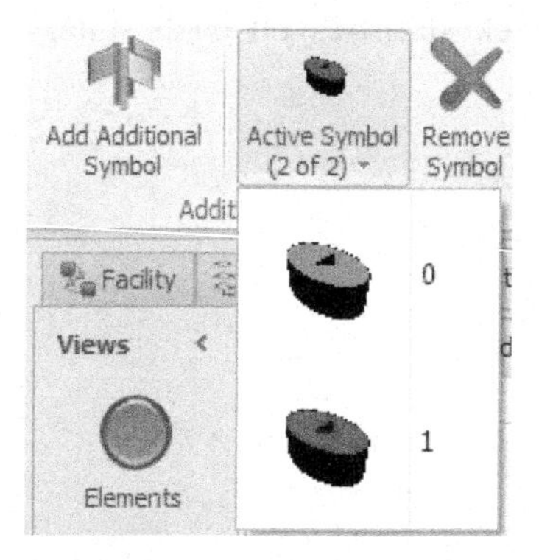

Figure 20.13: Symbols for the Two States Idle and Delaying

Step 7: To use the states to define which symbol to display, change the *Default Current Symbol Index* in the *Properties* window to the `SimpleDelay.StaDelayState`.

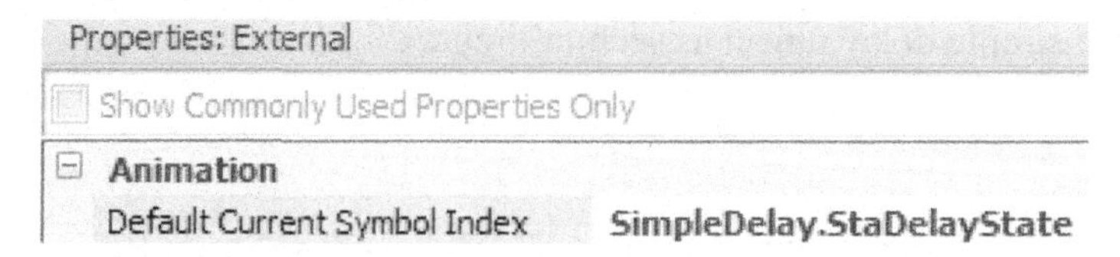

Figure 20.14: Using the Delay State to Change the Symbol

Step 8: Return to the main **Model** and delete the current **DelOperation** object from your SIMIO model since the external view is not updated with instances already defined. Insert a new **SIMPLEDELAY** using the same name, making sure to set the delay time to Random.Pert(1, 2, 4, 1) minutes. Save and run the model interactively.

Question 8: Does it change color when entities occupy the delay?[294]

Question 9: What other embellishments to this simple delay object do you think would be useful?

Part 20.3: Adding Defined Add-on Process Triggers

Perhaps the greatest flexibility you can add to your objects is to allow your users to add on processes that change the behavior of the object to meet their specific needs. This addition does assume your user will know how to create those processes. Many of the SIMIO standard objects allow the user to augment the processing of an object via add-on process triggers (e.g., *Run Initialized, Entered, Exited, Processing, Processed*, etc). To allow our new object to mimic the standard objects, these user defined add-on process triggers need to be specified.

Step 1: Select your **SIMPLEDELAY** from the [*Navigation*] panel and navigate to the *Definitions→ Properties* section. Properties associated with the add-on process triggers have to be defined so the user will be able to specify them. Five process triggers will be added *(Initialized, Entered, Exited, BeforeDelaying*, and *AfterDelaying*) using a "Process" property under the "*Element Reference*".

Step 2: Insert a PROCESS ELEMENT REFERENCE property for the "***Run Initialized***" with the following properties. Like the other objects we will have this process run at the start of a simulation. New category names can be created by just typing in the category into the entry box.

Property	Value
Name	RunInitializedAddOnProcess
Description	Executes when the simulation run is initialized.
Required Value	False
Display Name	Run Initialized
Category Name	Add-on Process Triggers
Default Value	Empty String

- Add the "***Entered***" PROCESS ELEMENT REFERENCE property with the following properties that will be executed as soon as an object enters the **DELAYOBJECT**.

Property	Value
Name	EnteredAddOnProcess
Description	Occurs immediately after an entity has entered this object and before the delay.
Required Value	False
Display Name	Entered
Category Name	Add-on Process Triggers
Default Value	Empty String

- Add the "***Exited***" PROCESS ELEMENT REFERENCE property with the following properties that will be executed once an object leaves the **DELAYOBJECT**.

---

[294] Note, the user has the ability change the *Current Symbol Index* in the model just like any other standard SIMIO object.

Property	Value
Name	ExitedAddOnProcess
Display Name	Exited
Description	Occurs immediately after an entity has exited this object
Category Name	Add-on Process Triggers
Required Value	False
Default Value	Empty String

- Add the "***Before Delaying***" PROCESS ELEMENT REFERENCE property that will run immediately before the ENTITY starts the delay process similar to the SERVER' S *Before Processing* trigger.

Property	Value
Name	BeforeDelayingAddOnProcess
Display Name	Before Delaying
Description	Occurs immediately before an entity is to be delayed
Category Name	Add-on Process Triggers
Required Value	False
Default Value	Empty String

- Finally, insert the "***After Delaying***" PROCESS ELEMENT REFERENCE property that will run once the ENTITY has finished delaying similar to the SERVER' S *After Processing* trigger.

Property	Value
Name	AfterDelayingAddOnProcess
Display Name	After Delaying
Description	Occurs immediately after an entity has been delayed
Category Name	Add-on Process Triggers
Required Value	False
Default Value	Empty String

Return to the *Processes* tab and insert an `Execute` step into the **OnEnteredDelay** process which allows processes to be executed as seen in Figure 20.15. Specify the process to be the referenced property (`EnteredAddOnProcess`). The *Token Wait Action* property should be "WaitUntilProcessCompleted" under the *Advanced Options* which forces the token to wait until the add-on process has completed running before it continues to the next step.

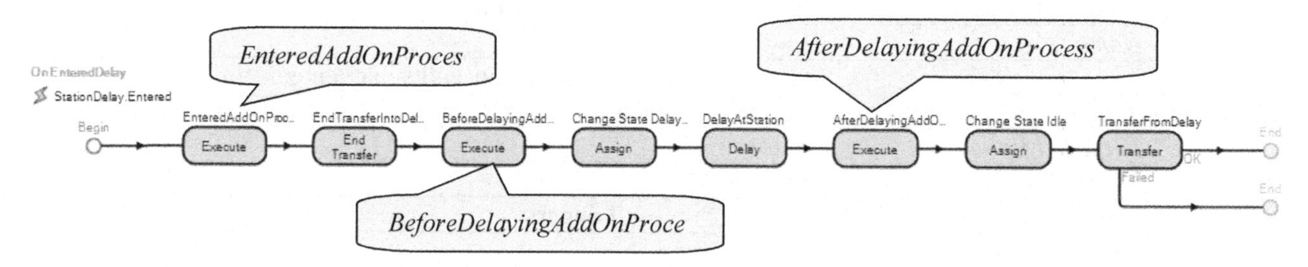

Figure 20.15: Specifying the Entered and Delayed Add-on Process Trigger

Step 3: Before the `Assign` step (see Figure 20.15) that changes the state in the **OnEnteredDelay** process, insert another `Execute` step that will run the process associated with the `BeforeDelayingAddOnProcess` trigger with the same (default) *Action* property value.

Step 4: After the `Delay` step, insert another `Execute` step that will run the process associated with the `AfterDelayingAddOnProcess` trigger with the same (default) *Action* property value.

Step 5: Add the **OnRunInitialized** Process which is automatically defined for all `FIXED CLASS` objects by selecting it in the "*Select Process*" dropdown box. Insert an `Execute` step into the process as before but specify the *RunInitializedAddOnProcess* referenced property as seen in Figure 20.16 and Figure 20.17.

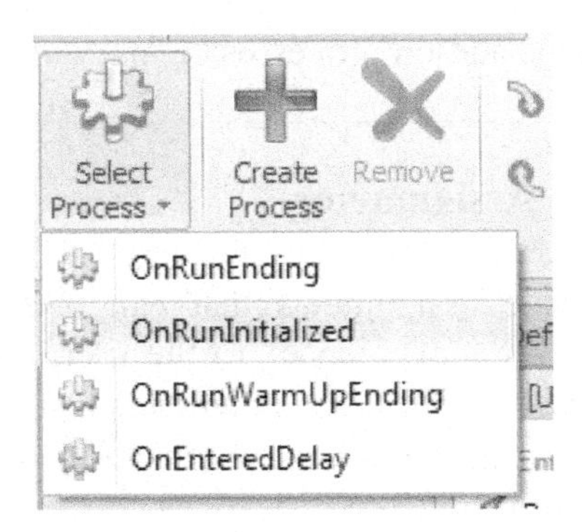

Figure 20.16: Using a Built-in Process for all Fixed Objects

Step 6: Create a new Process named **OnExited** and specify the *TriggeringEvent* property to be the `StationDelay.Exited` event. Again, insert an `Execute` step that will execute the process for the user defined *ExitedAddOnProcess* property.

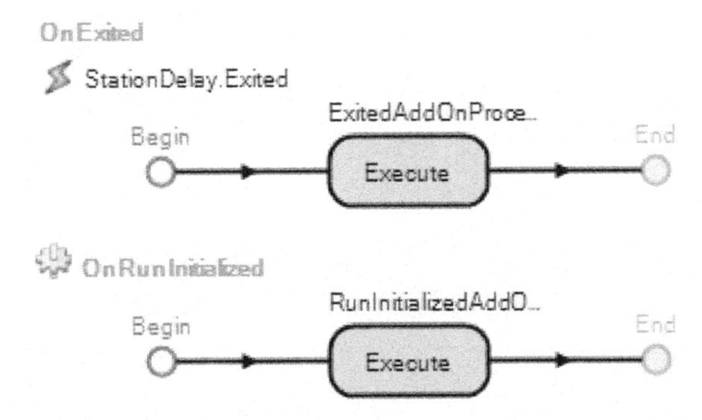

Figure 20.17: Adding an OnRunInitialized and OnExited Process

Step 7: Let's utilize our new add on process triggers, as a "user" of our object, to allow the delay time to vary for each entity (i.e., 25% of the time it will be four minutes, 25% of the time it will be six minutes and 50% of the time it will be eight minutes).

- Insert a new `MODELENTITY` real valued state variable named **EStaDelayTime** whose *Unit Type* is **Time** with default units of "**Minutes**" which will be used to store the delay time.
- Specify the *Delay Time* of the **DelOperation** be `ModelEntity.EStaDelayTime`.
- Navigate back to the **Model** and create a new *Entered Add-on Process* trigger for the **DelOperation** as seen in Figure 20.18.
- Add an `Assign` step that assigns **ModelEntity.EStaDelayTime** a value from a `Random.Discrete(4, 0.25, 6, 0.5, 8, 1.0)` minutes.

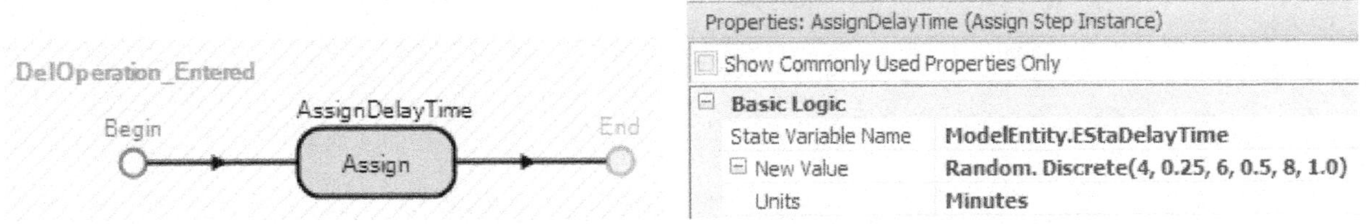

Figure 20.18: Assigning a Delay Time for Each Entity

Step 8: Save and run the model for 100 hours designating all entities to the new object.

Question 10: Theoretically, the time in station should be 6.5 minutes (why?). What did you get for the "Holding" time at the delay station in **DelOperation**?

____

Question 11: How important might the add-on processes be in a simple object?

____

Part 20.4: Embellishing with State Assignments

The SIMIO standard objects have state assignments that can perform assignments without the use of processes. This addition is considered a "convenience" to the user, especially since assignments are easily implemented in processes. Nonetheless, similar state assignments can be added to the new **SIMPLEDELAY** to illustrate how to add this capability. Recall that an `Assign` step may consist of several assignments. As a result, this addition must accommodate repeating specifications.

Step 1: State assignments are specified by a "repeating group property" to the new object so the number of assignments can vary. Add a *Repeat Group* property named **AssignmentsOnEntering** in the **SIMPLEDELAY** object with the properties specified in Figure 20.19. Again, type in the new category name **State Assignments** so it will be similar to the other standard objects. The repeat group allows you to specify multiple rows (i.e., repeating) of a set of properties (i.e., the state variable and the value expression).

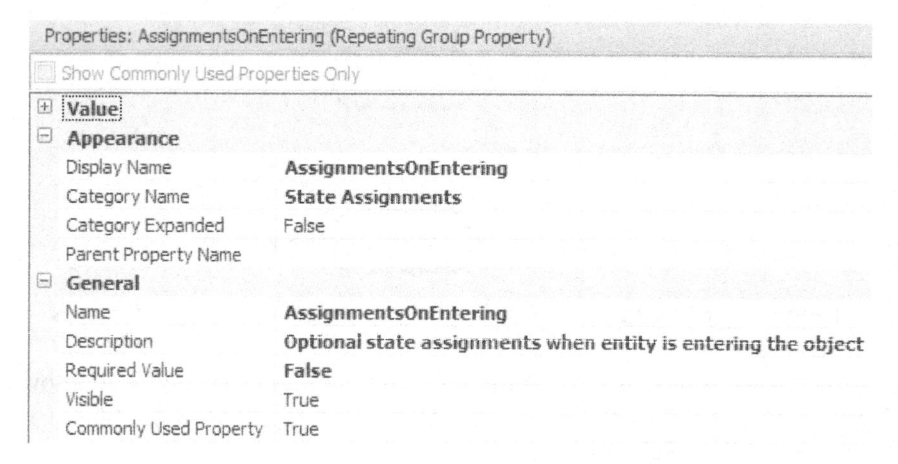

Figure 20.19: Adding a Repeating Group Property

Step 2: To create the properties that are to be repeated (i.e., grouped), first select the **Assignments OnEntering** group property, and then insert a new *State Standard Property* named **AssignmentsOnEnteringStateVariableName** since a state variable is used in an assignment with the properties specified in Figure 20.20.

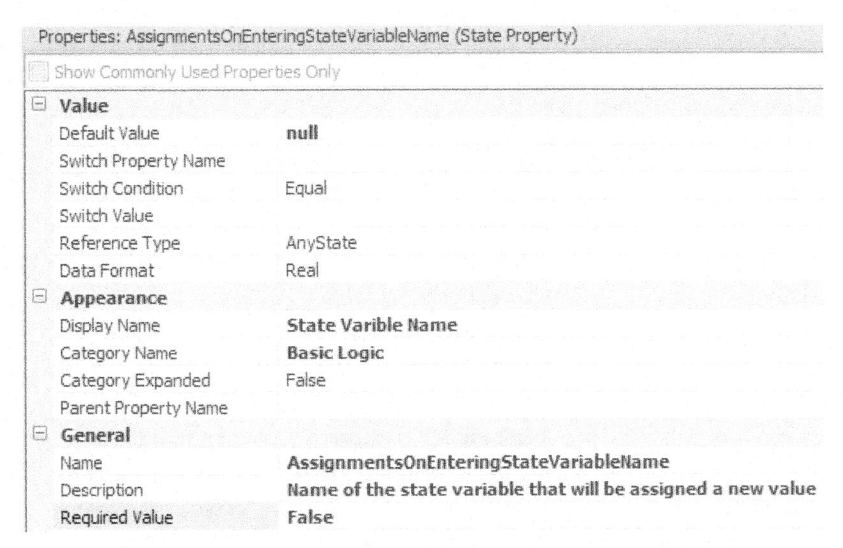

Figure 20.20: Specifying a State Variable Property

Step 3: Next, insert the *expression* property that will specify the new value to be assigned the state variable. Again, select the **AssignmentsOnEntering** group property first, and then insert a new *Expression Standard Property* named **AssignmentsOnEnteringNewValue** with the properties specified in Figure 20.21. Note carefully the specification of the *Unit Type*, so units of the value can be specified based on the state variable.

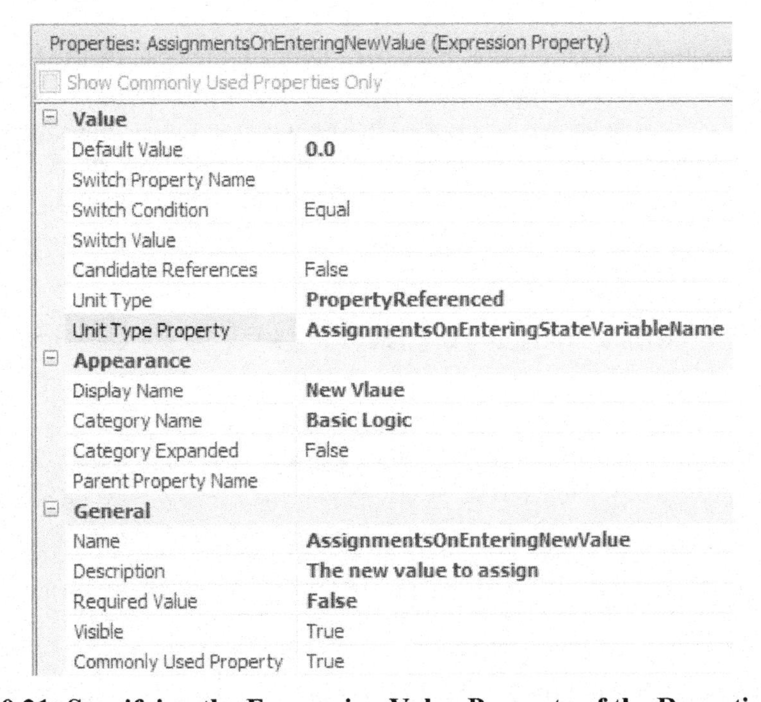

Figure 20.21: Specifying the Expression Value Property of the Repeating Group

Step 4: Repeat the process for another *Repeat Group Property* named **AssignmentsBeforeExiting** as seen in Figure 20.22 which shows all the properties. Notice how the group properties have their own separate properties section. You can copy the **AssignmentsOnEntering** and just change the name.

Name	Object Type	Display Name	Category
> **Properties (Inherited)**			
> **WorkDayExceptions.Properties (Inherited)**			
> **WorkPeriodExceptions.Properties (Inherited)**			
˅ **Properties**			
DelayTime	Expression Property	DelayTime	Process Logic
RunInitializedAddOnProcess	Process Element Property	Run Initialized	Add-on Process Triggers
EnteredAddOnProcess	Process Element Property	Entered	Add-on Process Triggers
ExitedAddOnProcess	Process Element Property	Exited	Add-on Process Triggers
BeforeDelayingAddOnProcess	Process Element Property	BeforeDelaying	Add-on Process Triggers
AfterDelayingAddOnProcess	Process Element Property	AfterDelaying	Add-on Process Triggers
AssignmentsOnEntering	Repeating Group Property	AssignmentsOnEntering	State Assignments
AssignmentsBeforeExiting	Repeating Group Property	AssignmentsBeforeExiting	State Assignments
˅ **AssignmentsOnEntering.Properties**			
AssignmentsOnEnteringStateVariableName	State Property	State Variable Name	Basic Logic
AssignmentsOnEnteringNewValue	Expression Property	New Value	Basic Logic
˅ **AssignmentsBeforeExiting.Properties**			
AssignmentsBeforeExitingStateVariableName	State Property	State Variable Name	Basic Logic
AssignmentBeforeExitingNewValue	Expression Property	New Value	Basic Logic

Figure 20.22: Properties Associated with the SIMPLEDELAY

Step 5: Once the *Repeating Group* properties have been defined, they can be used in `Assign` steps. Insert two `Assign` steps as seen in Figure 20.23 before the "*OnEntering*" and "*BeforeExiting*" state assignments. One `Assign` step should be inserted before the execution of the entered add-on process and the other one right before the entity is transferred out of the node. Figure 20.23 shows how the reference properties are used to specify the *Before Exiting Assignments* property repeating group.

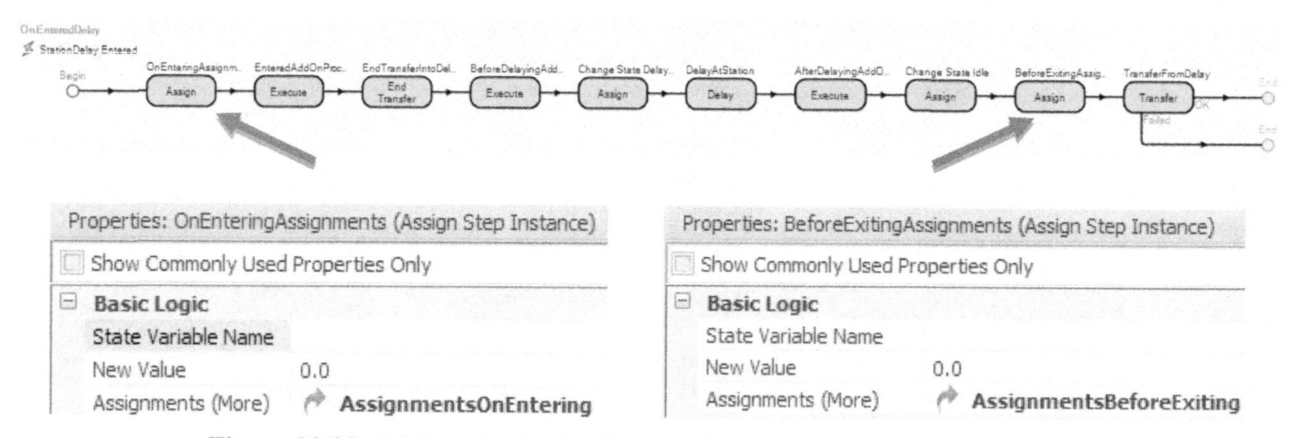

Figure 20.23: Adding the Assign Process Steps for the Assignment States

Step 6: For each of the Assignments (More) properties, the state variable and new value properties have to be associated with the reference properties as well. Bring up the *Repeating Property Editor* by clicking the more button **AssignmentsOnEntering** [...] as seen in Figure 20.24.

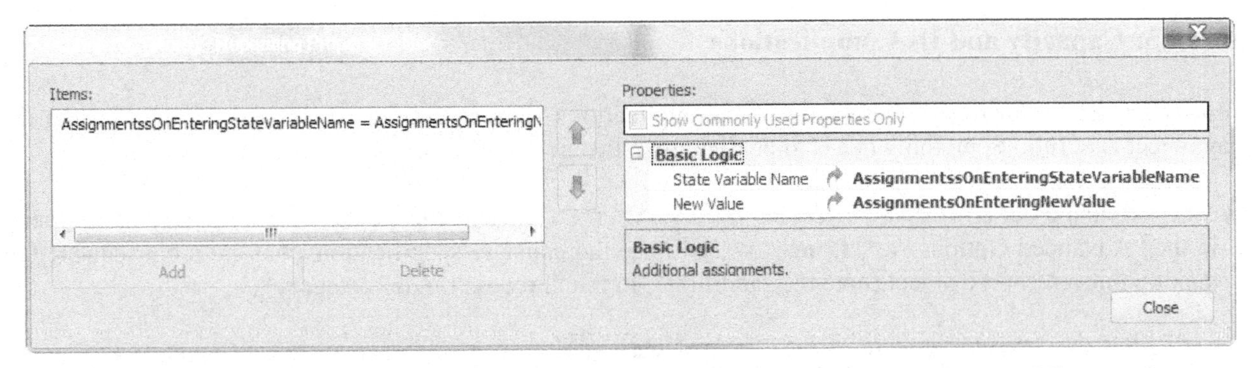

Figure 20.24: Setting up the State Variables and New Values to the Referenced Properties

Step 7: For both the ***State Variable Name*** and ***New Value properties***, set the reference property to the top property for each property. Figure 20.25 shows selecting the *State Variable Name* referenced property.

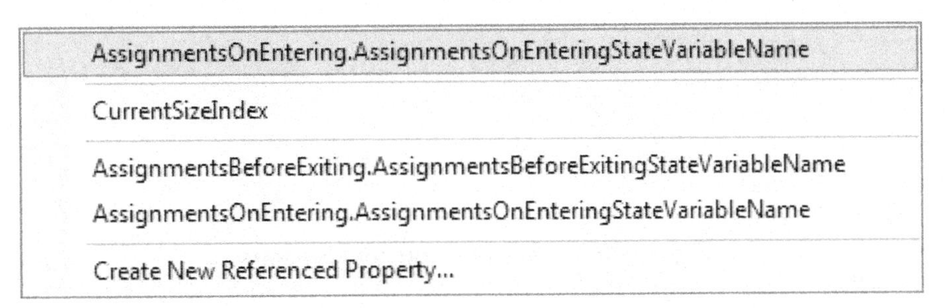

Figure 20.25: Select the Top Reference Property for the State Variable Name

Step 8: Let's utilize our new processes and assignments by specifying the delay time of each of the entities using the *OnEntering State Assignments*.

- First remove the *Entered Add-on process* trigger by resetting it (reset or Null) so the delay time will not be specified by the add-on process.
- Repeat the same assignment as was done in Figure 20.26 using the ***AssignmentsOnEntering*** *State Assignments*. Assign the **ModelEntity.EStaDelayTime** a value from a `Random.Discrete(4, 0.25,6,0.5,8,1.0)` minutes.[295]

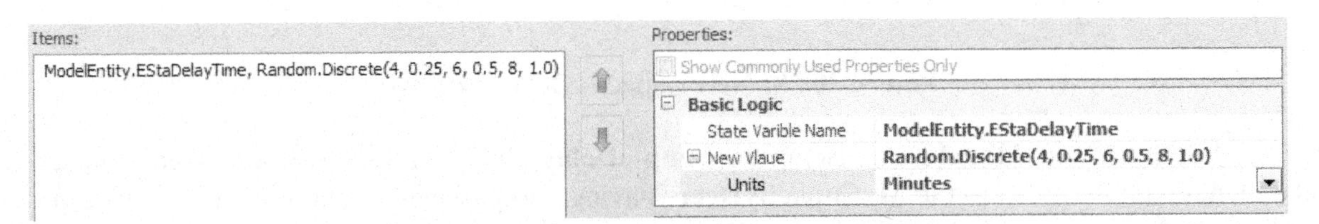

Figure 20.26: Using the State Assignments Property

Step 9: Save and run the model for 100 hours designating all entities to the new object.

Question 12: What did you get for the Holding time in the **DelOperation?**

____

Question 13: How important are the state assignments in a simple object?

____

Question 14: Is the state assignment needed when there are add-on triggers?

____

295 If the *Units* do not show, then the New Value property was not setup correctly (see Figure 20.21).

Part 20.5: Capacity and Its Complications

Adding capacity to our delay object adds new complications because entities cannot enter a delay when the delay object is at full. Something needs to be done with the arriving entities

Step 1: Rename our SIMPLEDELAY object[296] to CAPACITYDELAY and change the "*Resource Object*" property in the "Advanced Options" to "**True**". We will use the capacity specifications that are a consequence of making this a resource object (note the "additions" to the "Process Logic" category).

Step 2: Set the *Initial Capacity* of the **StationDelay** station to a reference to InitialCapacity so that the capacity of the station is identical to the CAPACITYDELAY object.

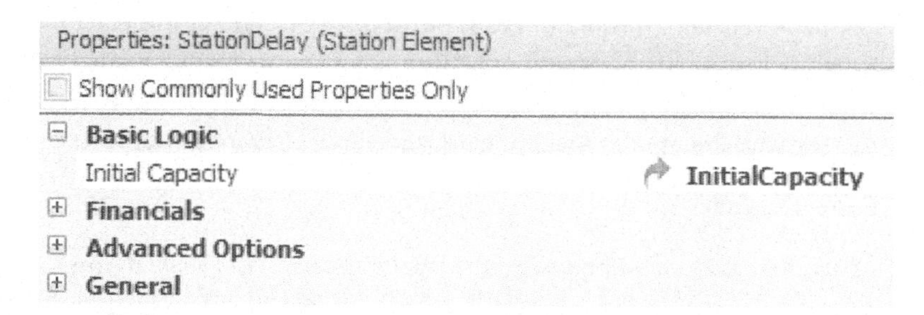

Figure 20.27: Setting Station Capacity

Step 3: The **OnEnteredDelay** process can now be modified as shown in Figure 20.28. To seize capacity from the object, use PARENTOBJECT as the *Object Type* within the `Seize` step. The same object type will be used at the `Release` step. We will use only one unit of capacity from this object for any delayed entity. We will choose to put the `Seize` step before the change in state assignment and the `Release` step directly after the `Delay` since we know the object is no longer busy

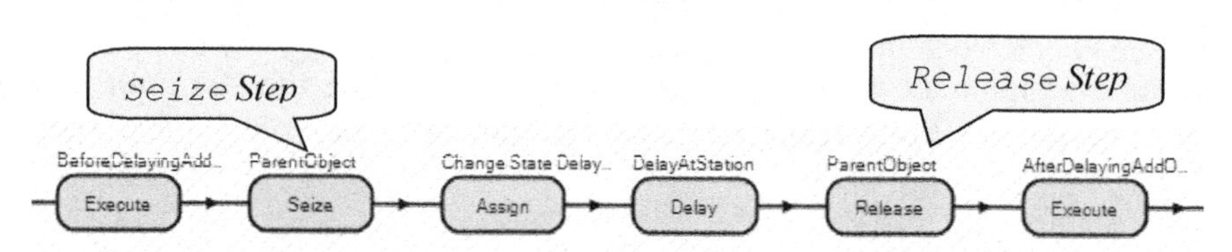

Figure 20.28: New CapacityDelay **Process**

Step 4: Use this in the **Model** by replacing the **SimpleDelay** object with the **CapacityDelay** object. Run the model with the PATH to the **MyCapacityDelay** having "no passing" Run it again to "allow passing" to that path[297].

Question 15: What happens to the entities that can't enter the object's station?

____

Step 5: When you run the model, say for 24 hours, ignoring the build-up in the path, examine the "Results". Look at the output of the **StationDelay** and notice there are now queuing statistics on the *EntryQueue* of the station.

Step 6: The *EntryQueue* is a SIMIO defined queue that exists for each station, in case the station is full when the entity arrives. Thus the station element has two queues associated with each station: `Contents`

---

[296] You could start at the beginning and create a new Fixed Class object.

[297] The connections need to be paths to have the "no passing" option. Other types of paths do not have this feature.

and *EntryQueue*. (If you specify that the input buffer is ten for a SERVER and the eleventh entity arrives to the server, the entity will be placed into the *EntryQueue* and not be allowed to enter the server - it will stay at the input node of the server).

Step 7: Now let's modify our object so that the EntryQueue is visible as shown below in Figure 20.29. Then replace the existing **CapacityDelay** with this new **CapacityDelay** and execute the model. Use a capacity of 1 and a delay time of 2 minutes. Observe the behavior of the entities entering the station.

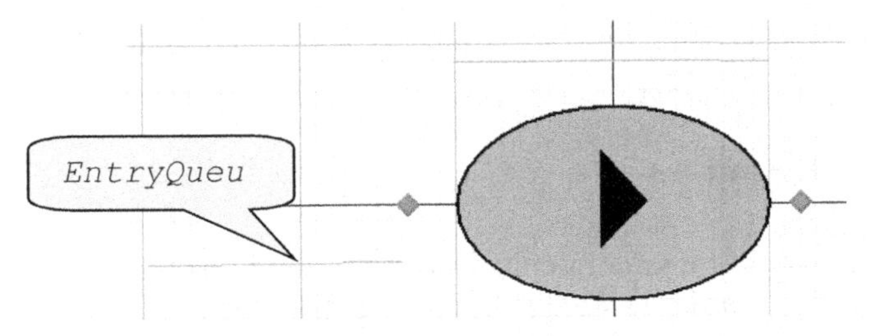

Figure 20.29: Add Entry Queue

Question 16: Is there anything about the animation or the statistics that trouble you?

Step 8: Although the statistics are ok, the animation shows an entity waiting on the path that is duplicated within the visible entry queue. This duplication occurs because the delay station is full and there are entities waiting.

Step 9: You will recall that the standard SERVER object uses a station for waiting – namely the *InputBuffer*. So let's use a second station in our object to account for waiting to enter the delay.

- Insert a new STATION named **StationWaiting**. Let's make its *Initial Capacity* `Infinity` so there is no limit on waiting especially since we do not want to deal with another set of capacity concerns.
- In the *External* view change the *Input External Node* to reference the **StationWaiting** instead of the **StationDelay** to route entities to this station first.
- Change the queue representing the *EntryQueue* to the `StationWaiting.Contents` queue.

Step 10: Finally, we need to insert an **OnEnteredWaiting** process to handle entering the waiting station. Move the `Assign`, `Execute`, and `Seize` steps from the **OnEnteredDelay** to the **OnEnteredWaiting**. Doing this will separate the time spent waiting for capacity from the time in the delay – see Figure 20.30.

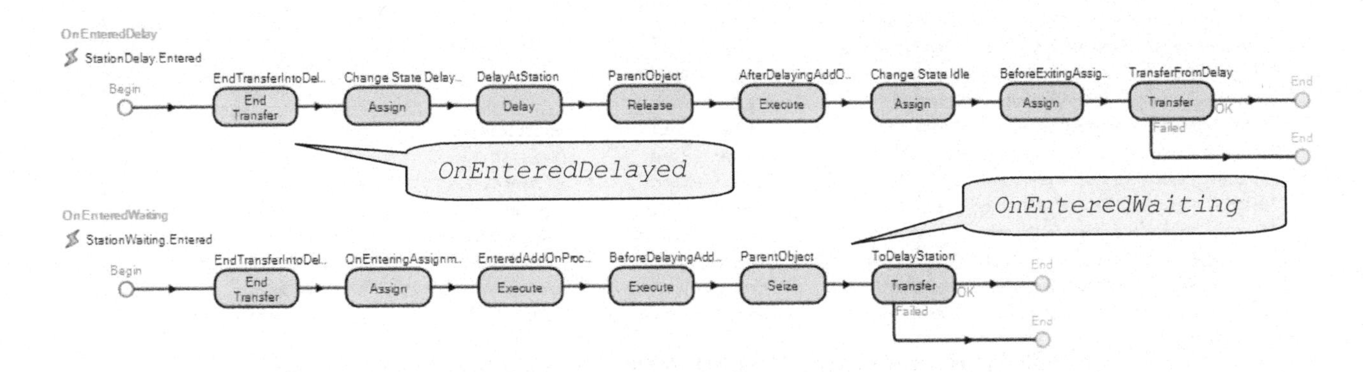

Figure 20.30: OnEntered Processes

Step 11: Configure the **DelOperation** in the **Model** to have a capacity of "1" and a delay time of "2" minutes. Run the model interactively to observe the behavior of the new object.

Question 17: Does the observed queuing seem appropriate?

____

Step 12: Now run the Simulation for 100 hours.

Question 18: What is the average time (in minutes) spent waiting to enter the station?

____

Question 19: What is the average holding time spent in the station (in minutes)?

____

Part 20.6: Adding Multiple Resources

The capacity of the new object would be greatly enhanced if multiple resources could be required before the entity is delayed. We would to like require resources before the delay occurs and then release them right after the delay, as is done with *Secondary Resources* in the SERVER. Again a *Repeat Group* property similar to the assignments has to be utilized since the `Seize` and `Release` steps within the secondary resources can request/release more than one type of capacity.

Step 1: Recall, the assignments had just two properties (i.e., the state variable and the new value) associated with the *Repeat Group*. That is not the case for the `Seize` and `Release` steps which have many properties. The easiest way is to sub-class the SERVER object and then copy and paste the secondary resources (e.g., *SecondaryResourcesSeizesBeforeProcessing* and *SecondaryResourcesReleasesAfterProcessing*) into the **SIMPLEDELAY**. Therefore, sub-class the SERVER by right clicking to create a **MYSERVER** and then expand the "Inherited Properties."

Step 2: Select the **MYSERVER** from the [*Navigation*] window. Under the *Definitions→Properties* section, expand the **Properties(Inherited).** Copy the **SecondaryResourceSeizesBeforeProcessing** and then paste this *Repeat Group* property into the **CAPACITYDELAY** properties.

Step 3: Repeat the process for the **SecondaryResourceReleasesAfterProcessing** *Repeat Group* property (see Figure 20.31).

AssignmentsBeforeExiting	Repeating Group Property	AssignmentsBeforeExiting	State Assignments
SecondaryResourceSeizesBeforeProcessing	Repeating Group Property	Before Processing	Secondary Resources/Other Resource Seizes
SecondaryResourceReleasesAfterProcessing	Repeating Group Property	After Processing	Secondary Resources/Other Resource Releases
> **AssignmentsOnEntering.Properties**			
> **AssignmentsBeforeExiting.Properties**			
> **SecondaryResourceSeizesBeforeProcessing.Properties**			
∨ **SecondaryResourceReleasesAfterProcessing.Properties**			
SecondaryResourceReleasesAfterProcessingObjectType	Enumeration Property	Object Type	Basic Logic
SecondaryResourceReleasesAfterProcessingObjectName	Object Property	Object Name	Basic Logic
SecondaryResourceReleasesAfterProcessingObjectListName	Object List Property	Object List Name	Basic Logic
SecondaryResourceReleasesAfterProcessingNumberOfObjects	Expression Property	Number Of Objects	Advanced Options
SecondaryResourceReleasesAfterProcessingUnitsPerObject	Expression Property	Units Per Object	Advanced Options
SecondaryResourceReleasesAfterProcessingReleaseOrder	Enumeration Property	Release Order	Advanced Options
SecondaryResourceReleasesAfterProcessingReleaseCondition	Expression Property	Release Condition	Advanced Options
SecondaryResourceReleasesAfterProcessingKeepReservedCondition	Expression Property	Keep Reserved If	Advanced Options
SecondaryResourceReleasesAfterProcessingReservationTimeout	Expression Property	Reservation Timeout	Advanced Options
SecondaryResourceReleasesAfterProcessingOnReleasedProcess	Process Element Property	On Released Process	Advanced Options

Figure 20.31: Repeat Group Properties for Allowing Secondary Resources

Step 4: From the **OnEnteredProcessing** process (see Figure 20.32) inside the **MYSERVER** object, copy the `Seize` and `Release` steps before and after process.

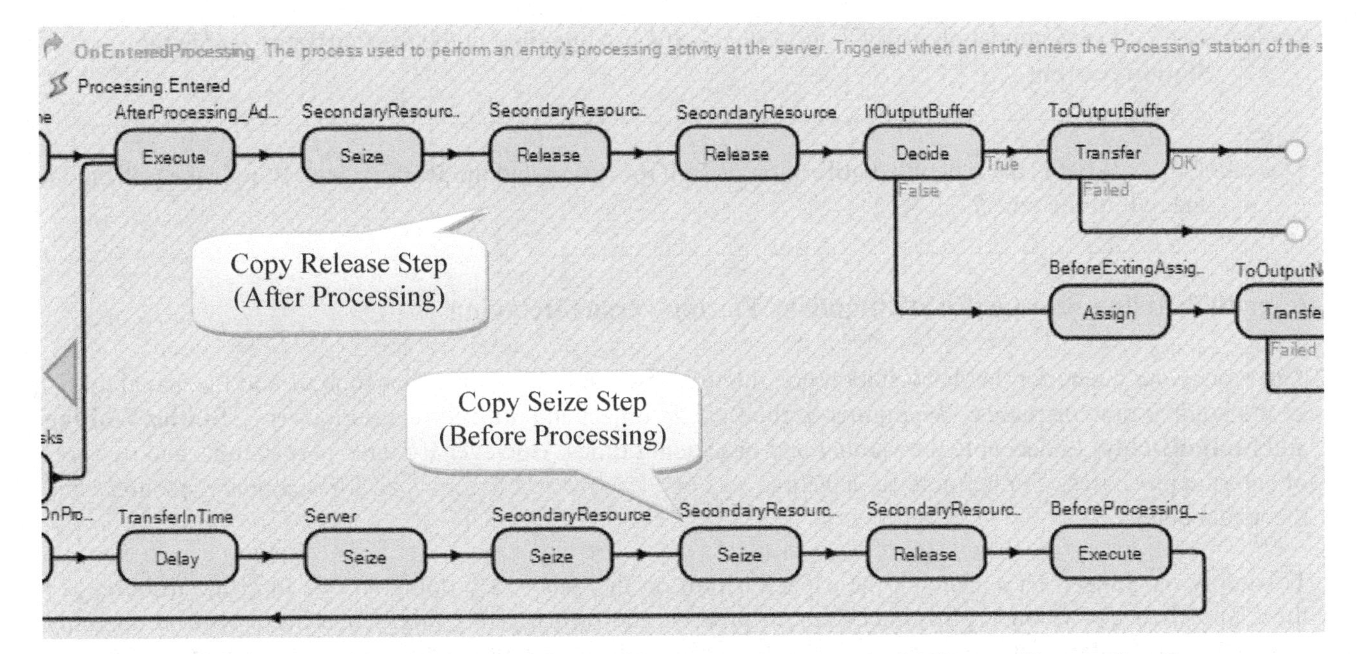

Figure 20.32: Copy Seize and Release Steps from the MyServer's OnEnteredProcessing Process

Step 5: Now in the **OnEnteredWaiting** process of the SIMPLEDELAY object, add the `Seize` secondary resources before the `Seize` step which seizes the parent object as in Figure 20.33.

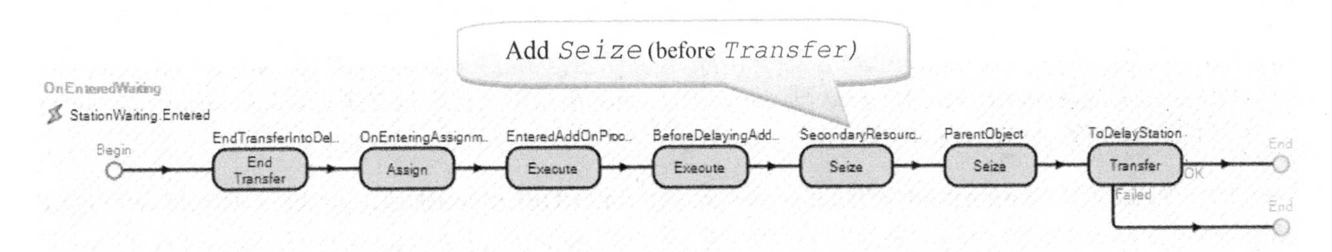

Figure 20.33: Adding the Seize into the OnEnteredWaiting

Step 6: Next in the **OnEnteredDelay** process, add the `Release` right after the `Delay` step as seen in Figure 20.34.

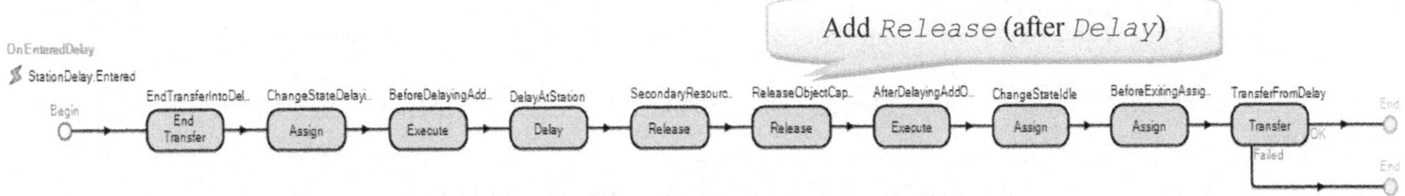

Figure 20.34: Adding the Release to the OnEnteredDelay

Step 7: Return to the **Model** and insert a RESOURCE named **ResService** with a capacity of two, of which one unit is required before the delay can occur. The **DelOperation** also has a capacity of one (it only makes sense for the delay capacity to be one or two, since the secondary resource has capacity of two).

Step 8: In the *"Secondary Resources"* of the **DelOperation**, seize the **ResService** in the *Other Resource Seizes→Before Processing* and release it in the *Other Resource Releases→After Processing.*

Step 9: Save and run the model interactively observing the animation.

Question 20: Does the animation behave as you expect?

____

Question 21: For a simulation of 100 hours, what is the holding time in the **StationDelay** *and* in the **StationWaiting**

Question 22: What is the scheduled utilization of **DelOperation** and the **ResService**? Why is the utilization one half of the other?

Part 20.7: Using Storages to Distinguish Waiting versus Delaying

The processing queue for the delay station now includes waiting for secondary resources and the actual delay, so it's holding time increases. We mimicked the SERVER object and utilized two stations (i.e., **StationWaiting** and **StationDelay**) to decouple the waiting and processing times. However we have two stations and two sets of entering processes. Furthermore we have two sets of `Seizes` and `Releases` for secondary resources and the actual delay.

To simply our object, let's abandon the capacity limit on the delay. By doing so, we limit the resources to those specified as "secondary resources" or in process add-on triggers. In essence these will be our only resource specifications and in turn we have a simpler object. Further, rather than two stations (locations) within the object, it would be better to have one station, so we avoid the complications of transferring between stations (these transfers produce events in the current events calendar, meaning they may not be executed immediately). Instead we will utilize STORAGE elements which also automatically calculate statistics and which we have used earlier.

Step 1: Select the **CAPACITYDELAY** object from the [*Navigation*] panel. Right-click to bring up its "*Properties*" and under the "*Advance Options*" make the *Resource Object* **False**. This means our object will no longer have capacity limitations.

Step 2: Since the **StationWait** will be removed, reset the **OnEnteredDelay** process as seen in Figure 20.35.

- Move the "OnEnteringAssignment" `Assign` and the "EnteredAddOnProcess" `Execute` from the **OnEnteredWait** before the `End Transfer`.
- Move the "SecondaryResources" `Seize` and the "BeforeDelayingAddOnProcess" `Execute` from the **OnEnteredWait** right after the `End Transfer`.
- Finally, remove the `Release` step that releases the parent object in the **OnEnteredDelay**.

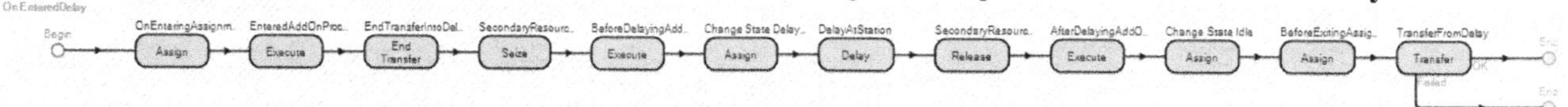

Figure 20.35: Resenting the OnEnteredDelay

Step 3: In the *Definitions→ Elements* section, delete the **StationWaiting** station. Next, insert two STORAGE elements named **StorageWait** and **StorageDelay** as in Figure 20.36. Finally, set the ***InitialCapacity*** of the **StationDelay** back to `Infinity` so the station has no capacity limit.

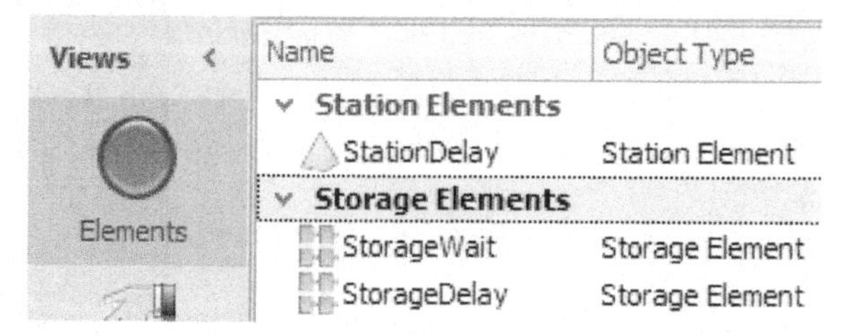

Figure 20.36: Adding Two STORAGE Elements to Keep Track of Waiting and Delaying

Step 4: To calculate how long entities waited for secondary resources, we need to insert the entities into the **StorageWait** before we `Seize` and then `Remove` them after the BeforeDelaying procces-trigger. We need to repeat the process for keeping track of the actual delay time by inserting the entities into the **StorageDelay** right before the `Delay` and then removing them right after the `Delay` as seen in Figure 20.37

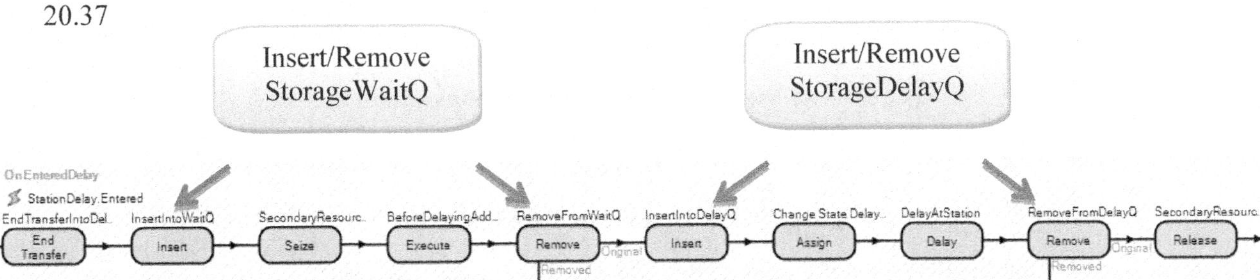

Figure 20.37: Utilizing the StorageWait and StorageDealy to Track Waiting Times

Step 5: Finally modify the CAPACITYDELAY object "External" view. Insert a new animated queue that references **StorageWait.Queue** while the processing queue should reference the **StorageDelay.Queue**. Change the *Input* "External Node" to send entities entering (i.e, its *Station Name* is **StationDelay**).

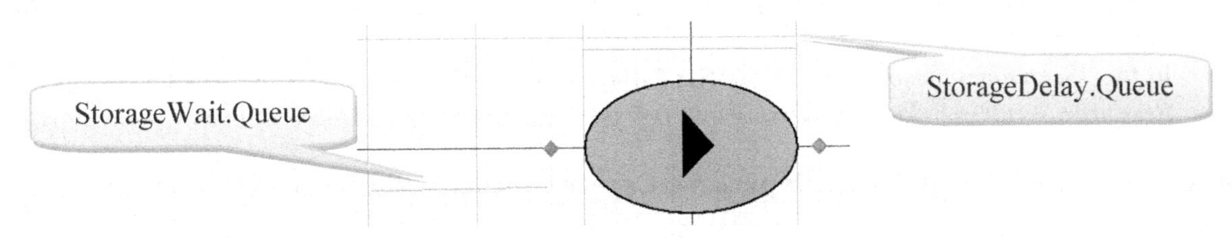

Figure 20.38: Adding the Delay and Wait Queues to the External View

Step 6: Save and run the **Main** model for 100 hours. Be sure your model seizes one unit of **ResService** before processing releases one unit of **ResService** after processing. The delay time is 3 minutes. You will need to delete the current CapacityDelay and reinsert it for the new external view to be used.

Question 23: When you switched back to the **Main** model, are the two new queues visible?

____

Question 24: On average, how long did parts wait for the **MyCapacityDelay**?

____

Question 25: How long were parts actually delayed in the **MyCapacityDelay**?

____

Question 26: Comment on the simplicity and transparency of this object.

____

Step 7: At this point, we have replicated a lot of the SERVER object, which is a complex object (i.e., a heavy weight object), versus our CAPACITYDELAY (i.e., light weight) object. Therefore, we will check our created object with that of the SERVER again.

- For the SERVER **SrvComparison**, *make sure the Initial Capacity* to "Infinity" and set the *Processing Time* property to `ModelEntity.EStaDelayTime`.
- For the **DelOperation**, set the *Delay Time* property to `ModelEntity.EStaDelayTime`.

Step 8: In the **SrcParts_CreatedEntity** add-on process, add the `Assign` step to set the entities delay time to `Random.Pert(1,2,4,1)`[298], as in Figure 20.39.

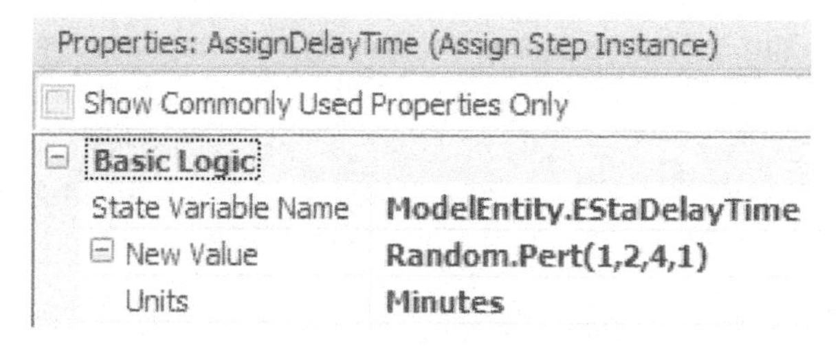

Figure 20.39: Delay Time Assignment

Step 9: Next for the **SrcParts**, *set* the *Interarrival Time* to `Random.Exponential(7,2)`.

Step 10: To compare the two systems, we need to add the use of secondary resources to both objects. We will require one unit of **ResService** at each.

- In the "*Secondary Resources*" of the **CapDelayComparison**, seize one unit of the **ResService** in the *Other Resource Seizes→Before Processing* and release it in the *Other Resource Releases→After Processing.*
- In the "*Secondary Resources*" of the **SrvComparison**, seize one unit of the **ResService** in the *Other Resource Seizes→Before Processing* and release it in the *Other Resource Releases→After Processing.*

Step 11: Set the *DestinationNode* to **Input@SepCopy** and run the simulation for 1000 hours.

Question 27: What did you get for the *TimeWaiting* in the **StorageWait** of the **CapDelayComparison** and the *HoldingTime* for the *InputBuffer* of the **SrvComparison**?

____

Question 28: Why are these identical?

____

Question 29: What corresponds to the *TimeWaiting* in the **StorageDelay** of the **CapDelayComparison** and relative to the **SrvComparison?**

____

Step 12: Employ a SIMIO experiment where the *DestinationNode* is the only control. First run the experiment for 1000 minutes with the *DestinationNode* being **Input@DelOperation**. Then run the experiment with *DestinationNode* being **Input@SrvComparison**.

Question 30: What did you get for the execution time for each of the experiments? Do you think the difference is important?

____

Step 13: Now in the Experiment, set up two scenarios that specify **Input@DelOperation** and **Input@SrvComparison** respectively with 100 replications. Change the run length back to 100 hours.

Step 14: Save and run the experiment.

---

[298] Note we are using the fourth parameter which specifies the random stream so we induce as much correlation into comparing scenarios.

Question 31: What is the average and half width of total time in system as well as the time waiting for the two scenarios?

____

Part 20.8: Some Observations on the Design of Objects

In this chapter we carefully and deliberately created an object from one of the SIMIO base classes. Since the goal was to create a "queuing-like" object, our SIMIO reference was the SERVER object. But we started where the SIMIO developers may have started, namely with a base class. Our initial object was a simple delay based on a station, but we were able to show how the object could be enhanced with assignments and add-on processes. In the midst of our design concerns, we introduced capacity into the simple object through the station and that brought several queuing complications, which the SIMIO developers also had to resolve for the SERVER. Although we considered several options, we eventually abandoned the capacity of the station in favor of simply using "secondary resources" as the primary resources. And we obtained statistics on waiting and in delay from storages.

It is useful to contrast the simplicity and power of the created object with the SERVER. Without doubt the SERVER provides a wide range of powerful features. For example the SERVER has a variety of queuing options, it has input and output buffer capacity, and the processing can handle tasks and task sequences. There is reliability logic and financials associated with the SERVER and the new object doesn't have buffer locations for either input or output.

Nevertheless, the new object has the benefits of simplicity and transparency. The SERVER has three stations and numerous processes, including three entering process. The new object has only one station and one entering process. Further, looking at the SERVER processes, notice the proliferation of steps, add-on processes, and assignments. This proliferation creates a challenge to understand just when something is done in the SERVER and further, it now becomes a challenge as to when each of these features are invoked. The new object features are much clearer The fact there is only one entering process, it is easy to see when add-on processes, assignments, and resources seize/release occur. Also in the SERVER, the processes must account for when the stations have zero capacity. In the new object, the only resource constraint are the secondary resources.

While stations provide the possibility of a capacity constrained location, the capacity must be seized and released while the location is obtained by transfer. Leaving a station fires a transfer event that must be resolved from the current events. In fact, these current events emanate from the set of processes in the object. If you examine the SERVER carefully, you will notice the many processes which cause any number of current events and we have seen in prior examples how these "zero-time" events can dramatically affect the simulation execution.

So our advice relative to creating you own objects is to focus on simplicity and transparency so using your object does not become an error-prone challenge. The easier it is to use an object and the more transparent its behavior, the more success you will have in creating object-oriented simulations

Part 20.9: Commentary

- Creating your own specialized objects can really enhance your models, which is one of the important advantages of using SIMIO. However, if you are only going to use the object once and in one particular model, it may not make sense to spend the time creating a specialized object that could be handled in processes. The advantage is the reuse with-in the same model or different models.
- Instead of the need for a capacity on the station (location), considered the capacity restriction of the object as being a separate concern from the resources needed. It greatly simplified the object, especially in contrast to the SIMIO server.

- But as we developed the object we realized that there is a strong interest to adding "features" and the appropriate stopping point is not at all clear.

Chapter 21
Continuous Variables, Reneging, Interrupt, Debugging: A Gas Station

Continuous variables are variables whose value can change continuously during the simulation (as opposed to discrete variables). Continuous variables in SIMIO can be "Level" or "Level with Acceleration" state variables – see "*States*" under the *Definitions*"tab in the "*Continuous*" section. When a LEVEL state variable is created, you may also specify the rate of change, called the *Initial Rate Value* property as well as its initial state called the *Initial State Value property.* LEVEL WITH ACCELERATION variables in addition to specifying the rate of change and the initial state, you may also specify acceleration rate, called *Initial Acceleration Value* property as well as the length of the acceleration via the *Initial Acceleration Duration Value*. Unlike discrete state variables, continuous state variables can change continuously based on the rate and acceleration.

Part 21.1: Simple Tank[299] Manual Process

We are going to manage a simple gasoline tank by changing its rate value. We will assume the tank volume is between 0 and 100 gallons.

Step 1: Create a new model. Add a fixed RESOURCE object to the model named **ResTank** which will represent the tank.

Step 2: Create a new symbol named **TankSymbol** to associate with the tank, (from the *Project Home→New Symbol→Create Symbol*) such as the one in Figure 21.1. The symbol which represents the interior is a square with a height of ten meters (whose height determines the content/level of the tank). Add a *Gold Comb* texture to give it the appearance of gasoline for both the top and the sides.

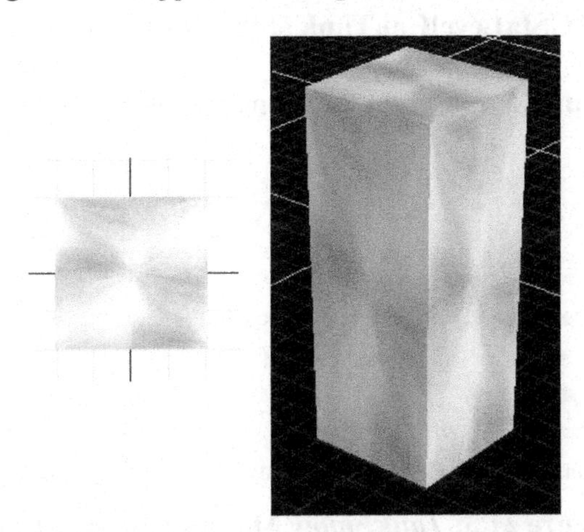

Figure 21.1: 2D and 3D Representation of the Tank Symbol

Step 3: Next, use the **TankSymbol** as the symbol for the **ResTank** RESOURCE.[300] Once this is done, draw a static "case" around it so you can see the level of the tank in 3D as shown in Figure 21.2. The case is made from a thick "Polyline", leaving a small opening in the front, so as to see the level of the tank. The polyline also needs a height of ten meters.

---

[299] This "simple tank" example is adapted from the SIMIO "SimBits" which are another way help you learn about SIMIO.

[300] Select the RESOURCE object and then choose the symbols from the "*Symbols*" tab.

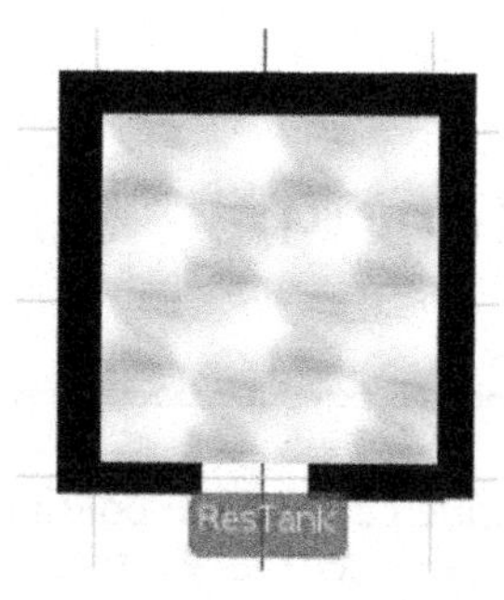

Figure 21.2: Tank Symbol

Step 4: Now add a LEVEL continuous state variable named **GStaLevelConTank** from the *Definitions→States* section. Let's set the *Initial State Value* to 0 and the *Initial Rate Value* to 0. A negative rate value means the tank is being drained while a positive value means it is being filled.

Step 5: When we need to modify a parameter for a model, it is convenient to specify a model property rather than specifying a constant. Doing so also helps in eliminating mistakes by using the same symbolic name for a particular value. Therefore define a minimum and maximum level for the tank as a standard real property named **TankMin** and **TankMax.**[301] For these two properties, the *Unit Type* should be "**Volume**" and *Default Units* equal to "**Gallons**." Add two "Expression" standard properties named **RatePositive** and **RateNegative** to represent the rate at which the tank will fill up and the rate at which the rate will be emptied. For these two properties, the *Unit Type* should be "**VolumeFlowRate**" and *Default Units* equal to "**Gallons per Hour**." Place each property into a new **Tank** *Category*.

Step 6: Next, right click on the **Model** in the [*Navigation*] panel to access the model properties and set the values of the minimum level property (***TankMin***) to 0, the maximum level property (***TankMax***) to 100, the fill rate (***RatePositive***) to 100, and the emptying rate (***RateNegative***) to -200.

Step 7: Now add two MONITOR *Elements* named **MonTankEmpty** and **MonTankFull**. These MONITORS will detect the crossing of the **GStaLevelConTank** at the ***TankMin*** and ***TankMax*** levels respectively.

Step 8: For the **MonTankEmpty** MONITOR, set the properties as shown in Figure 21.3.

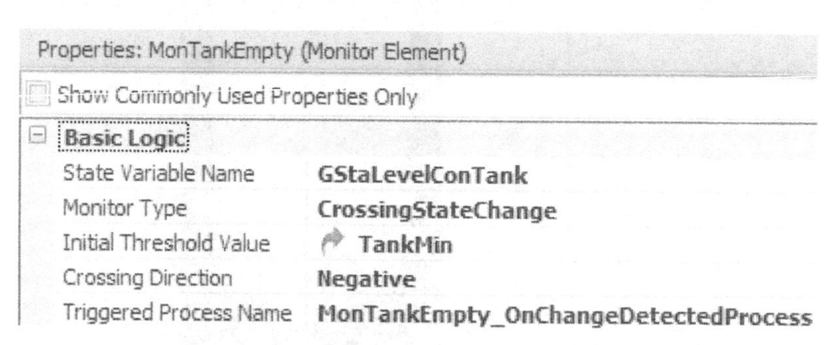

Figure 21.3: *TankEmpty* Monitor Specification

Step 9: And for the **MonTankFull** MONITOR, see Figure 21.4 for the property settings.

---

[301] This is done under the *Definitions* tab in the *Properties* section..

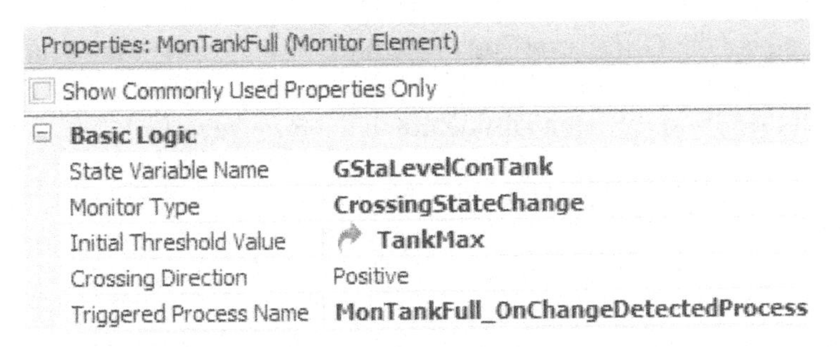

Figure 21.4: MonTankFull **Monitor Specification**

Step 10: Create new processes for the two *On Change Detected Process* add-on process triggers which are defined in Figure 21.5 and Figure 21.6.[302]

Step 11: The ***MonTankEmpty_OnChangeDetectedProcess*** process is called when the tank empties (i.e., the **GStaLevelConTank** reaches zero). We will need to stop emptying the tank, have a small delay before beginning to fill, and then start filling as shown in Figure 21.5.

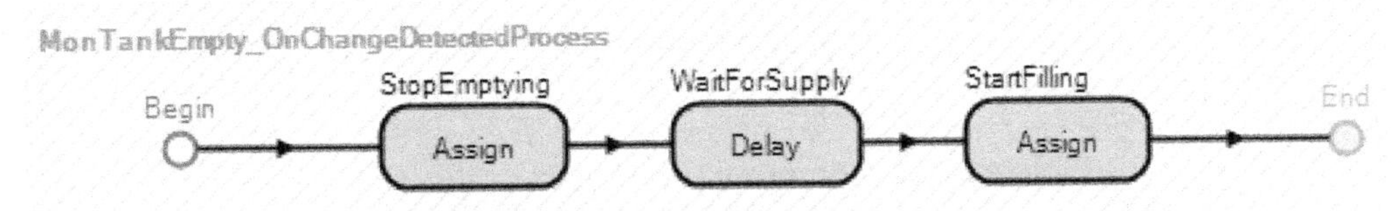

Figure 21.5: When the Tank Empties

- The first `Assign` step sets both the **GStaLevelConTank.Rate** to "0" which stops the LEVEL state variable from changing and sets the **ResTank.Size.Height** to "0".
- Use the `Delay` step to force a 0.2 hour delay before the tank begins to fill again.
- The final `Assign` step sets the **GStaLevelConTank.Rate** to its maximum value of ***RatePositive*** which will allow the tank to start to fill since it is a positive value.

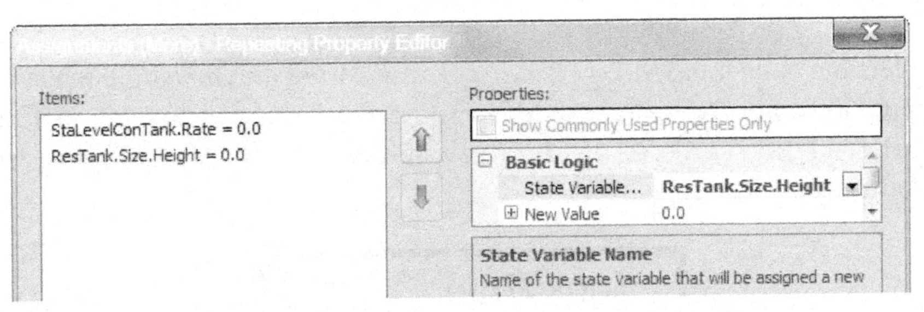

Figure 21.6: Assigning the Rate and the Height

Step 12: The ***MonTankFull_OnEventProcess*** is called when the tank fills to the maximum level. Once it reaches this level, stop filling, force a small delay before allowing it to empty, and then allow the emptying to start as seen in Figure 21.7.

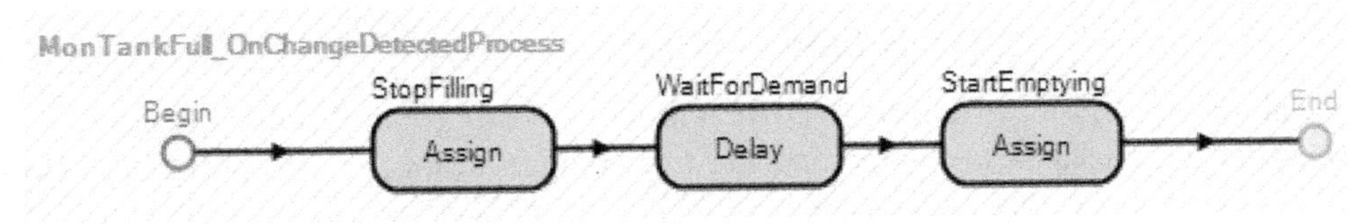

Figure 21.7: When the Tank Fills

---

[302] You can either create the process here that will respond to the MONITOR event or you can create a generic process with the *Triggering Event* property set to the MONITOR event (e.g., `MonTankFull.Event`).

- The `Assign` step sets the **GStaLevelConTank.Rate** to 0. A 0.1 hour delay is assumed before the tank can start emptying and then the last `Assign` step will set the rate to a negative value to allow emptying (i.e., set the **GStaLevelConTank.Rate** to the ***RateNegative***).

Step 13: Finally, we need to take care of the "height" of the tank animation. We will do that in the *OnRunInitialized* process for the model which will run when the simulation starts. From the "*Processes*" tab, select the *OnRunInitialized* process from the "*Select Process*" drop-down.

- This process should `Execute`[303] the ***MonTankEmpty_OnEventProcess*** once to initialize the tank for filling. Then, it will `Delay` for one minute and then `Assign` the **ResTank.Size.Height** to `GStaLevelConTank /(TankMax/10)` as in Figure 21.8.

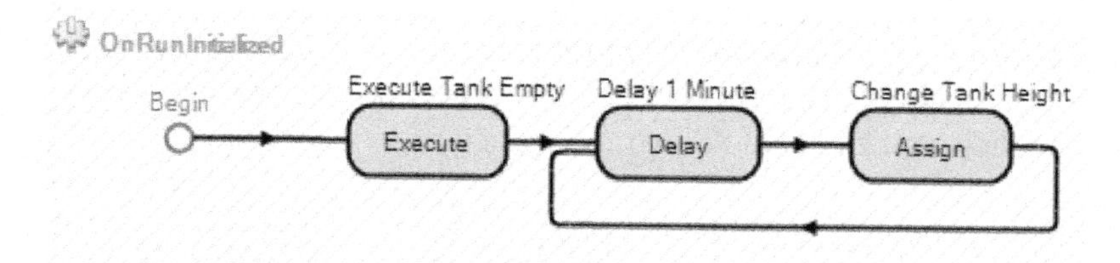

Figure 21.8: Changing the Height of the Tank

- Notice, the system loops continuously every minute updating the tank height for animation purposes. Do this by grabbing the "*End*" point and dropping it on top of the `Delay` step. Therefore the TOKEN will continually loop through these steps until the simulation stops.

Step 14: For the tank level to show in the animation, the *Current Symbol Index* in the "Animation" section of the **ResTank** resource must be set to zero, so the *OnRunInitialized* process is the only specification to change the height. Select the **ResTank** RESOURCE and add an additional symbol and then set the *Current Symbol Index* property to zero.

Step 15: To complete our animation, add six STATUS LABELS from the *Animation* tab as seen in Figure 21.9 where the expressions are `GStaLevelConTank.Rate*264.1720524`, `GStaLevelConTank *264.1720524`, and `ResTank.Size.Height` respectively[304]. Also, for the word Status labels, change the background to white.

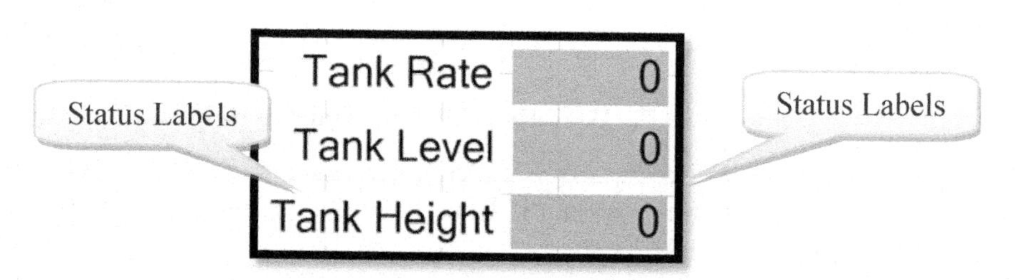

Figure 21.9: Adding Status Labels

Step 16: Next, add a Status Plot of **GStaLevelConTank** vs. Time (hours) – see Figure 21.31. Change the Title, *x-axis* label, *y-axis* label, *text scale* and the time range to plot (i.e., 2 hours) from the *Appearance* tab.

---

[303] The `Execute` step will run any process that has been defined.

[304] The default unit for volume and flow rate is cubic meters and cubic meters per hour which we are converting back to gallons and gallons per hour.

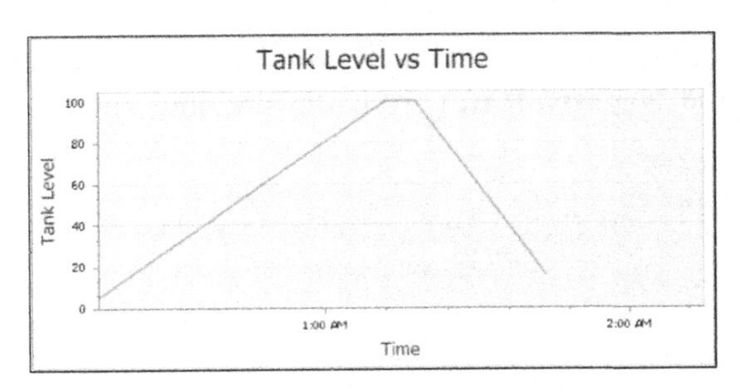

Figure 21.10: Status Plot

Step 17: Also, a Circular Gauge (from the "Animation" panel) may be placed on top of the tank as seen in Figure 21.11.[305] It is a way to visualize the tank level in 2D. After adding the circular gauge with the correct style, under the *Scale* section' set the *Minimum* property to zero, the *Maximum* property to 100 and the *Ticks* to 11. The *Expression* should be the level state **GStaLevelConTank** .

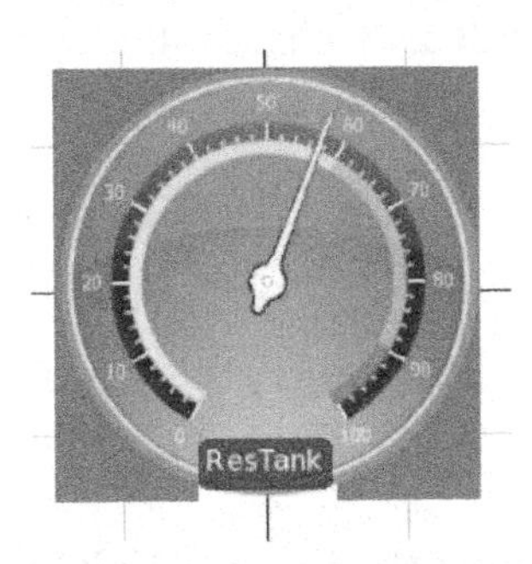

Figure 21.11: Adding a Circular Gauge the Current Tank Level

Step 18: Save and run the model observing the status labels and plots. Also, switch into 3D to see the tank content rise and fall.

Question 1: Do the animations and plots appear appropriate?

____

Question 2: Why does the tank empty faster than filling up?

____

Part 21.2: Simple Tank Revisited using the Flow Library

SIMIO provides a [*Flow Library*] to assist in handling pipes (i.e., flows) and tanks. The library provides a FLOW SOURCE for creating flow entities, FLOW SINK, TANK, FLOWNODE (i.e., regulator), and FLOW CONNECTOR used to connect FLOWNODES. We will model the simple tank again using the [*Flow Library*] as seen Figure 21.12

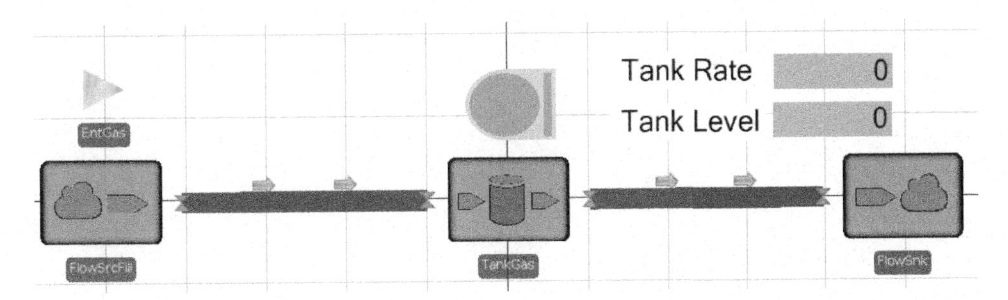

Figure 21.12: Simple Tank using the SIMIO Flow Library

[305] Move to the 3D view and use the shift key to move the gauge up in space to place it directly on top.

Step 1: Insert a new fixed model named **TankFlow**.[306] Copy the four properties (i.e., **TankMax**, **TankMin**, **PositiveRate**, and **NegativeRate**) from the previous model and set their values as seen in Figure 21.13.[307]

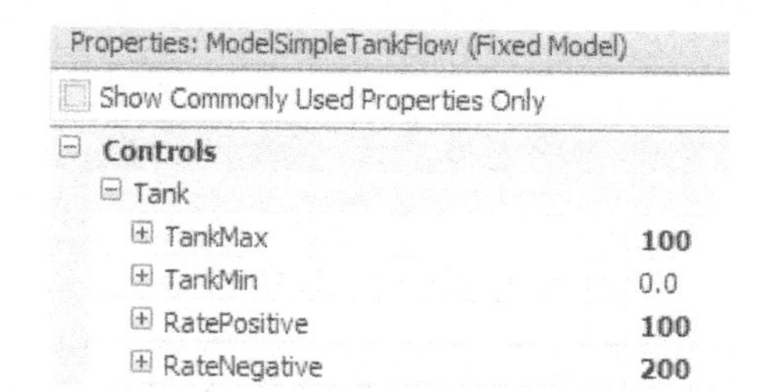

Figure 21.13: Setting up the Prosperities

Step 2: Insert a new FLOWSOURCE named **FlowSrcFill** which will be used to fill the tank.[308] Select the output FLOWNODE **Output@FlowSrcFill** and change the *Initial Maximum Flow* to **RatePositive** (see Figure 21.14).

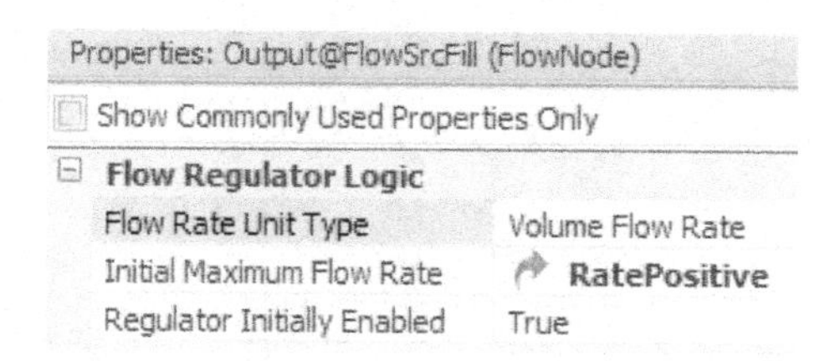

Figure 21.14: Setting the Correct Unit of the Output Flow

Step 3: Insert a new MODELENTITY named **EntGas**. Change the texture of the entity to be the *Gold Comb* texture as before to cause the tank to fill up with gas.

Step 4: Next, add a TANK named **TankGas** which will hold the gasoline flowing from the FLOWSOURCE. Set the *Initial Volume Capacity* and *High Mark* to **TankMax** as well as the *Low Mark* to **TankMin**. Tanks have the ability to specify the capacity as well as have five different level marks (i.e., low-low, low, middle, high, and high-high marks) which will trigger processes when the tank reaches these levels. You should shrink the size of the tank animation to be smaller so you can see the tank filling and emptying easier.

---

[306] You can also just delete the two MONITOR elements, the three processes, the circular gauge, the ResTank and polyline from the current model rather than copying the objects over to the new model.

[307] Note, the rate negative is positive this time because of the way the flow library works.

[308] The regulator inside the **Output@FlowSrcFill** that embodies the continuous level state variables and thus the rates as was done before.

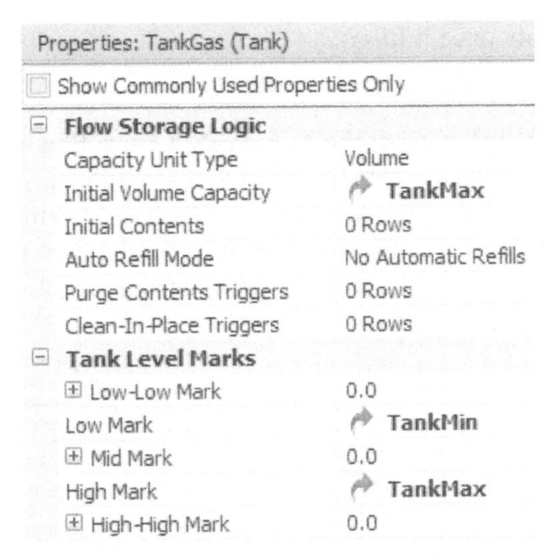

Figure 21.15: Setting up the 100 Gallon Tank

Step 5: Select the **Output@TankGas** FlowNode and set the *Initial Maximum Flow Rate* to 200 gallons per hour which represents the emptying rate.[309] Also, set the *Regulator Initially Enabled* to "False" to indicate the regulator is closed and therefore not flowing (i.e., emptying) as seen in Figure 21.16.

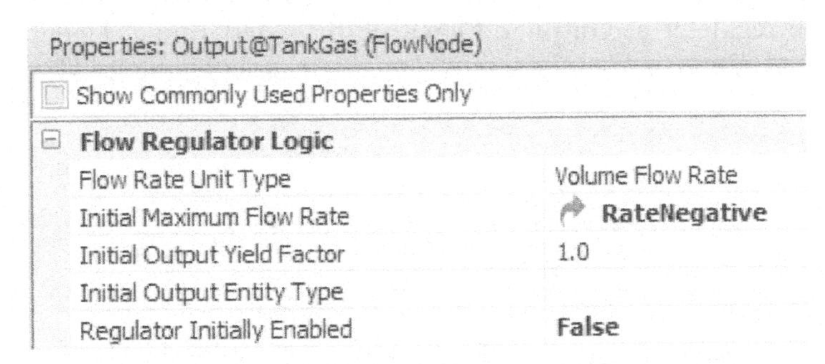

Figure 21.16: Specifying the Tank Emptying Rate

Step 6: Add a FlowSink named **FlowSnk** to destroy the flow of gas emptying the tank.

Step 7: Connect the **FlowSrcFill**, **TankGas**, and **FlowSnk** using FlowConnectors.

Step 8: Copy the Tank Rate and Tank Level status labels as well as the status plot from the previous model. Change the expressions to `TankGas.FlowContainer.Contents.Volume.Rate * 264.1720524` and `TankGas.FlowContainer.Contents.Volume * 264.1720524` for the rate and tank level respectively. Let the plot expression be the same as the tank level expression.

Step 9: Save and run the model. You may need to modify the speed factor to 100 to see the tank level rising.

Question 3: What happens when the simulation runs with regard to the tank and the two FlowConnectors?

____

Question 4: What did you notice about the tank rate when the tank reached the full capacity?

____

[309] Notice we do not specify it to be negative as was done in the manual process of the previous section.

Question 5: Does the tank empty once it fills?

Step 10: Once the tank fills up, the flow from the **FlowSrcFill** stops flowing but the tank does not empty as the **Output@TankGas** regulator is not enabled. When the tank fills up we need to delay for 0.1 hours and then open up this regulator while shutting down the regulator at **Output@FlowSrcFill**. Under the *Tank Level Rising* category, insert the *Tank Full* Add-on Process Trigger as seen in Figure 21.17.[310] The FLOWNODES contain a REGULATOR element which can be opened and closed to control the flow.

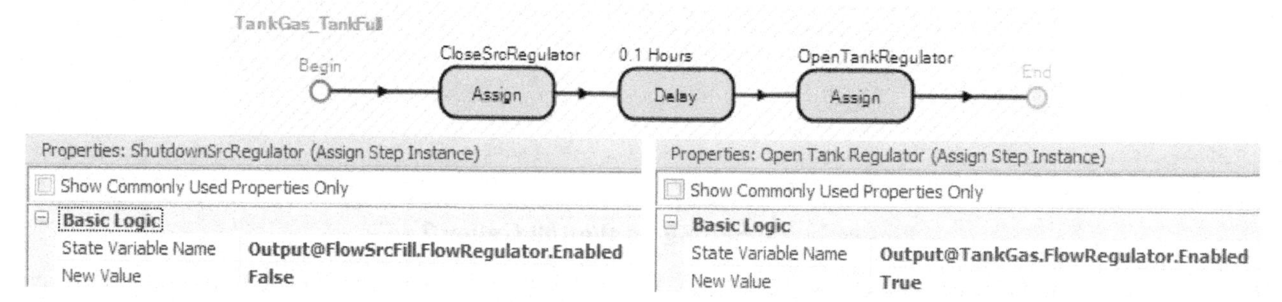

Figure 21.17: Shutdown the Flow Source and then Open up the Tank Output

Step 11: Once the tank empties, we need shut down the outflow regulator of the tank, delay for 0.2 hours and then open up the regulator at **Output@FlowSrcFill** to start filling. Under the *Tank Level Falling* category, insert *the Tank Empty Add-on Process Trigger as seen in* Figure 21.17

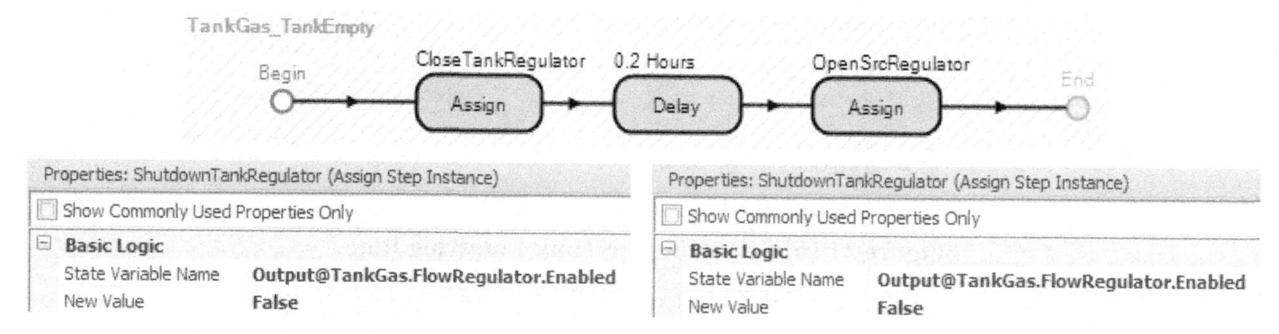

Figure 21.18: Shutdown the Outflow from Tank and Start the Source Flow

Step 12: Save and run the model. You may need to modify the speed factor to 100 to see the tank level rising.

Question 6: What happens when the simulation runs with regard to the tank and the two FLOWCONNECTORS?

Part 21.3: The Gas Station

A gas station has four gas pumps, which dispense the same grade of gas. All the gas pumps pump their gas from a single tank in the ground. Cars arrive and obtain gas from any of the four pumps. If the ground tank is depleted to 100 gallons, the pumps are closed, although any car currently being filled finishes. Any cars in line will stay in line, but newly arriving cars will balk (i.e., leave). A tanker truck comes by periodically to refill the ground tank. As soon as the ground tank is full (i.e., 10,000 gallons), the pumps are open again to deliver gas.

---

[310] You will have to type in these state variables as they will not show up in the dropdown list of possible state variables.

This problem can be approached in several ways. Again, a continuous variable (i.e., tank) will be used to model the ground tank and we will let the cars and the tanker truck change the rates of inflow and outflow to the tank. The cars and the tanker truck will be our active entities. The amount of gas any car needs is assumed to be `Triangular(4,9.5,25)` gallons. The rate that any car pumps gas is `Triangular(125,150,200)` gallons per hour. The tanker will fill the tank at 20,000 gallons per hour.

Step 1: We will modify the model of the simple tank from the previous section. Add two SOURCES, one SERVER, three SINKS and two MODELENTITIES named **EntCars** and **EntRefillTrucks,** as illustrated in Figure 21.19.

- **EntCars** arrive Exponentially every 1.5 minutes beginning at time 0.1 while the **EntRefillTrucks** arrives Uniformly between 6.75 and 8.25 hours apart.
- Use CONNECTORS to link all of the objects together associated with the cars. Make the link between **SrcTanker** and **SnkTankerLeaves** a PATH.[311]
- Make sure each of the sources creates the correct entity type. Also change the picture of the **EntRefilTrucks** to be a tanker[312]. The **EntCars** should be given six different car pictures randomly chosen (i.e., set the *Random Symbol* property to "True" under the *Animation* section).

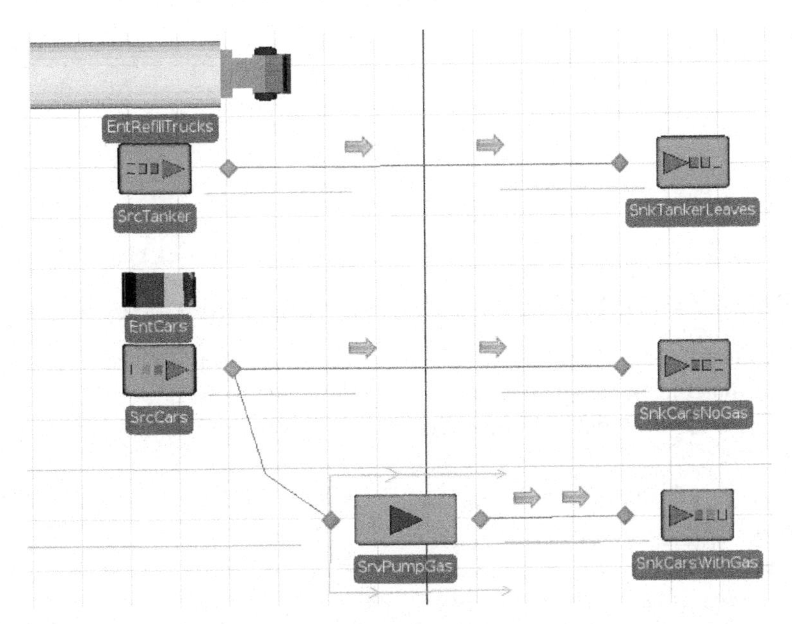

Figure 21.19: Model Structure

Step 2: Define a new DISCRETE INTEGER state variable for the main Model named **GStaPumpOn** to model whether the pumps are on or off. A value of one means "on" and zero means "off". Set the *Initial State Value* property to zero.

Step 3: Set the *Initial Contents* property of the **TankGas** to be **TankMin**.

---

[311] The one path will allow the tanker to be seen as it fills up the tank. Also, we have two sinks for the cars to automatically collect statistics on cars that balk and those do not.

[312] You might want to make the tanker a little smaller.

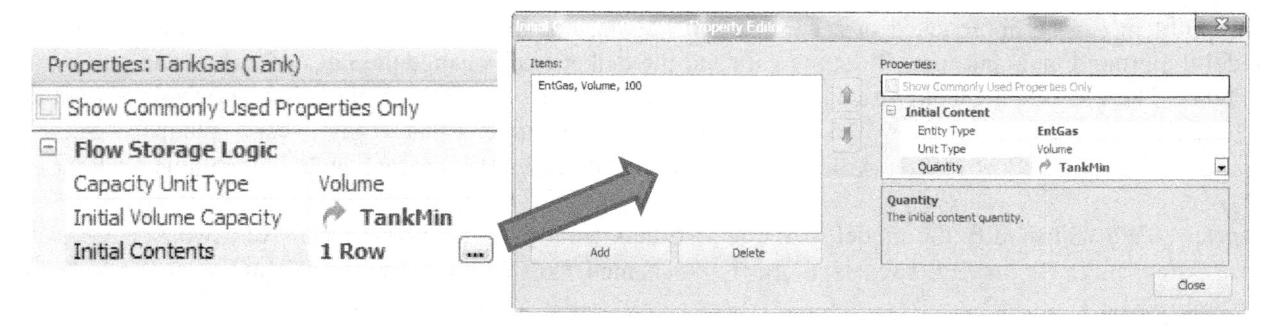

Figure 21.20: Setting the Initial Amount of Gas in the Tank

Step 4: For the **Model**, change the *TankMax* and *TankMin* properties to be 10,000 gallons and 100 gallons respectively. By setting the low mark value to 100, we will not allow our tank to empty completely since cars may still be getting gas when the tanker arrives. Change the ***PostiveRate*** and the ***RateNegative*** model properties to 20,000 and `Random.Triangular(100,125,175)` gallons per hour to represent the rate a car will empty the tank.

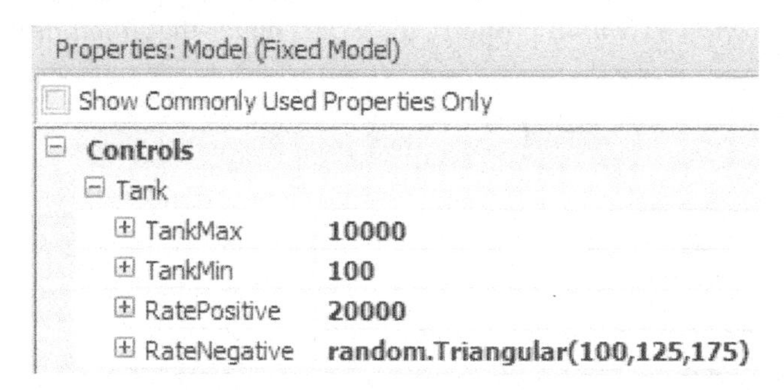

Figure 21.21: Setting the Rates and Initial Values of the Tank

Step 5: Now for the cars (i.e., MODELENTITY) let's define two state variables to represent the amount of gas needed and the rate that the person can pump gas. Select the MODELENTITY in the [*Navigation*] panel. Insert the following two states which will be characteristics of the individual cars as they arrive.

- **EStaGasNeeded** is a DISCRETE REAL STATE variable with a *Unit Type* of "Volume" and an *Initial State Value* of "0" gallons.
- **EStaCarPumpingRate** is a DISCRETE REAL STATE variable with a *Unit Type* of "VolumeFlowRate" an *Initial State Value* of "0" gallons per hour.

Step 6: Since the cars arriving will set the outflow rate, we need to set *Initial Maximum Flow Rate* of the **Output@TankGas** to "0" Gallons per hour.

Step 7: When cars are created by the source, define a process in the *Created Entity* add-on process trigger as in Figure 21.22 which assigns the car characteristics.[313]

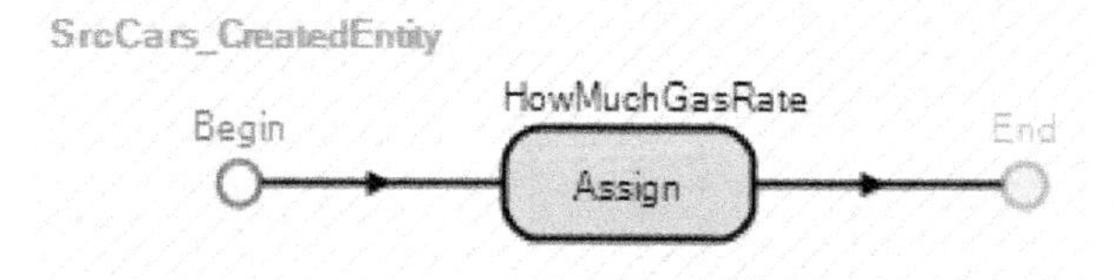

Figure 21.22: Cars Leaving the Source

[313] Throughout this example we will use the `Assign` process step rather than using *the State Assignments* in the modeling objects. Ultimately the `Assign` steps offer more generality since the *State Assignments* are only available on entry to and exit from the object.

Where the `Assign` step sets the following two state variables making sure the rate is negative.

```
ModelEntity.EStaGasNeeded = Random.Triangular(4,9.5,25)
ModelEntity.EStaCarPumpingRate = RateNegative
```

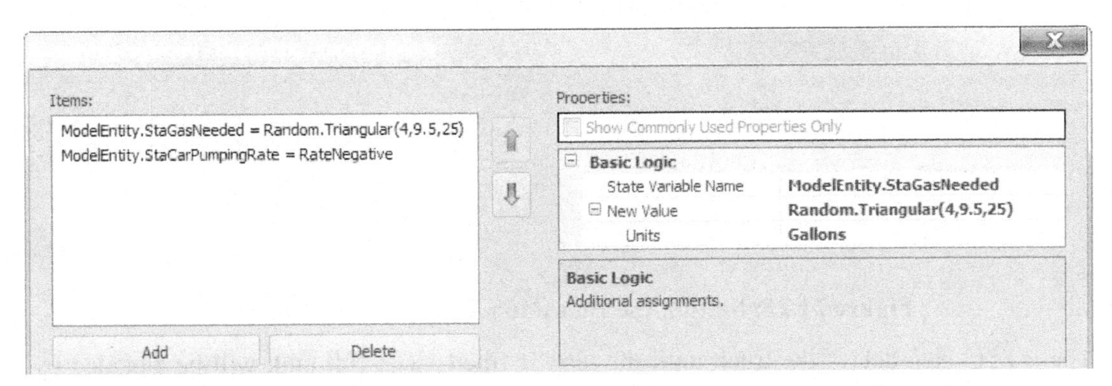

Figure 21.23: Specifying Gas Needed and the Pumping Rate of Each Car

Step 8: Set the *Processing Time* property (in Hours) in the **SrvPumpGas** server to be the following.

```
ModelEntity.EStaGasNeeded/ModelEntity.EStaCarPumpingRate
```

This expression takes into account the amount gas needed divided by the rate the customer can pump it. Make sure to change the *Units* to "Hours" and do not use the default of "Minutes". Be sure that the *Initial Capacity* property of **SrvPumpGas** server is set to "**4**" to indicate room for four cars.[314]

Step 9: If the pumps are on, cars will be routed to the pumps and, if they are off, they will be sent directly to the **SnkCarsNoGas**. The selection link weights on the paths leaving the source are defined as follows to ensure that one path will have a weight of one and the other one a zero.

- For the connection to **SnkCarsNoGas**, set the *Selection Weight* property to `1 - GStaPumpOn.`
- For the connection to **SrvPumpGas**, the *Selection Weight* should be `GStaPumpOn.`

Step 10: All discrete event simulation languages use events and the user can define their own events in SIMIO. Under the "*Definitions*" tab add a new EVENT named **EvntStartPumps**, which will allow entities to wait until the pumps are started up after they have been turned off to start pumping again.

Step 11: Consider what happens when the tanker truck arrives, as shown in Figure 21.24. The tanker truck will turn off the pumps and start filling the tank at 20,000 gallons per hour. The tanker truck will continue to fill until the tank is full (i.e., the monitored **MonTankFull** reaches its threshold). Use the *Entered* add-on process trigger for the input node of the **SnkTankLeaves** (i.e., when the truck entity enters the sink).[315]

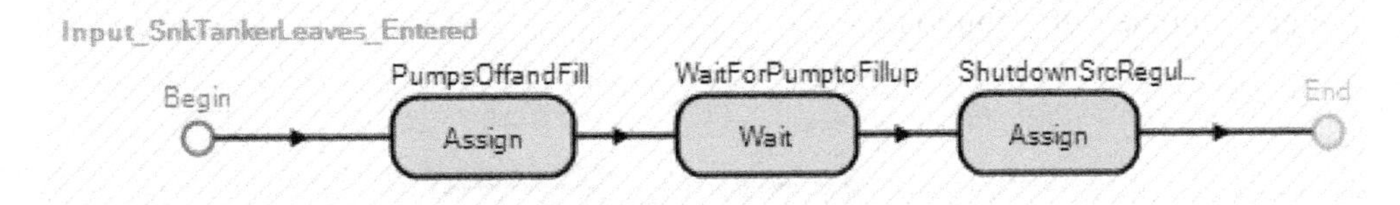

Figure 21.24: Tanker Truck Leaves

[314] You may want to arrange the processing station animated queue to be oriented with four points around the server like a gas pump.

[315] It is necessary to invoke this process after the truck leaves the output at the source so that a time in system can be computed for the truck. If this process is invoked before the truck leaves the output, then its time in system is not updated.

- In the `Assign` step, set the **GStaPumpOn** state variable to zero, turn off the flow REGULATOR of the TANK and turn on the REGULATOR associated with the FLOWSOURCE as seen in Figure 21.25.

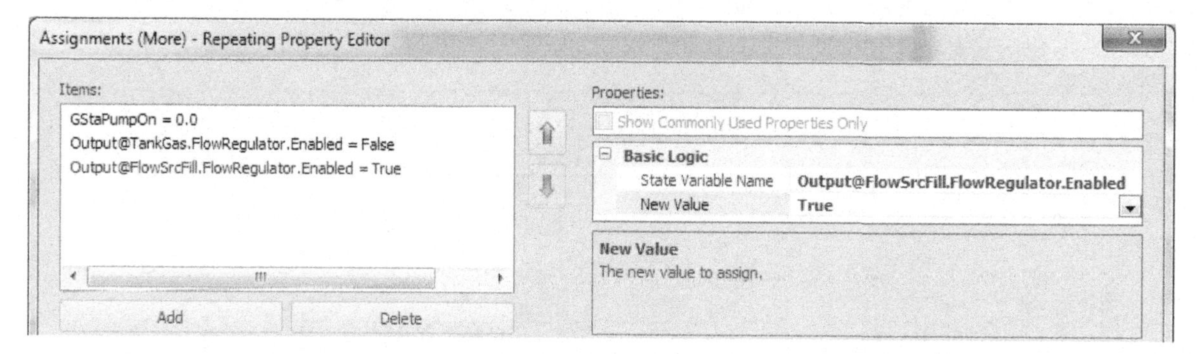

Figure 21.25: Setting the Pump to Off and Filling the Tank

- The `Wait` step delays the truck until the tank is filled. The full tank will be signaled by an event called **EvntStartPumps**.

Figure 21.26: Waiting for EvntStartPumps to Continue

- Once the tank has filled up we need to turn the pumps on and shut down the FLOWSOURCE REGULATOR as well enable the tank regulator to start again.

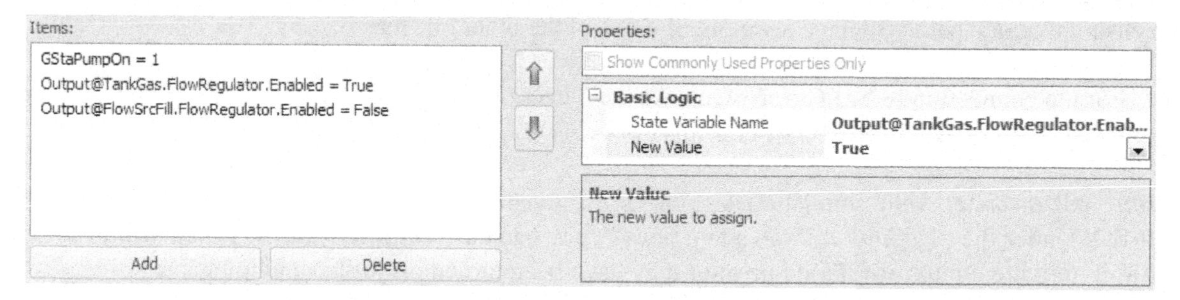

Figure 21.27: Turning On Pumps and Shutdown Filling the Tanks

Step 12: When the pumps are off new cars that arrive will leave the system. However, cars that are currently in line and/or filling up will remain. Those that are currently filling up will finish but no other car should be allowed to start filling. Therefore, within the **SrvPumpGas** server, consider first when the processing of a car filling their individual tank begins (see Figure 21.28). Insert the *Before_Processing Add-On Process Trigger* with the following logic.

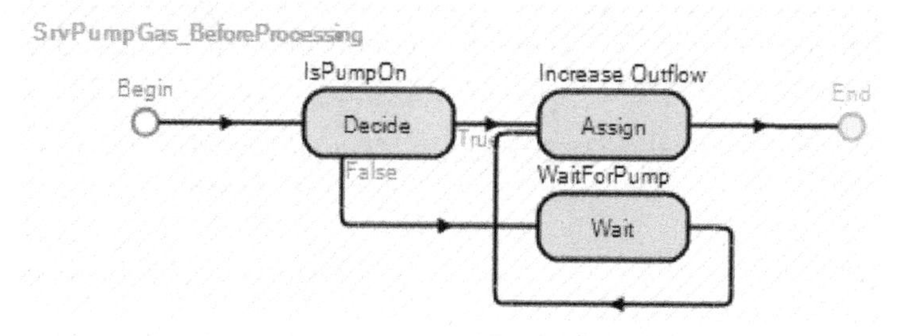

Figure 21.28: Processing Begins at the Pump

- When the *Before_Processing* add-on process trigger starts: Decide if `GStaPumpOn == 1`

- If the pumps are off (i.e., "False" branch), then "`Wait`" for the event **EvntStartPumps.**
- Otherwise, if the pumps are on, enable the **TankGas** FLOWREGULATOR (i.e., `Output@TankGas.FlowRegulator.Enabled = True` and change the tank empting rate via an *`Assign`* step by adding the outflow due to the car being filled to the current rate:
  ```
  Output@TankGas.FlowRegulator.CurrentMaximumFlowRate =
  Output@TankGas.FlowRegulator.CurrentMaximumFlowRate +
  ModelEntity.EStaCarPumpingRate
  ```

Step 13: When the car finishes (i.e., processing) at the pump (i.e., *After_Processing* add-on process trigger for the **SrvPumpGas**), refer to Figure 21.29, we need to remove the outflow due to the car being filled (i.e., reduce rate). Use the `Math.Max` to ensure that you never go below zero.

```
Output@TankGas.FlowRegulator.CurrentMaximumFlowRate = Math.Max(0,
Output@TankGas.FlowRegulator.CurrentMaximumFlowRate -
ModelEntity.EStaCarPumpingRate)
```

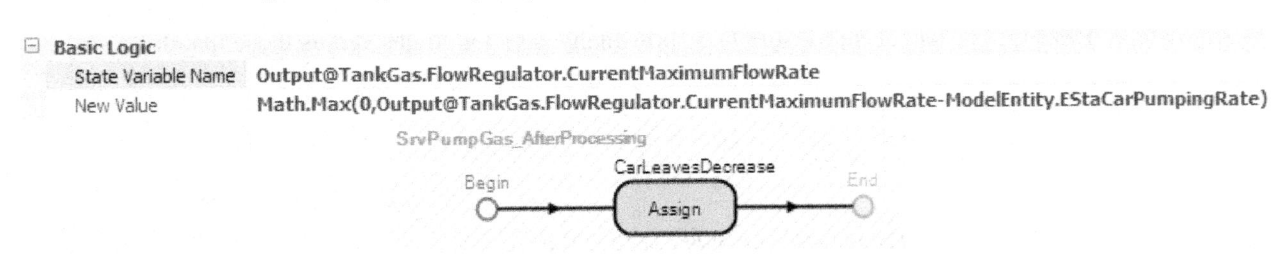

Figure 21.29: Car Finishes at Pump Decrease Rate

Step 14: Next, take care of the process when the **Tank** reaches the low mark of 100 gallons. For the **TankGas**, insert the *TankGas_TankLevelFallingBelowLowMark* add-on process trigger and an *`Assign`* step that will set the `GStaPumpOn` to "0".

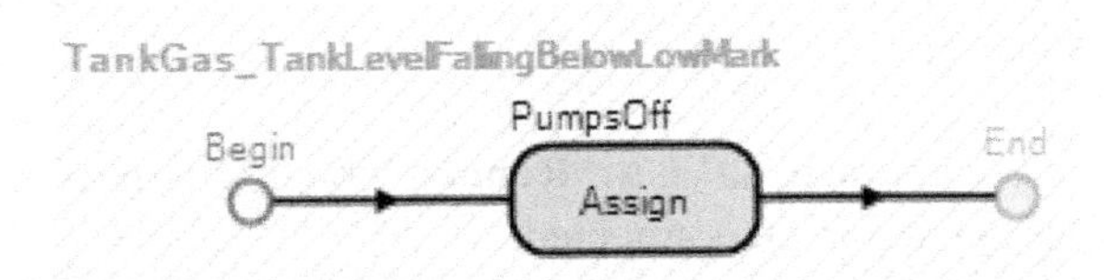

Figure 21.30: Turn off the Pumps so Cars will Balk

Step 15: For the **TankGas_TankEmpty** (see Figure 21.32) we need to disable the tank's out flow (i.e., set `Output@TankGas.FlowRegulator.Enabled` to "False) and then set the rate to zero (i.e., `Output@TankGas.FlowRegulator.CurrentMaximumFlowRate`=0.0).

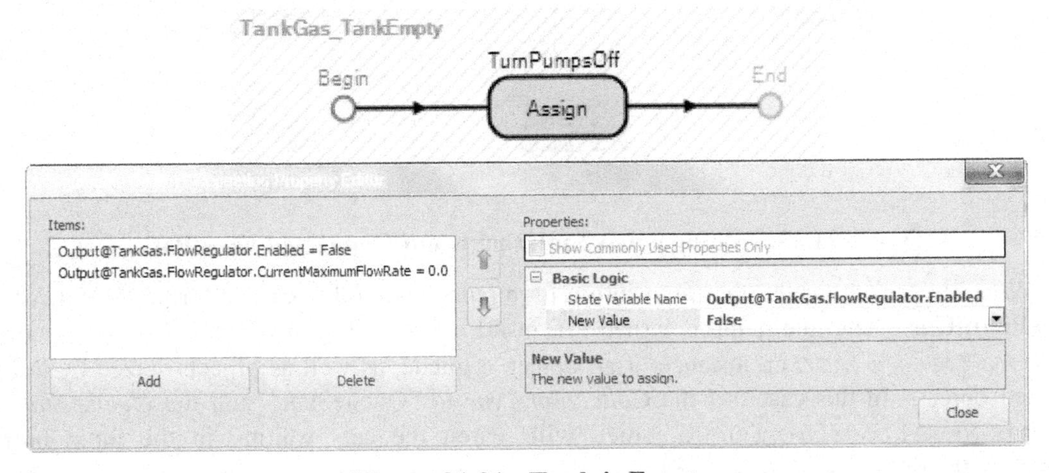

Figure 21.31: Tank is Empty

Step 16: And for the **TankGas_TankFull** event (see Figure 21.32), it should turn the pump on and let any cars waiting as well as the refill truck know that the filling process has completed. Use an `Assign` step to set the `GStaPumpOn` to "1" and then use the `Fire` step, to "Fire" the **EvntStartPumps** event.

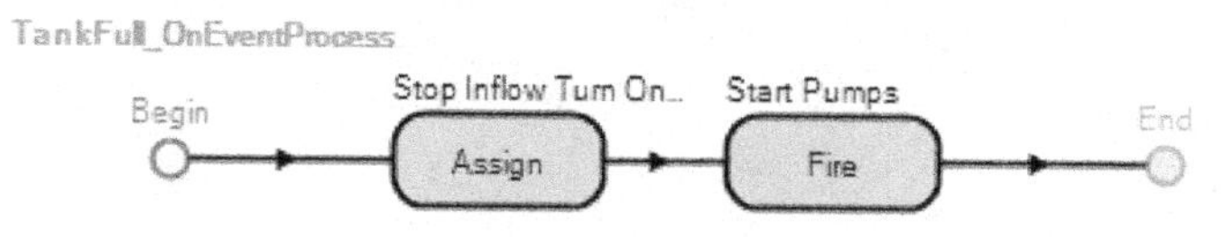

Figure 21.32: Tank is Full

Step 17: When the event **EvntStartPumps** is fired, the truck waiting for the tank to fill is activated and the cars waiting at the pumps will be activated. If these should be activated differently, then the process needs to be modified.

Step 18: Save and run the current model for 48 hours.

Question 7: Does this system seem to work (i.e., how many cars get no gas versus those that do)?

____

Question 8: What happens when the tanker comes and their cars pumping gas and waiting in line?

____

Question 9: What is the average level in the tank?

____

Part 21.4: Reneging the Cars When the Pump Goes Off

The previous model assumed that when the tanker arrived and shut the pumps off, the cars at the pumps would continue to wait until the pumps are turned back on. Let's modify the model such that all of the cars waiting at the pumps would renege (that is leave).

Step 1: Add a SINK to your model named **SnkCarsRenege** which will automatically generate statistics for the reneging cars but does not needed to be connected.

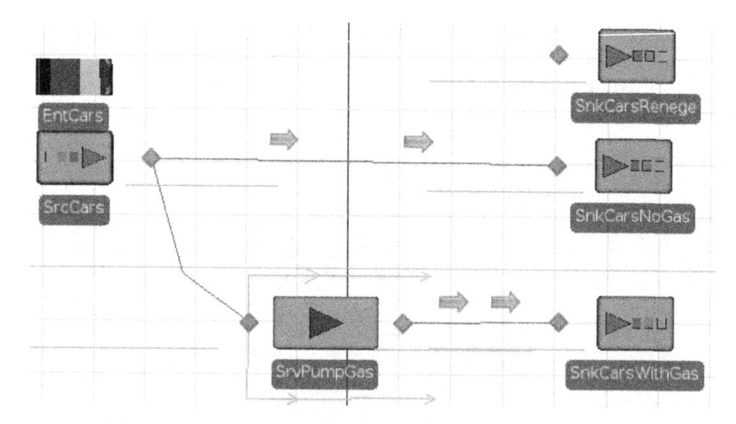

Figure 21.33: Adding a new Sink that is not Connected via a Path

Step 2: When the tanker arrives, the cars waiting for a pump should be removed from the `input` queue of the **SrvPumpGas**. After the pump is turned off, insert a `Search` step which can search a collection of objects in a TABLE, a LIST, an instance of an object, a queue, as well as objects seized by the parent or associated object. In this case, set the *Collection Type* to "QueueState" and the *Queue State Name* to `SrvPumpGas.AllocationQueue` which will search the cars waiting in the input buffer to be processed at the pump. See Figure 21.34.

- The search needs to return all the cars waiting, therefore set the *Limit* property to "Infinity". The *Match Condition* will allow searching for certain objects that has to be true for the item to be selected. In this case we could have been left blank to return all entities in the queue. However, if you chose paths to link the objects and a car was on the link and arrived at the same time the tanker arrived then there could be a car that has seized the pump but is waiting in the "Wait" state for the pump to come on and therefore we don't want to remove any cars that have already seized a pump. For each car (i.e., **EntCars**) found, a new TOKEN associated with the found car is sent down the "Found" path while the original TOKEN will eventually continue down the "Original" path once all items have been found[316].

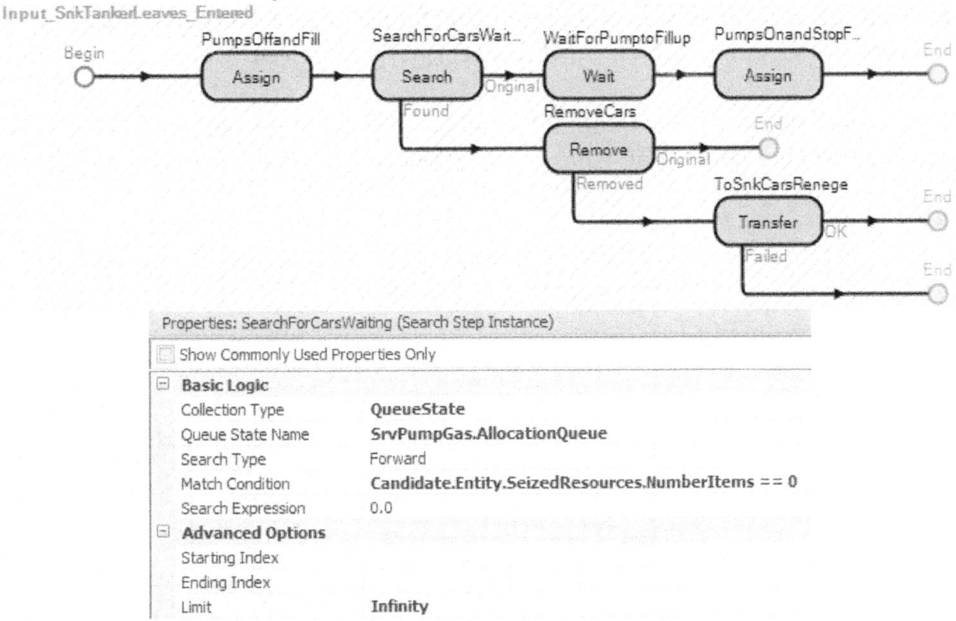

Figure 21.34: Reneging all Cars Waiting for Gas when the Pump goes Off.

Step 3: Insert a *`Remove`* step that will remove the current car associated with the "Found" TOKEN from the `SrvPumpGas.AllocationQueue`. A TOKEN associated with the removed item will exit out the "Removed" path.

Step 4: Now insert a *`Transfer`* step that will move the car (i.e., modelentity) to the **SnkCarsRenege** by specifying the *Node Name* as `Input@SnkCarsRenege` as the node to send these cars. The car is currently in the **InputBuffer** STATION so this step needs to transfer the entity from the current station to the node specified as seen in Figure 21.35.

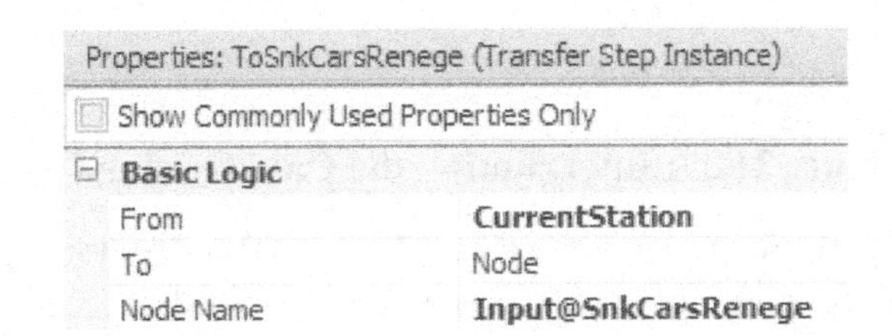

Figure 21.35: Transferring the Cars to the Renege Sink for Statistic Collection

Step 5: Save and run the model observing what happens when the Tanker arrives.

Question 10: Do the waiting cars renege (you may need to look at results)?

___

[316] Note that two tokens are now active.

Question 11: How many cars reneged in 48 hours of the simulation?

Part 21.5: Interrupting the Cars When Pump Goes Off

In the previous model, the cars waiting at the pump reneged when the pump was turned off and we had already not allowed other cars to arrive to the gas pumps while the pumps are off. The cars already in service getting gas when the pumps go off were allowed to continue to fill up their car and then leave. Suppose now we want the model to interrupt the cars in service and deny them further use of the pump when the tanker arrives. These interrupted cars will leave through the **SnkCarsRenege** EXIT as well.

Step 1: To interrupt a process that is currently being delayed, insert an `Interrupt` step before the `Search` step as shown in Figure 21.36 with specifications in Figure 21.37.

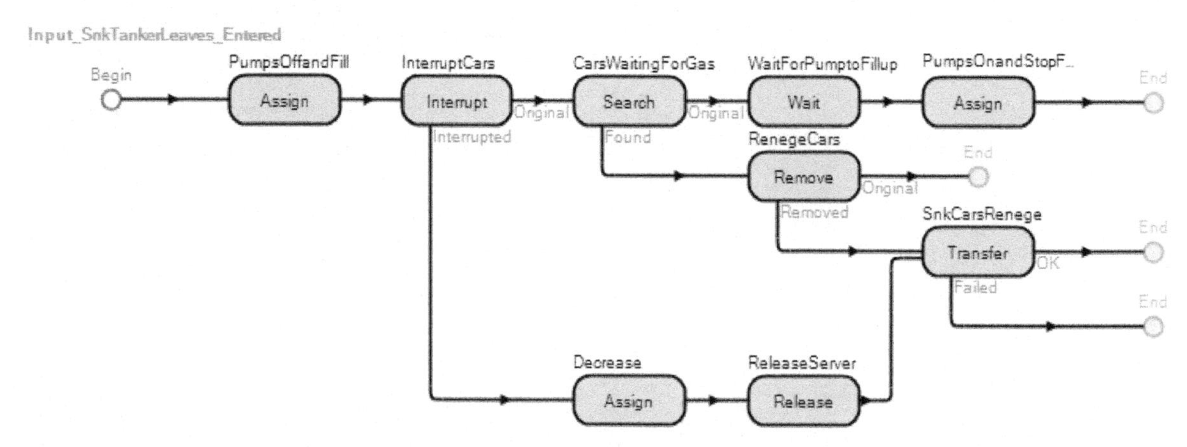

Figure 21.36: Adding the Interrupt to Force Cars Currently Processing to Renege

- Use the *Process Name* **SvrPumpGas.OnEnteredProcessing** which is the process that is running the `Delay` step (i.e., processing of the cars). Notice that we "EndProcess" for the interrupted processes and we establish a *Limit* of "4" since there are four pumps potentially in operation.

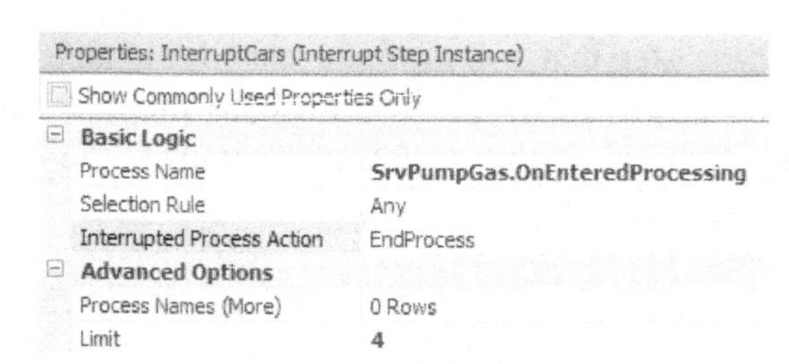

Figure 21.37: Interrupting the Cars Getting Gas

- Copy the `Assign` step from the step **SrvPumpGas _AfterProcessing** to the "Interrupted" branch which will decrease the flow rate of each car. Then connect the `Assign` step to the `Transfer` step that is used previously to force the cars to exit to the reneging sink (i.e., **SnkCarsRenege**).
- Insert a `Release` step that will release a Specific resource (i.e., **SrvPumpGas**).

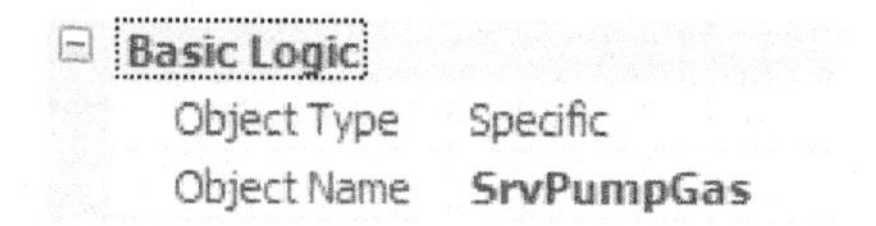

Figure 21.38: Releasing the Server SrvPumpGas after Interruption

Step 2: Save and run the model observing what happens.

Question 12: How can you tell if the interrupt is working (Hint put a break point on Search and then single step the simulation)?

____

Question 13: How many cars exited the renege sink now (i.e., how many cars were interrupted)?

____

Part 21.6: Using Entities that Carry a Tank

In our gas scenario, there are really three distinct systems (i.e., the cars and the pump, the refill tanker, and finally the continuous gas tank system). The first two systems are discrete systems that interact with the continuous system by turning on and off the regulators as well as changing the flow rate. The [*Flow Library*] provides for several other flow objects that combine both discrete and continuous together as seen in Table 21.1. Currently the tanker truck arrives and causes the FLOWSOURCE to fill up the tank at certain rate till it is full. We assume the tanker has enough gas to fill the tank until it is full. However, the tanker truck can only carry about 4500 gallons of gas. Now, we are going to assume that we own one tanker that has a volume capacity (i.e., actually caries the gas) and that it will travel between the refill station and the gas station where it fills the tank to its capacity. The tanker can move faster when it is empty (returning to the refill station).

Table 21.1: Additional Flow Objects Linking Discrete and Continuous Entities

Flow Object	Description
CONTAINERENTITY	A MODELENTITY that contains a TANK which is transported with the entity.
FILLER	This object behaves in a similar fashion as the COMBINER. Instead of combining a parent and member entity together, the object combines (i.e., fills up the tank) a discrete container entity with a flow
EMPTIER	This object behaves in a similar fashion as the SEPARATOR does for entities by separating (i.e., emptying the tank) the flow from the container entity.
ITEMSTOFLOWCONVERTER	Converts discrete model entities into continuous flows (e.g., large drinking bottles are delivered and placed into the water machine which flows out water to people until the bottle is empty)
FLOWTOITEMCONVERTER	Converts continuous flows into discrete entities (i.e., continuous process of spinning of fibers to produce a bobbin of yarn which is transferred to the next station).

Step 1: Since we are going to model the truck physically transporting the gas back and forth from the refill station, we can delete unneeded modeling components. Delete the **SnkTankerLeaves** as well as the **EntRefillTrucks** since we need now to use a CONTAINERENTITY for the tanker.

Step 2: From the *Definitions* tab, insert a new EXPRESSION PROPERTY named **TankerCapacity** that has a *Unit Type* of "Volume" with *Default Units* set to "Gallons" as seen in Figure 21.39.

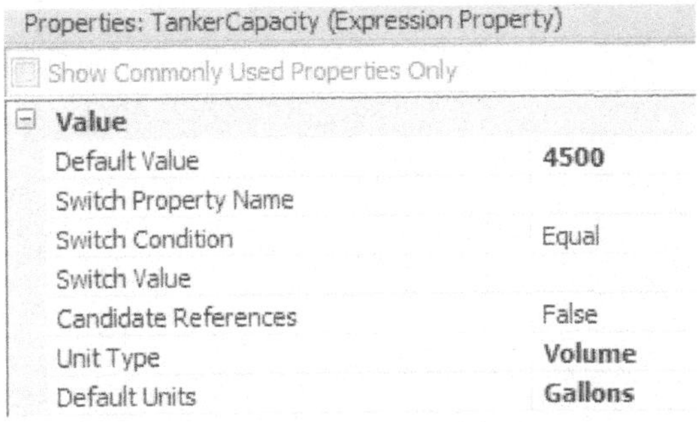

Properties: TankerCapacity (Expression Property)

Show Commonly Used Properties Only

Value	
Default Value	**4500**
Switch Property Name	
Switch Condition	Equal
Switch Value	
Candidate References	False
Unit Type	**Volume**
Default Units	**Gallons**

Figure 21.39: Properties of the TankerCapacity Property

Step 3: From the [*Flow Library*] insert a CONTANERENTITY named **CEntTankerTruck** which has the *Initial Volume Capacity* set to the **TankerCapacity** (i.e., a 4500 gallon tanker) as seen in Figure 21.40. Change the picture to be the same tanker resizing it to be smaller. Also, you may want to make the tank square and stretch it out.

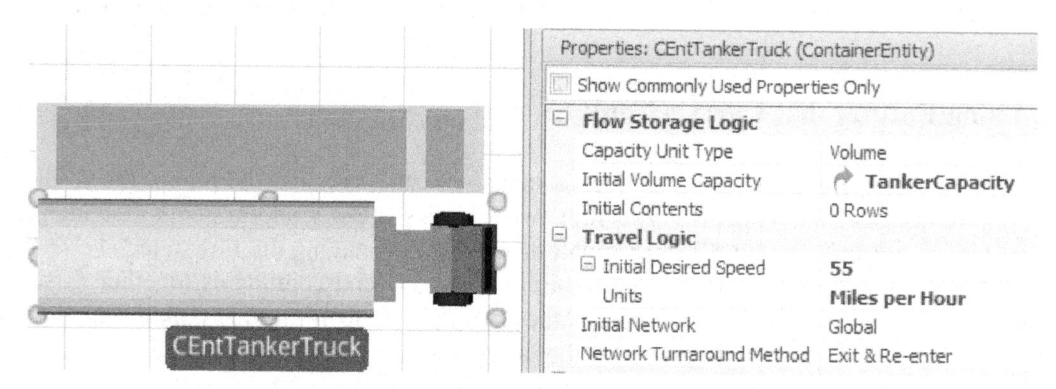

Figure 21.40: Setting up the CEntTankerTruck

Step 4: Select the **Srctanker** and specify the new **CEntTankerTruck** as the *Entity Type*. Since we only will have one tanker in the system set the *Maximum Arrivals* to "1" under the *Stopping Conditions* section.

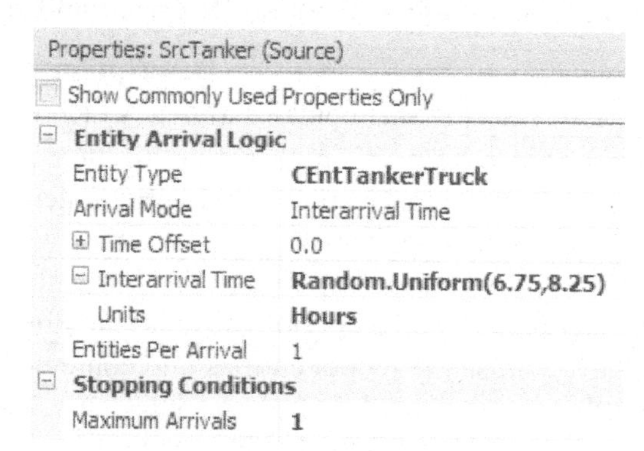

Figure 21.41: Modifying the SrcTanker to Create One Tanker

Step 5: Next, we will create the refill station where the tanker will be filled up and then allowed to transport the gas to the gas station as seen in Figure 21.42.

- Disconnect the **FlowSrcFill** from the **TankGas** as we will use the tanker to fill up the tank.
- Insert a new FILLER named **FillTanker** which allows us to fill up a container in an entity. Also, make the processing station queue oriented to allow for the large tanker to be seen while processing.
- Connect the **SrcTanker** via a path to a **ContainerInput** node of the **FillTanker**.[317]
- Finally Connect the **FlowSrcFill** to the **FlowInput** node of the **FillTanker**

[317] The simulation will warn you when you run the simulation and the entity that is entering the container input node does not have a tank associated with the entity (i.e., it is not a `ContainerEntity`).

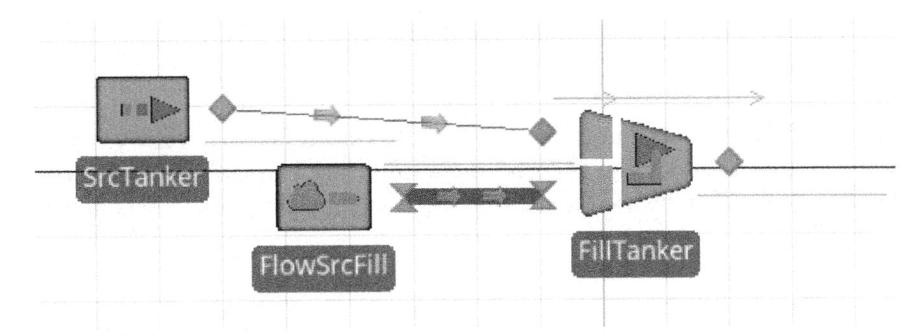

Figure 21.42: Refill Station Area

Step 6: Once the tanker arrives at the gas station, we need to pump the gas out of the tanker and into the tank. Insert a new EMPTIER named **EmptyTanker** and connect **FlowOutput** node to the **TankGas** via a FLOWCONNECTOR. The emptier can either empty all the contents of the container entity or a specific amount. In this case we will empty all of it until the **TankGas** is full (i.e., the **EventStartPumps** is fired when the tank is full).

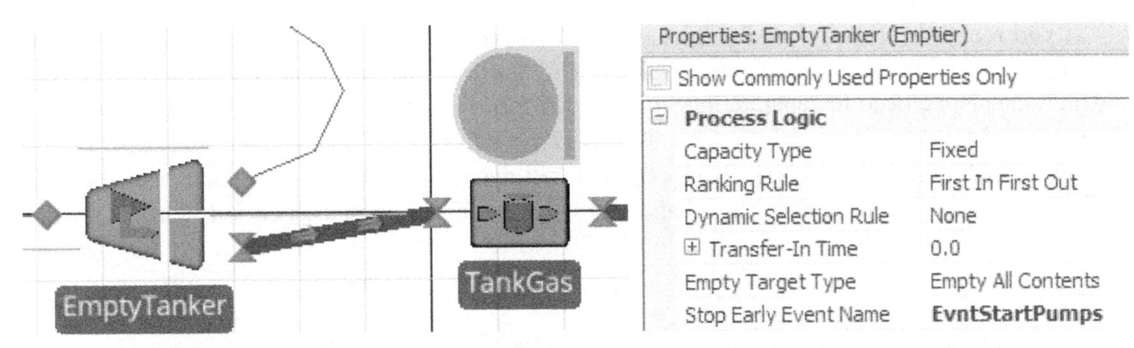

Figure 21.43: Using the Emptier to Fill up the Gas Tank from the Tanker

Step 7: Finally connect the **Output@FillTanker** to the **Input@EmptyTanker** via a time path as well as the **ContainerOutput@EmptyTanker** to the **ContainerInput@FillTanker**. Both time paths should have a Travel Time set to `Random.Uniform(3.37,4.125)` hours to travel between the gas and refill station.

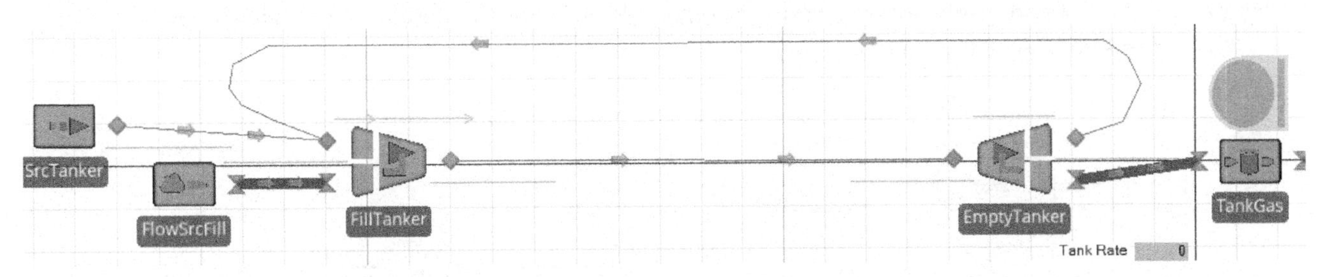

Figure 21.44: Tanker and Gas Station Refilling System

Step 8: Save and run the model.

Question 14: Are cars being allowed to get gas immediately?

____

Question 15: How many trips does it take before cars allowed into the gas station and why?

____

Question 16: Does the tanker always return empty?

____

Question 17: How many cars renege during the 48 hours?

____

Step 9: If we want the pumps to be shut off and the cars to renege like before, we will need to specify the *Before Processing* add-on process trigger to be the **Input_SnkTankerLeaves_Entered**. Create a new *After Processing* trigger that will be used to the turn the gas pumps back on as seen in Figure 21.45.

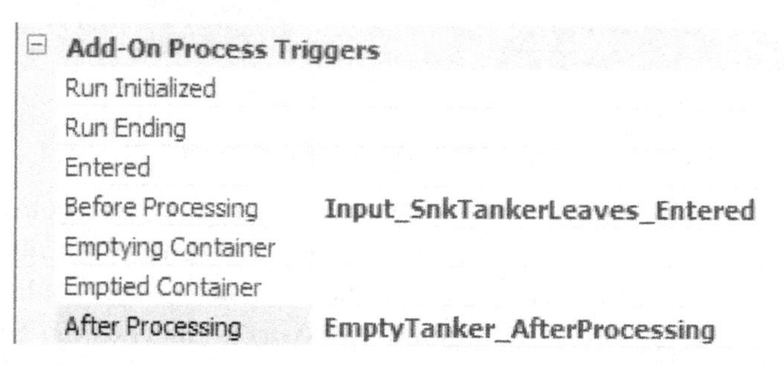

Figure 21.45: Using Process to Turn the Pumps On and Off when Filling up Tank

Step 10: If you recall the **Input_SnkTankerLeaves_Entered** process shuts off the pumps and then waited for the tank to fill up before turning them back on. The **EmptyTanker** object is performing the filling task now which means we need to separate the shutting down of the pumps from the starting of the pumps. For the **EmptyTanker_AfterProcessing**, copy the `Assign` step (i.e., "PumpsOnandStopFlow") that is after the `Wait` step which turns the pumps back on. You will need to delete the assignment that shuts down the FlowSrcFill output regulator since that is handled by the refilling station as seen in Figure 21.46.

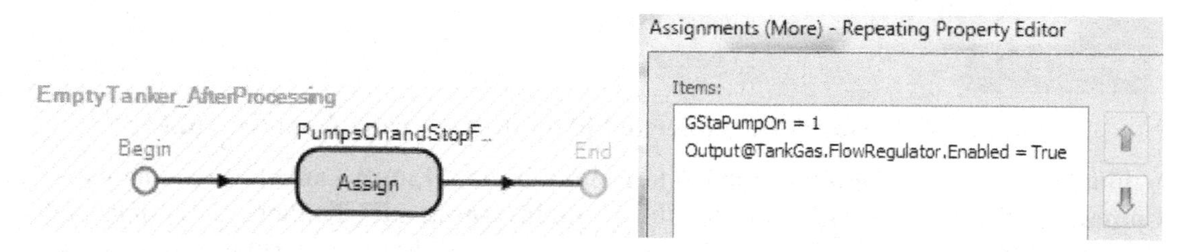

Figure 21.46: After the Tanker Finishes Empting Turn Pumps Back On

Step 11: For the **Input_SnkTankerLeaves_Entered**, delete both the Assign and Wait process steps after the *Search* step as seen in Figure 21.47.

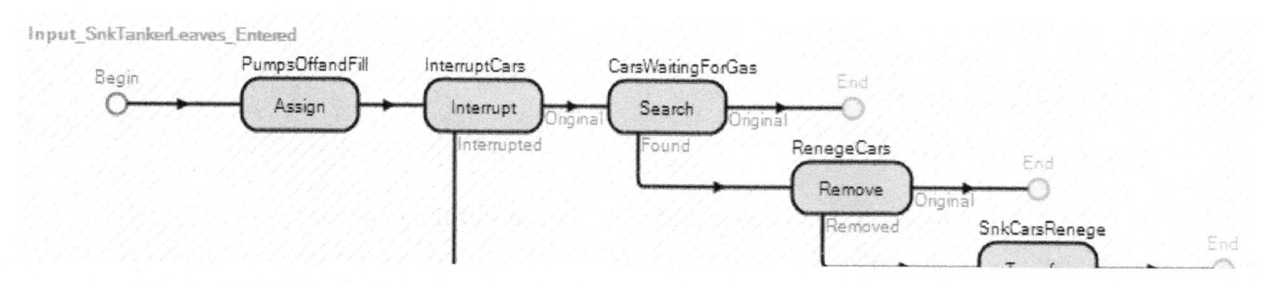

Figure 21.47: Only Shutting Down the Gas Pumps and Forcing the Cars to Renege

Step 12: Save and run the model.

Question 18: How many cars renege now?

____

Part 21.7: Commentary

- Continuous variables in SIMIO can be very useful, although the rates of change (second derivative) are limited to constants. The addition of the new flow objects has made it much easier to use these and integrate them into discrete systems.
- For our gas station, we could have combined the gas pumps in with the continuous system as well using the FILLER . Because the FILLER object can fill only one item (i.e., capacity of one) at a time, you would need a FILLER for each of four gas pumps. Then, we would need to employ the concepts from Chapter 8 to restrict flow entities to the pumps (i.e., a single queue using a RESOURCE object). Then the cars become container entities which when created would modify their gas tank size to specify the amount gas needed. You would connect the output of the **TankGas** to the four gas pump fillers while removing the FLOWSINK.
- The uses of reneging and interruption concepts were introduced in this chapter and should not be ignored. Reneging allows entities in queues to be removed and either destroyed or transferred. Processes may also be interrupted. These features are not often used, but when they are needed, they are extremely valuable. Notice also that the modeling is accomplished through processes, not though additional specifications on the objects, as one might see in other simulation languages. This approach grants considerable flexibility to the modeler.
- The debugging facilities in SIMIO are quite extensive. When combined with animation display options, the trace, breakpoint, watch, and dashboard window can provide needed detail about the behavior of the simulation.
- The value of the "Simbits" should not be underestimated. They contain numerous examples of highly useful modeling approaches on many various topics.

Chapter 22
More Subclassing: Advanced Modeling of Supply Chain Systems

Chapter 10 demonstrated the modeling of a very simple three tier supply chain system. However, only the DC (Distribution Center) component in the supply chain used an inventory model to make decisions. Figure 22.1 shows a more complicated supply chain where two different store chains place orders on our manufacture's DC. The two chain DCs run inventory models similar to the manufacturer's DC and also places replenishment orders. The manufacture can utilize two different suppliers for a particular product. The suppliers use an inventory model to replenish its inventory positions to satisfy demands from the manufacturer.

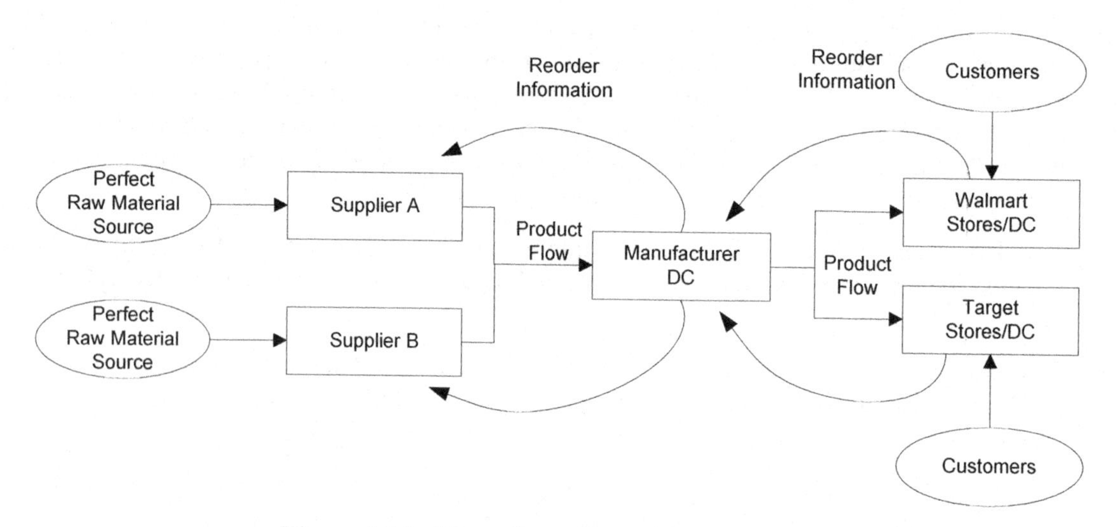

Figure 22.1: More Complicated Supply Chain

Part 22.1: Developing a Specialized Supply Chain Workstation Object

In Chapter 10, an inventory model system was built for a single inventory system. A similar approach could be used for each segment of the supply chain in Figure 22.1 by adding additional state variables and process logic. As the system gets more complicated, changes or the requirement of different inventory models this approach is not very efficient. However, the ability to create specialized objects will make it easier to model these systems with very little additional logic.

Step 1: Open up the final project from Chapter10. Now also create a new project in a separate instance of SIMIO. It will be convenient to copy items from the previous model in constructing the new one.

Step 2: The MODELENTITY is used to model the flow of orders as well as the products. Insert the following three discrete integer state variables into the new project model by selecting the MODELENTITY object in the [*Navigation*] panel.

Table 22.1: Three New Discrete State Variables for Order ModelEntities

Variable	Description
EStaOrderAmt	The amount of product that is needed or has been ordered.
EStaDeliverNodeID	The input node where the order should be delivered.
EStaOriginalOrderAmt	The amount of the original order.

Step 3: Our goal is to encapsulate the logic and parameters that were in the simple supply solution into an object that can be used more than once. Select the **Model** from the [*Navigation*] panel and subclass the WORKSTATION object[318] and rename the new object **SUPPLYINVWORKSTATION**.

Step 4: Select the **SUPPLYINVWORKSTATION** object in the [*Navigation*] panel. Under the *Definitions→Properties*, insert the same five *Expression Standard Properties* named **InitialInventory**, **ReorderPoint**, **OrderUptoQty**, **ReviewPeriod,** and **ReviewPeriodTimeOffset.** These should all be placed in the "Inventory" Category[319] and the two review period properties need to have units of time with default of unit of days.

Properties

InitialInventory	Expression Property	InitialInventory	Inventory
ReorderPoint	Expression Property	ReorderPoint	Inventory
OrderUptoQty	Expression Property	OrderUptoQty	Inventory
ReviewPeriod	Expression Property	ReviewPeriod	Inventory
ReviewPeriodTimeOffset	Expression Property	ReviewPeriodTimeOffset	Inventory

Figure 22.2: Adding the Four Inventory Properties to our Supply Object

Step 5: For the **SUPPLYINVWORKSTATION** object, insert a DISCRETE INTEGER STATE variable named **StaInventory** which will represent the current inventory amount of the product at the current segment and another one named **StaOnOrder** which represents the total amount this segment has been currently ordered from the supplier. The *Initial State Value* property should be zero for both the **StaInventory** and **StaOnOrder** variables respectively.

Step 6: Next, we need to insert the same ELEMENTS as was done for the simple supply chain model into our new object as seen in Figure 22.3 and Figure 22.4.[320]

- Insert/Copy two TALLY STATISTICS named **TallyStatSL** to track the service level performance and TallyStatAmtOrdered to track the amount ordered.
- Insert/Copy two STATESTATISTICS named **StateStatInv** and StateStatOnOrder to track the inventory level performance and the amount on order. Specify the *State Variable Name* property to be the **StaInventory** variable and **StaOnOrder** respectively.
- Insert a TIMER named **TimerReview** to model the periodic review. The supply workstation will review their inventory positions[321] based on the reference property **ReviewPeriod** and start reviewing based on the **ReviewPeriodTimeOffset**.

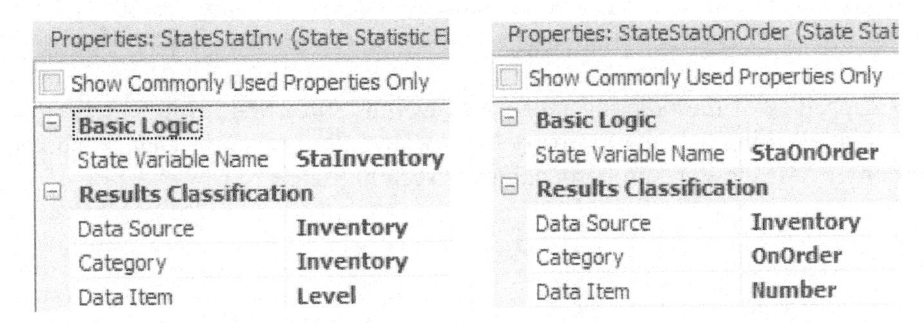

Figure 22.3: Setting up the State Statistics Track Inventory and On Order

[318] See Chapter 18 for more information on subclassing objects in SIMIO.

[319] The first time, you need to type the category name "Inventory" and then the next time it can be selected from the dropdown box.

[320] You can copy the elements, state variables, and properties from the Chapter 10 models.

[321] Remember that inventory position is the inventory "on hand" plus the inventory "on order."

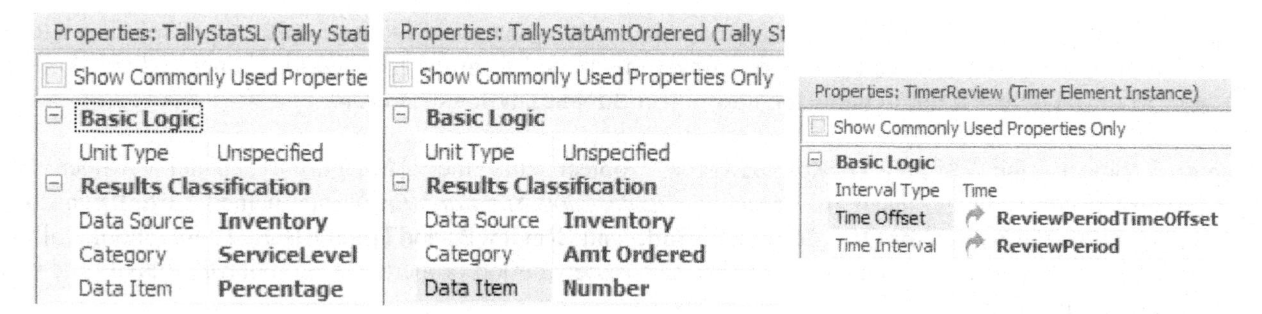

Figure 22.4: Setting up the Tally Statistics and Periodic Review Timer

Part 22.2: Adding the Ordering Station and Characteristics to Handle Orders

In the previous section, the same variables, properties, and elements were added to our new supply server object that was used for simple supply chain model of Chapter 10. Recall that the model included two separate systems (i.e., an ordering system to handle orders and one to produce the parts to fill the inventory). Figure 22.5 shows the anatomy of the new object. The original workstation will still act as the replenishment piece (i.e., convert raw material into a finished good) but it will not send the product on to the network but update the inventory level while the new order processing piece will ship out product based on customer orders.

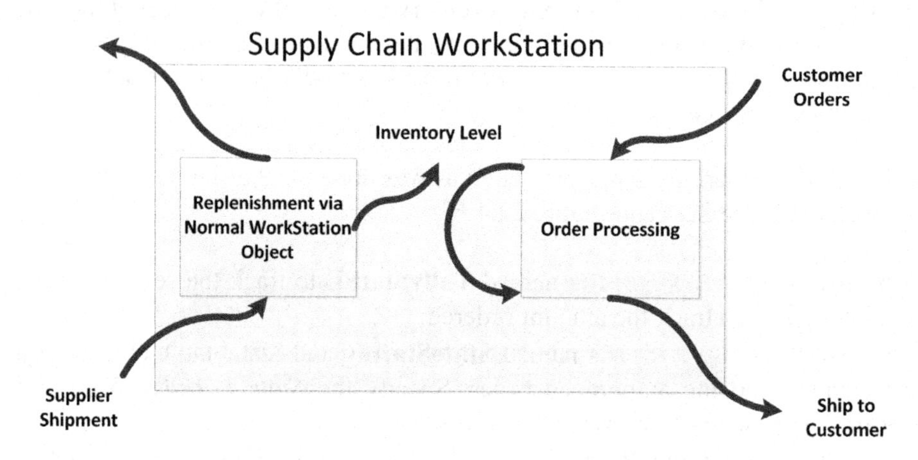

Figure 22.5: Anatomy of the Supply Chain WorkStation

Step 1: Our new object will need to receive orders and process them. Therefore, a location (i.e., a STATION) is needed to house these order entities while they are being processed. From the *Definitions→Elements→General* section, insert a new STATION named **OrderProcessing.**[322]

Step 2: To set capacities of the ordering process as well as sort orders waiting, create a new referenced property by right clicking on the property except the *Parent Cost Center* should be set to the **ParentCostCenter**. Use the same property names as seen in Figure 22.6.

[322] See Chapter 20 for more information on STATIONS.

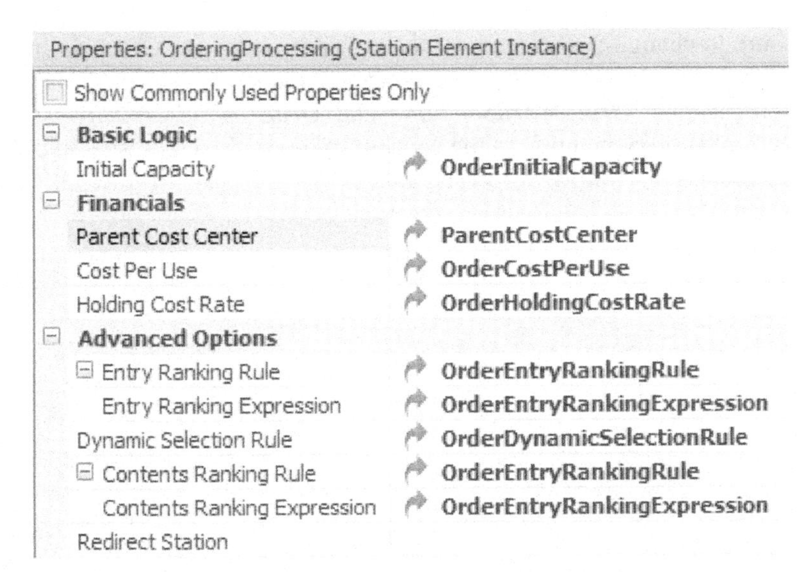

Figure 22.6: Creating Properties to Allow User to Change the Ranking of Orders

Step 3: Switch to the *Definitions→Properties* panel and modify the new properties by placing the *OrderCostPerUse* and *OrderHoldingCostRate* into the "Financials/Order" category and the rest in the "Ordering Information" category.

OrderInitialCapacity	Expression Property	OrderInitialCapacity	Order Processing
OrderEntryRankingRule	Enumeration Property	OrderEntryRankingRule	Order Processing
OrderEntryRankingExpression	Expression Property	OrderEntryRankingExpression	Order Processing
OrderDynamicSelectionRule	Selection Rule Property	OrderDynamicSelectionRule	Order Processing
OrderCostPerUse	Expression Property	OrderCostPerUse	Financials/Order
OrderHoldingCostRate	Expression Property	OrderHoldingCostRate	Financials/Order

Figure 22.7: All of the Properties

Step 4: To separate the processing of the ordering and production systems from the "*Facility*" tab, insert a new RESOURCE named **ResOrder** which will be used in the ordering system while the workstation will be used for the production system. Also, the resource should not be visible externally in the model.[323]

Step 5: Next, we will link the new resource capacity and ranking to the ordering station. For the *Capacity Type* and *Work Schedule* properties create new reference properties, while for the others just choose the ones already created.

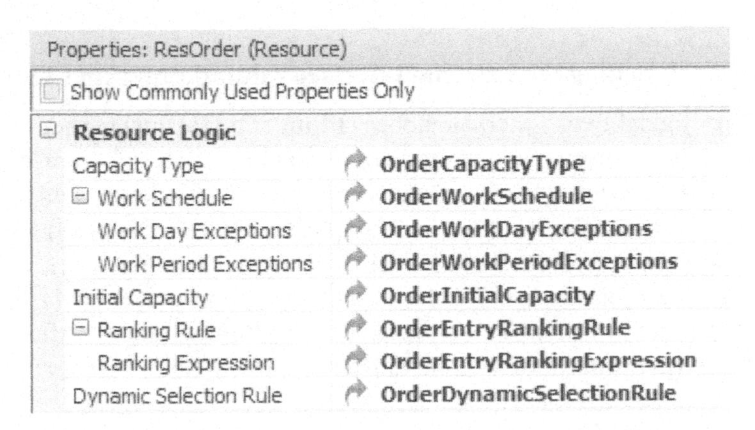

Figure 22.8: Process Logic Properties of the New Resource

[323] To make sure an object is not visible in the external view, right click the object and toggle the "*Externally Visible*" button.

Step 6: Also, make sure to change the category of the new properties to "Ordering Information" as well as move the two new properties up above the ***OrderInitialCapacity*** as seen in Figure 22.9. Let's create a Switch Condition on the new ***OrderWorkSchedule*** and ***OrderInitialCapacity*** which will not show it unless the user selects a "WorkSchedule" as the capacity type or "Fixed" respectively. This will allow the ordering processing system to follow work schedules. Do the same for work day and work period exceptions.

Properties
- OrderCapacityType
- OrderWorkSchedule
- OrderInitialCapacity
- OrderEntryRankingRule
- OrderDynamicSelectionRule
- OrderEntryRankingExpression
- InitialInventory
- ReorderPoint
- OrderUptoQty
- ReviewPeriod
- ReviewPeriodTimeOffset

Properties: OrderWorkSchedule (Schedule Property)

Show Commonly Used Properties Only

Property	Value
Logic	
Default Value	
Switch Property Name	**OrderCapacityType**
Switch Condition	Equal
Switch Value	**WorkSchedule**
Schedule Type	**CapacitySchedule**
Appearance	
Display Name	**OrderWorkSchedule**
Category Name	**Order Processing**
Category Expanded	False
Parent Property Name	
General	
Name	**OrderWorkSchedule**
Description	
Required Value	**False**
Visible	True

Properties: OrderInitialCapacity (Expression Property)

Show Commonly Used Properties Only

Property	Value
Logic	
Default Value	**Infinity**
Switch Property Name	**OrderCapacityType**
Switch Condition	Equal
Switch Value	**Fixed**
Candidate References	False
Unit Type	Unspecified
Appearance	
Display Name	**OrderInitialCapacity**
Category Name	**Order Processing**
Category Expanded	False
Parent Property Name	
General	
Name	**OrderInitialCapacity**
Description	
Required Value	True

Figure 22.9: Modifying the Order Capacity Type and OrderWorkSchedule Properties

Step 7: Next, we need to add an entry point (i.e., input node) for orders coming into the supply inventory server as well as an exit point for reorders leaving the supply server. For the **SUPPLYINVWORKSTATION** object, navigate to the *Definitions→External* section. Let's add a Status Label with an expression "Sup Inv" as on top of the picture as seen in Figure 22.10 to distinguish from the original WORKSTATION object.

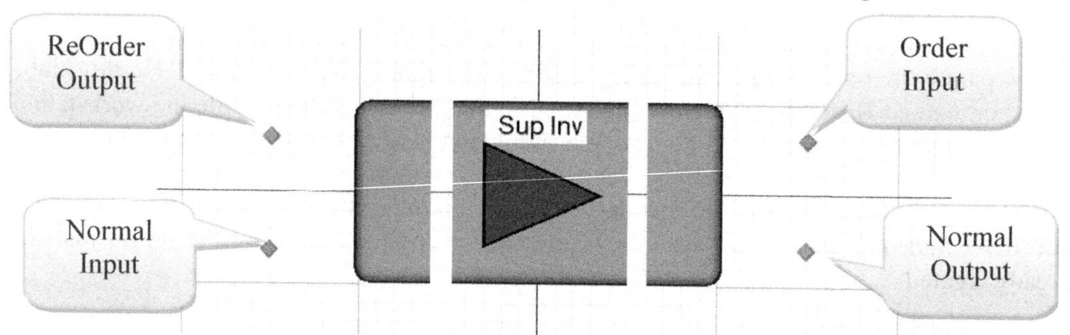

Figure 22.10: Creating the External Interface Adding Entry and Exit Nodes

Step 8: Now, move the original two "External" nodes (**Input** and **Output**) to the bottom third of the object as seen in Figure 22.10. Insert two additional "External" nodes[324] named **ReorderOutput** and **OrderInput**. The **ReorderOutput** node is a TRANSFERNODE and should be right above the normal **Input** node since orders will be transferred out to their supplier while the **OrderInput** node is a BASICNODE placed right above the normal **Output** node which will accept arriving orders. This will facilitate the notion of products flowing from the back of the chain to the front of the chain (i.e., left to right) while orders will flow the opposite direction. The properties for both nodes are specified in Figure 22.11. Once order entities enter the **OrderInput** they will be transferred directly to the **OrderProcessing** STATION.

[324] See Chapter 17 and Chapter 20 for more information on inserting external nodes into the external view of the object.

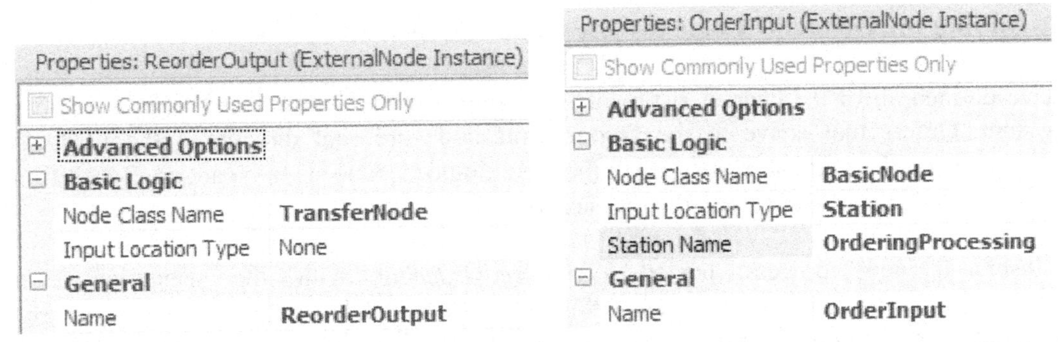

Figure 22.11: Specifying the Properties of the ReorderOutput and OrderInput Nodes

Step 9: Next add an additional animation queue associated with the contents queue of the OrderProcessing station as seen in Figure 22.12.[325] Add the same six status labels as before which represent the inventory level (**StaInventory**), average service inventory (`StateStatInv.Average`), and average service level (`TallyStatSL.Average`).[326]

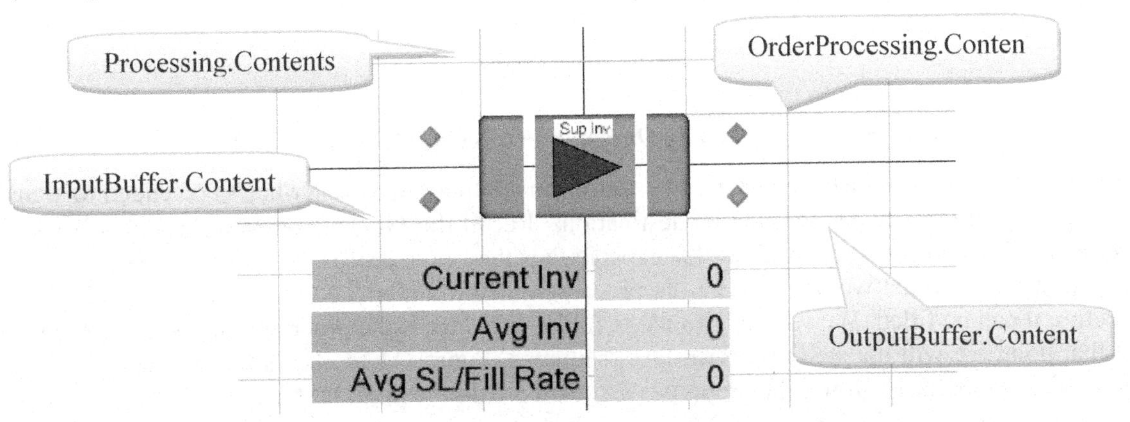

Figure 22.12: Adding the Animation Queues and Status Labels

Step 10: As part of the ordering system, three additional properties will be needed, as seen in Table 22.2 below. Also, the **OrderProcessingTime** property should have units of "Time" with the default units being "Days" placing these properties in the "Ordering Information" category.

Table 22.2: Three Properties to Allow for the Ordering Process to Take Place

Property	Type	Description	Required
OrderProcessingTime	*Standard Property→Expression*	The time spent processing order before it is filled	Yes
DeliverNodeID	*Standard Property→Expression*	The input node id where the order should be delivered	Yes
OrderEntityType	*Object Reference Standard Property→Entity*	The type of order entity to send out	Yes

The *OrderEntityType* will be used in a similar fashion as the *EntityType* property for the SOURCE object. This allows the object to create Reorders based on a user defined type (i.e., allows the modeler to make more changes to the orders without having to modify the **SUPPLYINVWORKSTATION** object).

---

[325] For each queue, set the *Alignment* to "None" under the *Appearance* tab.

[326] Just copy the six labels from the Chapter 10 model that is currently open to get all the expressions correct.

Part 22.3: Adding the Behavior Logic for the Ordering System

We now have defined all of the characteristics of the supply inventory object and now we need to add the ordering system. Orders that arrive to the **OrderInput** node are sent directly to the **OrderProcessing** STATION. At this point, the logic necessary to fill the incoming orders needs to be added. Recall from Chapter 20 on the **DELAY** object, STATIONS respond to entities arriving and exiting the particular station.

Step 1: Insert a new process named *OnEnteredOrdering* with the triggering event being `OrderProcessing.Entered` as seen in Figure 22.13 to handle the processing of orders. The next few steps will be broken down into sections in order to build the process.

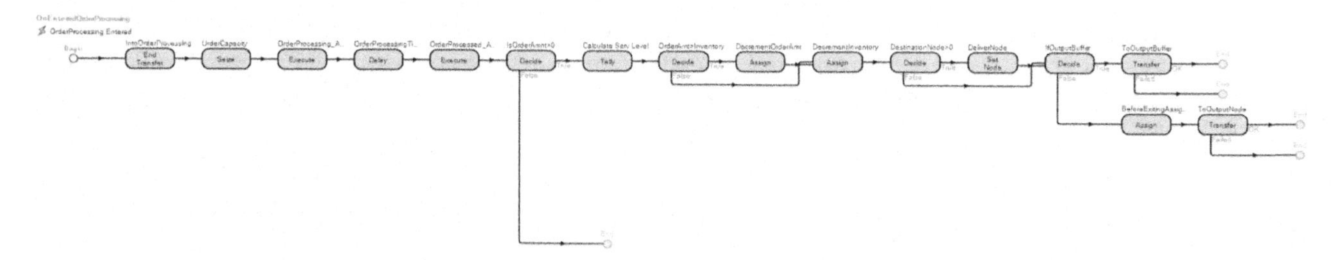

Figure 22.13: The *OnEnteredOrdering* Process

Step 2: When entities are transferred from one location to another, the transfer has to be ended to signal to the transferring step that is has reached its destination. Recall the **OrderProcessing** STATION's capacity was set as a reference property to allow for capacitating the order processes. Also, we added an order processing expression that would simulate the processing of the order (i.e., checking information, credit, etc.,) before it can be filled. We will only to try to fulfill the order if the order amount is greater than zero and at that point we will update the service level statistics. Figure 22.14 shows the first few steps of the process while Figure 22.18 shows the properties associated with the process steps.

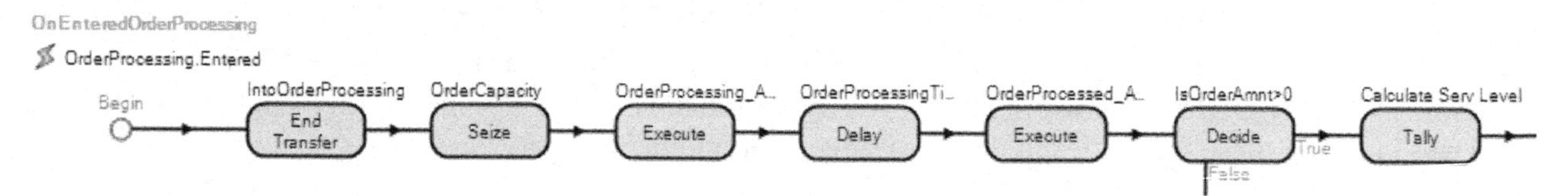

Figure 22.14: First Five Steps of Handling the Arriving of Orders

- Insert an `End Transfer` step to end the transfer of the arriving order entity from the **OrderInput** external node into the **OrderProcessing** STATION.
- Insert a `Seize` step that will seize one unit of capacity from the "Specific" resource which is the **ResOrder as shown in** Figure 22.15.

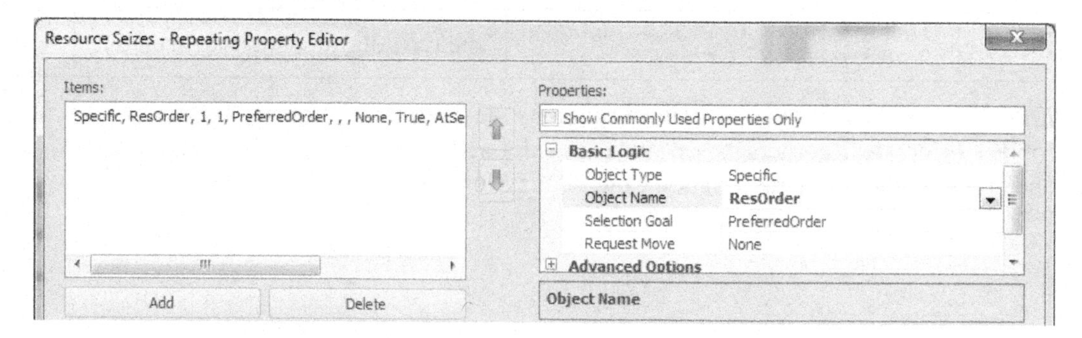

Figure 22.15: Seizing a Unit of ResOrder Capacity

- Like the normal WORKSTATION we would like to provide the capability for the user to execute add-on process triggers before processing the order and after the order has been processed, insert two

`Execute` steps. Create a new reference property named **OrderBeforeProcessingAddOn** and **OrderAfterProcessingAddOn** as the *Process Name* property as seen in Figure 22.16. In the *Definitions→Property* section, change the category of this new reference property to "Add on Process Trigger" and the display name to "Order Processing" and "Order Processed" respectively.

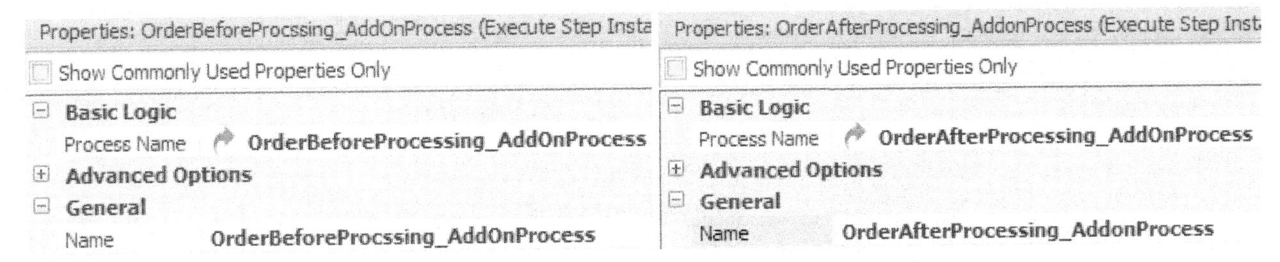

Figure 22.16: Order Processing and Processed Add-on Process Trigger

- Next, insert a `Delay` step between the two `Execute` statements to represent the processing of the order before filling the order using the existing *OrderProcessingTime* property.

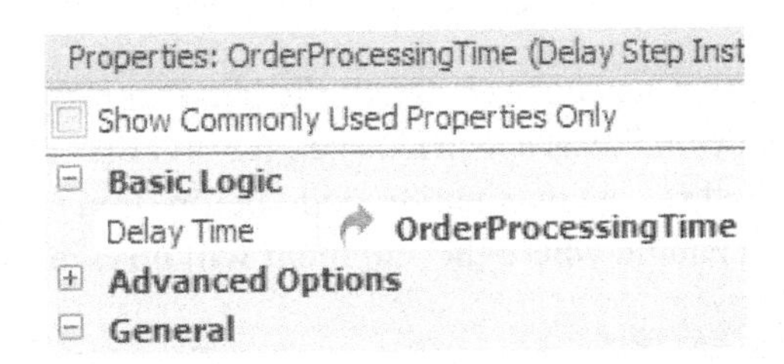

Figure 22.17: Setting up the `Delay` Step

- Add a `Decide` step to determine if the order amount is greater than zero before continuing on to the fulfillment process.
- Update the service level statistic using a `Tally` step using the same logic from .

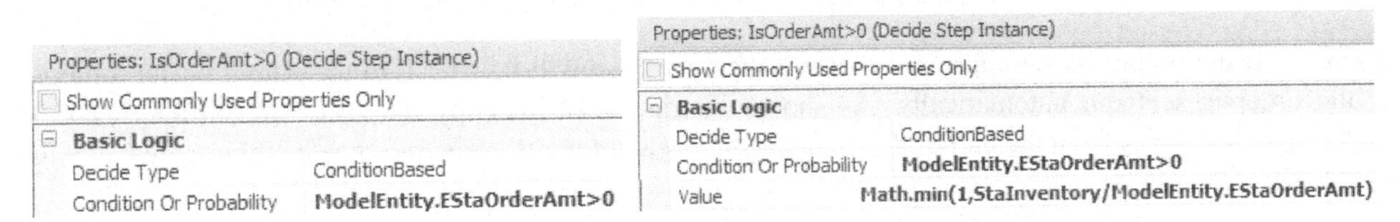

Figure 22.18: Properties Associates with Beginning of the Order Process

Step 3: The previous steps were similar to the ones used in the simple supply chain model in Chapter 10. These were the only ones associated with the ordering process since the DC was the object of interest and orders did not flow to the downstream customer in the supply chain. The actual filling and shipping of the order to the customer will be handled by the last steps of the process as seen in Figure 22.19.

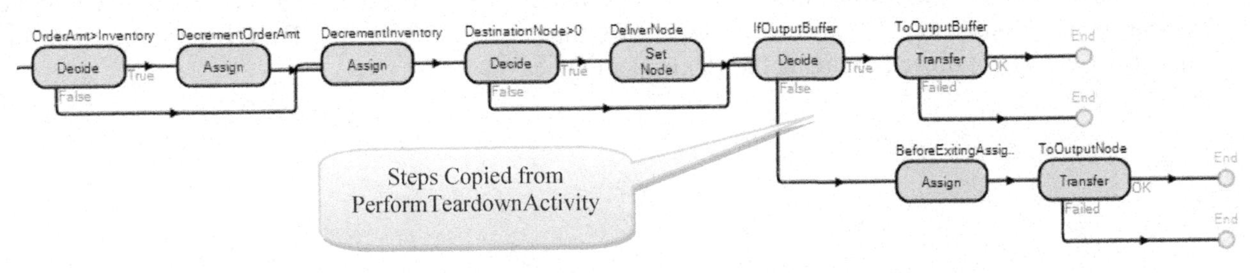

Figure 22.19: Steps Used to Fill the Order and Ship it out to the Next Element

Step 4: The first section (i.e., the first three steps) will update the inventory position by the amount being shipped out. If the current inventory position cannot handle the order amount (i.e., `Decide` step) then set the order amount equal to the current inventory level.[327] If the inventory is large enough (i.e., "*False*" branch) go directly to decrementing the current inventory position by the order amount. Insert the `Decide` and the two `Assign` steps with the properties in Figure 22.20. Otherwise, we need to change the order amount of the entity to let the customer know how much we were able to ship to them.

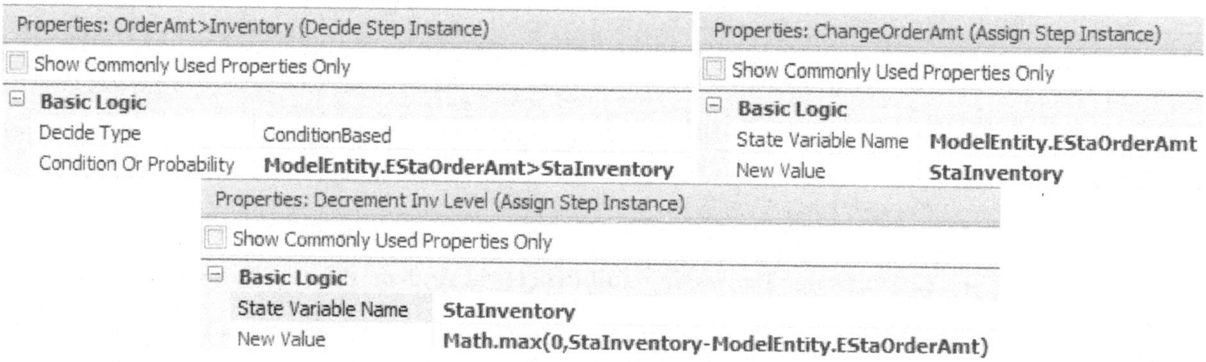

Properties: OrderAmt>Inventory (Decide Step Instance)

☐ Show Commonly Used Properties Only

⊟ **Basic Logic**	
Decide Type	ConditionBased
Condition Or Probability	**ModelEntity.EStaOrderAmt>StaInventory**

Properties: ChangeOrderAmt (Assign Step Instance)

☐ Show Commonly Used Properties Only

⊟ **Basic Logic**	
State Variable Name	**ModelEntity.EStaOrderAmt**
New Value	**StaInventory**

Properties: Decrement Inv Level (Assign Step Instance)

☐ Show Commonly Used Properties Only

⊟ **Basic Logic**	
State Variable Name	**StaInventory**
New Value	**Math.max(0,StaInventory-ModelEntity.EStaOrderAmt)**

Figure 22.20: Updating the Inventory Position

Step 5: After the inventory position has been updated, the shipment has to be sent out the normal **Output** external node. If the deliver node "id" was specified (i.e., `ModelEntity.EStaDeliverNodeID >0`) in the MODELENTITY then we will set the destination node before transferring the shipment out. Otherwise just ship it out, and let the links handle where the shipment will go (see Figure 22.21).

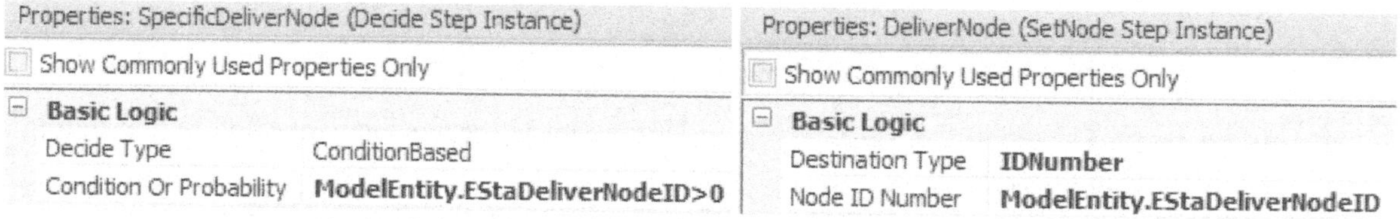

Properties: SpecificDeliverNode (Decide Step Instance)

☐ Show Commonly Used Properties Only

⊟ **Basic Logic**	
Decide Type	ConditionBased
Condition Or Probability	**ModelEntity.EStaDeliverNodeID>0**

Properties: DeliverNode (SetNode Step Instance)

☐ Show Commonly Used Properties Only

⊟ **Basic Logic**	
Destination Type	**IDNumber**
Node ID Number	**ModelEntity.EStaDeliverNodeID**

Figure 22.21: Shipping the Order Out to Downstream Customer

Step 6: If the output buffer capacity is greater than zero, we will transfer it to the output buffer, otherwise the order is sent out automatically. As shown in Figure 22.19, copy the `Decide`, `Assign` and two `Transfer` steps from the end of the "*PerformTeardownActivity*" process since it is the same (see Figure 22.28).

Step 7: Next, link the "False" branch from the `Decide` on whether they specified a deliver node to the `Decide` of the output buffer greater than zero.

Step 8: Insert a new process named "*OnExitedOrderingProcess*" with the triggering event being `OrderProcessing.Exited` as seen in Figure 22.22 which it executes the "*ExitedAddOn*" process trigger if it is specified. Copy all three steps from the "*OnExitedProcessing*" process. This process will run and execute all the steps when the output buffer capacity is zero and the orders are directly sent to the external node. Modify the `Release` step to release the "Specific" resource **ResOrder** rather than the capacity of the parent object, changing the "Units per Object" to 1.

---

[327] Before we just set the inventory level to 0 if the order amount was greater than zero but we are now shipping it out.

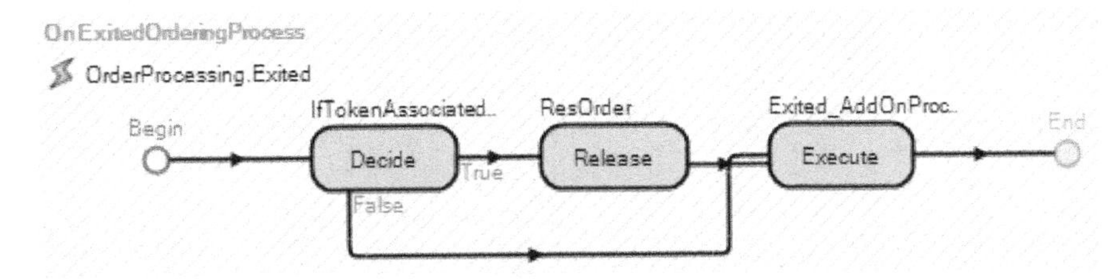

Figure 22.22: Process that Executes when Entities leave the OrderProcessing STATION.

Step 9: Since the orders are being shipped out through the output buffer, we need to override the "*OnEnteredOutputBuffer*" so it releases the **ResOrder** rather than the parent object like the previous step. Click on the process and override and modify the `Release` step to release one unit of **ResOrder**.

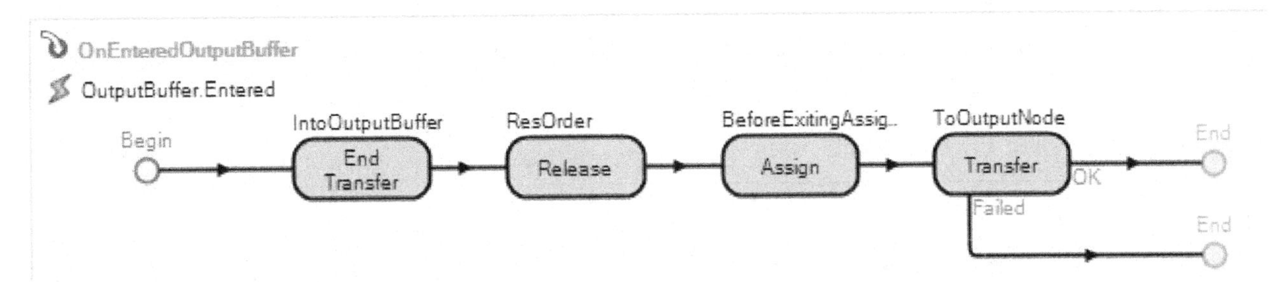

Figure 22.23: Modifying the OutputBuffer Entered Process

Part 22.4: Adding the Behavior Logic for the Inventory Replenishment System

We now have defined the logic for the ordering and fulfillment system. Next, the logic necessary to implement the inventory replenishment system will be added to the new object.

Step 1: Setup the periodic review by creating a process to respond to the timer **TimerReview**. From the "*Processes*" tab of the **SUPPLYINVWORKSTATION** object, create a new process named **InventoryReview.** Set the *Triggering Event* property to respond to the `TimerReview.Event` which causes the process to be executed every time the timer event fires, as seen in Figure 22.24. This is almost identical to the same process created in Chapter 10. Copy the entire process and all steps can be just copied and then the *Create, Assign* and *Transfer* steps need to be modified. You will need to modify all the GSta variables to Sta since they are part of the workstation.

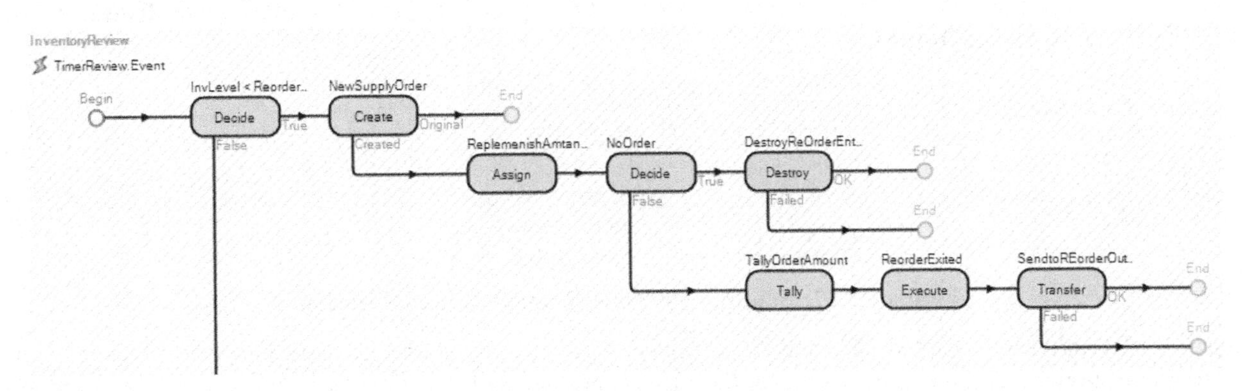

Figure 22.24: Process to Reorder

- The `Decide` step does a "ConditionBased" check to see if the current inventory is less than the reorder point (i.e., `StaInventory < ReorderPoint`)
- If a new reorder is needed, use the `Create` step to create a new entity as seen in Figure 22.25 which creates the entity based on the *OrderEntityType* reference property this time.

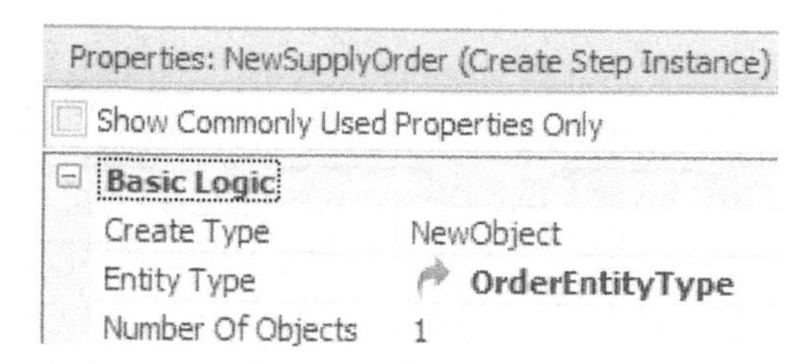

Figure 22.25: Creating a New Reorder to Send the Supplier

- For the newly created order, use an `Assign` step to assign the order amount and then increase the **WIP** value based on the order amount. Additionally, we will also set the original amount to help model reliability of the suppliers as well as the deliver node, if it has been specified, as in Table 22.3.

Table 22.3: Determining the Order Amount and Updating the WIP

State Variable	New Value
`ModelEntity.EStaOrderAmt`	`Math.Max(0,OrderUpToQty - StaInventory-StaOnOrder)`
`StaOnOrder`	`StaOnOrder + ModelEntity.EStaOrderAmt`
`ModelEntity.EStaDeliverNodeID`	`DeliverNodeID` (This is a property)
`ModelEntity.EStaOriginalOrderAmt`	`ModelEntity.EStaOrderAmt`

- Insert an `Execute` step that will run a user defined add-on process trigger which will be executed before the reorder exits the system to be sent to the supplier (see Figure 22.26). In the *Definitions→Property* section, change the category of this new reference property to "Add-On Process Triggers" and the display name to "ReOrder Exited."

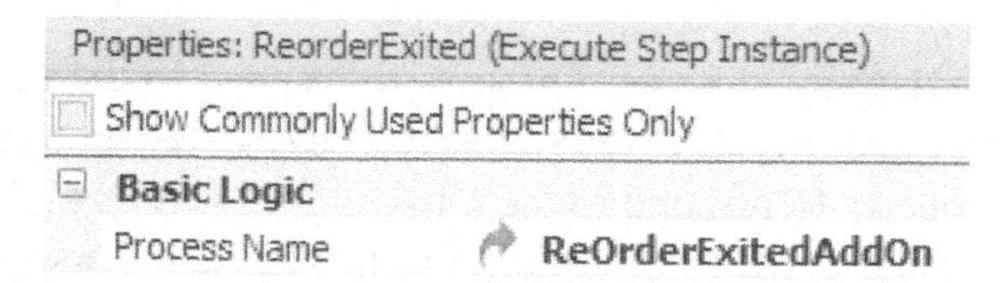

Figure 22.26: Inserting a User-Defined Add-on Process Trigger

- Once the *OrderEntityType* has been set, it needs to be sent to the supplier by using a `Transfer` step to move it from "FreeSpace" to the **ParentExternalNode** of the **ReOrderOutput**.[328]

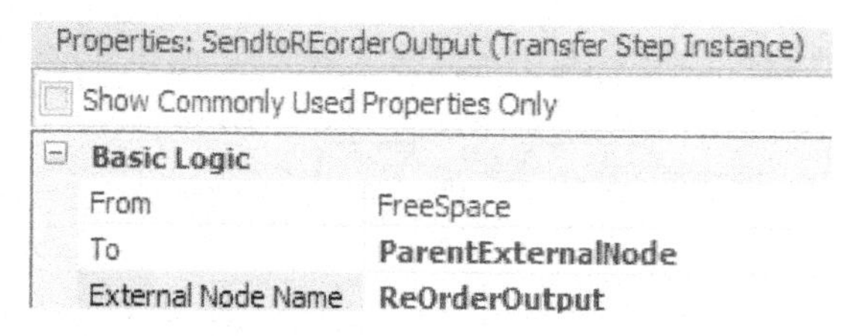

Figure 22.27: Sending the New Reorder to the Supplier

Step 2: Recall from Chapter 10 when the products arrived back at the DC and had been packaged, the **StaInventory** variable was increased by the order amount, while the **StaOnOrder** variable was reduced by the value in the "*Processed*" add-on process trigger for the **WrkDC**. Since we are encapsulating this same logic inside our new **SUPPLYINVWORKSTATION** object, the same logic needs to be added right after the "*Finished Good Operation*" trigger executed in the *PerformTeardownActivity* process. This process is run when the batch of parts has finished processing and the tear down activity has finished for the batch.

[328] Now the modeler will indicate where the reorders should travel via a link from the **ReOrderOutput** node.

Select the *PerformTeardownActivity* process from the "*Processes*" tab and then click the override button (Override) which allows you to modify it as seen in Figure 22.28.[329]

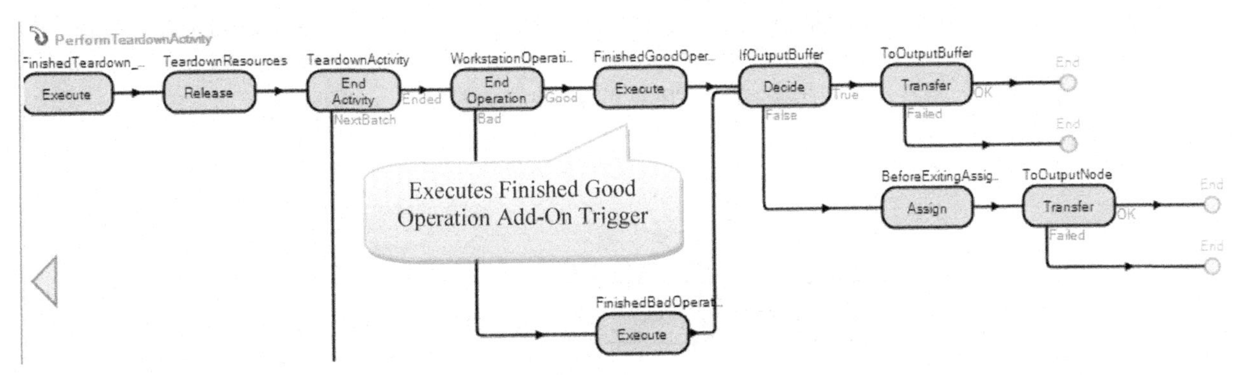

Figure 22.28: The Default *OnEnteredProcessing* Process

Step 3: After the execution of the "*Finished Good Operation*" add-on process trigger as seen in Figure 22.28, the entity will be transferred to the output buffer if the buffer capacity is greater than zero (i.e., "True" branch) or it will execute any before exiting assignments and then transferred to the external output node. In our make-to-stock model, the product is placed into inventory and not sent on to the next process. Therefore, you need delete all the steps after the `Execute` step and then insert three new process steps (i.e., `Assign`, `Release`, and `Destroy`) as seen in Figure 22.29.

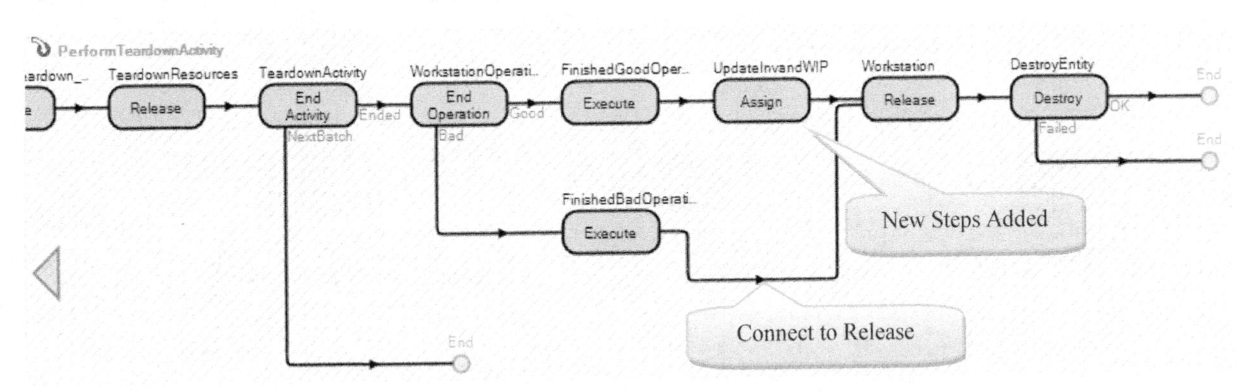

Figure 22.29: The New *PerformTeardownActivity* Process

- The `Assign` step updates the **Inventory** and **WIP** values as specified in Table 22.4.

Table 22.4: Updating Inventory and DC once Product Arrives

State Variable	New Value
`StaInventory`	`StaInventory + ModelEntity.EStaOrderAmt`
`StaOnOrder`	`StaOnOrder - ModelEntity.EStaOriginalOrderAmt`

- Since the entity is no longer sent to the Output Buffer (i.e., it will not enter the *OnEnteredOuputBuffer* process) or to the external node which will not run the *ExitedProcessing* process, the workstation capacity is never released. Therefore, copy the `Release` step from the *OnExitedProcessing* process after the `Assign` step and make sure *Exclusion Property* is empty (i.e. reset).
- Insert a `Destroy` step to destroy the entity order that has arrived since it will not continue.

[329] See Chapter 19 and Chapter 20 for more information on overriding and restoring processes.

- Make sure to connect the end of the *FinishedBadOperation* add on process execution step to the `Release` step.

Recall, the initial inventory property was used to set the beginning **StaInventory** position in the *OnRunInitialized* process. Override the *OnRunInitialized* process of our new object and insert the same `Assign` step as before as seen in

Step 4: Figure 22.30.

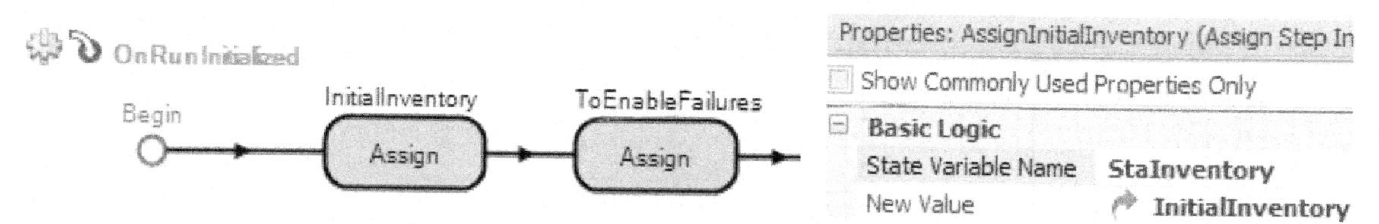

Figure 22.30: Setting the Beginning Inventory Position

Part 22.5: Using the New to Model the Complex Supply System

The new **SUPPYINVWORKSTATION** object has been defined and can now be used to build the model (see Figure 22.31) of the complex supply chain system of Figure 22.1. [330]

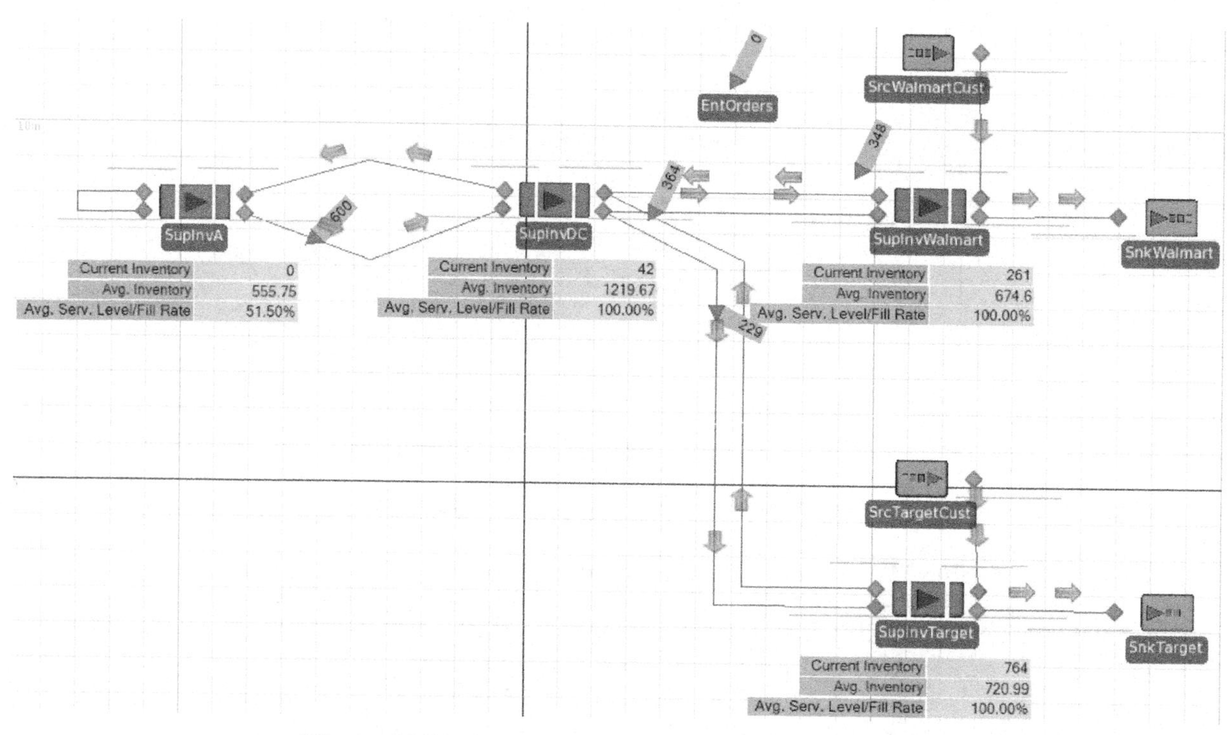

Figure 22.31: SIMIO Model of the Complex System

Step 1: In the model "Facility", add a new MODELENTITY named **EntOrders**. Add a status label that is connected to the entity in similar to the one in Chapter 10 with the *Expression* property set to `EStaOrderAmt` as seen in the top right of Figure 22.31.

[330] You should spread out the objects until the path direction arrows show. Having paths and connectors in the wrong direction can create some "hard to find" errors. When you are sure the paths and connectors are in the proper direction, place the objects back into their appropriate places.

Step 2: Insert two SOURCES named **SrcWalmartCust** and **SrcTargetCust** that will represent the daily number of customers arriving at the stores to purchase the products (interarrival times are one each day). As was done before, set the *Before Exiting State Assignments* that will set the `ModelEntity.EStaOrderAmt` to `Random.Poisson(70)` and `Random.Poisson(50)` in days, for the Walmart and Target stores respectively to represent the number of daily orders for each customer type. Here is a case where if we had done the assignment in a tokenized process it would have been easier.

Step 3: Next add a new **SUPPYINVWORKSTATION** named **SupInvWalmart** that will represent the inventory position object of Walmart.

- Set the *Processing Batch Size* to two, *Operation Quantity* to `ModelEntity.StaOrderAmt` and the *Processing Time* property to `Random.Triangular(1,2,3)` in minutes to reflect the processing of the product into the stores and on the shelves as seen in Figure 22.32.

Properties: SupInvWalmart (SupplyInvWorkStation)

☐ Show Commonly Used Properties Only

⊟ **Process Logic**	
Capacity Type	Fixed
Ranking Rule	First In First Out
Dynamic Selection Rule	None
⊞ Transfer-In Time	0.0
Operation Quantity	**ModelEntity.EStaOrderAmt**
⊟ Setup Time Type	Specific
⊞ Setup Time	0.0
Processing Batch Size	**2**
⊞ Processing Time	**Random.Triangular(1,2,3)**
⊞ Teardown Time	0.0

Figure 22.32: Setting up the Processing of the Batches

- Under the *Inventory* property category, set the *Review Period* to every five days, the *Initial Inventory* to 1100, the *Reorder Point* to 900 with the *Order Up to Qty* to 1100 as seen in Figure 22.33.
- For the *Order Information*, it takes a ¼ of day to process an incoming order. Also, set the *Deliver Node ID* to `Input@SupInvWalmart.ID`[331] which will allow orders sent to the supplier to be sent back to the Walmart store since two different stores are ordering on the same supplier.

⊟ **Inventory**	
InitialInventory	**1100**
ReorderPoint	**900**
OrderUptoQty	**1100**
⊟ ReviewPeriod	5
Units	Days
⊟ ReviewPeriodTimeOffset	0
Units	Days

⊟ **Order Processing**	
OrderCapacityType	Fixed
OrderInitialCapacity	Infinity
OrderEntryRankingRule	FirstInFirstOut
OrderEntryRankingExpression	Entity.Priority
OrderDynamicSelectionRule	None
⊞ OrderProcessingTime	**0.25**
DeliverNodeId	**Input@SupInvWalmart.ID**
OrderEntityType	**EntReplenishments**

Figure 22.33: Inventory and Ordering Information Properties of the Walmart Store

Step 4: Repeat Step 3 for the Target store by copying the Walmart **SupInvWalmart** naming the object **SupInvTarget** changing the properties seen in Figure 22.34 making sure to change the *Deliver Node ID* to `Input@SupInvTarget.ID` and the inventory properties since Target does not service as many customers.

---

[331] Recall each object in SIMIO has a unique identification number which is accessed by the `ID` function.

Inventory	
InitialInventory	1100
ReorderPoint	700
OrderUptoQty	1100
ReviewPeriod	5
Units	Days
ReviewPeriodTimeOffset	0
Units	Days

Order Processing	
OrderCapacityType	Fixed
OrderInitialCapacity	Infinity
OrderEntryRankingRule	FirstInFirstOut
OrderEntryRankingExpression	Entity.Priority
OrderDynamicSelectionRule	None
OrderProcessingTime	0.25
DeliverNodeId	Input@SupInvTarget.ID
OrderEntityType	EntReplenishments

Figure 22.34: Inventory and Ordering Information Properties of the Target Store

Step 5: Connect the two sources to their appropriate **OrderInput** node at the **SupInvWalmart** and **SupInvTarget** via connectors as seen in Figure 22.31.

Step 6: Insert two SINKS into the model named **SnkWalmart** and **SnkTarget**. Connect the **Output** nodes of the **SupInvWalmart** and the **SupInvTarget** **SUPPYINVWORKSTATIONS** to the respective sinks via CONNECTORS.

Step 7: Next we need to create the distribution center which will service the two stores. Copy the **SupInvWalmart** object and name the new **SUPPYINVWORKSTATION** object **SupInvDC** to represent the DC.

- Set the *Processing Time* property to the following expression in minutes.
  `Random.Triangular(5,10,20)`
- Set the *Initial Inventory* property to 1800, the *Reorder Point* property to 1400 and the *Order Up to Qty* property to 1800.
- Set the *Delivery Node ID* to 0.

Step 8: Next we need to connect the DC to each of the two stores.

- For products to flow to the stores, connect the **Output** node of **SupInvDC** as seen in the whole model of Figure 22.31 to the **Input** nodes of **SupInvWalmart** and **SupInvTarget** via TIMEPATHS. It takes generally four days to transport product from the DC to each store. However, it can take a minimum of three or a maximum of seven days to travel from the DC to each store. Therefore set the *Travel Time* property to `Random.Pert(3,4,7)days.`
- For orders to flow from the stores to the DC, connect the **ReorderOutput** node of both stores to the **OrderInput** node of the **SupInvDC** via connectors to represent EDI (Electronic Data Information) order transmissions.

Question 1: With our new object, how can we model the situation where Walmart orders are satisfied before Target orders, if they arrive at the same time?

____

Step 9: In the first portion, we have utilized one supplier to supply the DC, which is a make-to-stock type supplier. Therefore, copy the **SupInvDC** and paste in a new **SUPPYINVWORKSTATION** object **SupInvA** as seen in Figure 22.35. Only a few things will actually change.

- Set the *Processing Time* property to `Random.Triangular(10,20,30)` (minutes) since it takes longer to convert the raw material into a product.
- Set the *Initial Inventory* property to 600, the *Reorder Point* property to 300 and the *Order Up to Qty* property to 600.
- Set the *Delivery Node ID* to 0.

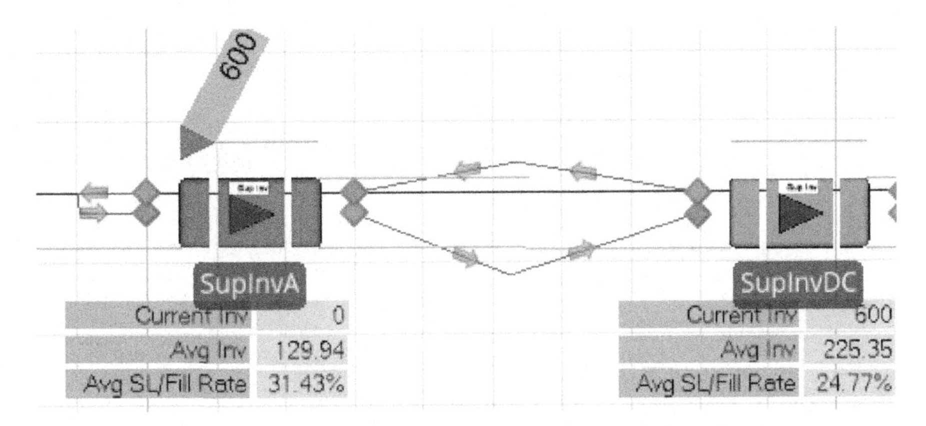

Figure 22.35: Connecting a Supplier to the DC.

Step 10: Similar to the stores, connect the **ReorderOutput** node of the **SupInvDC** to the **OrderInput** node of the **SupInvA** via a connector to have orders flow back to the supplier.

Step 11: Like the DC, products take the same amount of time to ship from the Supplier A to the DC. Therefore, connect the **Output** node of **SupInvA** to the **Input** node of the **SupInvDC** via TIMEPATH with the same *Travel Time* property set to `Random.Pert(3,4,7)` days.

Step 12: Supplier A has a perfect raw material supplier (i.e., we assume the material is always available). Therefore, use a connector to connect the **ReorderOutput** of **SupInvA** directly to the **Input** node of **SupInvA** which allows the supplier to order a particular amount of product from its perfect raw material supplier to replenish its inventory position through its own manufacturing.

Step 13: Save and run the model for 52 weeks observing the output.

Question 2: What are the average service levels of each of the segments in the supply chain?

____

Question 3: What is the current and average inventory level of each segment in the chain?

____

Question 4: Do you think the current system is adequate and what seems to be the problem?

____

Part 22.6: Adding a Secondary Supplier for Overflow Orders

By reviewing the results, most of the inventory positions were zero which led to very low service levels. The central issue is the lead time length and variability of getting product from the back of the chain to the front of the chain. There are many ways this situation can be fixed by optimizing reorder points and order up to quantities. Also process improvement projects might reduce the processing times or shipping times. Let's add a second supplier (Supplier B) which is a make-to-order operation to handle orders from the DC as well. In many situations multiple sourcing strategies exist and can be modeled.

Step 1: Insert a new WORKSTATION named **WrkSupB** to act as a make-to-order supplier (i.e., when orders arrive they are produced and not stored ahead of time). Modify the model as shown in Figure 22.36.

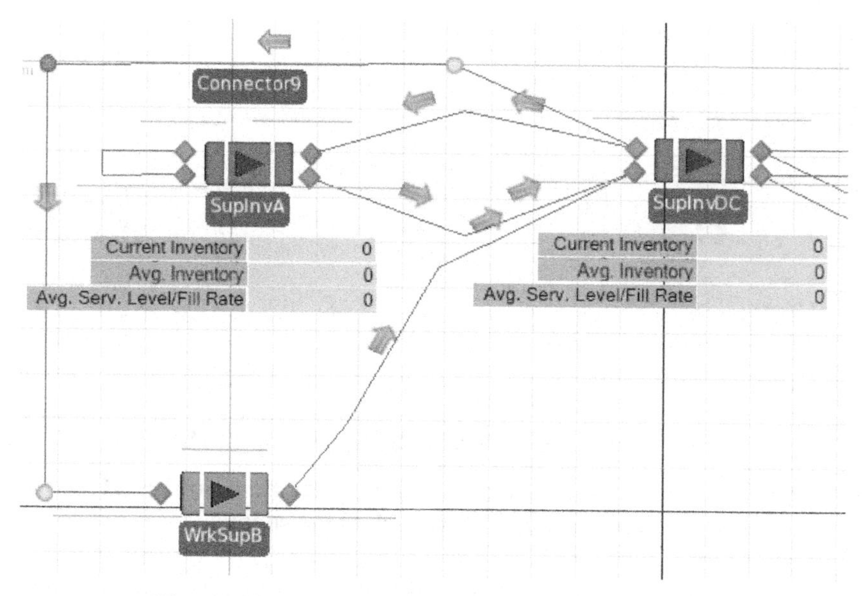

Figure 22.36: Add Make-to-Order Supplier B

- Connect the **ReorderOutput** of **SupInvDC** to the **Input** of the **WrkSupB** via a CONNECTOR link to allow orders to be sent to the new supplier electronically.
- To allow products to flow, connect the **Output** node of the **WrkSupB** to the **Input** node of the **SupInvDC** via a TIMEPATH. It takes `Random.Pert(1,2,3)` days to ship then product to the DC from this supplier.
- Set the *Processing Time* property of the **WrkSupB** to produce the products to `Random.Triangular(8,10,20)` (minutes). Again set the *Operation Quantity* to `ModelEntity.EStaOrderAmt` and the *Processing Batch Size* property to two.

Step 2: In this example, the DC has access to the Suppliers A current inventory level. If the order amount is greater than the current inventory level, the DC will order only up to the inventory position and then order the remaining portion from Supplier B. Insert a new "*ReOrderExited*" add-on process trigger that is executed when the order exits the **WrkSupB** as seen in Figure 22.37.

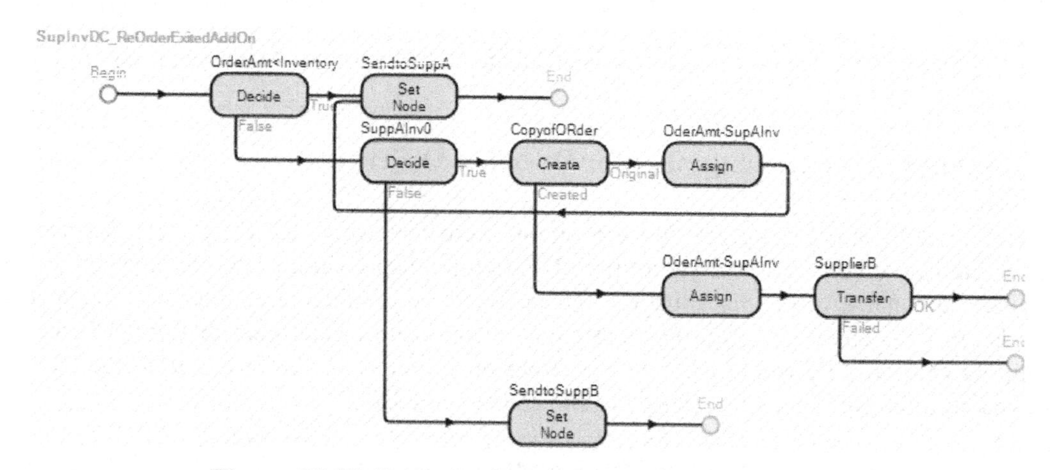

Figure 22.37: ReOrder Exit Add-on Process Trigger

- Insert a *`Decide`* step that checks to see if the `ModelEntity.EStaOrderAmt` is less than or equal to the **SupInvA.Inventory** position. If it is (i.e., the "True" branch) then use a *`Set Node`* step to specify that the order should be sent to the Supplier A (i.e., **OrderInput@SupInvA**).

Properties: OrderAmt<Inventory (Decide Step Instance)	
Show Commonly Used Properties Only	
Basic Logic	
Decide Type	ConditionBased
Condition Or Probability	**ModelEntity.EStaOrderAmt<SupInvA.StaInventory**

Properties: SendtoSuppA (SetNode Step Instanc	
Show Commonly Used Properties Only	
Basic Logic	
Destination Type	Specific
Node Name	**OrderInput@SupInvA**

Figure 22.38: The Properties of the `Decide` and `Set Node` for Reordering Process

- If the order amount is bigger than the current inventory position then decide if supplier A has any inventory at all (i.e., `SupInvA.StaInventory >0`). If it does not (i.e., "False" branch) then send the current order to supplier B via the *`Set Node`* with *Node Name* property set to `Input@WrkSupB`. Otherwise, *`Create`* a copy of the current order and set the "Original" order amount to the current inventory level as seen in Figure 22.39.

Properties: SuppAInv0 (Decide Step Instance)	
Show Commonly Used Properties Only	
Basic Logic	
Decide Type	ConditionBased
Condition Or Probability	**SupInvA.StaInventory>0**

Properties: CopyofORder (Create Step Instance)	
Show Commonly Used Properties Only	
Basic Logic	
Create Type	**CopyAssociatedObject**
Entity Type	
Number Of Objects	1

Figure 22.39: Create Step to Copy the Current Order

- In the assignment, as shown in Figure 22.40 set the inventory position at **SupInvA** to the **StaOrderAmt** and the original order's **StaOriginalOrderAmt** needs to be set equal to the new **StaOrderAmt.** Then go ahead and send the order on to the **SupInvA** by moving the "*End*" point on top of the *`Set Node`* as seen in Figure 22.37.

Basic Logic	
State Variable Name	**ModelEntity.EStaOrderAmt**
New Value	**SupInvA.StaInventory**

Basic Logic	
State Variable Name	**ModelEntity.EStaOriginalOrderAmt**
New Value	**ModelEntity.EStaOrderAmt**

Figure 22.40: Set Order Amount and the Original Order Amount

- For the newly created order which is a copy, set the **StaOrderAmt** to the remaining portion and also update the **StaOriginalOrderAmt** as shown in Figure 22.41.

Basic Logic	
State Variable Name	**ModelEntity.EStaOrderAmt**
New Value	**ModelEntity.EStaOrderAmt-SupInvA.StaInventory**

Basic Logic	
State Variable Name	**ModelEntity.EStaOriginalOrderAmt**
New Value	**ModelEntity.EStaOrderAmt**

Figure 22.41: Creating an Order to Send to the Second Supplier

Step 3: Then transfer from "Free Space" to **Input@WrkSupB** as seen in Figure 22.42.

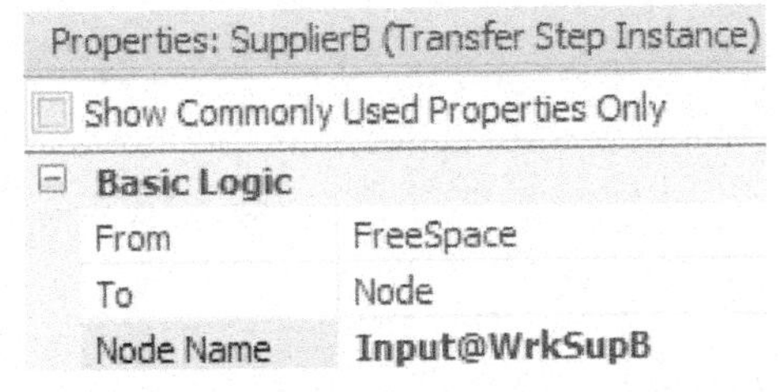

Properties: SupplierB (Transfer Step Instance)	
Show Commonly Used Properties Only	
Basic Logic	
From	FreeSpace
To	Node
Node Name	**Input@WrkSupB**

Figure 22.42: Transfer Input at WrkSupB

Step 4: Save and run the model for 52 weeks observing the output.

Question 5: What are the average service levels of each of the segments in the supply chain?

Question 6: What is the current and average inventory level of each segment in the chain?

Part 22.7: Commentary

- This model provides a framework for many interesting embellishments and much added realism.
- Failure information could be linked with the **ResOrder** resource. The ability to suspend the ordering process when failures occur should also be added.
- Cost information can easily be added to the orders and objects using the cost parameters under the financials and one can choose suppliers as well as have different shipping options (i.e., air, rail, and trucking) to assist in meeting consumer demand to make it even more realistic.
- The reliability of the supplier can be added to the new **SUPPLYINVWORKSTATION** object such that shipping or picking errors can be modeled (suppose a customer asks for 100 but only 98 are shipped, even though there is sufficient inventory).
- The object can be modified to handle multiple products rather than a single product type.
- Additional inventory policies can be built into the model that would allow the user to choose the particular inventory policy for each segment of the supply chain.
- We could also add a MONITOR event to create a continuous review system rather than just a periodic review system. Users could choose between the systems.
- The WORKSTATION was chosen as the object to subclass rather than the SERVER since it can process a batch of parts correctly. However, you cannot increase the capacity of WORKSTATIONS like you can SERVERS by changing the capacity. You can change the batch size but that is not quite the same as the average flow time for the parts would be correct but the variance would not. One would have to insert multiple workstations to achieve the same thing.

Chapter 23
More Subclassing: Process Planning/Project Management

Part 23.1: Process Planning

A new product (called Product 3) is being introduced. One unit of Product 3 is produced by assembling one unit of Product 1 and one unit of Product 2. Before production begins on either Product 1 or Product 2, raw materials must be purchased and workers must be trained. Before Products 1 and 2 can be assembled into Product 3, the finished Product 2 must be inspected. The table below gives the list of activities and their predecessors and the average duration of each activity.

Table 23.1: Activity Precedence and Duration

Activity	Predecessors	Duration (days)
A = train workers	--	6
B = purchase raw materials	--	9
C = produce Product 1	A, B	8
D = produce Product 2	A, B	7
E = test Product 2	D	10
F = assemble Products 1 and 2	C, E	12

The typical project management diagram for the example can be seen in Figure 23.1.

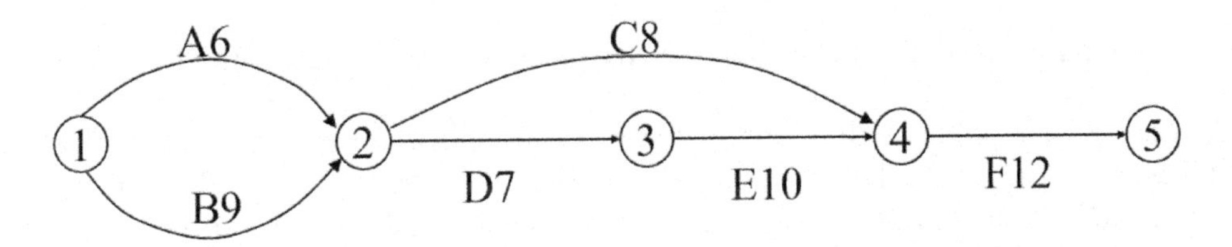

Figure 23.1: Sample Project Planning Example

Clearly the project will take an average of 38 days and activities B, D, E, and F are critical ("on the critical path") – meaning if any of these increases, then the time to completion is increased. But in reality, activity completion times are uncertain (random) and different activities may become critical. So we can't rely on deterministic values for the activity times and instead need to represent the times with random variables.

Also, activities might share resources to complete the task. Therefore, the goal will be to determine the critical path and the probability that a path will be on the critical path.

Table 23.2: Project Activity Statistics

Earliest Finish	The time the current task finishes (i.e., entity leaves the path).
Latest Finish	The time the current task could finish without changing the critical path or causing the successive tasks to start later.
Slack	The time difference between the Earliest Finish and the Latest Finish
Percent CP	The percentage of the time the current path is on the critical path.

Earlier chapters have demonstrated the ability to specialize/subclass existing objects which allows additional characteristics and behaviors to be added or existing ones to be modified, but none have specialized one of the link objects (i.e., PATHS or TIMEPATHS). We will model the project management problem using the TIMEPATHS to consume time. We will also use one replication to represent a single completion of the project, so multiple replications will be used to statistically characterize the project.

Part 23.2: Creating a Specialized TIMEPATH to Handle Activities

The ability to create new objects is one of the advantages of SIMIO and the TIMEPATH link is an attractive object to use for the activity time, since the entities can be seen visually moving along the paths. The paths represent the important objects (i.e., critical path determination). Each path needs to keep track of its own statistics (i.e., Earliest Finish, Slack, Percent CP, etc.). The path also needs to know when an entity reaches the end of the path, and also restrict only one entity to travel on each path. This encapsulation of the time path gives the object the "intelligence" to automatically calculate the statistics, as well as allow us to utilize it for many different project management problems.

Step 1: Create a new model and subclass the TIMEPATH and change the *Model Name* property to **PRJTIMEPATH**[332].

Step 2: Select the **PRJTIMEPATH** from the [*Navigation*] panel. Insert two DISCRETE REAL STATE variables named **StaEarliestFinish** and **StaCalculateSlack**. The first state variable will be used to keep track of the time the activity finishes while the second one will be used to signal when to calculate the slack.

Step 3: To keep track of critical path calculations, add three TALLY statistics named **TallyStatEarliestFinish**, **TallyStatSlack**, and **TallyStatPercentCP** via the *Definitions→Elements* section.

- Change the "*Category*" property to "Critical Path" for all three TALLY statistics under the "Results Classification.
- Change the *Data Item* property to "Earliest Finish", "Slack", and "Percent CP" for the appropriate statistic.
- Also, for the **TallyStatEarliestFinish** and **TallyStatSlack**, set the *Unit Type* to "Time."

Step 4: Recall, only one project will be completed for each replication. In order to model a "junction" from which multiple activities start, the paths should not allow more than one traveler and travel should be in one direction only. From the *Definitions→Properties*, click on the down arrow to reveal the inherited properties.

- Select the *IntialTravelerCapacity* property and change the *Default Value* from Infinity to **one** and the *Visible* property to "**False**" so a user cannot change it and it will automatically be set.
- Select the *InitialDesiredDirection* property and change the *Default Value* to **forward** and the *Visible* property to "**False**" so a user cannot change it and it will automatically be set.

[332] Right click on the object in the [*Navigation*] section and select properties.

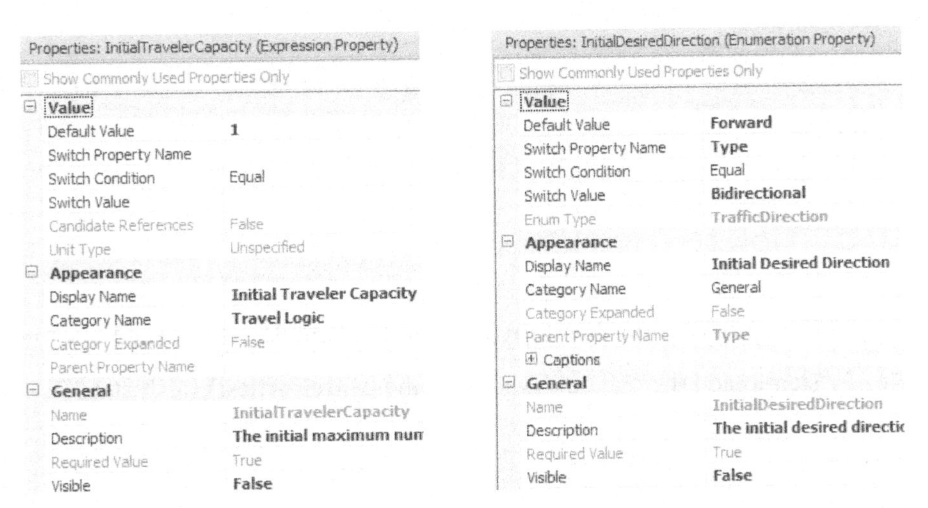

Figure 23.2: Setting the Default Values of the Travel Capacity and Desired Direction

Step 5: The TIMEPATH defines seven processes that determine the behavior of the link as seen in Table 23.3.

Table 23.3: Processes of the TimePath

Process	Description
OnEntered	Process runs when the leading edge of an Entity, Worker, or Transporter reaches the beginning of the TIMEPATH.
OnExited	Process runs when the trailing edge of an Entity, Worker, or Transporter leaves the TIMEPATH.
OnReachedEnd	Process runs when the leading edge of an Entity, Worker, or Transporter reaches the end of the TIMEPATH.
OnRunEnding	Process runs once at the end of the simulation.
OnRunInitialized	Process runs once at the beginning of the simulation.
OnTrailingEdgeEntered	Process runs when the trailing edge of an Entity, Worker, or Transporter enters the path.

Step 6: Once the project ENTITY reaches the end of the path, the activity has finished and the earliest finish time can be calculated. Override the "*OnReachedEnd*" process as seen in Figure 23.3 to perform the calculation.

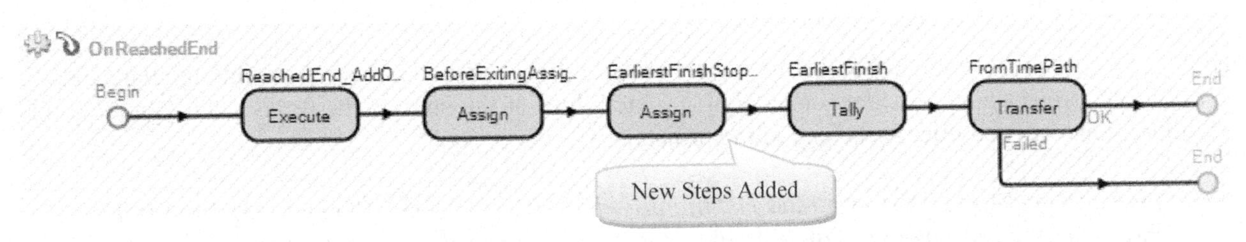

Figure 23.3: Calculating Earliest Finish and Shutting Down Path

- Insert an `Assign` step that will set the `DesiredDirection` to `Enum.TrafficDirection.None` which will shut down the path and not allow anymore entities to travel this path. Also, set the **StaEarliestFinish** state variable to the current time (`Run.TimeNow`). Refer to Figure 23.4.

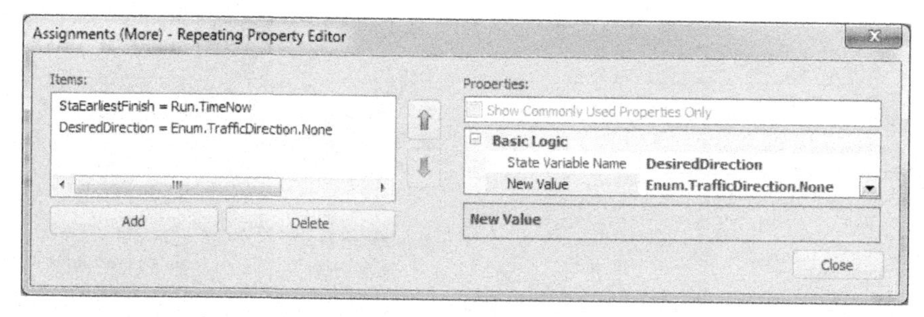

Figure 23.4: Shut down the Path and Set the Earliest Finish

- Insert a `Tally` step to add the one observation of the **TallyEarliestFinish** statistic. as seen in Figure 23.5.

Figure 23.5: Add an Observation of the Earliest Finish Statistic

Part 23.3: Creating a Junction Object to Handle Precedent Constraints

In the previous section, we modeled the project activities as specialized TIMEPATHS which automatically calculated all the statistics. However, they cannot enforce the precedence constraints or start up activities nor determine when the latest finish happens. The precedence constraints have to be enforced by the junctions which also initiate the next set of activities. Therefore, the junction has to wait until all the preceding activities have finished before starting all of the succeeding activities at the same time. Also, the junctions determine the latest finish of all the preceding activities based on the time the last of the preceding activities finishes.

Step 1: None of the standard objects can model this circumstance easily and most standard objects are too complicated to modify for this purpose. A new object similar to the **DELAY** object of Chapter 20 will be created. From the "*Project Home*" tab, insert a new "Fixed Class Model" named **JUNCTION** which will be used to model the junctions.

Step 2: We need our junction to act as a start node and an end node, as well as intermediate junctions within the project network. Start nodes will start activities at the start of the project while the end nodes will terminate all activities at the end of the project.

- Insert a new String LIST from the *Definitions→List→Create* section named **ListJunctionType** with three entries as seen below. This will allow the user to choose the Junction Type.

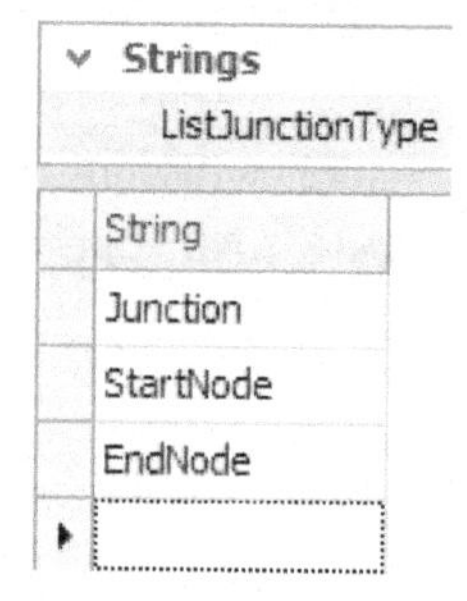

Figure 23.6: Defining Different List Types

- Insert a new standard "List Property" named **JunctionType** which allow the modeler to specify the junction type with "Junction" as the default type and place it in the "Project Management" category as seen in Figure 23.7.[333]

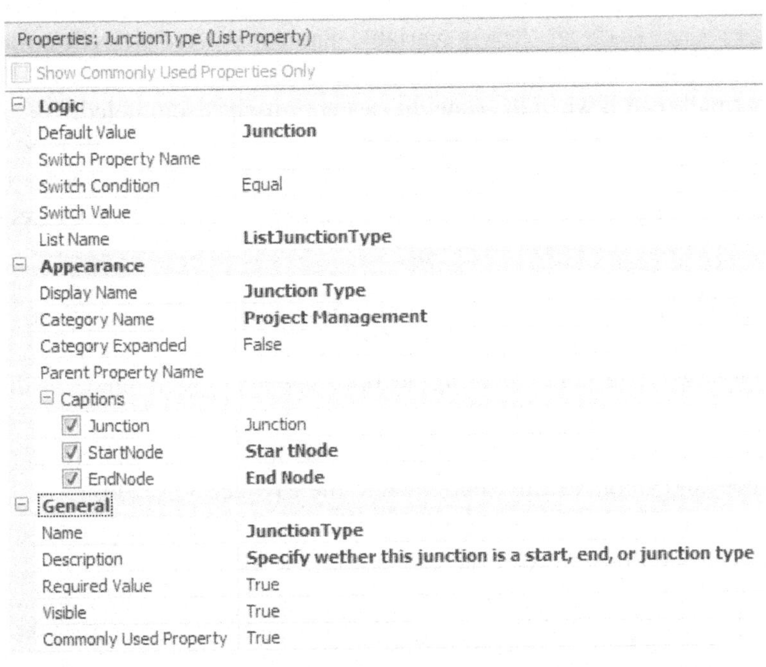

Figure 23.7: Specifying the List JunctionType Property

Step 3: Like the standard SOURCE, if the **JUNCTION** object is a "Start Node" it will need to create entities to be sent out into the project network. From the *Definitions→Properties→Add→Object Reference*, add an *Entity* property named **EntityType** making sure to place it into the "Project Management" category created in the step before. Also, you only want the property displayed if the ***JunctionType*** property is set to "StartNode".

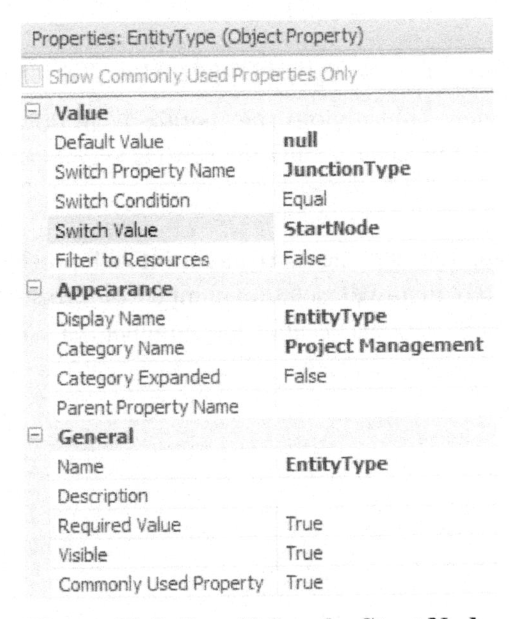

Figure 23.8: Specifying the Start Node

---

333 Note, that you can change the caption of the list object as a space was added between "*StartNode*" and "*EndNode*" options.

Step 4: Each junction needs to know the number of precedent activities to wait on and the number of successive activities to start. Insert two new DISCRETE INTEGER STATE properties named **StaPredecessors** and **StaSuccessors** which will be set at the beginning of the simulation run.[334]

Step 5: Insert a DISCRETE INTEGER STATE variable named **StaActivitiesCompleted** which will keep track of the number of activities that have completed and a DISCRETE REAL STATE variable named **StaLatestFinish** that will record when the time the last activity has completed.

Step 6: As was done in Chapter 20, the **DELAY** object, the incoming entities need a location where actions can occur. From the *Definitions→Elements* insert a new STATION named **Project** with all the default values.

Step 7: Next, let's define the external view seen by the user of the **JUNCTION** object.

- From the *Definitions→External→Drawing* section, use a Polygon shape to draw a red diamond as seen in Figure 23.9.
- Insert an EXTERNAL NODE that is a BASICNODE on the left edge of the red diamond that will accept incoming activities and automatically send them to the **Project** STATION.
- Next create the output of activities by inserting an External Node that is a TRANSFERNODE on the right edge of the red diamond that will route successor activities.

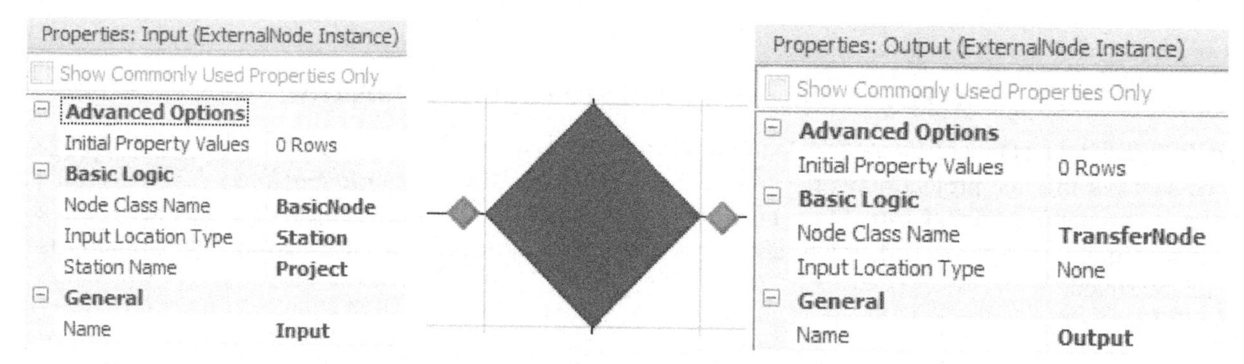

Figure 23.9: External View of the Junction Object Along with Input and Output Nodes

Now that we defined the external view and activities (i.e., entities) can flow in and out of our object, we need to add the process logic to handle the precedence and the starting of the activities as well as determining the number predecessors and successors.

Step 8: From the "Process" tab, selected the **OnRunInitialized** process from the *Select Process* drop down. Insert an *Assign* step that will set the number of predecessors (i.e., **StaPredecessors** to `Input.InboundLinks.NumberItems`) and the number of successors (i.e, **StaSuccessors** to `Output.OutboundLinks.NumberItems`).

[334] In Versions prior to 5 there was no way to access the number of links coming or going out of BASICNODES or TRANSFERNODES and these values had to be passed in as a number.

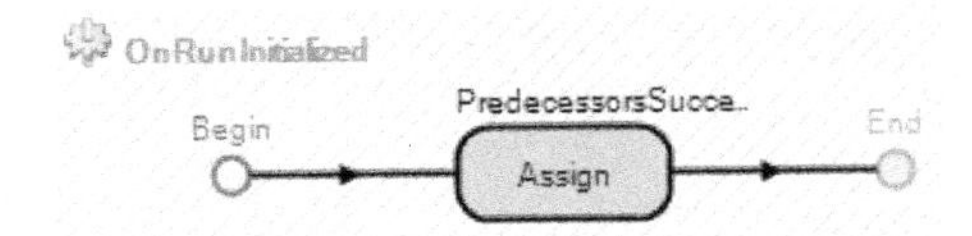

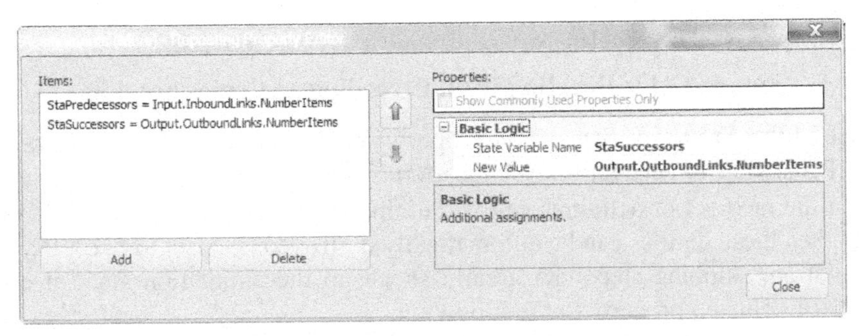

Figure 23.10: Initializing the Number of Successors and Predcessors

Step 9: Create a new process named "*OnEnteringProject*" that will execute every time an activity (i.e., ENTITY) enters the **Project** station by specifying the *Triggering Event* to be "Project.Entered" as seen in Figure 23.11. The process needs to update the number of activities finished. If the number of activities equals the number of predecessors, it will create the number of successors and send them out to the network otherwise there are still activities that need to complete and we will destroy the current activity entity.

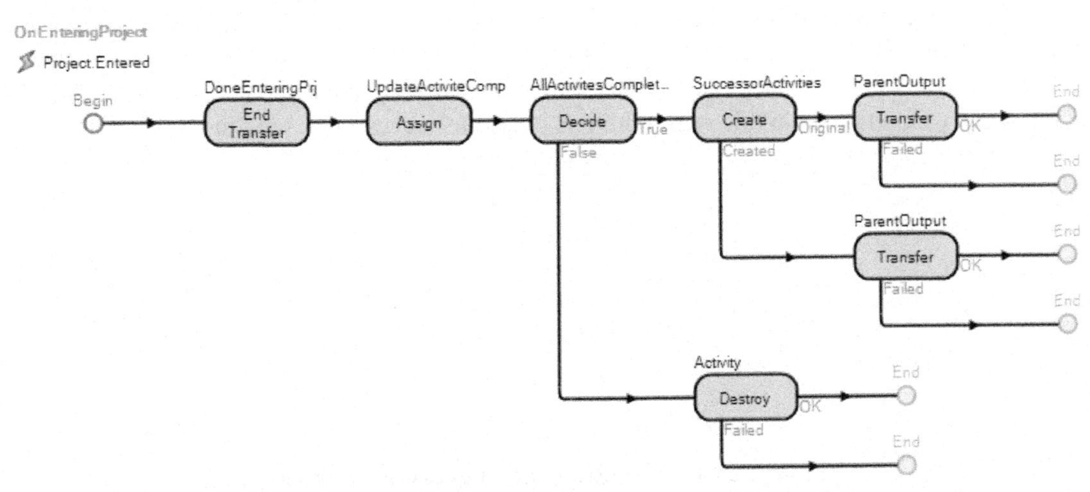

Figure 23.11: Process to Handle Activities that Enter the Junction

- As always, the first thing you need to do is to end the transfer of the entity object into the station by inserting an `End Transfer` step.
- Next update the number of activities that have completed by one.

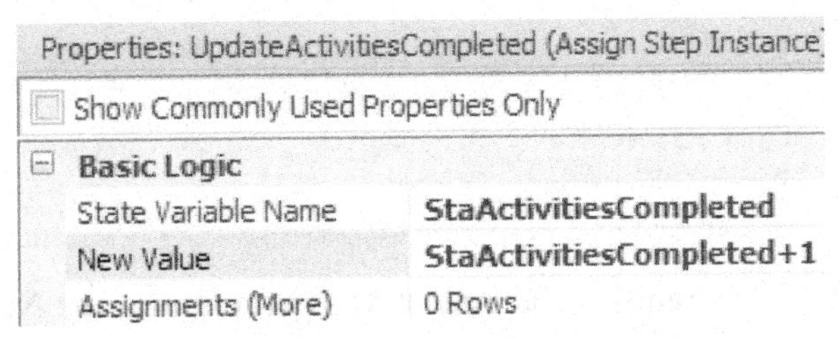

Figure 23.12: Updating the Number of Activities by One

- Now decide if all the precedent activities have completed (i.e., `StaActivitiesCompleted == StaPredecessors`) as seen in Figure 23.13.

Properties: AllActivitiesCompleted (Decide Step Instance)	
Show Commonly Used Properties Only	
Basic Logic	
Decide Type	ConditionBased
Expression	**StaActivitiesCompleted == StaPredecessors**
Advanced Options	
Exclusion Expression	**((StaPredecessors==1 &&JunctionType != List.ListJunctionType.EndNode)**

Figure 23.13: Deciding if All Preceding Activities are done

- Note the use of the *Exclusion Expression* property which can be used at the start of the simulation to determine if this step is necessary.[335] If the JUNCTION is a "*StartNode*" or only has one preceding activity then the next set of activities can be automatically started (it will be set to one) or this is an "*EndNode*" then these entities can be automatically destroyed (i.e., set to two). If you know at the start of the simulation, skipping steps can greatly speed up the simulation since it does not need to be evaluated. Set Exclusion property to the following.

```
((StaPredecessors==1 &&JunctionType != List.ListJunctionType.EndNode)
|| JunctionType == List.ListJunctionType.StartNode) + 2*(JunctionType
== List.ListJunctionType.EndNode)
```

- If the total number preceding activities has not been completed (i.e., still waiting on some activities to complete), then insert a `Destroy` step to kill the current activity entity.
- If this activity is the last one to complete or the junction is a "Start Node", then we need to start the next set of successor activities. Insert a `Create` step that will create copies of the current Entity (i.e., set the *Create Type* property to "CopyAssociatedObject"). The *Number of Objects* will be set to `StaSuccessors -1` since one activity ENTITY is already present. Note the use of the *Exclusion Expression* property to skip this step if there is only one successor as seen in Figure 23.14.

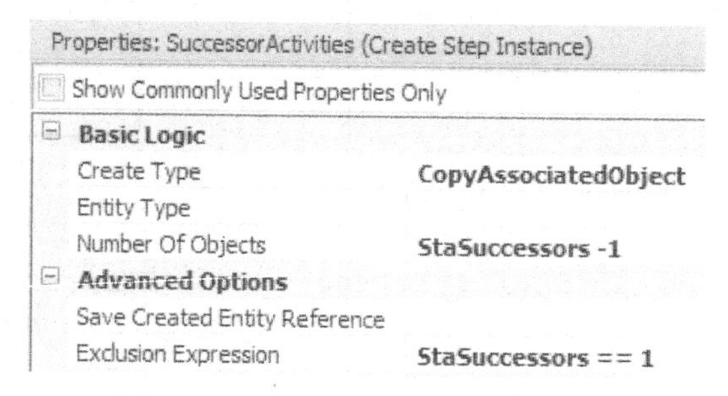

Figure 23.14: Creating the Successor Activities

- Using `Transfer` steps, the original activity ENTITY needs to be sent from the "Current Station" to the **"ParentExternalNode" Output** while the newly created activity entities need to be sent from "Free Space" to the same **ParentExternalNode Output** as seen in Figure 23.15.

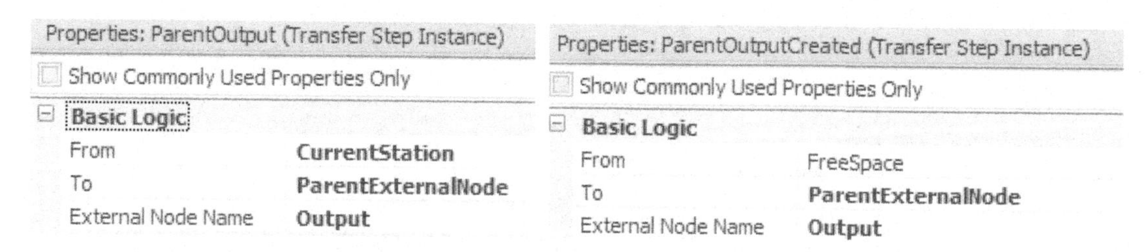

Figure 23.15: Sending Successor Activities Out to the Network

[335] If the expression evaluates to a one then then the step will not be included and the entity will flow directly out of the true or primary branch. If it evaluates to a two then it would be skipped and flow directly out of the false or secondary branch. All other evaluations will have the process step evaluated each time.

Step 10: If the **JUNCTION** is a “Start Node” then it must automatically create the start activity to start this branch of the network. Therefore, the “*OnRunInitialized*” process which is executed for each object at the beginning of the simulation will be used to handle this logic. Select the “*OnRunInitialized*” and add a `Create` and `Transfer` steps after the `Assign` sets the number predecessors and successors.

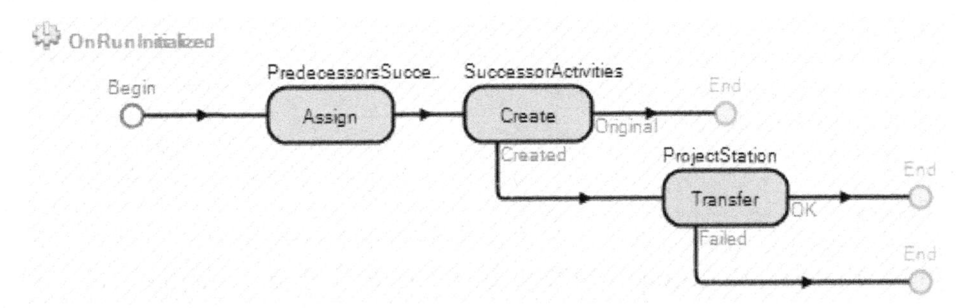

Figure 23.16: “*OnRunInitialized*” Process to Start Activities

- Insert a `Create` step that will produce one new object of ***EntityType*** property. Again, right click on the *Object Instance Name* property and select the referenced property. Again, note the use of the *Exclusion Expression* property to skip this step if it is not a “Start Node”. Refer to Figure 23.17.
- If it is a start node and an activity has been created, then insert a `Transfer` step to move it from “FreeSpace” to the **Project Station**.[336] This will then run the “*OnEnteringProject*” process which will produce the correct number of successors without having to repeat the code as seen in Figure 23.17.

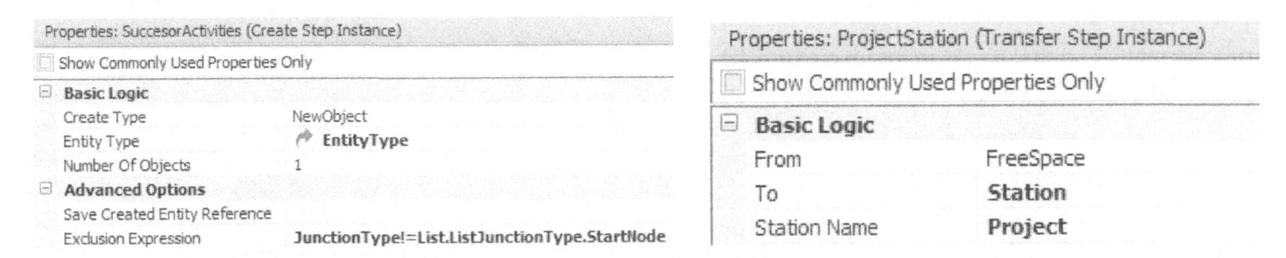

Figure 23.17: `Create` and `Transfer` Steps for the “*OnRunInitialized*” Process

Step 11: A warning may appear stating that there is nowhere for the entity to go once it exits the last **JUNCTION**. SIMIO now warns you when entities are sent to nowhere and will be automatically destroyed. This warning helps modelers make sure this was no modeling error. In this case, this is not an error but not a good idea to still have the warning. Therefore, if this is an “End Node” **JUNCTION** we do not want to transfer the entity out of the **Output ParentExternalNode**.

- Select the `Transfer` step connected to the “Original” path and add `2*(JunctionType == List.ListJunctionType.EndNode)` to the *Exclusion Expression* property which will force the model entity out of the “Failed” branch. A value of two sends the entity automatically out the failed.
- Next, drag the “Failed” branch to the `Destroy` step so entity will be destroyed now.

[336] Note you could have sent the entity from “FreeSpace” to the ***Input ParentExternalNode*** as well. Either way it runs the process.

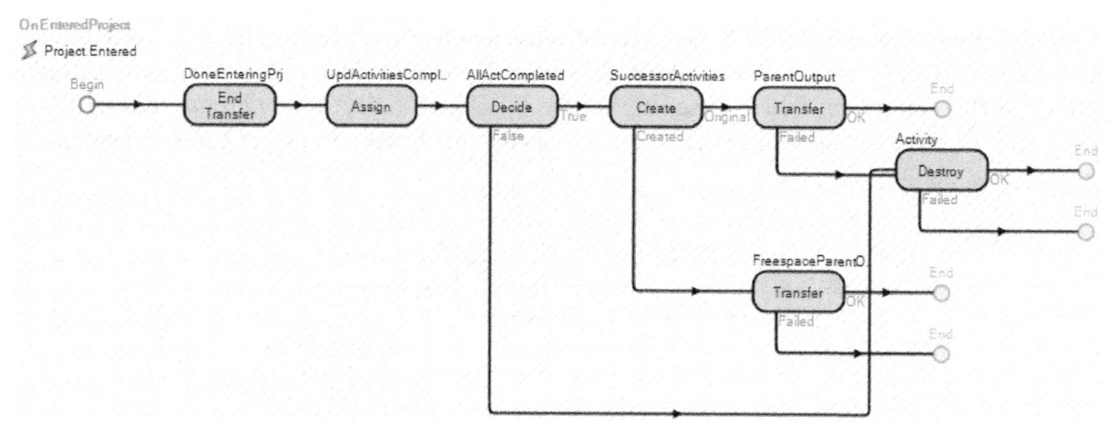

Figure 23.18: Updated Model that will Destroy the Entities if the Junction is an End Node

Part 23.4: Creating Small Network to Test the New Object

At this point our junction object seems to be setup to handle the precedence constraints and start new activities. Go back to the original model and create a very simple network to test this ability. At this point we will not worry about naming the objects.

Step 1: Insert a new MODEL ENTITY named **EntActivities** and four Junction objects leaving their names as the default. Place them in the configuration as seen in Figure 23.19.

String.format("The successor or predecessor {0} connected to {1} is not a PrjTimePaths. Change the link to PrjTimePath",Link.Name, Name)

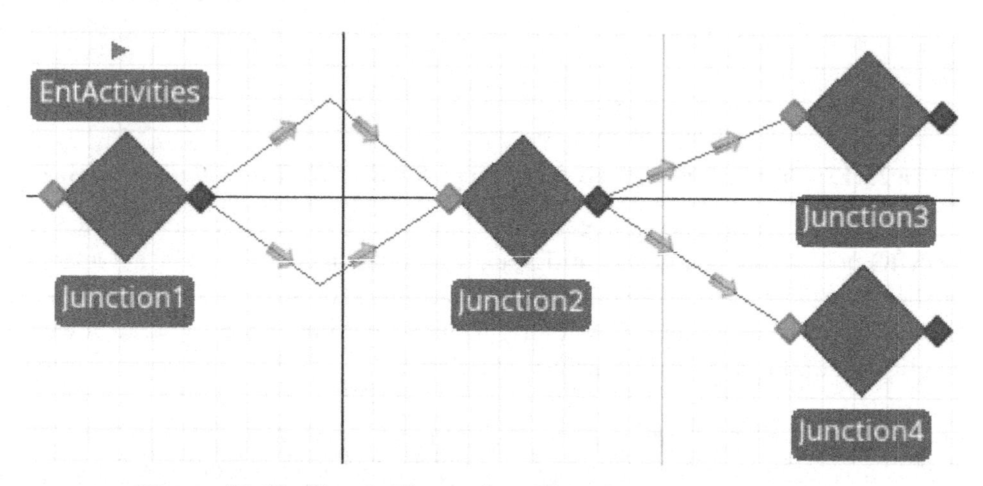

Figure 23.19: Simple Network to Test the JUNCTION **Object**

Step 1: Connect **Junction1** to **Junction2** via two **PRJTIMEPATHS** with one path taking one day and the other one taking two days and then connect **Junction2** to **Junction3** and **Junction4** via one day **PRJTIMEPATHS**.

Step 2: Set **Junction1** to be a "Start Node" while **Junction2** should be just a "Junction" as seen in Figure 23.20.

Properties: Junction1 (Junction)	
Show Commonly Used Properties Only	
⊞ **Process Logic**	
⊞ **Financials**	
⊞ **Advanced Options**	
⊟ **Project Management**	
Junction Type	**Star tNode**
EntityType	**EntActivities**

Properties: Junction2 (Junction)	
Show Commonly Used Properties Only	
⊞ **Process Logic**	
⊞ **Financials**	
⊞ **Advanced Options**	
⊟ **Project Management**	
Junction Type	Junction

Figure 23.20: Setting up the Start Junction and the Secondary Junction

Step 3: Save and run the model observing what happens.

Question 1: Did two activities get created from Junction1 and Junction2?

____

Question 2: Did Junction2 wait before both activities were completed before starting the next set of activities?

____

Step 4: It looks like only one was created when actually both were placed on the same path. Click on the **Output** node of **Junction2** and change the *Outbound Link Preference* from "Any" to "Available" which is used to select the next link. In this case we only want to select from the ones that are currently available.

Step 5: Save and run the model observing what happens.

Question 3: Did two activities get started from Junction2 now?

____

Step 6: At this point, the user will have to change the default property from "Any" to "Available" every time which is not acceptable since they may forget. Therefore, we will automatically set it for the modeler by specializing the TRANSFERNODE object. Subclass the TRANSFERNODE object and name it **PrjTransferNode**.

Step 7: Select the **PRJTRANSFERNODE** from the [*Navigation Panel*]. From the *Definitions→Properties*, click on the "Properties (Inherited)" down arrow to show all the inherited properties. Select the *OutboundLinkPreference* property and change the *Default Value* to "Available" and set the *Visible* property to "False" so the user cant accidently set it back.

Step 8: Return back to the **JUNCTION** object and select the external **Output** node in the *Definitions→External* section. Change the *Node Class* property from the TRANSFEROBJECT to the new **PRJTRANSFERNODE** as seen in Figure 23.21.

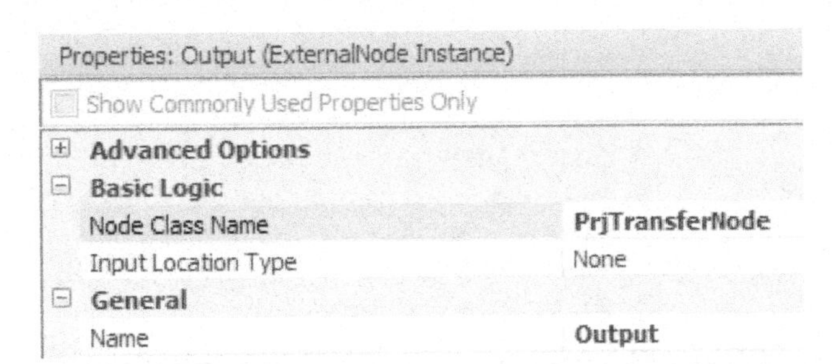

Figure 23.21: Changing the Output node to be PrjTransferNode

Part 23.5: Building the Example Network

Let's build the simple example from the beginning section.

Step 1: Rearrange the current **JUNCTIONS** in the model and insert a new **JUNCTION** using the default names as seen in Figure 23.22.

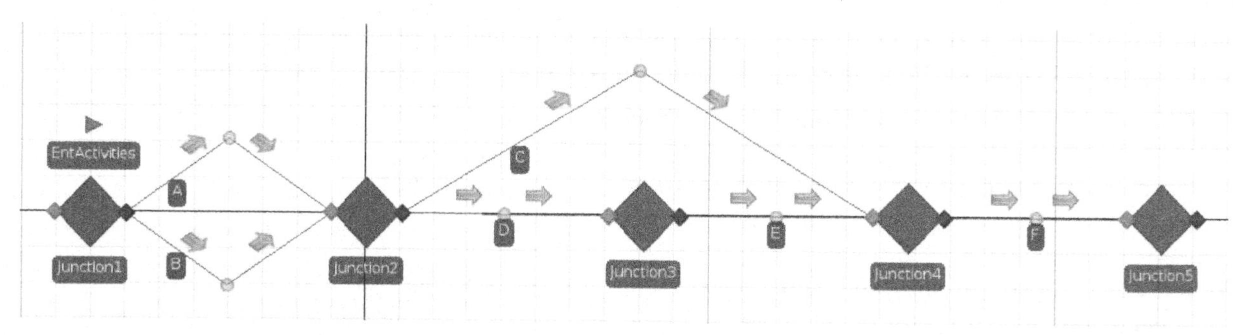

Figure 23.22: Example Project Management Model

Step 2: Use **PRJTIMEPATHS** to represent each of the activities by connecting two **JUNCTIONS** together. Table 23.4 should be utilized specify the name of each of the paths as well as to set the *Travel Time* and *Units* properties.

Table 23.4: Activity Times and Names

From	To	Name	Travel Time	Units
Junction 1	Junction 2	**A**	`Random.Pert(5,6,10)`	Days
Junction 1	Junction 2	**B**	`Random.Pert(6,9,9)`	Days
Junction 3	Junction 4	**C**	`Random.Pert(6,8,16)`	Days
Junction 2	Junction 3	**D**	`Random.Pert(4,5,7)`	Days
Junction 2	Junction 4	**E**	`Random.Pert(6,7,10)`	Days
Junction 4	Junction 5	**F**	`Random.Pert(8,12,16)`	Days

Step 3: Next, use the values in Figure 23.23 to specify the *Project Management* properties of each of the five junctions to make sure the precedent constraints are satisfied and the number successor activities are started.

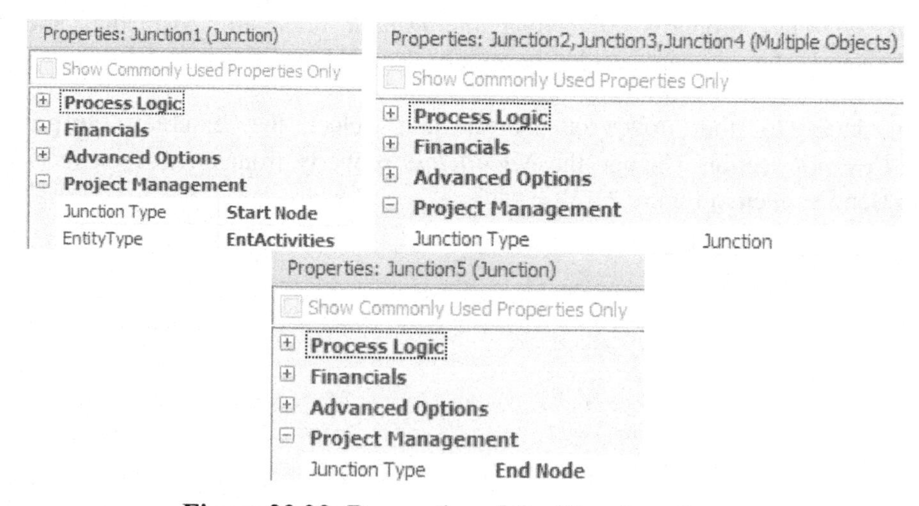

Figure 23.23: Properties of the Five Junctions

Step 4: Save and run the model, make sure the run length is "Unspecified" which will stop the simulation when no more entities are in the system.

Question 4: How many projects were completed and what was the total completion time?

Step 5: Remember, we decided to model one project completion with one replication. Therefore, insert a new experiment that will replicate the project management network 1000 times.

Step 6: Run the experiment and look at the results under the "*Pivot Grid* tab. Change the Unit Settings of time to Days.

Step 7: Filter the "*Category*" column by clicking on the funnel in the right hand corner (see Figure 23.24) and select only the "Critical Path" statistics and then filter "*Statistic*" column to only include the "Average" since we only have one observation per replication. Figure 23.25 shows the reduced statistics where the project will take between 27 and 38 days with the average taking 33 days plus or minus 1.8 days[337].

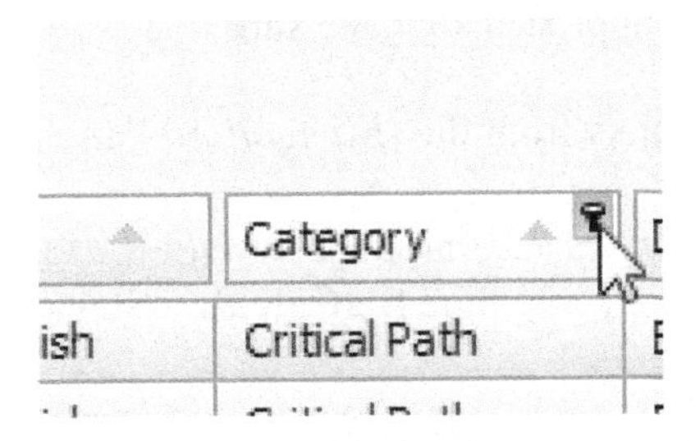

Figure 23.24: Filter the Category only to Include Critical Path Statistics

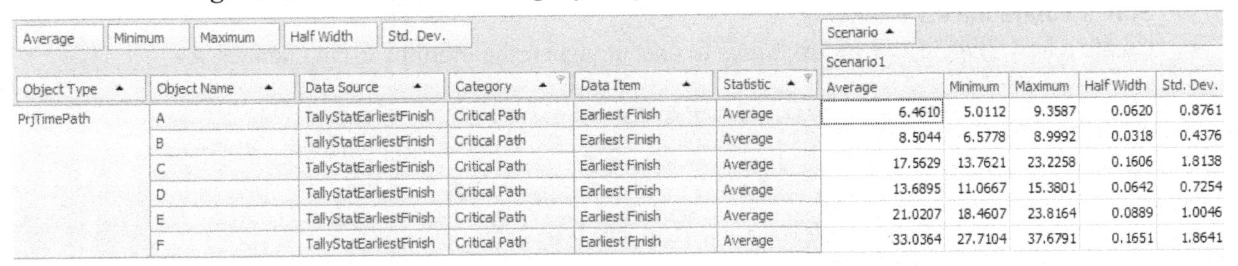

Object Type	Object Name	Data Source	Category	Data Item	Statistic	Average	Minimum	Maximum	Half Width	Std. Dev.
PrjTimePath	A	TallyStatEarliestFinish	Critical Path	Earliest Finish	Average	6.4610	5.0112	9.3587	0.0620	0.8761
	B	TallyStatEarliestFinish	Critical Path	Earliest Finish	Average	8.5044	6.5778	8.9992	0.0318	0.4376
	C	TallyStatEarliestFinish	Critical Path	Earliest Finish	Average	17.5629	13.7621	23.2258	0.1606	1.8138
	D	TallyStatEarliestFinish	Critical Path	Earliest Finish	Average	13.6895	11.0667	15.3801	0.0642	0.7254
	E	TallyStatEarliestFinish	Critical Path	Earliest Finish	Average	21.0207	18.4607	23.8164	0.0889	1.0046
	F	TallyStatEarliestFinish	Critical Path	Earliest Finish	Average	33.0364	27.7104	37.6791	0.1651	1.8641

Figure 23.25: Earliest Finish Statistics for 1000 Replications

Part 23.6: Adding the Slack and Percent of time on Critical Path Calculations

At this point we can determine the Earliest Finish (EF) time of all the paths and the total time to complete the project, but it is difficult to determine what paths constitute the critical path since path times are random. However, the slack time for each path and the percent of time this path is on the critical path can assist the project manager. The Earliest Finish (EF) time can be calculated without any other information from other objects. However, the computation of the Latest Finish (LF) requires knowing the slack and the number of times the path is on the critical path.

The Latest Finish time is only known at the junctions and is determined when the last activity finishes. When two SIMIO objects need to communicate with one another, there are two ways to approach this communication. In the first approach shown in Figure 23.26, the **JUNCTIONS** have to know the **PRJTIMEPATHS** that are connected to them. The **PRJTIMEPATHS** will monitor a state variable within their object that is set to one by the **JUNCTION** when the last activity finishes. In the second approach the **PRJTIMEPATHS** need to know the **JUNCTIONS** that they are connected too and they will monitor a state variable inside the **JUNCTION** object.

---

337 Your numbers may be a little different since the random numbers may be sampled differently in your model, due to the way the modeling components were added. However the results should be statistically identical.

We will illustrate the first approach in this section where the monitoring state variable resides inside the monitoring.

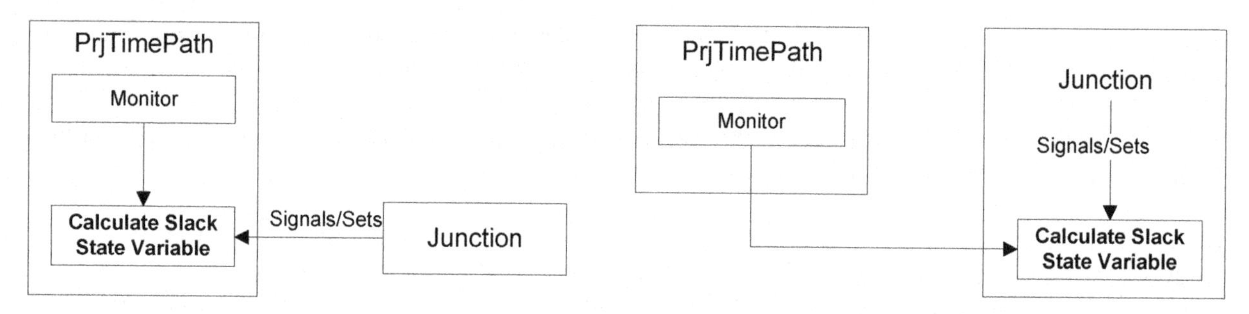

Figure 23.26: Approaches to Signaling from One Object to Another

Step 1: Save the current model as a **Chapter 23-6.spfx** because we will need to come back to the previous section model to create the second approach so make sure it is saved before renaming it.

Step 2: Select the PRJTIMEPATH object from the [*Navigation*] Panel.

Step 3: From the *Defintions→Elements* tab, insert a new MONITOR named **MonitorSlack** which will be used to calculate the slack calculation when the **StaCalculateSlack** state variable changes values to one as seen in Figure 23.27.

- Set the *Monitor Type* to "DiscreteStateChange" and the *State Variable Name* to the **StaCalculateSlack** variable.
- Create a new *Triggered Process Name* to execute when the monitor event changes.

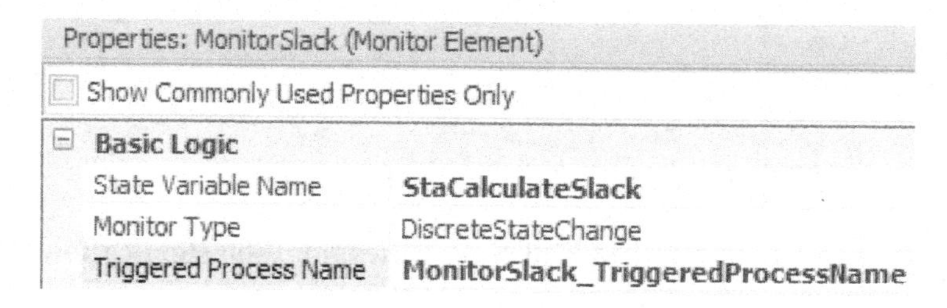

Figure 23.27: Setting up the Monitor to Calculate the Slack

Step 4: Once the last activity has happened, the associated JUNCTION will set the **StaCalculateSlack** state variable to one. Since this variable is being watched by the **MonitorSlack**, it will execute the "*Triggered Process*" as seen in Figure 23.28.

- Insert a `Tally` step that will calculate the slack (LF-EF) where the latest finish is the current time and add either a zero or one to the **TallyPercentCP** to indicate whether or not the path is on the critical path or not. You need to select the *Tallies (More)* repeating group to set both statistics.

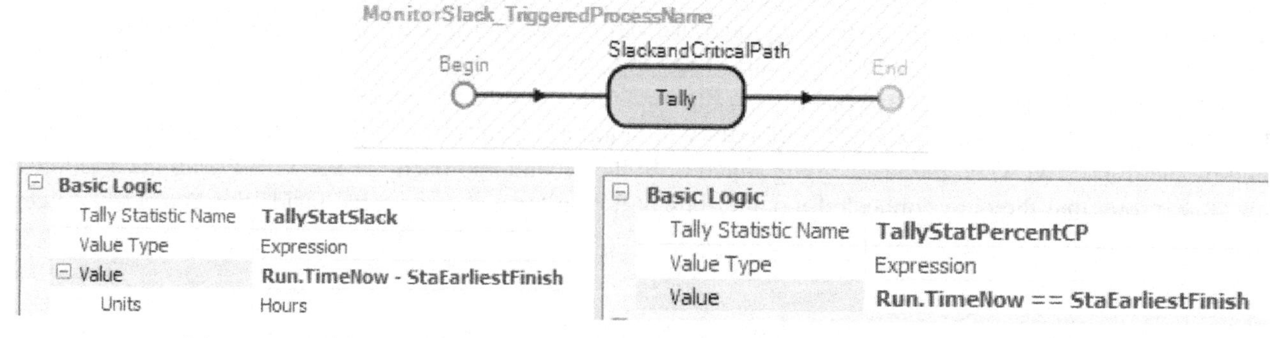

Figure 23.28: Calculating the Slack and Determining if it is on Critical Path

Step 5: The **PRJTIMEPATH** object is now setup to calculate the two new statistics when its **StaCalculateSlack** variable is changed. The **JUNCTION** object must now set the **StaCalculateSlack** variable for all of the precedent paths once the last activity has finished. Select the **JUNCTION** object from the [*Navigation*] panel.

Step 6: After the last precedent activity has finished (**JUNCTION** object's *OnEnteringProject* process), the **StaCalculateSlack** variable will need to be set to one for all the precedent paths. Insert a `Search` step that will use to search for the correct activity paths right after the `Decide` step. Recall the search step returns a TOKEN associated with the found object (i.e., **PRJTIMEPATH**) which can then be used to set its own variable.

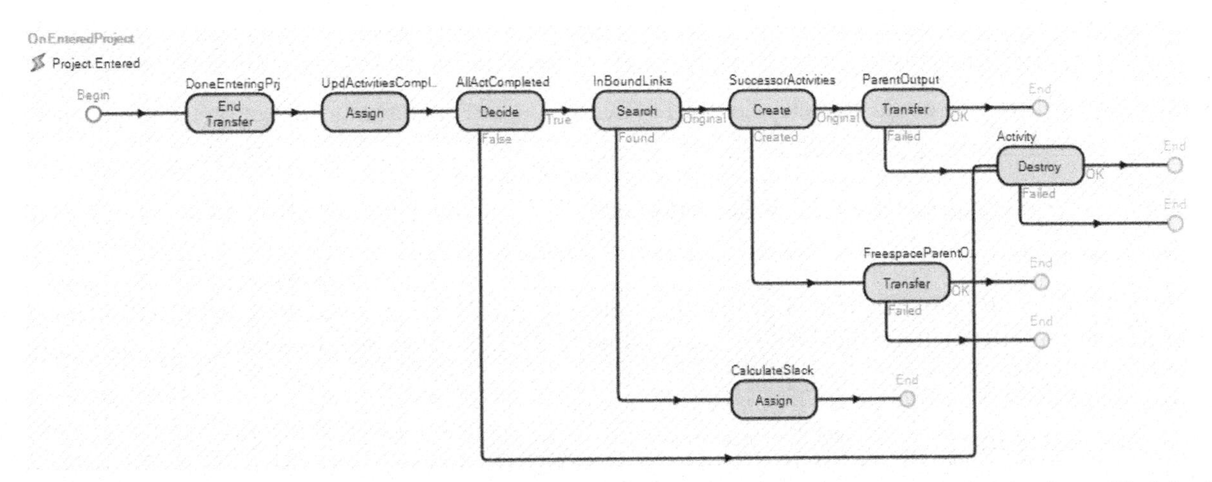

Figure 23.29: Setting the StaCalculateSlack Variable of the Paths When All Activities have Finished

- Figure 23.30 shows the `Search` step that will search a "NodeInBoundLinks" where the *Node Name* property is set to **Input** which is the parent input node. Note, this node will not be in the dropdown list but will need to be just typed into the field. The *Exclusion Expression* (i.e., `JunctionType ==List.ListJunctionType.StartNode`) specifies to skip this `Search` step if this **JUNCTION** is a "StartNode" and does not need to notify in predecessor links.

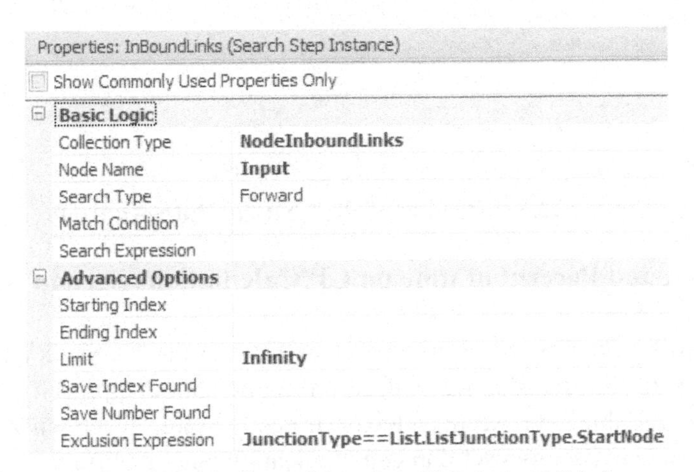

Figure 23.30: Searching for all Inbound Links

- Insert an `Assign` on the "Found" branch and specified with the values in Figure 23.29 which will set the `PrjTimePath.StaCalculateSlack` discrete variable to one.

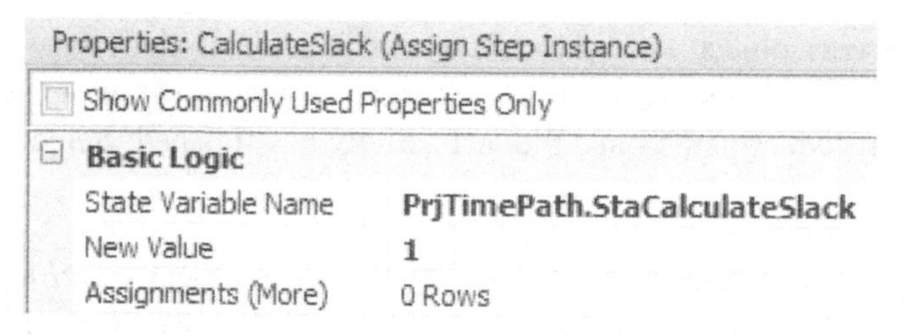

Figure 23.31: Setting the Calculate Slack Variable

Step 7: Save and rerun the experiment.[338]

Question 5: What percentage of the time is path A on the critical path?

____

Question 6: How long could we delay the start of activity A and not have any effect on the critical path?

____

Question 7: How long could we delay the start of activity C and not have any effect on the critical path?

____

Step 8: Looking at the results it is clear that the critical path is B D E F since these paths have the highest percentages of time on the critical path as seen in Figure 23.32 for Percent CP average.

Object Name	Data Source	Category	Data Item	Statistic	Average	Minimum	Maximum	Half Width
A	TallyStatEarliestFinish	Critical Path	Earliest Finish	Average (Days)	6.4610	5.0112	9.3587	0.0620
	TallyStatPercentCP	Critical Path	Percent CP	Average	0.0283	0.0000	1.0000	0.0147
	TallyStatSlack	Critical Path	Slack	Average (Days)	2.0230	0.0000	3.8161	0.0829
B	TallyStatEarliestFinish	Critical Path	Earliest Finish	Average (Days)	8.5044	6.5778	8.9992	0.0318
	TallyStatPercentCP	Critical Path	Percent CP	Average	0.9717	0.0000	1.0000	0.0147
	TallyStatSlack	Critical Path	Slack	Average (Days)	0.0119	0.0000	0.8448	0.0071
C	TallyStatEarliestFinish	Critical Path	Earliest Finish	Average (Days)	17.5629	13.7621	23.2258	0.1606
	TallyStatPercentCP	Critical Path	Percent CP	Average	0.0465	0.0000	1.0000	0.0187
	TallyStatSlack	Critical Path	Slack	Average (Days)	3.4986	0.0000	8.5482	0.1617
D	TallyStatEarliestFinish	Critical Path	Earliest Finish	Average (Days)	13.6895	11.0667	15.3801	0.0642
	TallyStatPercentCP	Critical Path	Percent CP	Average	1.0000	1.0000	1.0000	0.0000
	TallyStatSlack	Critical Path	Slack	Average (Days)	0.0000	0.0000	0.0000	0.0000
E	TallyStatEarliestFinish	Critical Path	Earliest Finish	Average (Days)	21.0207	18.4607	23.8164	0.0889
	TallyStatPercentCP	Critical Path	Percent CP	Average	0.9535	0.0000	1.0000	0.0187
	TallyStatSlack	Critical Path	Slack	Average (Days)	0.0407	0.0000	2.1276	0.0199
F	TallyStatEarliestFinish	Critical Path	Earliest Finish	Average (Days)	33.0364	27.7104	37.6791	0.1651

Figure 23.32: Results of the Critical Path Analysis

Part 23.7: Adding Slack and Percent of time on CP Calculations Second Approach

The previous section uses the **JUNCTION** objects to signal (i.e., set the **StaSlackVariable**) the activity time paths of the Latest Finish. In this section, we will take the alternative approach and let the **PRJTIMEPATH** monitor a **JUNCTION** state variable. This section demonstrates how an object can monitor state variables of other objects, which can prove to be very useful in communicating among objects.

Step 1: Save the current project (Chap23-6p2.spx) under a new name.

---

[338] Sometimes when adding properties to objects that are already in the model causes ghost errors of things not being specified. Save and reopen the model to eliminate them if this occurs.

Step 2: Remove the the `Search` steps from the "*OnEnteringProject*" process and insert a new `Assign` step that assigns the current time (`Run.TimeNow`) to the **StaLatestFinish** state variable.

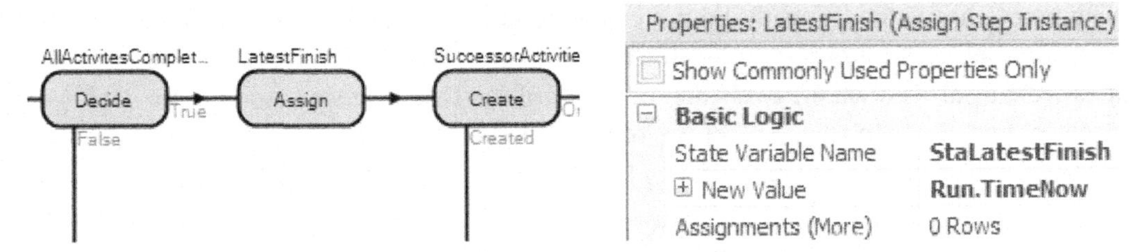

Figure 23.33: Setting the StaLatestFinish State Variable

Step 3: Next, select the **PRJTIMEPATH** object.

Step 4: From the *Definitions→Properties→Add→Standard Property* drop down, insert a new STATE property named **JunctionLF** which will allow the modeler to specify the **JUNCTION** state variable to monitor. Change the *Category Name* to "Project Management" by specifying it.

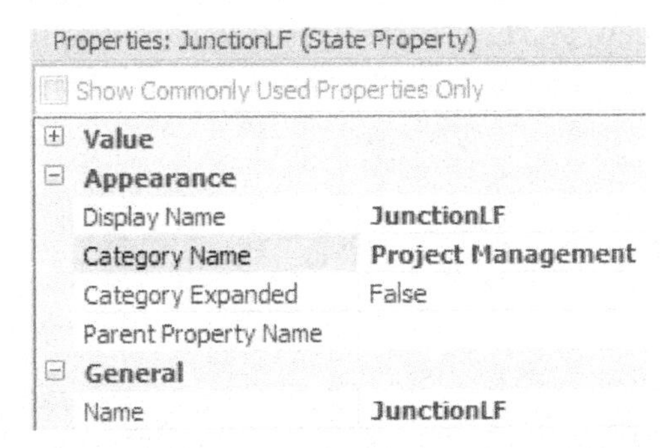

Figure 23.34: Create a State Variable Property

Step 5: Modify the MONITOR element **MonitorSlack** to monitor the new state reference property **JunctionLF** as seen in Figure 23.35.

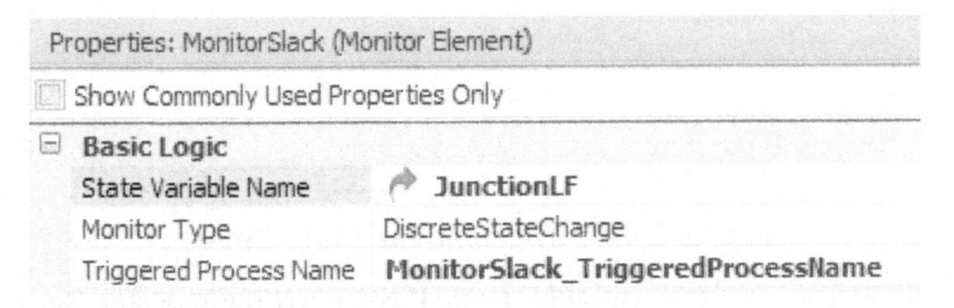

Figure 23.35: Using the Reference State Property to Monitor

Step 6: Return back to the model and for each **PRJTIMEPATH**, specify the appropriate **JUNCTION'S** **StaLatestFinish** state variable.

- For example (see Figure 23.36), *Junction LF State Variable* property would be set to `Junction2.StaLatestFinish` for both paths **A** and **B.**
- For paths **C** and **E**, **D**, and **F** it should be set to `Junction4.StaLatestFinish`, `Junction3.StaLatestFinish`, and `Junction5.StaLatestFinish` respectively.

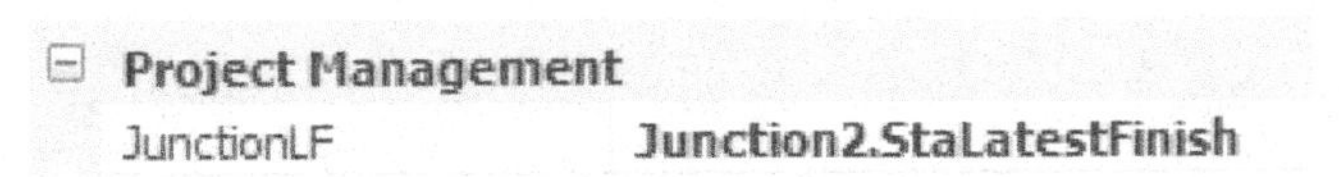

Figure 23.36: Specifying the State Variable

Step 7: Save the model and rerun the experiment comparing the results of the two systems.

Question 8: Did you get the same results?

Step 8: The problem was the last path did not calculate the latest finish statistics since it was heading to an "End" junction. Select the **JUNCTION** object and then inside the *OnEnteringProject* process. Recall that all the entities were destroyed via the exclusion property of the `Decide` step (see Figure 23.13) and therefore it is not sent to the `Assign` step. Therefore, remove the portion of the exclusion property associated with end node: `2*(JunctionType == List.ListJunctionType.EndNode)`.

Step 9: Save the model and rerun just the model.

Question 9: Did anything happen when the activity went through the end node?

Part 23.8: Error Checking to Make Sure Modeler Uses Junction Correctly

There are several assumptions the model has made. We are assuming the modeler will always connect up the junctions correctly using only the new **PRJTIMEPATH** links. To eliminate this assumption, we will modify the **OnRunInitialized** to check for these issues, warn the modeler, and then end the simulation run as seen in Figure 23.37.

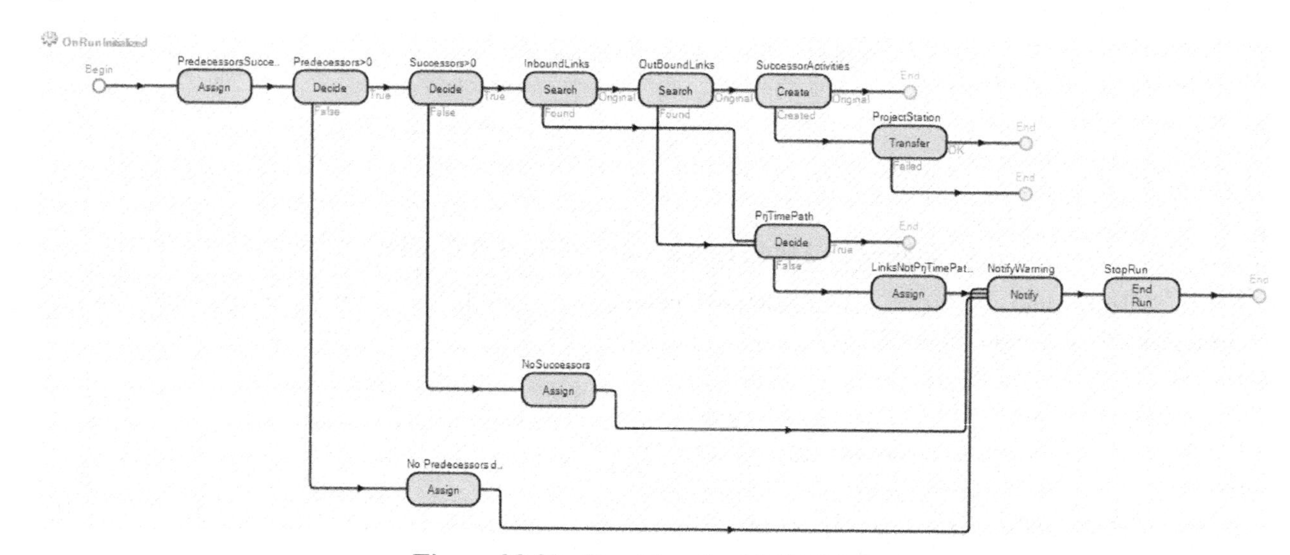

Figure 23.37: Checking for Valid Links

Step 1: Insert a new STRING STATE variable named **StaWarningMessage** to specify the particular message.

Step 2: Insert a conditional based `Decide` step to check to see if there are predecessor links connected to the Junction. If this is not a "Start Node," the step will be skipped automatically as specified by the *Exclusion Expression* property. Insert another `Decide` step that will check to make sure the successor links have been defined if this is not an "End Node" as seen in Figure 23.38.

Properties: Predecessors>0 (Decide Step Instance)

Show Commonly Used Properties Only

Basic Logic	
Decide Type	ConditionBased
Expression	**StaPredecessors >0**
Advanced Options	
Exclusion Expression	**JunctionType == List.ListJunctionType.StartNode**

Properties: Successors>0 (Decide Step Instance)

Show Commonly Used Properties Only

Basic Logic	
Decide Type	ConditionBased
Expression	**StaSuccessors > 0**
Advanced Options	
Exclusion Expression	**JunctionType == List.ListJunctionType.EndNode**

Figure 23.38: Checking to see if the Predecessors and Successors have been Defined

Step 3: If the **Predecessors>0** `Decide` steps fail (i.e., not start node but links are connected to the junction), insert an `Assign` step on the failed branch that will set the **StaWarningMessage** state variable to a string which states the **JUNCTION** name as well as what to do fix the issue.[339]

```
String.format("{0} is Not a Start Node and there are no predecessors (i.e, Project Path links) defined", Name)
```

Step 4: Next, if the **Successors>0** `Decide` steps fail, insert an `Assign` step on the failed branch that will set the **StaWarningMessage** state variable to a string which states the **JUNCTION** name as well as what to do fix the issue.

```
String.format(" {0} is Not an End Node and there are no successors (i.e, links) defined",Name)
```

Step 5: Connect both of the `Assign` steps to the same `Notify` step. The `Notify` step can either write a message in the Trace window or give a pop-up Warning to the user. We will use the **StaWarningMessage** as the *Message Content* property as seen in Figure 23.39.

Properties: NotifyWarning (Notify Step Instance)

☐ Show Commonly Used Properties Only

⊟ Basic Logic	
Notification Type	**Warning**
Message Heading	**"Junction Initialization Error"**
Message Content	**StaWarningMessage**

Figure 23.39: Sending a Warning to the User

Step 6: Insert an `End Run` step that will terminate the simulation since there is an error in the model and we don't want to continue.

Step 7: The first two `decide` steps determined if there were no predecessor or successor links defined. However, the user could connect the **JUNCTIONS** using normal PATHS, CONNECTORS, or TIMEPATHS which is also an error. Insert two `Search` steps that will search both the inbound and outbound links as seen in Figure 23.40. The inbound search will be skipped if this is a "Start Node" while the outbound search will be skipped if the **JUNCTION** is an "End Node" owing to the *Exclusion Expression* property.

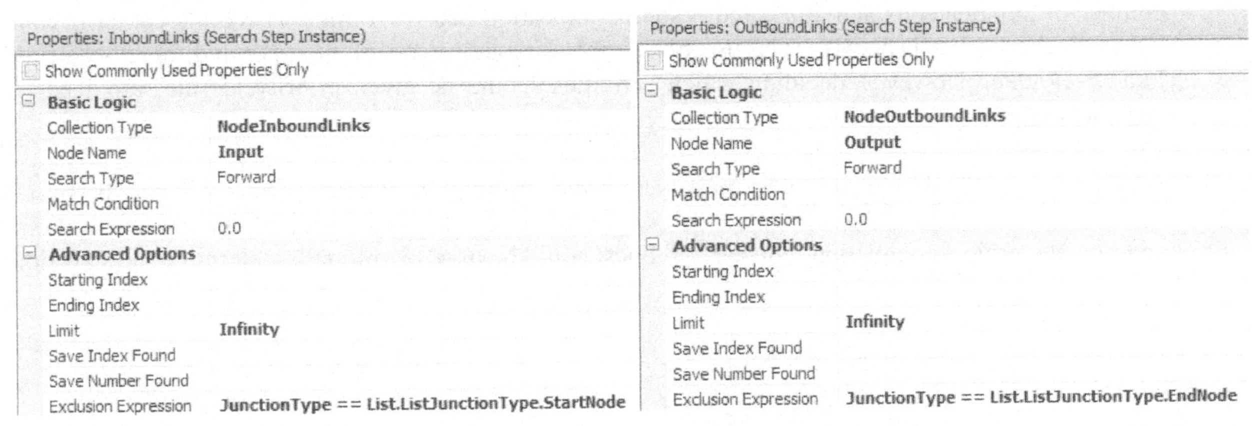

Figure 23.40: Searching to Make Sure the Predecessor and Successor Links are PrjTimePaths

---

[339] Note the use of `String.Format` function to create string that has one argument (i.e., "{0}") which is set to the value of the JUNCTION NAME.

Step 8: When links are found, we need to check to make sure they are **PRJTIMEPATH** links only. Connect both "Found" branches to a conditional `Decide` step that checks to see if these are **PRJTIMEPATH** using the `Is` keyword which returns **False** if these objects are not project time paths.

Properties: PrjTimePath (Decide Step Instance)	
☐ Show Commonly Used Properties Only	
⊟ **Basic Logic**	
Decide Type	ConditionBased
Expression	**Is.PrjTimePath**

Figure 23.41: Checking to see if the Links are Project Time Paths

Step 9: If any of the links are not project time paths, then set the warning message which will tell the modeler which link is not a project time path that is connected to which **JUNCTION** as seen below.

```
String.format("The successor or predecessor {0} connected to {1} is not a
PrjTimePaths. Change the link to PrjTimePath",Link.Name, Name)
```

Step 10: Next, connect the `Assign` step to the `Notify` step by dragging the End on top of the step.

Step 11: Select the Facility window and test the new error checking by inserting another Junction into the model.

Step 12: Save and then run the model with the new Junction not connected.

Question 10: Does the simulation stop and warn the user with the Junction in error?

____

Step 13: Next connect both the input and ouput to other nodes via paths and rerun the model.

Question 11: Does the simulation stop and warn the user with the Junction in error?

____

Part 23.9: Commentary

- Sometimes the activities within a project employ common resources. The resources can limit the number of simultaneous activities, and resource allocation and the timing of the allocation become limits on the project progress. The use of common resources can be incorporated directly into the **PRJTIMEPATHS**. However deciding which activities should be given priority for the resources can become a problem.

- Another possible option in project management is determining the amount of time to delay the start of an activity. By delaying the start of an activity, resources can be made available to more critical ones and thus the project completion reduced. Optquest could be used to optimally determine the start time.

Appendix A
Input Modeling

Any discussion of creating a "realistic" model of any operation eventually turns to the issue of variability or randomness. Incorporating "randomness" into your simulation model can be a significant challenge because the causes of randomness are difficult to understand completely. As a result we approximate this randomness with some kind of "randomness model", usually a statistical distribution. Because we must "approximate" the randomness, we view this as "modeling". And because the randomness is not explained by the simulation model, this randomness must be "input" to the simulation. Thus reason for the term "Input Modeling." There are several forms of input models for randomness.

One input model is the time-based arrival process, which in SIMIO is called the "Time Varying Arrival Rate" arrival mode at a SOURCE. Fundamentally, we are trying to model an arrival process that changes throughout time. The simplest model is a Poisson arrival process whose mean changes with time, which is generally referred to as a Non-Homogeneous Poisson Process (NHPP). Recall that the Poisson has only one parameter, namely its arrival rate (mean). A NHPP has a Poisson mean with an arrival rate change that can be any function of time. In SIMIO the rate function is a stepped function which provides an arrival rate during a specified time period. In SIMIO, the rate is always given in events per hour, while the time periods have a fixed interval size (and number). A more sophisticated NHPP would permit a more flexible rate function.

Another model of randomness is the specification of a statistical distribution for some observed randomness, usually an amount of time. The most prominent examples are Processing Times at SERVERS, WORKSTATIONS, COMBINERS, and SEPARATORS and Travel Time in a TIMEPATH. Other examples of times include Transfer-In Time, Setup Time, and Teardown Time in various objects. Also the "Calendar Time Based" failure type at working objects requires uptime between failures and time to repair. Creating input models for this kind of input may employ some way to find a statistical distribution to match the randomness behavior. Sometimes there is randomness in a number, such as the size of an arriving batch or the number of parts in a tote pan.

Part A.1 Random Variables

Randomness is thus represented in simulation by random variables. These random variables describe either a continuous value (such as time) or a discrete value (such as a quantity). While you may not know how the randomness occurs, you might know something about the distribution of these values such as some descriptive statistics. Descriptive statistics include measures of central tendency such as a mean, median, or mode. Maybe it's a measure of variability such as a variance, standard deviation, coefficient of variation, or range. Other descriptive statistics include the measure of asymmetry called the skewness and the measure of the peakedness called the kurtosis.

Part A.2 Collecting Data

More than likely you don't know much and will need to collect some data from the real input process to get some idea of how the randomness behaves. This data is simply a "sample" from which you might want to estimate how the population is distributed.

However before we begin to examine how to use "sample data", you need to remember the most important rule of data collection: *Build the simulation model before collecting data!*

The process of collecting data is almost always an onerous, frustrating, and expensive. You need to know that you absolutely need the input model before you mount a data collection effort. The best way to know that you need particular data is to construct the simulation model first. The simulation model doesn't know if the input models were obtained from real data. Simply use your best judgment to obtain input models.

One suggestion is to interview some people familiar with the process that is generating the randomness and try to elicit from them some general characteristics. If you can only estimate a minimum and a maximum, use a uniform distribution to match. However, generally, you can obtain three estimates: a mode, a minimum, and a maximum. For example, suppose we want to estimate the randomness that describes the amount of time it takes you to get to work or school. If I say, tell me the most likely or most common time it takes and you say 15 minutes. That estimate of 15 minutes becomes a mode. Now tell me how long it takes if you make all the lights and cruise here without any difficulty, namely the minimum amount of time. Suppose you say 10 minutes. Next give me the maximum, when everything goes wrong. You say the worst is 30 minutes. With these there estimates use a PERT distribution (sometimes called the BetaPERT) as shown in Figure A.1.

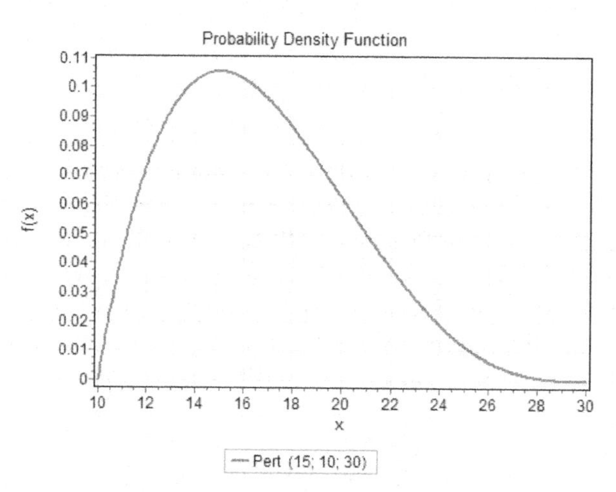

Figure A.1: Pert Distribution

As we you see, this distribution insures we bound our randomness to realistic values. In this case there are no negative values[340], the distribution is usually skewed to the right, and the mode is less than the mean. We highly recommend this means of obtaining estimates, especially in the early modeling stages. The Triangular distribution can be used also to model the case when given the minimum, maximum, and most likely point. However, it will be heavier in the tails while the Pert will have more probability around the mode[341].

With the liberal use of these kinds of estimates, build your simulation model. Vary some of the input model parameters (i.e., the mode, minimum or maximum) and see what impact they have on the simulation outputs. In particular, note how they impact the simulation outputs that are important to you. With a little experimentation of this type – called a "sensitivity analysis", you begin to get a sense of which data are most critical. You can marshal a data collection effort that is proportional to the importance of the input model. There are a number of hard-to-answer questions regarding data collection. Often you can't collect as much data as you would like, so you will need to consider the limitations of your data collection budget. You will want to collect as much data as you can, bearing in mind that not all data is equally important in the model.

Part A.3 Input Modeling: Selecting a Distribution to Fit to Your Data

Once you begin the data collection effort, you can begin to select distributions that fit your data. Why do we want to select a distribution to fit the data and not simply use the data directly in the simulation model? First, the data is only a sample from a population and not a total population, so we don't know exactly whether the sample completely represents the population from which it is sampled. Secondly, real data tends to be

---

[340] If a sample from a distribution for a time, like a processing time, produces a negative value, SIMIO will display an error. Distributions that can have negative values, like the Normal must be used with care. You can avoid negative values using and expression like Math.Max(0.0, Random.Normal(10.0,4.0).

[341] The PERT distribution does have some concerns. Its variance changes significantly as the mode approaches the end points. Some have suggested that it should not be used if the mode is within 13% of the end points.

incomplete in that it doesn't completely match the process being modeled. Third, we believe that there are enough standard statistical distributions from which to choose that will do a good job of modeling the "true" input process. So we will look for a distribution to fit the data.

Fortunately, once we have some data, the process of finding a distribution to fit the data is somewhat routine. Here are the steps we will follow:

1) Collect data and put into spreadsheet
2) Identify candidate distributions
3) For each candidate distribution
   a) Estimate parameters for the candidate distributions
   b) Compare candidate distribution to histogram
   c) Compare the candidate distribution function to the data-based distribution function
4) Select the best representation of the input (the input model)

Candidate Distributions

Before going into much detail, it is useful to do a SIMIO "Help" on "distributions". There will you see descriptions of all the distributions that SIMIO supports. Presently they include the standard "continuous" distributions of Beta, Erlang, Exponential, Gamma, JohnsonSB, JohnsonSU, LogLogistic, LogNormal, Normal, PearsonVI, Pert, Triangular, Uniform, and Weibull as well as the standard discrete distributions of Binomial, Geometric, NegativeBinomial, and Poisson. Also there are the empirical distributions called Continuous and Discrete, which represent a data-based continuous and discrete distributions. You should spend some time looking through these distributions so you have some sense of their "shape" and parameters. For instance, Figure A.2 displays the Erlang from the SIMIO help.

Random.Erlang(mean, K)

The Erlang distribution models an n-phase activity where the time for each phase is exponentially distributed. The Erlang has parameters that specify the mean and number of integer phases (K).

Erlang is the sum of K exponentials, each with an expected value of Mean/K. The Mean parameter is the mean of the Erlang, not the mean of the individual exponentials.

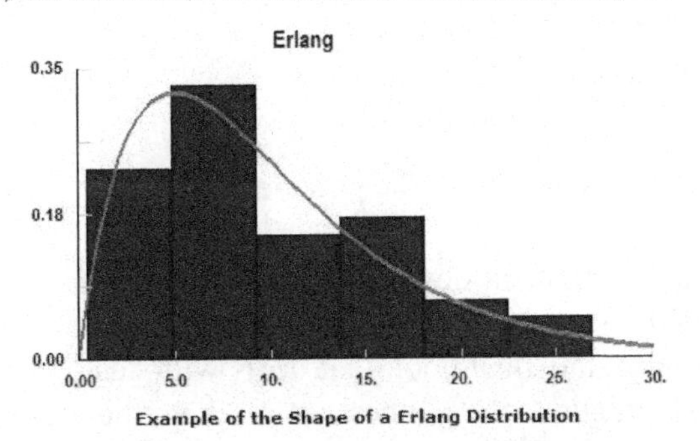

Example of the Shape of a Erlang Distribution

Figure A.2: Erlang Distribution

From the display, you see that the Erlang is a non-negative distribution specified by a mean and a number of phases (each phase being Exponentially distributed with a mean of the distribution divided by the number of phases).

In the prior figure, there is the display of a histogram. A histogram is simply a display of the frequency that data values fall within a given range. Typically the ranges or cells of a histogram have identical ranges (sometimes the end cells are longer) and the number of histogram cells range from 5 to 15. The histogram is an arbitrary display since the user determines the size of the cells and the number of cells. There are some "rules of thumb" used. For example, by dividing the range of observations (maximum – minimum) by the number of cells, you can determine the range for each cell. Although you would like to keep the number of cells, C, between 5 and 15, the following can be a good guide:

$$\text{Scott's Rule: } C \cong \frac{5}{3}(n^{1/3})$$

$$\text{Surges'Rule: } C \cong 1 + 3.3\log_{10}(n)$$

Here the number of observations in your data is n. With the range for each cell, you simply display the portion of the observations that fall within each range. By examining the histogram, you may want to select certain distributions to fit. But before you fit a distribution to the data you need the parameters for the distribution and they will need to be estimated from the data.

Estimating Distribution Parameters

The estimation of distribution parameters employ one of two methods: moment match and maximum likelihood. Moment matching requires that you know, for a two parameter distribution, the mean and variance of the distribution in terms of its parameters. If the Erlang mean is given by the symbol θ and there are k phases to the Erlang, then the

$$\text{Mean} = \theta$$

$$\text{Variance} = \theta^2/k$$

Therefore if you compute the average and the sample variance from your data as $\overline{x}$ and s^2, approximate mean

by the average and the variable by the sample variance. Solving, you can approximate the parameters of the Erlang by the following equations

$$\theta = \overline{x}$$

$$k = (\overline{x}/s)^2$$

The method of "matching" the sample statistics to the parameter values is the essence of moment matching.

Maximum likelihood requires more statistical development than we wish to devote here. However the basic idea is to find the parameters of the distribution so they have the greatest "likelihood" of coming from that parameterized distribution. Most any statistical book that deals with "estimation" will contain a description of maximum likelihood. In actual practice, the process of finding the maximum likelihood can become a computational challenge, if done by hand, so using some kind of software to perform the computations is highly desirable. We do need to note that most statisticians prefer the use of maximum likelihood to moment matching for parameter estimation. The reasons are somewhat technical, but the in general maximum likelihood will have the best parameter estimates (they are optimal in the sense they produce the best unbiased, minimum variance estimates).

Determining Goodness-of-Fit

Once a candidate distribution has been identified and its parameters determine, you can now compare the distribution you fit to the data. Here is an example of a histogram with a candidate distribution (Lognormal) being displayed in Figure A.3. The distribution parameters have been computed from the data.

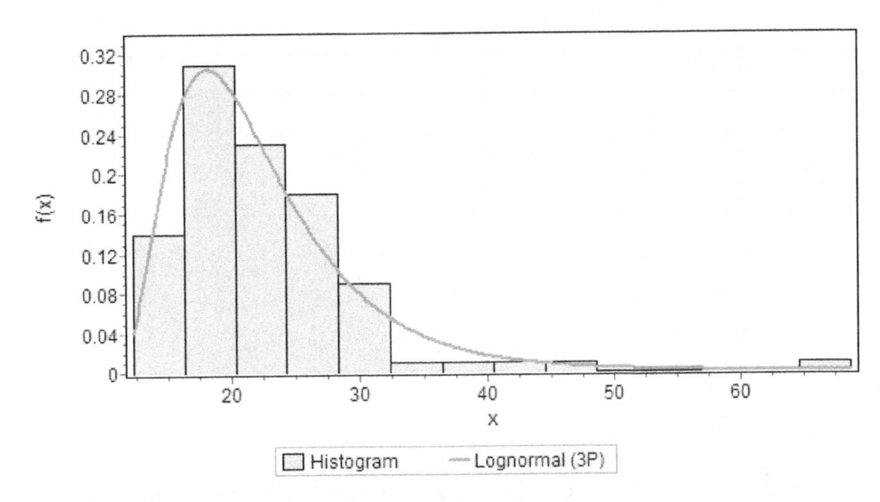

Figure A.3: Probability Density Function

The basic question is how good is the "fit" of the Lognormal to this data? Our answer is based on two approaches: visual inspection and statistical tests.

Visual Inspection: Visual inspection is the most important strategy because it visualizes the fit. Looking at the prior histogram and the fitted distribution, it's most important to see how the fitted distribution matches the observed histogram particularly in the tails. The "tails" are most important because the variation in the distribution is largest in the tails and it is the tails the most generally affects the queuing, utilization, and flow time statistics.

There are several other ways to "see" how well the fitted distribution matches the data. Most are based on the cumulative distribution function (CDF). For example, the comparison of the fitted CDF versus the data-based CDF is given in Figure A.4.

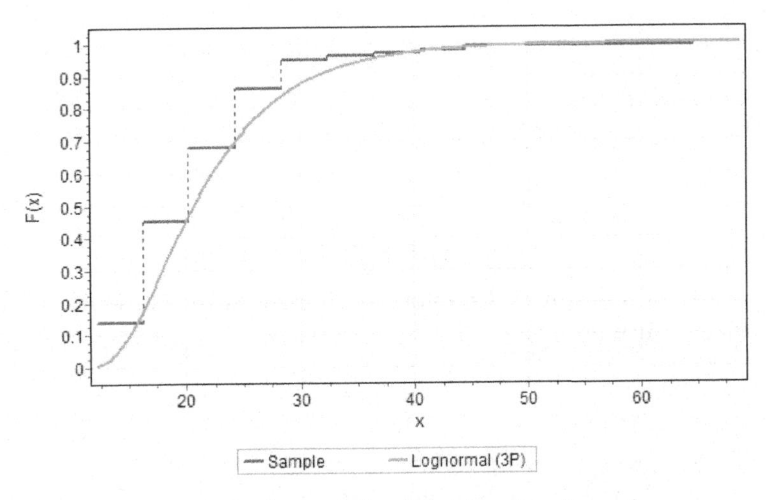

Figure A.4: Cumulative Distribution Function

Again, we are looked how well the fitted distribution represents the data. When looking at the difference, it is useful to plot the difference in the two CDFs as in Figure A.5.

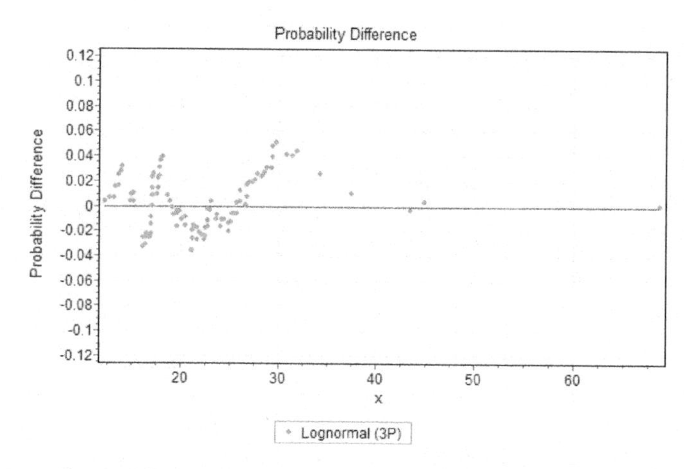

Figure A.5: Difference between Empirical CDF and Fitted CDF

In this plot we are looking to see if the difference is generally above or below 0, since that would indicate the fit overestimates or underestimates the data.

Two of the more useful visual plots is based on the "distance" between the fitted CDF and the data-based CDF. If you measure the difference vertically, you obtain the PP plot, which emphasizes the difference in the center of the distribution. The PP plot is shown in Figure A.6.

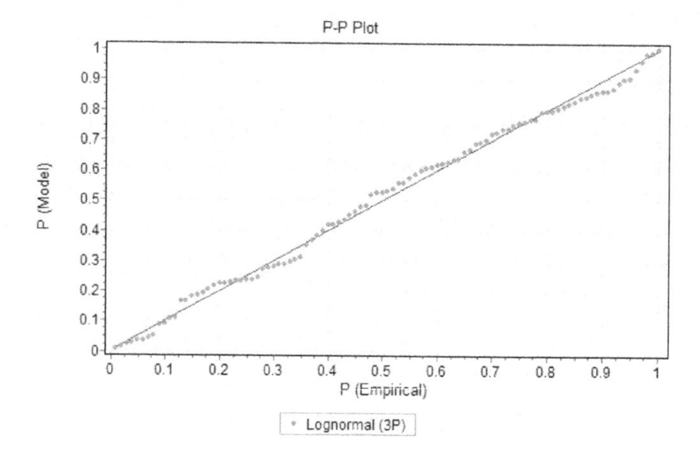

Figure A.6: P-P Plot Example

If you measure it horizontally, you obtain the QQ plot, which emphasizes the difference in the tails of the distribution. The QQ plot is shown in Figure A.7.

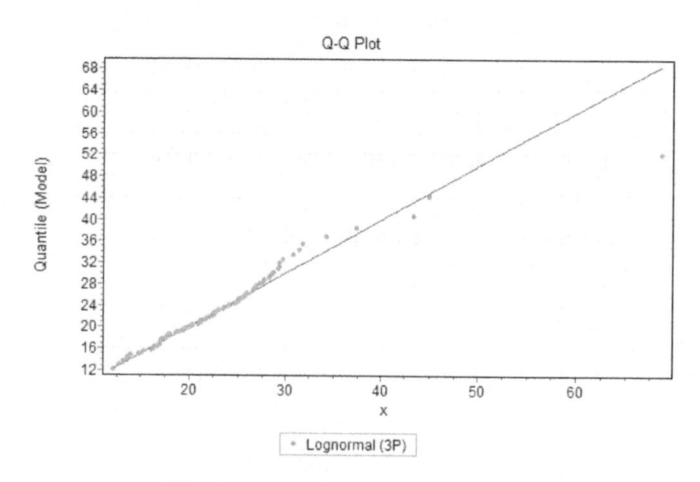

Figure A.7: Q-Q Plot Example

In both the PP and QQ plot, we want to see the points "close" to the dividing line. Deviations from the line indicate a potential for overestimation or underestimation of the data. Finally, we tend to have more interest in the QQ plot than in the PP plot since it emphasizes the difference in the tails of the distribution, where we believe the fit is most important.

Statistical Tests: Statistical tests are based on classical statistical testing in which a null hypothesis is compared to an alternative. The null hypothesis is that the data comes from the fitted distribution. But the "significance" of a statistical test is when the null hypothesis is rejected (Type 1 error). But in goodness-of-fit, we want to accept the null hypothesis so the power of the test is relevant (Type 2 error) and hence we would like are larger p-value. Thus statistical tests tend to only reject the poorest of fitted distributions. It is the reason we place so much emphasis on the visual inspection.

There are three standard goodness-of-fit statistical tests: the Chi-Squared test, the Kolmogorov-Smirnov test, and the Anderson Darling test. These tests are more thoroughly described in statistical texts. However we can provide a general description of each.

The Chi-Squared test is based on comparing the observed number of observations in a histogram cell with the number expected if the fitted distribution was the input model. Although, the test is often based on equal cell widths, it is generally better to use equal probability cells for the test.

The Kolmogorov-Smirnov (K-S) is a non-parametric test the looks at the largest (vertical) distance between the fitted CDF and the data-based CDF. The test is only approximate if parameters have to be estimated (which they usually are). A generalization of the K-S test is the Anderson Darling which puts more weight in the tails of the difference (the K-S uses equal weights).

Part A.4 Distribution Selection Hierarchy

The following is a rule of thumb and represents the distribution selection hierarchy one should follow.

1. Use recent data to fit the distribution using an input modeling software
2. Use raw data and load discrete points into a custom distribution (i.e., Empirical CDF)
3. Use the distribution suggested by the nature of the process or underlying physics
4. Assume a simple distribution and apply reasonable limits when lacking data (e.g.,Pert or Triangular)

It is always better to fit distributions to data collected about the process. If no good fits can be achieved the empirical distributions can be used to model the data. If there is some underling distribution suggested about the process owing scientific processes then one can utilize these. If you have very little or no data, then assuming a simple distribution and applying reasonable limits can be used as a start. Often we can obtain the minimum, maximum and sometimes the most likely. In this case we can use the Pert distribution.

Part A.5 Empirical Distributions in SIMIO

In contrast to the standard distributions, for which a functional form must be determined parameters must be estimated, empirical distributions are based totally on your data. You have two choices for empirical distributions: discrete and continuous. These distributions are represented in Figure A.8 which describes the probability mass/density function as well as the cumulative distribution function.

The SIMIO statements for the distributions in Figure A.8 are specified according to the following cumulative distributions:

$$\text{Random.Discrete}(v_1, c_1, v_2, c_2, \ldots, v_n, c_n)$$

$$\text{Random.Continuous}(v_1, c_1, v_2, c_2, \ldots, v_n, c_n)$$

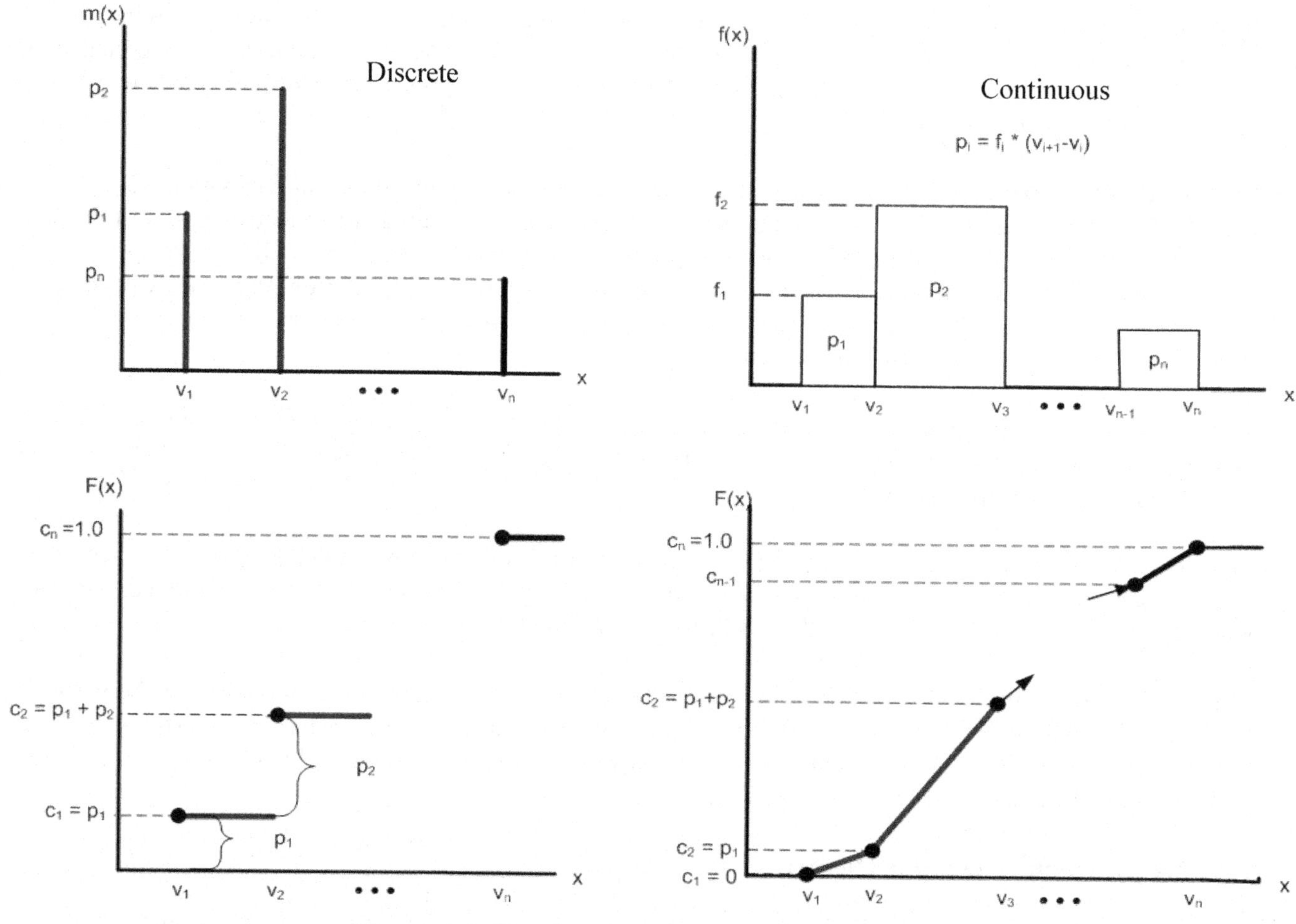

Figure A.8: Empirical Distributions in SIMIO

Of the two distributions the Discrete is the most often used. We might have a case where 30% have a value of 2.0, 50% have a value of 3.1, and 20% have a value of 6. A plot of this case is shown in Figure A.9.

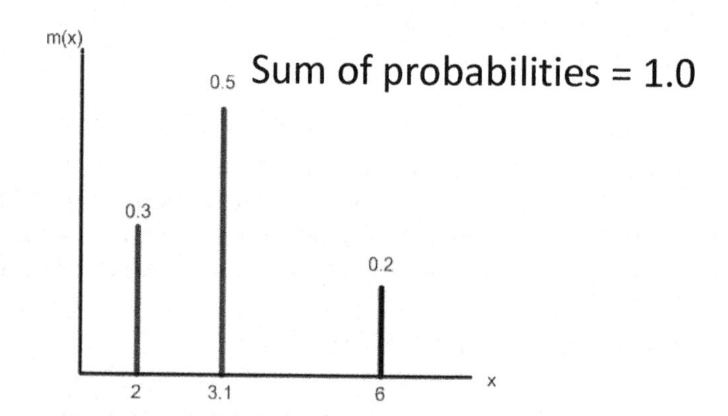

Figure A.9: `Random.Discrete(2, 0.3, 3.1, 0.8, 6, 1.0)`

Some other examples of the use of the empirical discrete distribution in SIMIO is:

```
Random.Discrete(2, 0.3, Math.Epsilon +5.2, 0.8, 6, 1.0)

Random.Discrete(Random.Exponential(0.5,0.3,Run.TimeNow,0.8,6, 1.0)

Random.Discrete("Red", 0.3, "Green", 0.8, "Blue", 1.0)
```

Of course the value/expression can be negative, although it occurs infrequently. Recall also the the cumulative probability parameters/expression must be in the range 0 to 1 and the last cumulative probability parameter must be 1.

An example of the use of continuous distribution might be where 30%of the time the value falls (uniformly) between 1 and 2.5, 60% between 2.5 and 4, and finally 10% between 7 and 8. The plot for this case is shown in Figure A.10.

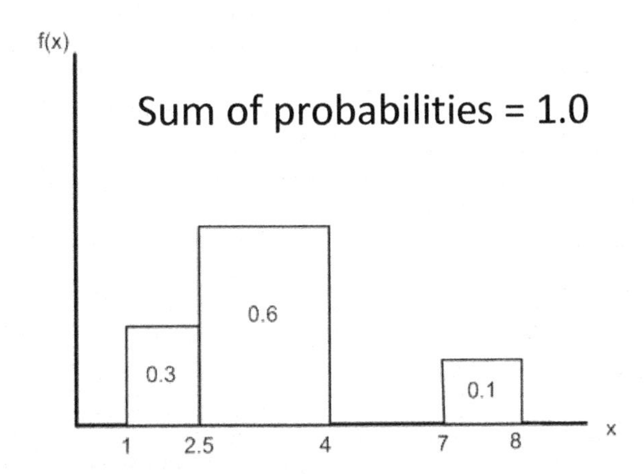

Figure A.10: SIMIO Specification: `Random.Continuous(1,0,2.5,0.3,4,0.9,7,0.9,8 1.0)`

The values/expression should not decrease, although SIMIO makes no check. Also cumulative probability parameters/expression values must be in range 0 to 1 and the last cumulative probability parameter must be 1.

Part A.6 Software for Input Modeling

The computational aspects of the distribution finding process have been incorporated into software. A number of software "packages" exist now to help in fitting a distribution to data (SAS Jmp, ExpertFit, EasyFit, Minitab). The software will do the work of finding candidate distributions that represent the data.

Typical of input modeling software is *EasyFit* from http://www.mathwave.com. It comes with a free trial version and a variety of licenses may be obtained. The software has a catalog of over 50 standard distributions. Data can be read from worksheets or entered and edited in the software's worksheet. The fitting system considers all the distributions as possible candidates. The user, may however, limit the fitting to (continuous) distributions which have an upper or lower bound or both and whether these bounds are open or closed. Maximum likelihood estimates are used for distribution parameter estimation for each candidate distribution. The visual displays include the probability density (or mass) function, the cumulative distribution function, the PP and QQ plots, and the CDF difference plot. The histogram of the data can also have the number of bins (cells) specified. The statistical goodness-of-fit tests include the K-S, the Anderson-Darling, and the Chi-Squared with equal probability or equal width cells. The candidate distributions can be ranking by the statistics produced from the statistical tests.

Whenever you use input modeling software, you need to be concerned with parameter correspondence between the software parameters and the SIMIO parameters. For example, the EasyFit parameters of Erlang (m,β) correspond to `Random.Erlang(m*β,m)` in SIMIO. Table A.1 shows how to convert EasyFit parameters to SIMIO parameters

Table A.1: Easy Fit to SIMIO Conversion of Input Parameters

Distribution	Easy Fit	SIMIO
Bernoulli	Bernoulli(p)	Random.Bernoulli(p)
Beta	Beta(α,β)	Random.Beta(α,β)
Binomial	Binomial(nt,p)	Random.Binomial(p,n)
Erlang*	Erlang(m,β)	Random.Erlang(m*β,m)
Exponential*	Exponential(λ)	Random.Exponential(λ)
Gamma*	Gamma(α,β)	Random.Gamma(α,β)
Geometric	Geometric(p)	Random.Geometric(p)
Johnson SB**	JohnsonSB(γ,δ,λ,ξ)	Random.JohnsonSB(γ,δ,ξ,ξ+λ)
Johnson UB(SU)**	JohnsonSU(γ,δ,λ,ξ)	Random.JohnsonUB(γ,δ,ξ,λ)
Log-Logistic*	Log-Logistic(α,β)	Random.LogLogistic(α,β)
Lognormal*	Lognormal(σ,μ)	Random.Lognormal(μ,σ)
NegativeBinomial	Neg.Binomial(n,p)	Random.NegativeBinomial(p,n)
Normal	Normal(stDev,mean)	Random.Normal(Mean,stDev)
PearsonVI*	Pearson6(α1,α2,β)	Random.PearsonVI(α1,α2,β)
Pert	Pert(m,a,b)	Random.Pert(a,m,b)
Poisson	Poisson(λ)	Random.Poisson(λ)
Triangular	Triangular(m,a,b)	Random.Triangular(a,m,b)
Uniform	Uniform(a,b)	Random.Uniform(a,b)
Weibull*	Weibull(α,β)	Random.Weibull(α,β)

*---Distributions that have a "Location" parameter, but was not included in the syntax. For an example on how to use a location parameter in a SIMIO random variable distribution, see below.

**---Distributions whose "Location" Parameters were indicated in the syntax, to show how to derive some of the parameters in SIMIO of the same distribution.

For many distributions EasyFit generates a "location" parameters along with the distribution parameters. This location parameter permits a distribution to be located at some arbitrary point. If EasyFit provides a location, say δ, for the Erlang then you will form the SIMIO expression:

```
δ + Random.Erlang(m*β,m)
```

The location parameter is used to shift a distribution so it will not be based at zero.

Part A.7 The Lognormal Distribution

The Lognormal Distribution may require some explanation in case you want to use it. The Lognormal is defined as the distribution you get by taking the logarithm of a normally distributed random variable. That normal distribution defined with a mean μ and a standard deviation σ are, in fact, the parameters of the SIMIO and the EasyFit Lognormal distribution. However they are not the mean and standard deviation of the Lognormal distribution which is given by:

$$\text{LogMean} = e^{\mu+\sigma^2/2}$$

$$\text{LogStd} = \sqrt{e^{2\mu+\sigma^2}(e^{\sigma^2}-1)}$$

If you have data from which you compute $\bar{x}$ and s_x, then you can compute SIMIO parameters (the Normal distribution parameters) from:

$$\mu \approx ln(\bar{x}^2 / \sqrt{\bar{x}^2 + s_x^{\ 2}}$$
$$\sigma \approx \sqrt{\ln(1 + s_x^{\ 2} / \bar{x}^2)}$$

Part A.8 Modeling the Sum of *N* Independent Random Variables

We are often faced with determining the processing time for a batch of parts to be processed individually. The processing time is, therefore, a sum of *N* independent random variables. For example, if there were 25 parts in a batch and each one took a Triangular(a, m, b) minutes to be processed, then the total time to process the batch is given by the summation of 25 independent triangular distributions as seen in the following equation.

$$Processing\,Time = \sum_{i=1}^{25} Triangular(a, m, b)$$

Often people will mistakenly model the processing time of summation of *N* independent random variables as *N* times the random variable of interest which is incorrect as it does not correctly create the appropriate distribution, but it is easy to specify.

$$25 * Triangular(a, m, b) \neq \sum_{i=1}^{25} Triangular(a, m, b)$$

To illustrate the point, let's take the Triangular distribution given a minimum value of *a*, mode of *m*, and a maximum value of *b*. The mean and variance of this distribution is given in the following table.

Mean	$\frac{a + m + b}{3}$
Variance	$\frac{a^2 + m^2 + b^2 - am - ab - bm}{18}$

Because it is much easier to just perform the multiplication rather than the summation, people will utilize the multiplication form which is given by the following equation.

$$N * \text{Triangular}(a, m, b) = \text{Triangular}(Na, Nm, Nb) \neq \sum_{i=1}^{N} \text{Triangular}(a, m, b)$$

As one can see in Table A.2 below the means of the two different versions are the same however the variances are not. In other words the modeling distribution is not what is desired. Figure A.11shows the PDF of each of the distribution and one can see that the true underlying distribution (i.e., summation distribution) is quite different. The reason is the variances of the summation of *N* independent variables add together rather than the standard deviations.

Table A.2: Calculation of the Mean and Variance of Summation and Multiplication Methods

Summation Mean	$\sum_{i=1}^{N}\left(\frac{a+m+b}{3}\right)=N\left(\frac{a+m+b}{3}\right)$
Summation Variance	$\sum_{i=1}^{N}\left(\frac{a^2+m^2+b^2-am-ab-bm}{18}\right)=N\left(\frac{a^2+m^2+b^2-am-ab-bm}{18}\right)$
Multiplication Mean	$\left(\frac{(Na)+(Nm)+(Nb)}{3}\right)=N\left(\frac{a+m+b}{3}\right)$
Multiplication Variance	$\left(\frac{(Na)^2+(Nm)^2+(Nb)^2-(Na)(Nm)-(Na)(Nb)-(Nb)(Nm)}{18}\right)=N^2\left(\frac{a^2+m^2+b^2-am-ab-bm}{18}\right)$

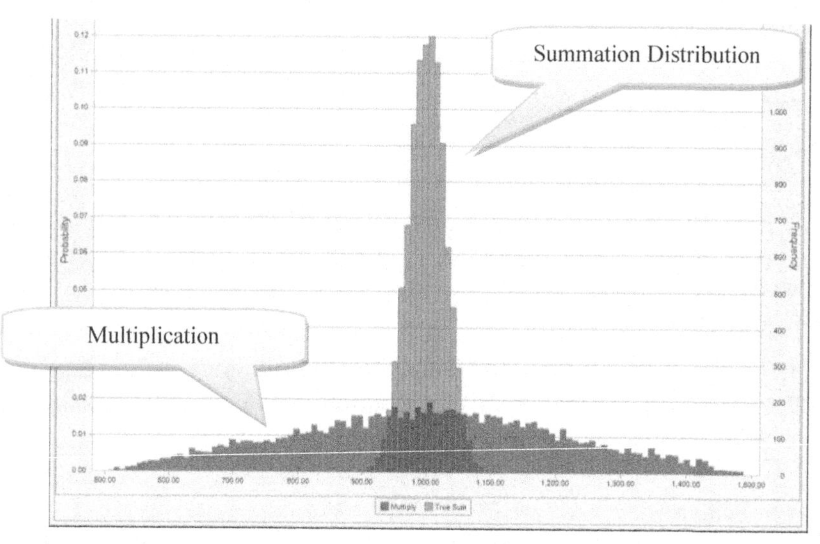

Figure A.11: Comparing True Summation Distribution with the Multiplication Distribution

Correct Handling of Summation of *N* Independent Variables

The most appropriate method is to sample *N* random variables and then sum these values to reach the correct sample. The WORKSTATION will model this exact scenario correctly since each order is made up multiple items and a processing time was given based an item. Figure A.12 shows a simplified version of the Processing process that handles the summation through batching. As you can see the `Delay` step will sample one random variable and then delay the associated entity. The `End Activity` step checks to see if there are any more batches. If there are more to perform, the TOKEN will loop back and delay the entity again. The `Delay` step will be invoked *N* times sampling an independent random variable each time. For example, the *Operation Quantity* property would be set to 25 for example with the *Processing Batch Size* set to one. Therefore, 25 separate batches or delays will be performed.

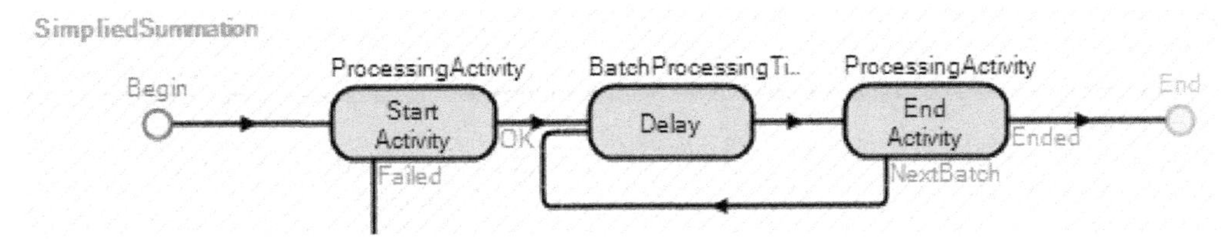

Figure A.12: Simplified Batch Processing similar to the Workstation

Approximation for the Summation of *N* Independent Variables

As *N* gets large, the distribution that results from the summation of *N* independent distributions becomes Normal owing to the *Central Limit Theorem*. Therefore, given a large enough *N*, the summation distribution can be approximated using a Normal distribution with the mean and standard deviation (STD) in Table A.3. Figure A.13 shows the exact summation distribution, the normal approximation and the incorrect multiplication distribution. As you can see the approximation is very good given *N* was 25.

Table A.3: Calculation of the Mean and Variance of Summation and Multiplication Methods

Normal Mean	$N * \textit{Single Distribution Mean}$
Normal STD	$\sqrt[2]{N * \textit{Single Distribution Variance}} = \sqrt[2]{N} * \textit{Single Distribution STD}$

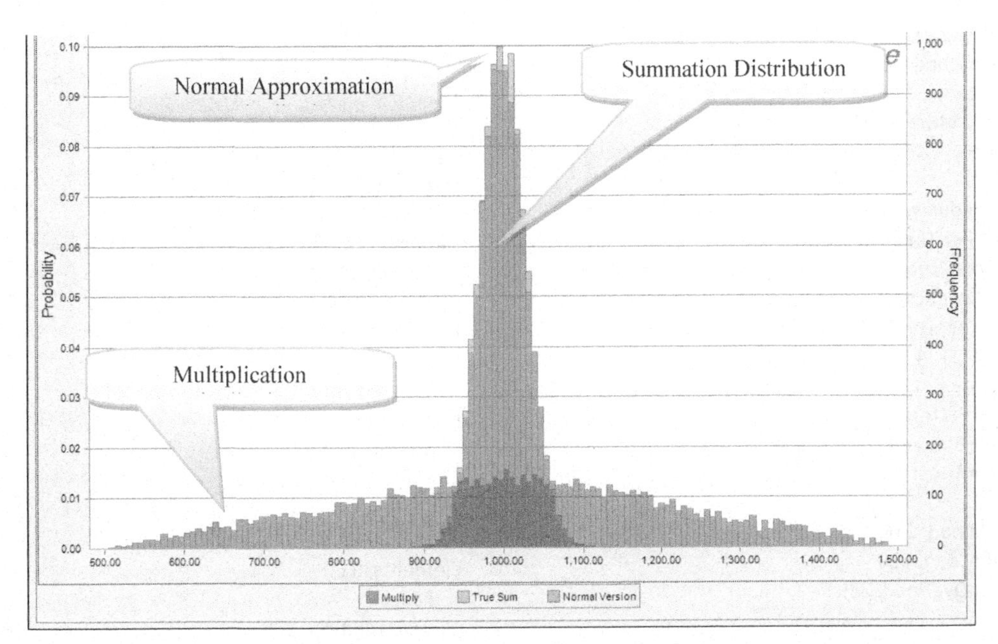

Figure A.13: Simplified Normal Approximation versus the True Summation Distribution

The disadvantages of using the normal approximation is the sampled values can become negative and the value *N* needs to be sufficiently large enough. Depending on the distribution, $N < 10$ will not yield a very good normal approximation and typically *N* larger than 20 is sufficient.

Index

F

G

H

I

K

L

M

N

O

P

Q

R

S

T

V

W

Z

Made in the USA
Columbia, SC
27 November 2018